The Physiology of Flowering Plants

Fourth Edition

This latest edition of *The Physiology of Flowering Plants* has been completely updated to cover the explosion of interest in plant biology. A whole-plant approach has been used to produce an integrated view of plant function, covering both the fundamentals of whole plant physiology and the latest developments in molecular biology. New developments in molecular techniques are explained within practical applications such as genetically modified plants. The book further examines:

- photosynthesis, respiration, plant growth and development;
- nutrition, water relations, photomorphogenesis and stress physiology;
- function, with particular attention to adaptations to different habitats.

Each chapter is fully referenced with suggestions for complementary reading including references to original research papers.

The Physiology of Flowering Plants is valuable to both undergraduate and postgraduate students studying plant biology.

HELGI ÖPIK was Senior Lecturer in the School of Biological Sciences at the University of Wales, Swansea until her retirement. Throughout her career she has taught plant physiology at all undergraduate levels, and since retiring has lectured in plant physiology for adult education. Her research interests have included plant respiration and ultrastructure, always aiming at integration of structure and physiological function.

STEPHEN ROLFE was awarded a European Molecular Biology Fellowship and undertook postdoctoral research on the phytochrome regulation of gene expression at the University of California, Los Angeles. He took up a post at the Department of Animal and Plant Sciences, University of Sheffield in 1991. His research interests include the study of photosynthesis and primary plant metabolism, with a special interest in non-invasive imaging techniques.

The Physiology of Flowering Plants

Fourth Edition

Helgi Öpik, *1936-*
Formerly Senior Lecturer,
School of Biological Sciences,
University of Wales,
Swansea

Stephen A. Rolfe
Senior Lecturer,
Department of Animal and Plant Sciences,
University of Sheffield

Academic Consultant Editor

Arthur J. Willis
Emeritus Professor,
University of Sheffield

CAMBRIDGE UNIVERSITY PRESS

Cambridge, New York, Melbourne, Madrid, Cape Town, Singapore, São Paulo

CAMBRIDGE UNIVERSITY PRESS

The Edinburgh Building, Cambridge CB2 2RU, UK

http://www.cambridge.org
Information on this title: www.cambridge.org/9780521662516

First published 2005

Printed in the United Kingdom at the University Press, Cambridge

A catalogue record for this book is available from the British Library

Library of Congress Cataloguing in Publication data

ISBN-13 978-0-521-66251-6 hardback
ISBN-10 0-521- 66251-6 hardback

ISBN-13 978-0-521-66485-3 paperback
ISBN-10 0-521-66485-3 paperback

Contents

Preface   *page* ix

Chapter 1 Introduction   1

1.1 Appreciating plants   1
1.2 What kind of plant physiology?   2
1.3 Molecular biology and plant physiology: the integration
    of disciplines   3
1.4 Outline of the text   5

Part I   Nutrition and transport

Chapter 2 Flow of energy and carbon through the plant:
              photosynthesis and respiration   9

2.1 Introduction   9
2.2 Energy flow and carbon turnover in the biosphere   9
2.3 Photosynthesis: light absorption and utilization   12
2.4 The fixation of carbon dioxide   18
2.5 Limiting factors for photosynthesis   30
2.6 The efficiency of energy conversion in photosynthesis   32
2.7 Photosynthesis and the increase in atmospheric
    carbon dioxide   36
2.8 Respiration: the oxidative breakdown of organic compounds   38
2.9 Terminal oxidation and oxidative phosphorylation   46
2.10 Anaerobic respiration   49
2.11 Respiration and plant activity   53

Chapter 3 Water relations   60

3.1 Introduction   60
3.2 Water movement and energy: the concept of
    water potential   61
3.3 Water potentials of plant cells and tissues   61
3.4 Water relations of whole plants and organs   68
3.5 The transport of solutes in the xylem   85
3.6 Water uptake and loss: control by environmental and
    plant factors   86
3.7 Water conservation: xerophytes and xeromorphic characters   95

Chapter 4 Mineral nutrition   100

4.1 Introduction   100
4.2 Essential elements   100

4.3 Ion uptake and transport in the plant 106
4.4 Nitrogen assimilation, fixation and cycling 122
4.5 Problems with mineral elements: deficiency
and toxicity 128

Chapter 5 | Translocation of organic compounds 133

5.1 Introduction 133
5.2 Phloem as the channel for organic translocation 133
5.3 The rate and direction of translocation 139
5.4 Phloem loading and unloading 142
5.5 Partitioning of translocate between sinks: integration
at the whole-plant level 146
5.6 The mechanism of phloem translocation 148

Part II   Growth and development

Chapter 6 | Growth as a quantitative process 161

6.1 Introduction 161
6.2 The measurement of plant growth 162
6.3 Growth, development and differentiation 163
6.4 Localization of growth in space and time 164
6.5 Conditions necessary for growth 165
6.6 Growth rates 167

Chapter 7 | Plant growth hormones 177

7.1 Introduction 177
7.2 Plant growth hormones 178
7.3 Detection and quantification of hormones in plants 191
7.4 How do plant hormones cause responses? 194

Chapter 8 | Cell growth and differentiation 205

8.1 Introduction 205
8.2 Meristems and cell division 205
8.3 Mitochondrial and plastid division 211
8.4 Cell expansion: mechanism and control 213
8.5 Cell differentiation 218

Chapter 9 | Vegetative development 221

9.1 Introduction 221
9.2 The structure and activity of the shoot apical meristem 221
9.3 Organ formation 225
9.4 Secondary growth 227
9.5 Development of the leaf 228
9.6 The structure and activity of the root apical meristem 239

Chapter 10 | Photomorphogenesis   246

10.1 Introduction   246
10.2 The switch from etiolated to de-etiolated growth   247
10.3 Phytochrome and photomorphogenesis   248
10.4 UV-A/blue light photoreceptors (cryptochrome)   255
10.5 Genes controlling etiolated growth   256
10.6 Unravelling photomorphogenesis   257
10.7 Phytochrome signal transduction   263

Chapter 11 | Reproductive development   270

11.1 Introduction   270
11.2 Juvenility and 'ripeness to flower'   270
11.3 The control of flowering by daylength and
     temperature   271
11.4 Plant size and flowering   277
11.5 The regulation of floral induction is a
     multifactorial process   279
11.6 Floral development   281
11.7 Pattern development in flowers   287
11.8 The formation of pollen   291
11.9 The formation of the embryo sac   293
11.10 Pollination   295
11.11 Embryo formation   301
11.12 Seeds and nutrition   303
11.13 Fruit development   308
11.14 Seed dormancy   310
11.15 Germination and the resumption of growth   315

Chapter 12 | Growth movements   318

12.1 Introduction   318
12.2 Nastic responses   318
12.3 Tropisms   320

Chapter 13 | Resistance to stress   344

13.1 Introduction   344
13.2 Terminology and concepts   344
13.3 Water-deficit stress   346
13.4 Low-temperature stress   354
13.5 High-temperature stress   362
13.6 Relationships between different types of stress
     resistance: cross-tolerance   366
13.7 Development of stress-resistant crop plants   368

Appendix 373

A.1 Naming genes, proteins and mutations 373
A.2 Units of measurement 373
A.3 Prefixes for units 375

Index 376

Preface

The history of this book dates back to the late 1960s, when the publishers Edward Arnold launched a series of student textbooks as the Contemporary Biology series, designed to provide up-to-date texts at elementary university and final-year school level. One of the first authors who was asked to contribute, on the topic of flowering plant physiology, was Professor H. E. Street, then Professor of Botany at the University of Wales, Swansea. He asked one of us (H.Ö.) to collaborate, and the first edition was duly published by Edward Arnold in 1970 under the authorship of H. E. Street and Helgi Öpik, and entitled *The Physiology of Flowering Plants: Their Growth and Development.* The emphasis of the text was on the 'whole plant' aspects of physiology. The second edition followed in 1976 and the third in 1984, although Professor Street sadly deceased in 1977.

While the second and third editions were still very much revisions of the original text, the longer time interval since the last edition, and the rapid pace at which biological knowledge has grown in the last few decades, have now necessitated a very thorough rewriting of large sections of the book, and the task has been quite challenging in the face of an accumulation of facts that on occasion has seemed quite overwhelming. It is not possible now to interpret many aspects of plant physiology without reference to molecular biology, even when one is basically interested in functioning at the organismal level. This applies particularly to the developmental aspects of physiology. Some reorganization of the text and shift of emphasis has accordingly been necessitated, though we have tried to retain the overall spirit of the original book.

One thing has remained unchanged during the preparation of this book from the first edition to the fourth: the unfailing encouragement and help from our editor, Professor A. J. Willis. Without him, the present text would not have been written. We are also grateful for the support of Dr Ward Cooper, Commissioning Editor, and Dr Alan Crowden, Editorial Director, of Cambridge University Press. Thanks are due for reading, and advising on, parts of the manuscript, to Professor Richard C. Leegood, Professor David Read and Dr Julie Gray of the University of Sheffield.

H.Ö. would like to acknowledge the generosity of Professor Ray Waters, Head of the School of Biological Sciences at the University of Wales, Swansea, for use of departmental facilities in preparing illustrations. H.Ö. also would like to thank Ken Jones of the School of Biological Sciences, Swansea, for printing figures; my nephew Kevin Miller and my niece, Heather Nagey, for help with word processing; and Professor Kevin Flynn and Dr Charles Hipkin of the University of Wales, Swansea, for helpful discussions.

We are grateful to all the people who have permitted us to reproduce their published data, and have provided material and helpful advice for figures; particular thanks are due to Professor Jane Sprent and Dr Euan James of the University of Dundee for supplying the original micrograph of bacteroids (Fig. 4.7).

Chapter 1

Introduction

1.1 | Appreciating plants

During a public open day at a university, a child trying to look at a botanical exhibit was dragged away by an impatient parent with the words 'Come on – we can't spend all day looking at dull green things!'

There is a tendency to consider plants as somewhat dull, passive and inactive. Yet plants face and overcome the same problems as animals: how to obtain nutrients and water, how to survive extreme environmental conditions, how to ensure reproduction and the survival of the next generation. The photosynthetic mode of life has conditioned plants to evolve as sessile organisms; their basic necessities – light, carbon dioxide, water and mineral ions – are ubiquitous and there has therefore been no selection pressure for mobility. An animal may obtain its nutrients and water by skilfully stalking its prey, and learning the path to a pool; this catches human attention as interesting behaviour. A flowering plant obtains nutrients and water by millions of minute root tips constantly growing through the soil, and by pumping ions across root cell plasma membranes with molecular-sized pumps. This is plant behaviour: **plant physiology is plant behaviour**. It need not be considered dull because it is less spectacular to the eye than what is called animal behaviour. The subsequent hauling up of the absorbed water and minerals to the top of a tree, 100 metres high, might indeed be considered a quite spectacular feat (imagine doing it with a bucket!). Without stirring from the spot, flowering plants are unceasingly monitoring their environment and responding to environmental signals. For them as for us, light is an information medium; they contain optical sensors (pigments) with which they perceive and respond to light direction, wavelength composition (i.e. colour) and the duration of daylight. They are sensitive to touch, with some responses as fast as animal movement: a Venus flytrap leaf snaps shut in a second or two to catch (and digest!) insects as big as wasps or small moths. Pea tendrils are a bit slower to react, but can be seen curling around a support within minutes of making contact. Plants respond also to the direction of gravity; it is the continuous responses to gravity and light that are largely responsible for plants growing 'the right way up', and with roots, branches and leaves orientated at various angles. Temperature changes are sensed. Signals with a regular annual variation – such as the changes in daylength – enable plants to synchronize their life cycles with the seasonal cycles in their environment. Even the embryo in a dormant seed, apparently quite inactive in its coat, is able to receive specific signals that stimulate it to commence germination at the appropriate season. Flowering plants, when looked at from the physiologist's point of view, are not merely alive, they are very lively. We hope that the readers of our book will be led to appreciate how marvellous and varied in their activities flowering plants are.

There are other reasons, too, for studying flowering plants. We are utterly dependent on them. Being the dominant plants in present-day terrestrial (land) vegetation, they are the primary producers, by photosynthesis, of organic material – food – on which all other terrestrial organisms including ourselves rely. We moreover need plants for wood, textiles, drugs, and hundreds of household chemicals; their very presence gives us joy, even comfort. Gardens are a main source of pleasure for many individuals; windowsills get filled with house plants, floral baskets are hung from balconies of high-rise flats. Animals not only feed on plants, but live in them, on them and under them. The oxygen we breathe is formed as a by-product of photosynthesis. Soil and climate, too, are influenced by plants; vegetation may for instance stabilize soil against physical erosion. Thus the importance of understanding the activities of flowering plants cannot be overemphasized. With the changes currently being imposed on the biosphere of this planet by human activities, it has become more imperative than ever to study the physiology of plants, if there is to be any hope of predicting how vegetation, all-important for life on earth, might respond to such changes in the environment. An understanding of flowering plant physiology is also vital for any attempts to improve plant productivity.

1.2 | What kind of plant physiology?

Plant physiology is the functioning of plants. This can be studied at several levels of complexity and organization, as indicated by such terms as *metabolic physiology*, *cellular physiology*, or *whole-plant physiology*. The aim of the current text is to give an account of the physiology of flowering plants mainly from the **whole plant** or **organismal** point of view. Moreover, since an organism functions within its environment, and by virtue of its physiological activities continuously interacts with its environment, plant–environment interactions are emphasized throughout the book. But many plant life processes are carried out at the level of the individual cell, even of the individual organelle. Physiological processes of plants can therefore be described only to a limited degree without reference to activities at the cellular or subcellular level. Water movement through an intact plant is a whole-plant level process; for some specified ecophysiological project, it may be adequate to discuss it, say, in terms of quantities of water absorbed by the roots, and the amounts lost by transpiration from the leaves. If, however, one wishes to understand the control exerted on the process by the plant, one is soon involved with the cellular physiology of the stomatal guard cells, which move to open and close the stomatal pores through which most of the transpiration occurs. Equally relevant to an understanding of physiological processes is a knowledge of cellular structure. To continue with the example of stomata, the opening and closing movements are also dependent on the shapes of the guard cells and on the pattern of their cell wall thickenings, right down to the precise arrangement of the cellulose microfibrils. Whilst the goal of this book is to promote the understanding of flowering plants as organisms, one cannot escape consideration of information from the realms of cellular physiology, biochemistry, cell structure and ultrastructure, and molecular biology. In this respect we have had to exercise our judgement about the depth to which such information should be presented, and the extent to which it might be taken for granted. In fact an elementary knowledge of basic metabolism, biochemistry, plant anatomy and plant cell structure has been assumed. Textbooks listed at the end of this chapter as **Complementary reading** will serve to fill in the background for readers who find this necessary, and will enable all interested readers to extend their knowledge of the more cellular aspects of plant physiology, and of plant structure, beyond this text.

1.3 | Molecular biology and plant physiology: the integration of disciplines

References to molecular biology will be found to figure liberally in parts of the text. The discipline of molecular biology embraces the study of **genes**: their isolation and identification, analysis (molecular sequencing) and their manipulation, i.e. modification and introduction of selected genes into cells at will. Gene expression, their 'switching on and off', can also be manipulated. This is a branch of science that is enabling biologists to identify and study the roles of specific genes in specified activities.

Molecular biology techniques have led to considerable advances in the understanding of plant physiology, particularly of development and differentiation. Every activity of a living organism can ultimately be traced to control by some gene(s) at some stage of its life history; the entire blueprint for everything that a flowering plant is capable of performing is inscribed in the DNA of the single-celled zygote from which it develops. Some physiological functions can be discussed with minimal reference to genetic activity. These include numerous aspects of the processes of nutrition and transport described in the first part of this book. Such processes may also show some short-term, quantitative, reversible responses to environmental factors, which are not mediated at the genetic level. When one comes to the physiology of growth and differentiation, however, as covered in the second part of the book, the need arises to refer constantly to genetic control. One tends to think of growth as a **quantitative** increase in mass. Growth is, however, almost always accompanied by **differentiation**, a change of form and/or physiological activity, and this results in the process termed development:

development = growth + differentiation

The growing seedling, for instance, does not become an enlarged version of the embryo in the seed, it develops into the adult form of the plant. Differentiation means **differential activity of genes**, i.e. activation and suppression of specific (sets of) genes at particular developmental stages. Developmental processes all involve qualitative as well as quantitative changes in response to environmental conditions. Environmental factors act as specific stimuli calling forth specific and generally irreversible changes in physiological activities and morphology. This applies to such events as the onset of reproductive growth, and reactions to environmental stresses. Such processes are the result of interactions between the external factors and the genome of the plant, and our understanding of such processes is dependent on studies of gene activity. Paradoxical as it may seem at first sight, it is the most conspicuously organism-oriented activities, such as the onset of flowering – which entails profound changes in the physiology of the whole plant – that need the molecular biology approach to the greatest extent. These are the processes which are most strongly under control at the genetic level.

We started this chapter with a very wide view, considering the overall role of flowering plants in the biosphere. We now have reached consideration of the finest, ultimate level of biological analysis – the individual gene. Flowering plants, like any other organisms, are complex entities and understanding of them as organisms can be achieved only by studying them at various levels of organization and integrating knowledge obtained by numerous biological disciplines, from molecular biology to ecology. The division of biology into various disciplines is an artificial one, made for the convenience of handling a vast field of knowledge; no such division exists in the plant. We hope that we have been able to keep the organism, the 'whole' plant, in sight throughout this text, even when interpreting events in terms of molecular biology.

Plant physiology also has something to offer to molecular biologists. The science of molecular biology has led to applications known under the general term of genetic engineering, i.e. the production of transgenic (transformed) organisms, containing selected genes which may come from other species. Basic gene structure being identical in all organisms on planet earth, genes from animals and even from prokaryotes can be inserted into the flowering-plant genome. Flowering plants possess great regenerative powers; single cells from vegetative organs are able to grow into fully formed plants. This makes them ideal subjects for genetic transformations. Some transgenic crop plants are already being grown commercially and many more potential transformations are being investigated. While some scientists see genetic engineering as a wonderful tool for improving plants as economic resources, many people regard transgenic operations with strong reservations, fearing that, accidentally or deliberately, new and dangerous organisms might be let loose upon the world. There can be no scientific safeguards against the deliberate misuse of genetic engineering. The defence against accidental disasters lies in knowledge and understanding. For successful engineering of 'plants for the future', it is imperative that the physiology of plants should be understood, and moreover understood at the organismal level. A particular transformation may be undertaken with one single activity in mind; but that activity needs to be studied in the context of the plant as a whole if one is to have a reasonable chance of predicting the result of the operation.

1.4 | Outline of the text

The book falls essentially into two parts: **Nutrition and transport** (Chapters 2 to 5) and **Growth and development** (Chapters 6 to 13).

Flowering plants are essentially autotrophic, photosynthetic organisms, with basic requirements of light, CO_2, water and some 17 elements as inorganic mineral ions. The survey of plant physiology accordingly begins with **photosynthesis** (Chapter 2), detailing how the flowering plants obtain and process the first two necessities listed above, light and CO_2. This is followed by **respiration**, the process providing energy and intermediates for metabolism. Next it is considered how they obtain and transport **water** (Chapter 3), and then how **mineral ions** are absorbed and assimilated (Chapter 4). The **translocation of organic materials**, the products of photosynthesis, is covered in Chapter 5, completing the first part of the book.

The second part of the text starts with treatment of **growth as a quantitative process** and covers some simple mathematical analysis of growth (Chapter 6). Then we proceed to discussions of the physiology of growth and development, beginning with surveys of **plant growth hormones** (Chapter 7) and **cell growth and differentiation** (Chapter 8). These are followed by consideration of **vegetative development**, **photomorphogenesis**, **reproductive development**, and **growth movements** (Chapters 9 to 12). Finally, in Chapter 13, the reactions of flowering plants to some **environmental stresses** are discussed.

Complementary reading

Anderson, J. W. & Beardall, J. *Molecular Activities of Plant Cells.* Oxford: Blackwell, 1991.

Dennis, D. T., Turpin, D. H., Lefebvre, D. D. & Layzell, D. B., eds. *Plant Metabolism*, 2nd edn. Harlow: Addison Wesley Longman, 1997.

Gunning, B. E. S. & Steer, M. W. *Plant Cell Biology, Structure and Function.* Sudbury, MA: Jones & Bartlett, 1996.

Lea, P. J. & Leegood, R. C., eds. *Plant Biochemistry and Molecular Biology,* 2nd edn. Chichester: Wiley, 1999.

Mauseth, J. D. *Plant Anatomy.* Menlo Park, CA: Benjamin/Cummings, 1988.

Mauseth, J. D. *Botany: an Introduction to Plant Biology*, 2nd edn. Sudbury, MA: Jones & Bartlett, 1998.

Taiz, L. & Zeiger, E. *Plant Physiology*, 3rd edn. Sunderland, MA: Sinauer, 2003.

Troughton, J. & Donaldson, L. A. *Probing Plant Structure.* London: Chapman & Hall, 1972.

Part I

Nutrition and transport

Chapter 2

Flow of energy and carbon through the plant: photosynthesis and respiration

2.1 | Introduction

All living organisms need a supply of raw materials from which their bodies can be constructed, and a supply of energy. This energy is needed for growth, i.e. for the formation of their bodies, also for the maintenance of their bodies, and for all the various types of work, chemical and mechanical, that are carried out by living systems.

Life as we know it is based on organic compounds of **carbon** (C). This element accordingly occupies a central place among the raw materials and it is found on earth abundantly in its inorganic forms as carbon dioxide (CO_2), carbonate (CO_3^{2-}) and bicarbonate (HCO_3^-). The ultimate energy source for most life forms on earth is the thermo-nuclear energy of the sun, transmitted to earth as electromagnetic radiation, light. **Photosynthesis** is the process by which the solar light energy is transformed into the chemical bond energy of organic carbon compounds. Photosynthesis is thus simultaneously a process of energy transduction, and a process by which inorganic carbon is converted to organic form and incorporated into living organisms. Chemically it is a reductive process. **Respiration** is the process of oxidative breakdown by which the energy stored in the organic products of photosynthesis is tapped for driving metabolism, for growth, for movements, and by which the C is returned to inorganic form again as CO_2.

2.2 | Energy flow and carbon turnover in the biosphere

2.2.1 Plants and the biosphere

The flow of energy and the turnover of C in the biosphere are illustrated in Fig. 2.1. Photosynthesis utilizes relatively high energy quanta, 'light', from the electromagnetic spectrum and these quanta energize the combination of hydrogen from water with CO_2 to organic compounds. In flowering plants, the first *stable* photosynthetic products are predominantly sugars. These sugars form the starting point of other plant constituents, including the

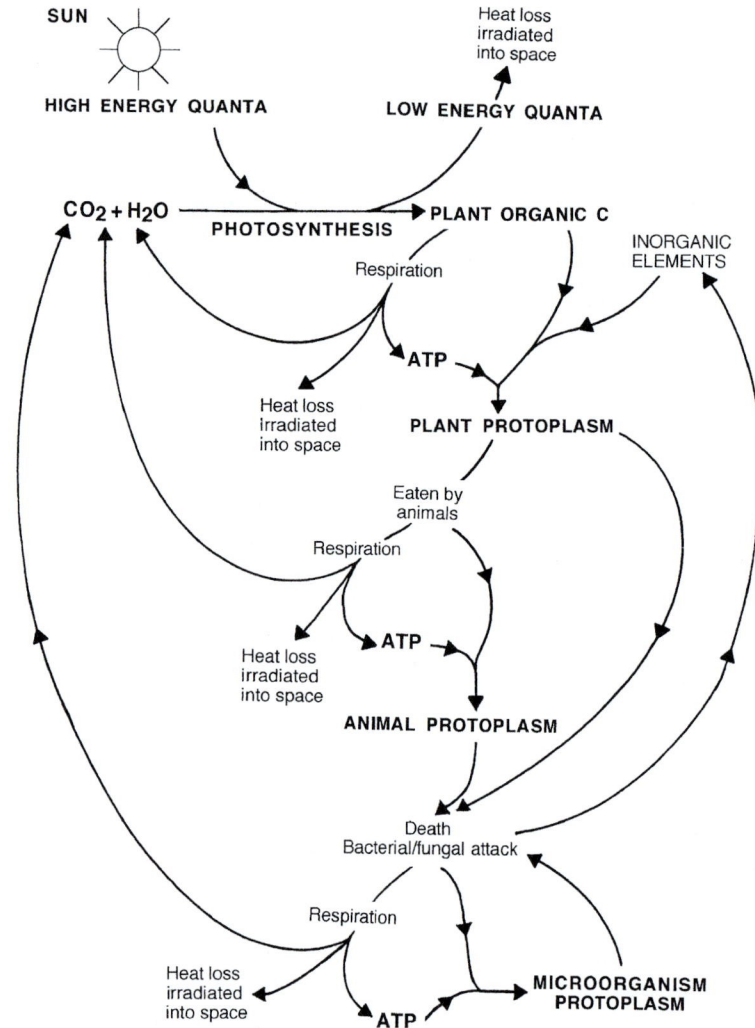

Fig. 2.1 The flow of carbon and energy through the biosphere. The term 'plant' is used here to denote all photosynthetic autotrophs and 'protoplasm' for all parts of an organism. The energy that is lost as heat is irradiated into space as infrared (long wavelength) radiation and cannot be recycled, but the C and other elements recycle as indicated.

macromolecules which are the units of cellular architecture. The sugars and polysaccharides (sugar polymers), sometimes also lipids, form a store of potential energy for plant cells. This store is continuously being drawn upon during respiration. The substrates are oxidized to CO_2 and water again, while some of the potential energy of the sugar molecules is transferred to molecules of ATP (adenosine triphosphate). This extremely reactive compound has been termed the energy currency molecule of living cells and its potential energy can be harnessed for cellular work – biosynthesis, growth, membrane transport, movement. During biosynthesis, other elements (Chapter 4) may be incorporated into organic combination. In the course of these activities the ATP is broken down to ADP or AMP (adenosine di- or monophosphate) respectively and it must be unceasingly resynthesized. In photosynthetic cells in the light, some cellular work may also be driven by photosynthetically formed ATP (see Section 2.3.2) without the intervention of respiration.

Photosynthesis is carried out by virtually all land plants. The only exceptions are some non-green parasitic or saprophytic species. The algae are also photosynthetic and land plants plus algae are sometimes collectively referred to as 'the green plants'. The green plants are **photoautotrophic** (literally 'light-self-feeding'), needing only light and inorganic compounds: CO_2, water and mineral ions. There are also photosynthetic bacteria (although some of these are photoheterotrophic, i.e. they need a supply of organic C compounds). The photoautotrophs are the **primary producers** and they form the basis of food chains. Plant organic matter is ingested as food by animals, fungi and other microorganisms. Any organic material synthesized by these non-photosynthetic organisms is regarded as **secondary** production, being derived from the photosynthetic products. As indicated in the lower part of Fig. 2.1, the cycles of biosynthesis and respiration are repeated in the secondary producers. Secondary producers in turn feed on each other and the cycles are multiplied, until ultimately microorganism action, decay, returns the last of the elements again to their inorganic forms of CO_2, water and mineral ions. The chemicals are recycled, although the recycling may take a long time. A living tree may retain organic materials in its body for thousands of years. Carbon fixed by the Carboniferous forests over 300 million years ago is only now being returned to the atmosphere by the burning of coal. *The energy, however, cannot be recycled.* In every process illustrated in Fig. 2.1, there occurs some energy loss as heat; this is in due course irradiated back into space as infrared radiation, low-energy quanta which can never drive photosynthesis again. The laws of thermodynamics make this energy loss inevitable: reactions are possible only if they proceed with an overall free energy loss.

Photosynthesis means a steady input of energy into the biosphere, vital for the continuance of most life. Hence photosynthesis occupies a central position, not only in the life of the green organisms which carry it out, but for life on earth in general. (An exception is provided by chemosynthetic bacteria.) On dry land, flowering plants are responsible for the major proportion of photosynthesis.

The global turnover of carbon

It is estimated that around 10^{11} tons of C are fixed annually in photosynthesis, representing a primary production of some 1.7×10^{11} tons of dry matter, while the total C held in the earth's biomass at any one time amounts to about 5.6×10^{11} tons. This, however, is a minute fraction of the C present on earth. Much of this C is present as carbonates in rocks and as sedimentary organic material in the deep oceans, unavailable for photosynthesis. But even the available stores are enormous: atmospheric CO_2 holds 7.5×10^{11} tons of C, and the oceans, which equilibrate with the atmosphere, contain in their upper, accessible, layers some 420×10^{11} tons as dissolved CO_2, carbonate and bicarbonate.

The site of photosynthesis

Any part of the shoot can be green and photosynthetic to some degree – stem, leaf, flowerbud, or young fruit; but the photosynthetic organ par excellence is the **leaf**, an outgrowth from the stem; by far the greater part of land plant photosynthesis is achieved in the leaves. The variety of leaf shapes and sizes may seem bewildering, but they mostly share a common basic structure which is adapted to obtain light and CO_2. Sunlight is diffuse and atmospheric CO_2 is present at a low concentration; a large aerial surface is required to collect these necessities efficiently. Essentially, the typical leaf is a thin, flat structure, often less than a millimetre thick. The photosynthetic cells make up the **mesophyll** ('mid-leaf') enclosed by an **epidermis**, usually non-photosynthetic, on both surfaces. Where the leaf has distinct upper and lower surfaces (a dorsiventral leaf), the mesophyll is generally differentiated into an upper palisade layer of elongate, cylindrical cells and a lower spongy mesophyll of more rounded, loosely packed cells. A transparent **cuticle**, relatively impervious to water, water vapour and other gases, covers the epidermis. The **vascular bundles** (veins) run through the mesophyll, supplying water and minerals, and translocating away the photosynthate. Even superficially such a leaf presents a large surface : volume ratio. From the point of view of *cell* surface, the ratio is even greater, for most of the leaf surface is inside, the mesophyll containing much air space, which communicates with the external air via stomatal pores in the epidermis. This structural arrangement enables the photosynthetic cells to function in an internal environment protected against excessive water loss. The physiological efficiency of the leaf may be inferred from the fact that it has arisen independently on numerous occasions during the evolution of land plants, and in nearly all except the most primitive groups.

Not all leaves conform to the 'typical' pattern. There are succulent leaves where the thickness may exceed a centimetre; there are plants with vestigial leaves, the stems taking over the function of photosynthesis. These variations represent adaptations to extreme environmental conditions, and are considered in connection with water conservation in Chapters 3 and 13.

Illustrated accounts of leaf structure may be found in the texts by Mauseth listed under *Complementary reading* at the end of this chapter.

2.3 | Photosynthesis: light absorption and utilization

2.3.1 The capture of light

The photosynthesis of green plants requires light in the visible range of the spectrum, with wavelengths approximately in the range of 400–700 nm; this range is called the **photosynthetically active radiation** (PAR). The photosynthetic pigments have their absorption peaks in the blue and red wavelengths (chlorophylls *a* and *b*) or in the blue (carotenoids, comprising carotenes and carotenols), i.e. at the two ends of the visible spectrum (Fig. 2.2). Since sunlight that has penetrated the atmosphere has its maximum energy output almost in the middle of the visible spectrum, in the green and blue-green range (Fig. 2.2; see also Fig. 10.1), the photosynthetic pigments may seem at first sight to be somewhat poorly adapted to capture solar energy. When, however, the action spectrum of photosynthesis for a whole leaf is considered, i.e. the rate of photosynthesis for the same number of incident photons (quanta) is plotted against the wavelength, only a moderate dip in the rate is seen in the green region of the spectrum (Fig. 2.3). A photon of green light which is not absorbed immediately is likely to be reflected and refracted between many internal leaf surfaces. If it spends long enough inside the leaf it may eventually be absorbed by a photosynthetic pigment molecule in spite of the low absorptivity of pigments in the green region. Carotenoid absorption is shifted towards green wavelengths through their association with chloroplast membranes, and this improves absorption in the green wavelengths. With high irradiance, it may be advantageous for pigments not to absorb too strongly in the green, for excessive energy absorption can damage the photosynthetic system.

2.3.2   The utilization of light

It is convenient to divide photosynthesis into two stages: (1) the reactions which achieve the transduction of light energy to chemical bond energy; and (2) the reactions in which the chemical energy is utilized – in flowering plants mainly for CO_2 fixation. In the living plant these stages proceed simultaneously, but they can be separated experimentally.

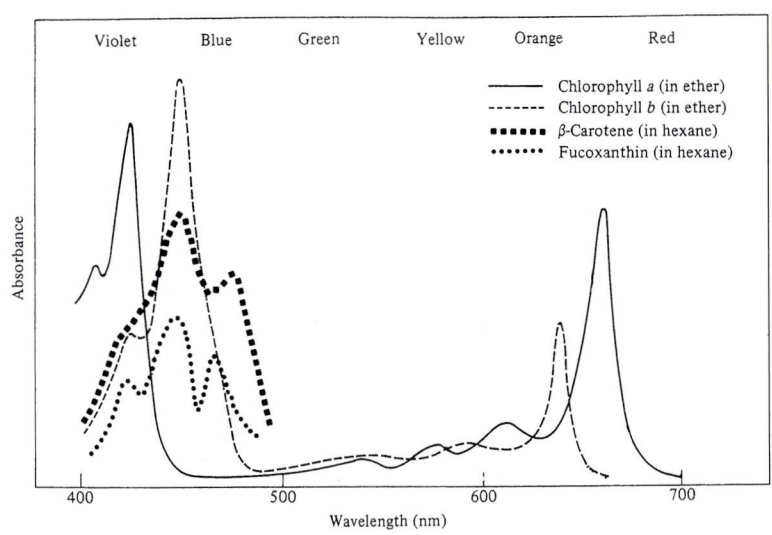

Fig. 2.2 Absorption spectra of chlorophyll *a*, chlorophyll *b*, β-carotene and fucoxanthin (a carotenol/xanthophyll). Maximal energy of sunlight lies in the green and yellow regions of the spectrum, the minimum region of absorption by chlorophylls and carotenoids. Adapted from Goodwin & Mercer (1972).

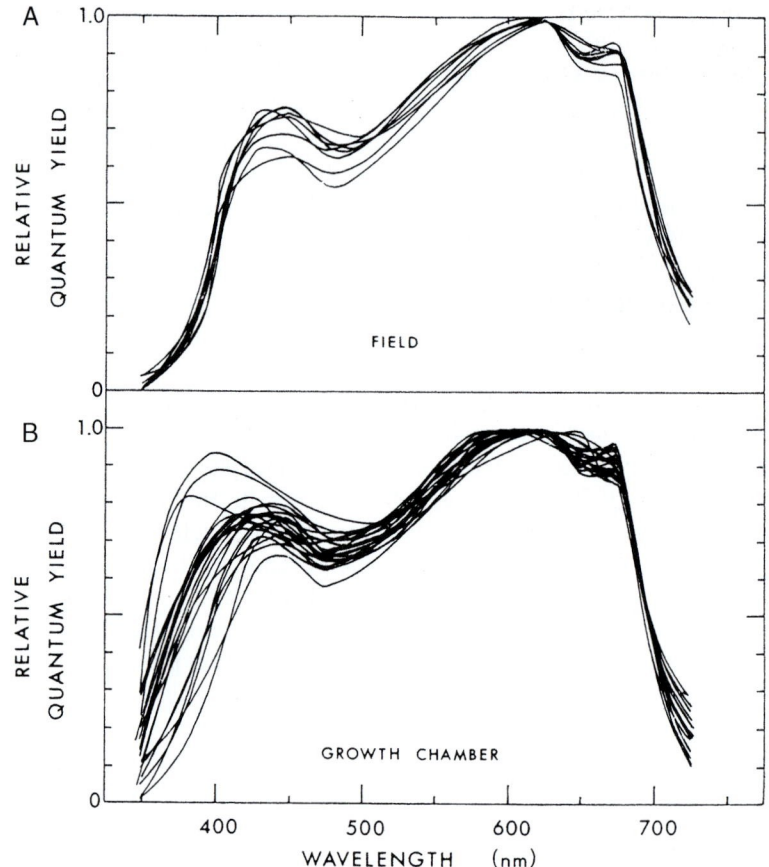

Fig. 2.3 The efficiency of energy utilization at different wavelengths in leaves of 8 species of crop plants grown in the field (A), and 20 species grown in a growth chamber (B). The efficiency is expressed as quantum yield, i.e. amount of C fixed for the same number of quanta, setting the maximum yield at unity. From McCree (1972). © Elsevier Science. Reprinted with permission.

The entire process of photosynthesis is carried out within the chloroplasts. The light-driven reactions occur in the thylakoid membranes (Fig. 2.4) which provide a very large surface area for light absorption and associated reactions. The thylakoid membranes contain a number of multimolecular protein complexes including the **photosystems I** and **II** (PSI, PSII), each consisting of several hundred pigment molecules and a number of specific proteins. One 'average' chloroplast may contain some 2 million photosystems. Each photosystem has a reaction centre containing a pair of chlorophyll *a* molecules in a special position. Only the reaction centre chlorophyll *a* molecules can undergo the photochemical reactions which are at the heart of photosynthesis. The remaining pigment molecules make up the light-harvesting pigment complexes, LHPC (also known as antenna pigments) and their function is to absorb photons and to channel the energy to the reaction centres. The LHP include carotenoids as well as most of the chlorophyll. The presence of LHPC enhances the efficiency of light capture very greatly over what it would be if the reaction centres had to rely on direct hits.

The energizing of the reaction centres by quanta of light energy results in a flow of electrons from water, the source, to the coenzyme

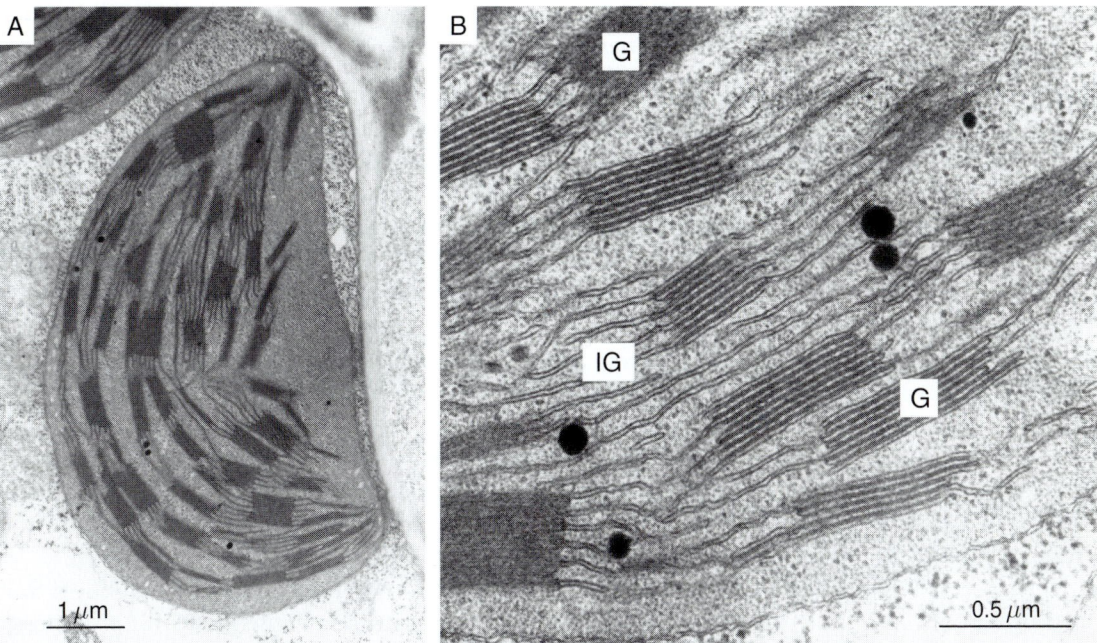

Fig. 2.4 (A) Low-power and (B) high-power electron micrographs of chloroplasts from the grass *Agrostis stolonifera* sectioned mainly at right angles to the thylakoid membranes. In three dimensions, the thylakoids are flattened membrane-bound sacs enclosing a narrow lumen. In a granum, G, the thylakoids are like stacked hollow disks; grana are joined by larger intergrana thylakoids, IG, so that the lumen within the thylakoids is probably a continuous if tortuous compartment through the whole chloroplast. The ground material or stroma contains some very densely stained lipid globules (plastoglobuli).

$NADP^+$ (nicotinamide adenine dinucleotide phosphate) along a precise path via the multimolecular complexes in the thylakoid membranes:

water-splitting complex → PSII → PSI → NADP-reducing system

The electrons together with the protons from the water reduce the $NADP^+$ to NADPH, while the oxygen (O_2) of the water is released as a by-product. In both PSI and PSII the electron receives a boost of energy. This means that **two** quanta of energy are used per electron, enabling green plants to use water as a reductant. Photosynthetic bacteria (cyanobacteria excepted) possess only PSI, use one quantum per electron, and utilize reductants which require a lower energy input than water. Concurrent with electron flow there is a synthesis of ATP from ADP and inorganic phosphate (Pi), the process of **photophosphorylation**, by a mechanism known as **chemiosmosis**, which can be summarized as follows:

(1) Inside the thylakoid lumen, hydrogen ions (H^+) accumulate from water splitting and from a coupling of the electron flow with an inward transfer of H^+ from the stroma. The H^+ being positively charged, this also makes the lumen more electropositive and the stroma more electronegative.

(2) The combined concentration gradient and electric potential gradient make a free energy gradient for the H^+, favouring their *outward* movement from the lumen into the stroma.

(3) The thylakoid membrane is highly impermeable to H^+. But it contains ATP-synthase enzyme complexes, with a proton channel through which the H^+ move to the stroma, and this movement, down their free energy gradient, is coupled with ATP synthesis.

The net result of the light reactions is thus the synthesis of ATP and NADPH, which can be utilized in CO_2 fixation (but can also be channelled into other processes, e.g. nitrate reduction). Normally all photosynthetic reactions occur simultaneously: ATP and NADPH cannot be stored and the cessation of illumination results in a stoppage of CO_2 fixation within a second or two. Several enzymes of CO_2 metabolism need light activation.

2.3.3 Levels of irradiance and rates of photosynthesis

Photometric units

In view of the basic role of light in photosynthesis, the rate of photosynthesis would be expected to vary with the amount of light available. Here it is appropriate to consider what exactly is meant by the 'amount' of light.

Since light is the energy source for photosynthesis, one's first instinct might be to measure it in *energy* units, say $J\ m^{-2}\ s^{-1}$ (energy per unit area, as joules per square metre per second). For energy balance sheets this may be appropriate. However, photochemical reactions are energized by individual quanta, the units of light energy carried by individual photons of light. **One** pigment molecule absorbs **one** quantum of energy at a time, to undergo **one** photochemical reaction. Hence, for many studies, the most meaningful measure of the 'amount' of light is the **number of photons** (or quanta), this number being given in **moles**. One mole of quanta can energize one mole of pigment molecules. The number of moles of photons of PAR, per unit area and unit time, is called the **photon flux density** or **PFD**. It is the PFD that exhibits the most direct relationship with the rate of photosynthesis. Bright sunlight has a PFD of 2000–2300 $\mu mol\ m^{-2}\ s^{-1}$.

Effects of varying the PFD: reactions of sun and shade plants

As the level of irradiance on a photosynthetic organ is increased, the rate of photosynthesis at first rises linearly, then levels off to a steady rate as light saturation is reached (Fig. 2.5). But the absorption of light does not fall proportionately, so that at increasing PFD, *less* CO_2 fixation takes place per photon absorbed: **photoinhibition** occurs. This was originally interpreted as due to photochemical damage. Excess light-excited chlorophyll can energize the formation of reactive oxygen species, ROS, from O_2: singlet oxygen $^1O_2{}^*$ and the superoxide radical $O_2{}^{\bullet-}$ (oxygen with an extra unpaired electron). These chemicals and their derivatives (Box 2.1) can destroy components of the photosystems. There is good evidence now that photoinhibition is in fact a *protective* process, during which excess energy is dissipated

---

Box 2.1

ROS, reactive oxygen species (alternative: AOS, active oxygen species) are extremely unstable, reactive and potentially destructive; they attack membranes by lipid peroxidation, and degrade DNA, RNA and proteins. From the superoxide radical $O_2{}^{\bullet-}$, reactions with cellular protons and electrons produce further ROS: the perhydroxyl radical $HO_2{}^\bullet$, the hydroxyl radical $OH^\bullet$ and hydrogen peroxide, H_2O_2. The symbol $^\bullet$ denotes an unpaired electron. *Small amounts of ROS are inevitably produced during photosynthetic and respiratory electron transport, and continually removed.* Superoxide is broken down by the enzyme **superoxide dismutase**, SOD, and hydrogen peroxide by **catalase**, two very fast-acting enzymes. SOD exists in several forms with different metal cofactors, FeSOD, MnSOD and Cu-ZnSOD, specific to subcellular locations. Cells also produce reductive antioxidants which react with ROS, including ascorbic acid (vitamin C) and glutathione. Numerous stresses stimulate the formation of ROS to levels which can be dangerous (Chapter 13).

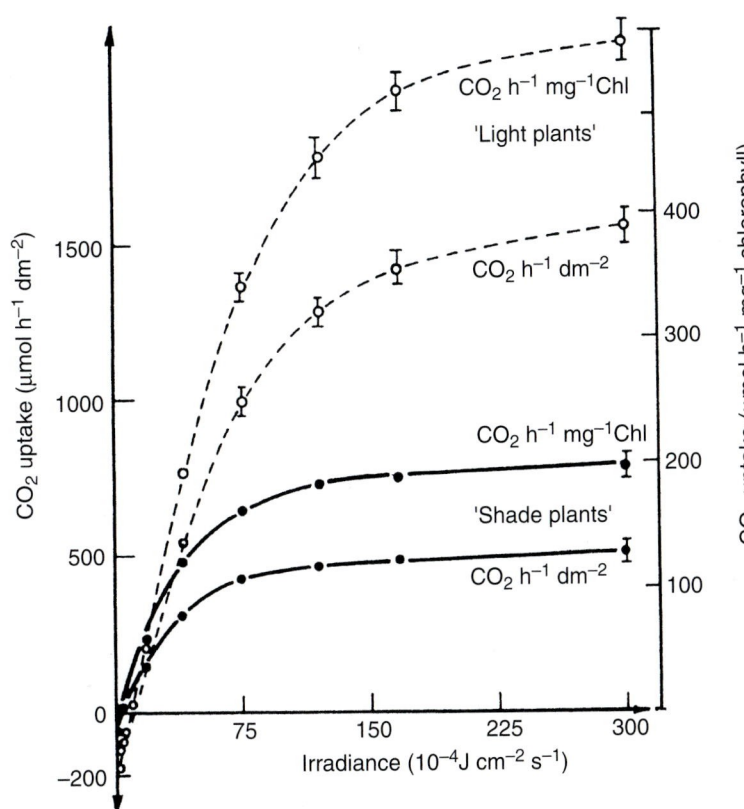

Fig. 2.5 Light saturation curves of photosynthesis for plants of *Sinapis alba* grown either under strong illumination, 'light (sun) plants' (dashed lines), or weak illumination, 'shade plants' (solid lines). The rate of photosynthesis changes similarly with change of irradiance whether expressed per unit of leaf area or unit of chlorophyll, showing that the differences in rates between sun and shade plants do not just result from a difference in total chlorophyll per unit area of leaf. Where the curves cut the *x*-axis is the *light compensation point*, below which respiration exceeds photosynthesis (negative CO_2 uptake = CO_2 output); this point lies at a lower irradiance for the shade plants. From Grahl & Wild (1972).

harmlessly (ultimately to heat) by reactions involving carotenoids in the thylakoids and preventing the buildup of ROS (Bartley & Scolnick 1995). When the PFD is too high to be counteracted by this process, then damage does occur, though it may still be repairable if not too extreme.

The irradiance level at which saturation occurs depends on a number of factors. If the temperature is very low, for instance, light saturation is reached at a low PFD: the rate of thermochemical reactions soon becomes limiting. Similarly, at low CO_2 concentrations, light saturation is reached once CO_2 has become limiting. Conversely, at higher temperatures and higher levels of CO_2, light saturation is reached at a higher PFD. However, other conditions being equal, significant differences in light saturation values are shown by individual photosynthesizing systems. Some species, e.g. plants of forest floors, are obligate **shade plants**, able to live only at low irradiance levels; examples include dog's mercury (*Mercurialis perennis*) and the enchanter's nightshade (*Circaea lutetiana*). There are also obligate **sun plants**, such as the aptly named sunflower (*Helianthus annuus*) and the daisy (*Bellis perennis*), plants of open habitats. Such species are genetically adapted for extremes of sun or shade. But for many species individuals can adjust appreciably to the light levels to which they are exposed during growth, as shown in Fig. 2.5. Trees commonly produce 'sun leaves' on the outside of the canopy, and 'shade leaves' within it. Shade plants (or leaves) become light-saturated at a much lower PFD than sun plants (or leaves). At low levels of light flux they have higher rates of photosynthesis than sun plants/leaves, whether the rate is measured per unit leaf area or per unit weight of chlorophyll. Shade plants have a low *light compensation point*, the value of irradiance at which photosynthesis exactly equals respiration, and below which respiration exceeds photosynthesis, leading to a net loss of organic matter. Numerical values of PFD at the light compensation point have been quoted as 20 μmol m^{-2} s^{-1} for shade plants, 80 μmol m^{-2} s^{-1} for sun plants. The shade plants can therefore survive at levels of light too low to support the growth of sun plants. In deep shade the PFD can fall below 50 μmol m^{-2} s^{-1}. Adaptations to growth in the shade include thin leaves (see Fig. 9.8) and very pigment-rich chloroplasts; there is a high proportion of LHPC to reaction centres, which increases the efficiency of light capture. In bright light the shade plants are relatively inefficient with respect to photosynthesis (Fig. 2.5) because of their low density of reaction centres, and therefore they are likely to be outcompeted by sun species. The shade plants are also highly susceptible to photochemical damage by bright light. The capacity for energy dissipation in shade plants is limited, whereas adaptable plants grown at a high PFD show increased levels of carotenoid pigments.

For whole plants, light saturation requires much higher levels of irradiance than for single leaves, because in an intact plant outer and upper leaves shade inner and lower ones. This shading is kept to a minimum by the arrangement of leaves in 'leaf mosaics', leaves arranging themselves so as to shade each other minimally (seen easily by looking up through the foliage of a tree!). Nevertheless, whereas a single leaf may be light-saturated with *c.* 25% of full sunlight, an entire plant may not reach light saturation even with the PFD of the full midsummer sun. Heavy clouding may bring a plant as a whole to its light compensation point.

2.4 | The fixation of carbon dioxide

2.4.1 The absorption of carbon dioxide

Gaseous diffusion

In the atmosphere CO_2 is present at an average concentration of about 370 μmol mol^{-1} (see Box 2.2). The leaf provides a large absorbing surface, and in this surface the stomata provide pores for entry. Within the leaf, the abundant air spaces permit gaseous diffusion between the cells and the large internal surface of the leaf is the main area for absorption of the gas into cells. The internal CO_2 concentration is kept below the atmospheric by photosynthesis.

The driving force for the inward movement of CO_2 is the concentration gradient, ΔCO_2, between the sites of fixation and the external atmosphere:

$$\Delta CO_2 = [CO_2]_{external} - [CO_2]_{internal} \qquad (2.1)$$

The ΔCO_2 is equivalent to a gradient of free energy: higher concentration is equivalent to higher free energy. The steepness of the gradient depends on both the external and internal concentrations. Under field conditions, photosynthesizing plants do not deplete the CO_2 supply in their vicinity greatly, for air mixes rapidly (though see below on boundary layers); however, CO_2 concentrations of about 270 μmol mol^{-1} have been measured within a crop.

During diffusion to the photosynthetic sites, the CO_2 molecules encounter resistances at various points: at the boundary air layer just outside the leaf; at the cuticle; at the stomata; and in the mesophyll. Since CO_2 concentration is often the limiting factor in photosynthesis, these resistances can determine the photosynthetic rate.

The **boundary-layer resistance** is the result of a layer of relatively still air, also known as the unstirred layer, immediately adjacent to the outside of the leaf. In this layer the CO_2 concentration is *lower* than in the bulk atmosphere owing to depletion by the leaf. Its presence has the effect of decreasing the effective ΔCO_2. In still air, a relatively thick boundary layer builds up over a plant surface and this slows down the rate of diffusion of CO_2 into the leaf; but usually there is sufficient air movement to keep the boundary-layer resistance low.

The **cuticle**, which forms a continuous layer over the epidermis, presents a very high resistance to CO_2 diffusion. As long as the stomata are open at all, the proportion of CO_2 entering through the cuticle is very small.

The **mesophyll resistance** is a combination of all the resistances that a CO_2 molecule meets while diffusing through the mesophyll air spaces, the cell walls, the plasma membrane, the cytosol and the chloroplast envelope, until it finally reaches the carboxylation sites within the chloroplast. This resistance accordingly depends on leaf structure and is more or less fixed once growth of the leaf has ceased.

The **stomatal resistance** depends on stomatal density (number per unit area) and the size of the stomatal pores. The stomatal density, like mesophyll structure, is fixed during leaf development. But the size of the pore is variable: stomata respond to several stimuli by 'stomatal movements', i.e. by opening or closing (partly or fully). The stomatal resistance is under physiological control.

Box 2.2

There are several ways of expressing the atmospheric concentration of CO_2 or other gases. The SI unit is used here, μmol mol^{-1}, micromoles per mole. Another unit in frequent use is ppm, parts per million, numerically equal to μmol mol^{-1} (since 1 μmol = 1 millionth of a mole). Other alternative units are % (per cent, parts per 100), or partial pressure as Pa, Pascals. Thus 370 μmol mol^{-1} = 37 Pa = 370 ppm = 0.0370%.

Box 2.3 | Diffusion of CO_2 and resistances

The rate of diffusion is inversely proportional to the resistance, R. If we denote the rate of entry of CO_2, which equals the rate of photosynthesis, by P, then

$$P = \frac{\Delta CO_2}{R} \tag{2.2}$$

R comprises components contributed respectively by the stomata, R_s; the cuticle, R_c; the boundary air layer, R_a; and by the mesophyll tissue, R_m. (Some authors refer to *conductances* rather than resistances; conductance is the reciprocal of resistance, $1/R$.) Because the stomatal and cuticular resistances act in parallel rather than in series, the mathematical relationship between them is

$$\frac{1}{R_{(s+c)}} = \frac{1}{R_s} + \frac{1}{R_c} \tag{2.3}$$

But since values of R_c are 500–1000 times higher than values of R_s, $1/R_c$ is negligible compared with $1/R_s$ and is consequently often omitted in calculations. The boundary layer, stomatal and mesophyll resistances all act in series and consequently can be added up to make R. If cuticular resistance is ignored, we can now expand Equation 2.2 :

$$P = \frac{\Delta CO_2}{R_a + R_s + R_m} \tag{2.4}$$

Usually there is sufficient air movement to keep R_a low relative to R_s and R_m, and variation in wind speed, once above a minimum, does not have much effect on CO_2 uptake. As stated in the text, mesophyll resistance R_m does not vary once growth has ceased. Hence R_s, the stomatal resistance, becomes the critical one.

Stomata

Most of the entry of CO_2 into photosynthetic tissues occurs through the **stomata** (singular: stoma). These are minute structures in the epidermis, consisting of two highly specialized elongate **guard cells** enclosing a pore between them (Fig. 2.6). The guard cells are often flanked by a few subsidiary (accessory) cells differing morphologically from the remaining epidermal cells. The shape of the guard cells and the arrangement of their cell-wall thickenings ensure that when the guard cells are more turgid than the subsidiary cells, the guard cells bulge outwards into the subsidiary cells and separate in the middle, opening the pore. When the guard cell turgor equals or is less than that of the adjacent cells, the guard cells shrink together and the pore closes. All intermediate stages between maximal opening, as permitted by the elasticity of the walls, and complete closure are possible. At full opening, the stomatal apertures of *Phaseolus vulgaris* measure only 3×7 µm, while fully open stomata of *Zebrina pendula* reach pore sizes of 12×31 µm. In the grass family, Poaceae, stomata are very elongate; a fully open stomatal pore of *Avena sativa* (oat) measures 8×38 µm. The stomatal frequency per cm^2 of leaf surface usually ranges from *c.* 1000 to 200 000. The apertures are so small that at the most 3% of the total leaf surface is occupied by the pores. Yet an illuminated leaf absorbs CO_2 from the atmosphere with great efficiency. A leaf can maintain a steep diffusion gradient for the gas, and many small pores have a large amount of *edge* in relation to their surface area. Gas diffusion through a hole is more rapid round

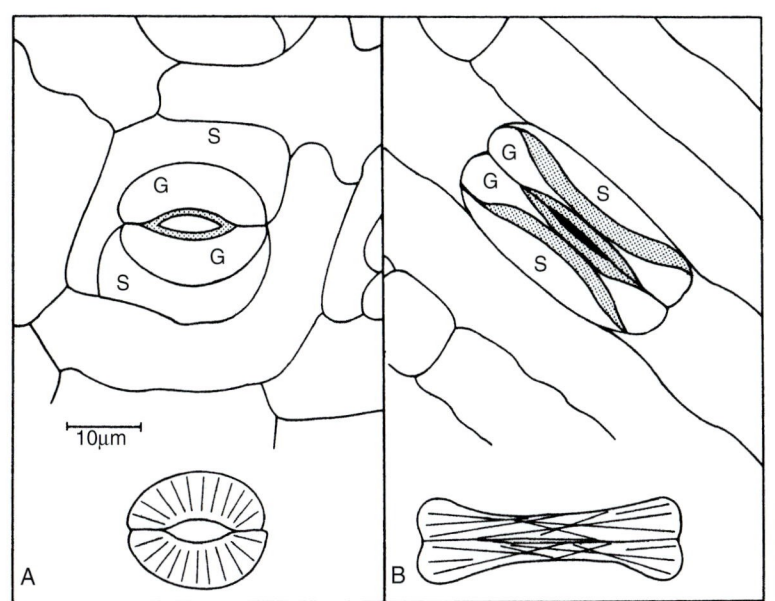

Fig. 2.6 The structure of stomata. The drawings show surface views of stomata. (A) Open stoma in a leaf of mung bean (*Vigna radiata*) and (B) closed stoma from a grass leaf. G = guard cell; S = subsidiary cell. The diagrams at the bottom of each drawing indicate the direction of cellulose microfibrils in the guard cell walls. When the guard cells are sufficiently turgid, they bulge apart (A), since the microfibrils cannot stretch, and open the pore; in the case of the grass stoma, only the bulbous cell ends are able to expand, again forcing the cells apart in the centre. See also Fig. 3.15.

the edges, where the molecules can fan out into the region of lower concentration.

The stomatal resistance R_s is nevertheless appreciable and the rate of photosynthesis depends very much on stomatal aperture; stomatal closure can reduce photosynthesis almost to zero. It is generally agreed that stomata have evolved in response to the need of land plants to permit the entry of CO_2 without excessive water loss. The high surface : volume ratio of a leaf, which makes it an efficient absorber of CO_2 and light, is equally conducive to water vapour loss. The cuticle protects against water loss, but, as stated, is also highly impermeable to CO_2. The stomatal pores which provide for CO_2 entry obviously act also as channels for the exit of water vapour, but they can open or close according to circumstances. Water stress has an overriding effect over all other stimuli, causing stomatal closure. This occurs at quite moderate levels of water stress, before wilting, i.e. the stomata do not close because of flaccidity of the guard cells, but the decline in the plant's water status acts as a specific stimulus for closure. A hormonal signal appears to be involved: the concentration of the hormone ABA (abscisic acid; see Section 7.2.7) rises rapidly in response to water stress and ABA causes stomata to close. The associated inhibition of photosynthesis is less detrimental than desiccation would be.

In many species stomata show a diurnal rhythm, opening by day and closing by night, light stimulating stomatal opening with blue wavelengths being most effective. Accordingly the pores are open during the period when light is available for photosynthesis; in the dark, when CO_2 cannot be assimilated, water loss is minimized. Stomatal opening is also promoted by low concentrations of CO_2 within the leaf – a feedback type of control.

Complex physiological mechanisms underlie the changes in guard cell turgor. The current view is that, basically, increases in guard cell turgidity follow from an active pumping of K^+ (potassium) ions into the guard cells, water then entering by osmosis. Decreases in guard cell turgor are attributed to an outward leakage of K^+ because of an opening of K^+ channels (Chapter 4) in the plasma membranes. This opening is promoted by increases in cellular Ca^{2+} concentrations caused by the closing stimulus. Chloride and malate anions accompany the K^+ cations so that ionic balance is maintained.

2.4.2 The pathways of carbon dioxide fixation

The mechanisms of energy transduction and the synthesis of ATP and NADPH are essentially the same not only in all flowering plants, but in all land plants, algae and cyanobacteria. The mechanism of CO_2 fixation, however, shows variations. The flowering plants can be divided into three categories according to their strategies for CO_2 fixation: the C_3 plants, the C_4 plants and the CAM (crassulacean acid metabolism) plants.

The C_3 cycle and C_3 plants

The C_3 cycle is the universal CO_2-fixing cycle present in all CO_2-fixing photosynthetic organisms (Fig. 2.7). It is located in the chloroplast stroma. The CO_2-fixing enzyme of this cycle is **Rubisco** (ribulose-1, 5-bisphosphate carboxylase-oxygenase). Rubisco catalyses the reaction of CO_2 with the 5-carbon (5-C) phosphorylated sugar, ribulose-1, 5-bisphosphate (RuBP), the CO_2 acceptor. The resulting 6-C compound immediately splits into two molecules of phosphoglycerate, PGA (see Box 2.4) and the PGA is reduced to triose (3-C, or C_3) sugar in reactions utilizing the products of the light reactions, ATP and NADPH:

$$CO_2 + RuBP \rightarrow 2PGA \qquad (2.5)$$

and

$$2PGA + 2ATP + 2NADPH \rightarrow 2C_3 + 2ADP + 2NADP^+ \qquad (2.6)$$

The name of the cycle derives from the fact that the first stable products are **three**-carbon compounds. Some of the sugar formed is withdrawn as C gain from the cycle and may be further processed to starch; the rest is recycled to replenish the acceptor RuBP, using one more molecule of ATP for every molecule of CO_2 fixed. The overall stoichiometry is therefore

$$1\ CO_2 : 3\ ATP : 2\ NADPH$$

A plant which fixes CO_2 exclusively by the C_3 cycle is known as a C_3 plant. The great majority of flowering plant species are C_3 plants. Export of carbohydrate from the chloroplast is mainly as triose phosphate, which is exchanged across the inner envelope membrane

Box 2.4

'Phosphoglycerate' is the anion of phosphoglyceric acid, which can dissociate to phosphoglycerate ions and H^+. The organic acids found in cells are weak acids and at the cytosolic pH they are largely dissociated. Hence they are often referred to by their anionic names – phosphoglycerate, pyruvate, glutamate, etc., although in formulae the acid form is commonly given; this terminology is followed in the present text.

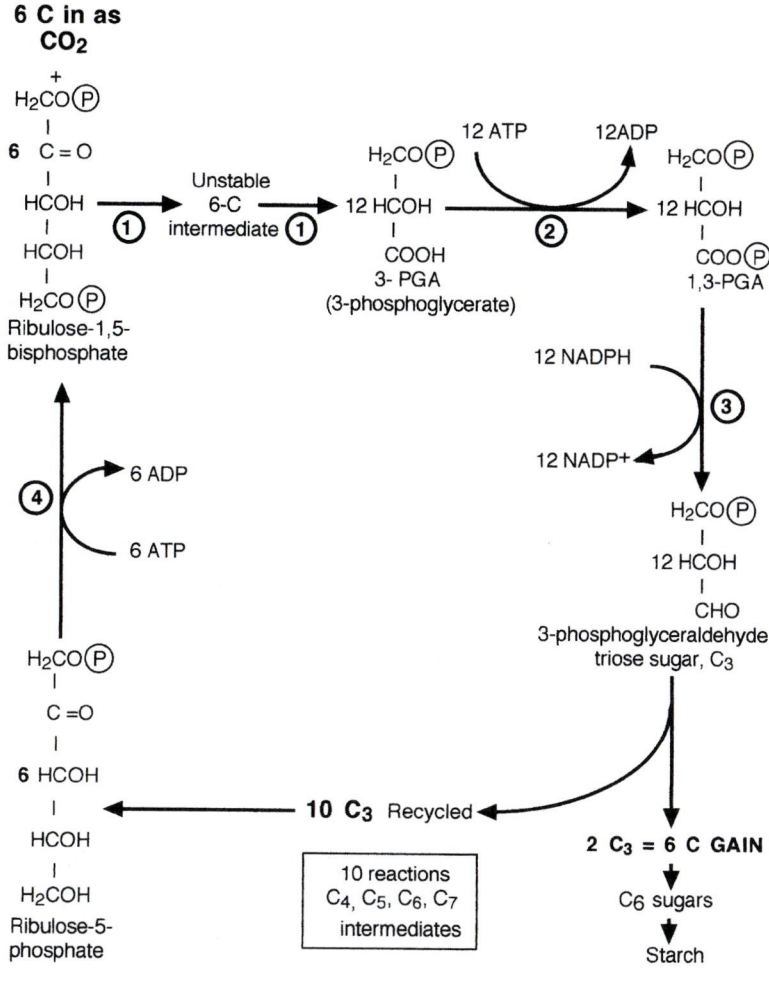

6 C in as
CO_2

Fig. 2.7 Summary of the C_3 cycle of photosynthetic CO_2 fixation. The enzymes catalysing the numbered steps are: ① Rubisco; ② phosphoglycerokinase; ③ glyceraldehyde phosphate dehydrogenase; ④ phosphoribulokinase. ℗ = phosphate group.

for Pi from the cytosol by a specific transport protein, the phosphate translocator.

Whilst carbohydrate is the main photosynthetic product in flowering plants, a proportion of the C_3 cycle intermediates is channelled into the synthesis of amino acids and lipids.

Photorespiration

The net rate or amount of CO_2 fixation is given by the difference between photosynthetic fixation and respiratory loss:

net photosynthesis = total (gross) photosynthesis − respiration

When leaf respiration rates are measured in the *dark*, they are low compared with the net rate of photosynthesis under favourable conditions of PFD, CO_2 concentration and temperature. Accurate values for the respiration rate of a photosynthesizing organ in the *light* are very difficult to obtain, for while respiration utilizes O_2 and evolves CO_2, photosynthesis utilizes CO_2 and evolves O_2. Estimates can be made by the use of isotopes; for instance, radioactive $^{14}CO_2$ can be supplied for photosynthesis, while respiration is evolving unlabelled CO_2. By such means it has been found that in photosynthetic tissues of C_3 plants, respiration is strongly stimulated by light. The light-stimulated respiration does not follow the same biochemical pathways as the 'dark' respiration proceeding in all organs in darkness and light. It is a distinct **photorespiration**, involving a special organelle, the **peroxisome**, in addition to chloroplasts and mitochondria (Fig. 2.8). The substrate is the newly fixed organic carbon, and a considerable loss of photosynthate can occur. Photorespiration can reduce net photosynthesis by up to 50%; 15–25% reductions are commonly quoted.

Photorespiration results from the potential for dual action by Rubisco. In addition to carboxylating RuBP with CO_2, the enzyme can also oxygenate RuBP with O_2 to give one molecule each of PGA and **phosphoglycolate**, a two-C compound (Fig. 2.9). The phosphoglycolate becomes the substrate for photorespiration, with eventual loss of CO_2, the intermediates reacting in turn with enzymes in chloroplasts, peroxisomes and mitochondria. The two gases CO_2 and O_2 compete in the reaction catalysed by Rubisco. Hence high levels of O_2 and low levels of CO_2 favour photorespiration, while high CO_2 concentrations suppress it. Photorespiration is stimulated by high irradiance, which may create a CO_2 shortage by promoting rapid assimilation. High temperatures also promote

Fig. 2.8 Electron micrograph from leaf of *Agrostis stolonifera*, showing a 'photorespiratory assembly' of closely adpressed organelles: chloroplast, C; mitochondrion, M; and peroxisome, P. The crystal in the peroxisome is catalase, an enzyme involved in photorespiration.

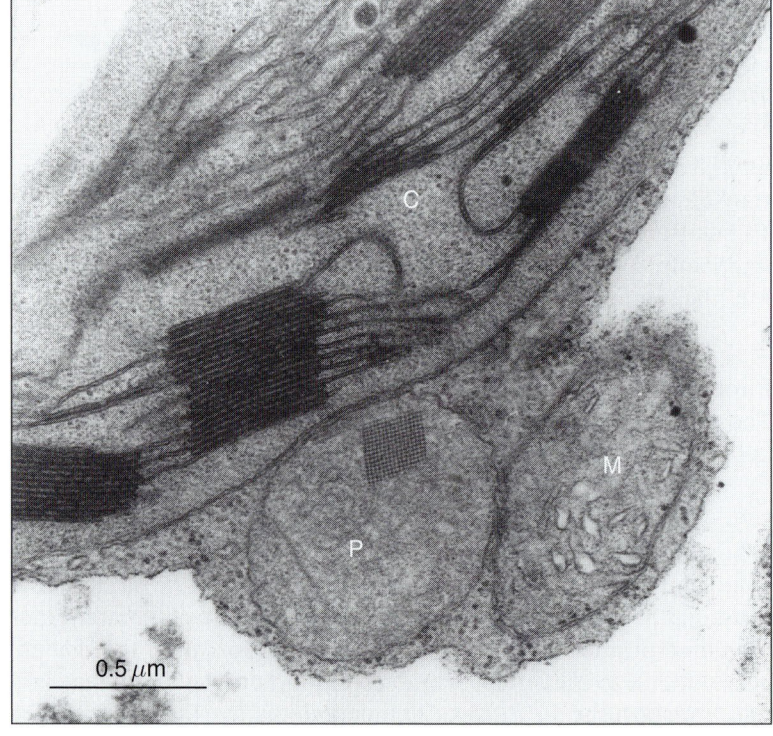

0.5 μm

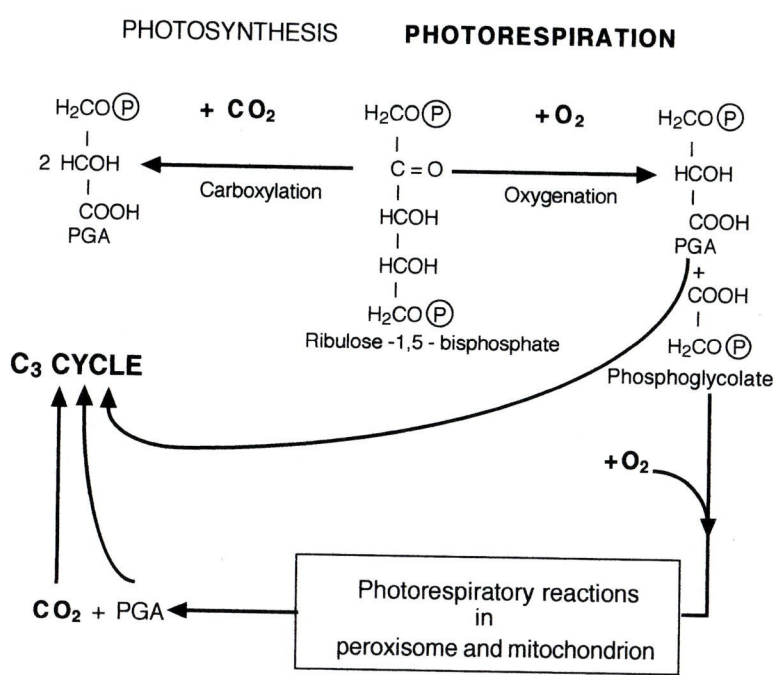

Fig. 2.9 Photorespiration: initial reactions and summary. (P) = phosphate group; PGA = phosphoglycerate. The oxygenation and carboxylation reactions are both catalysed by Rubisco.

photorespiration through rapid assimilation of CO_2. Additionally, with rising temperature the affinity of Rubisco for O_2 relative to CO_2 becomes greater, whilst the solubility of CO_2 relative to O_2 falls; this is relevant, for both gases must dissolve in cell water to reach the reaction sites.

Photorespiration seems to be a wasteful process, dissipating newly fixed C to CO_2 again and although some ATP is produced, there is no net ATP gain. One view regards photorespiration as an evolutionary accident. We know that photosynthesis originally evolved in an atmosphere rich in CO_2 but devoid of O_2. If Rubisco by chance evolved with an oxygenase activity, this would have been latent, with no evolutionary pressure against it. Then, when O_2 accumulated in the atmosphere – as a result of photosynthesis – the series of reactions we term photorespiration evolved in response to the need to metabolize the phosphoglycolate, which must be removed or it accumulates indefinitely. A mutant of *Arabidopsis thaliana* is unable to metabolize the phosphoglycolate and in normal air it dies; it can survive in an atmosphere of 1% O_2, which suppresses the oxygenase activity of Rubisco. Alternatively, photorespiration has been assigned the function of protecting the photosynthetic apparatus from damage under conditions of high PFD and high temperature. Under these conditions, the pigments may absorb more energy than can be utilized in CO_2 reduction. As stated earlier (Section 2.3.3), if photoexcited chlorophyll is unable to pass on electrons to normal acceptors, ROS such as the superoxide radical are formed and react destructively with photosynthetic pigments and other

Box 2.5

Strictly speaking, the substrate for PEP carboxylase is bicarbonate, HCO_3^-, derived from CO_2 by the action of carbonic anhydrase (Section 4.2.2, p. 105). The cellular concentration of bicarbonate depends on that of CO_2 and it is customary to give the Km value of the enzyme in terms of CO_2 concentration.

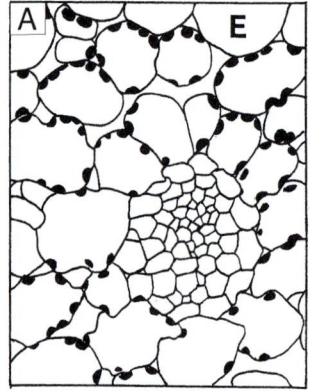

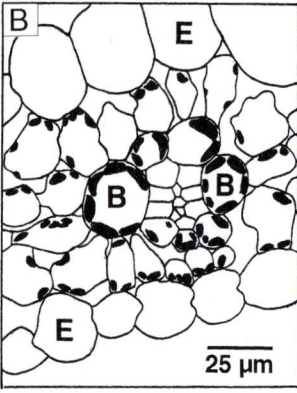

Fig. 2.10 (A) C_3 plant (barley, *Hordeum vulgare*) and (B) C_4 plant (maize, *Zea mays*) leaf cross-section, both to same scale. The C_4 plant is distinguished by large bundle sheath cells, B, rich in chloroplasts. E = epidermis; chloroplasts shown black.

components of thylakoid membranes. Photorespiration can dissipate surplus energy by recycling the CO_2.

Photorespiration also results in a considerable cycling of nitrogenous compounds; one can talk of the photorespiratory nitrogen cycle. The processing of the glycolate involves amino acids as intermediates, and for every molecule of CO_2 released, a molecule of ammonia (NH_3) is also produced. This is assimilated to glutamate in the usual way (see Chapter 4), and the glutamate then returns to the cycle.

The C_4 cycle and C_4 plants

Rubisco is not the only plant enzyme capable of fixing CO_2. In some species application of radioactive ^{14}C-labelled CO_2 to photosynthesizing leaves results in the tracer appearing first in the 4-C acids malate and/or aspartate. Only after a time-lag does the radioactivity appear in PGA and other C_3 cycle intermediates. As these experiments indicate, these species run an additional reaction series, the C_4 cycle, *preceding* the C_3 cycle in which the final CO_2 fixation takes place. The C_4 cycle by itself cannot achieve a net fixation of CO_2. These plants are known as the C_4 plants. In such plants there is typically a division of labour between two types of photosynthetic cells, mesophyll and bundle sheath. Both cell types are concentrically arranged round the vascular bundles. In the mesophyll, which contacts the epidermis, the cells are loosely packed with air spaces into which the stomata open. The bundle sheath cells are tightly packed around the vascular bundles and have large chloroplasts (Fig. 2.10). This leaf structure is known as 'Kranz' anatomy, from the German word for wreath; the appearance of the bundle sheaths as seen in cross-section is reminiscent of wreaths. The primary CO_2-fixing enzyme of the C_4 cycle is **PEP carboxylase** (phosphoenolpyruvate carboxylase) which catalyses the reaction of CO_2 (see Box 2.5) with a 3-carbon acceptor, **phosphoenol pyruvate**, to produce the 4-C acid oxaloacetate (Fig. 2.11). The PEP carboxylase is confined to, or mainly concentrated in, the mesophyll cells and is present in their cytosol. The oxaloacetate does not accumulate but is immediately either reduced to malate or transaminated to aspartate, according to species. These 4-C acids then travel to the bundle sheath cells where they are decarboxylated to yield CO_2 again, which is fixed by Rubisco and the C_3 cycle proceeds in the normal way. The Rubisco is confined to, or concentrated in, the bundle sheath cells. The 3-C fragment left after the decarboxylation of malate or aspartate returns to the mesophyll where it is reconverted to PEP, ready for another carboxylation. There is much chemical traffic between the mesophyll and bundle sheath cells, believed to occur by diffusion through plasmodesmata. There is no evidence for active transport; the plasmodesmata are very abundant and calculations indicate that, over the distances involved, diffusion should be adequate.

The above may seem to be a roundabout way of fixing CO_2. Extra ATP, too, is needed; there are three biochemical variants of the C_4

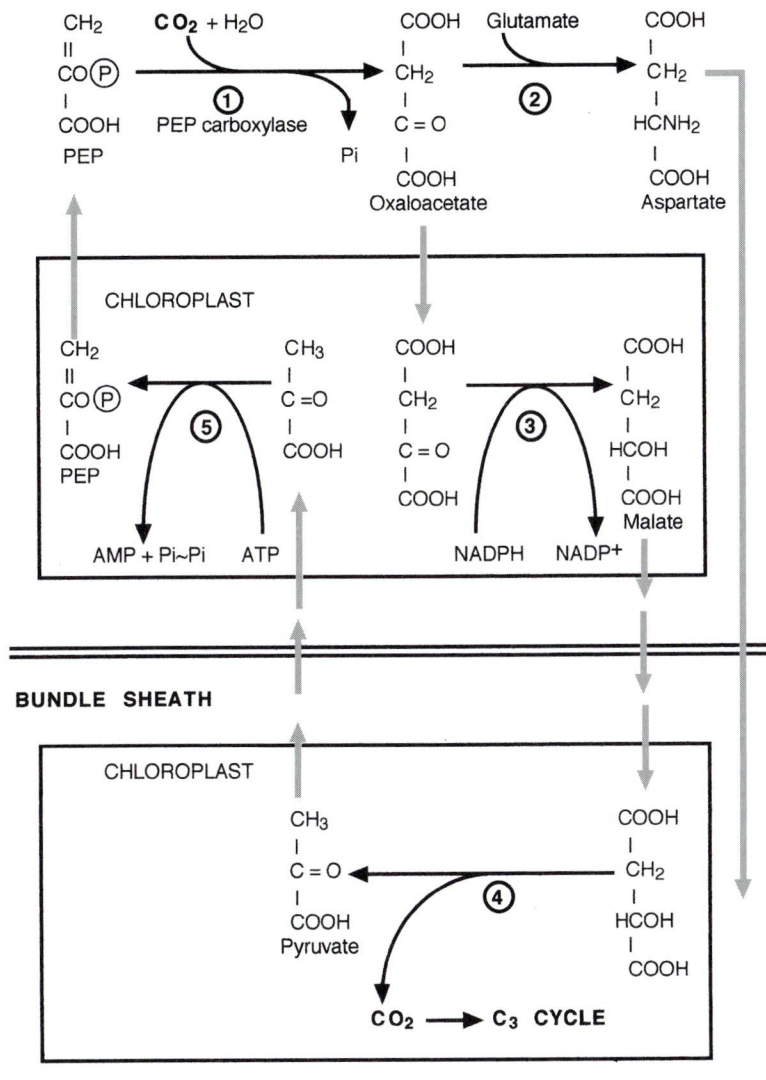

Fig. 2.11 Summary of the C_4 cycle of photosynthetic CO_2 fixation with particular reference to the most common malate pathway. Black arrows denote chemical reactions, grey arrows movement of chemicals between cells and subcellular compartments. The enzymes catalysing the numbered steps are: ① PEP carboxylase; ② glutamate, 2-oxoglutarate aminotransferase; ③ malate dehydrogenase; ④ malic enzyme; ⑤ pyruvate, inorganic phosphate dikinase. ⓟ = phosphate group; Pi = inorganic phosphate; Pi~Pi = inorganic pyrophosphate. Details of reactions involved in the decarboxylation and recycling of the malate to pyruvate have been omitted. Aspartate, if formed, is decarboxylated in the bundle sheath cytosol and also eventually yields pyruvate.

cycle, requiring, according to the variant, four or five molecules of ATP for every molecule of CO_2 finally fixed, compared with the three ATP per CO_2 as needed for the C_3 cycle by itself. Actually the C_4 plants are very efficient at fixing CO_2. Firstly, PEP carboxylase has a higher substrate affinity than Rubisco and, being in the outer cells, PEP carboxylase captures the gas efficiently from its low concentration in the air. But even more importantly, the C_4 cycle acts as a mechanism for **concentrating CO_2** at the site of Rubisco; the cycle has been called a CO_2 pump. The decarboxylation of the 4-C acids in the bundle sheath cells results in a localized concentration of CO_2 10–20 times higher than the CO_2 concentration in photosynthetic cells of C_3

Table 2.1 Maximum rates of dry matter production by C_3 and C_4 crop and pasture plants in the field, per unit leaf area. Energy conversion efficiency is given as the percentage of intercepted light energy utilized. From Hatch (1992).

Plant type	Dry matter production ($g\ m^{-2}\ day^{-1}$)		Energy conversion efficiency (%)
	Range	Average	Average
C_4	37–57	49	4.1
C_3	13–43	26	2.6

plants. Back-diffusion of the CO_2 is prevented by a very low permeability to the gas at the bundle sheath/mesophyll interface. The high concentration of CO_2 suppresses the oxygenase activity of Rubisco and consequently suppresses photorespiration in C_4 plants, in which photorespiration is virtually undetectable. The photorespiratory enzymes are present in small amounts, and probably a low level of photorespiration does occur, but the CO_2 thus released would be captured by the PEP carboxylase in the mesophyll cells before it could diffuse outside and be detected. Thus C_4 plants can synthesize efficiently at high PFD and high temperature; the optimum temperature for net photosynthesis in normal air is 30–40 °C for C_4 plants, against 10–25 °C for many C_3 species. Under optimal conditions of light and temperature, C_4 plants can attain, in the field, rates of net photosynthesis averaging about twice those of C_3 plants (Table 2.1). Sometimes the C_4 species have been termed 'high photosynthesis plants' contrasted with the 'low photosynthesis' C_3 species, but this is too sweeping a statement. When photorespiration is suppressed in C_3 plants by high external levels of CO_2, they can reach rates of net photosynthesis as high as C_4 plants at high irradiances and high temperatures. At lower levels of irradiance and temperature, when CO_2 concentration is not limiting, C_4 plants do not have an advantage. It might be more appropriate to term C_4 species 'low photorespiration plants' than 'high photosynthesis plants'.

C_4 plants also are more efficient at using water; they can photosynthesize efficiently without full opening of the stomata, and thereby suffer less water loss per unit of C fixed.

Crassulacean acid metabolism (CAM) plants
There is a third strategy of fixing CO_2, **crassulacean acid metabolism or CAM**. The name is derived from the family Crassulaceae in which CAM was first observed, although it is by no means confined to this family. In CAM plants the stomata typically open at night and close by day. During darkness, CO_2 is fixed by PEP carboxylase to form malate, as in C_4 plants, and the malate is stored in vacuoles. The PEP is derived from stored carbohydrate. The photosynthetic tissues of CAM plants are succulent, with very large vacuoles, and the

vacuolar pH falls sharply at night (dark acidification). The nightly fixation is promoted by low temperatures. During the day, the PEP carboxylase is inactivated by light, and the malate is transported to the cytosol and decarboxylated to provide CO_2 for Rubisco and the C_3 cycle. Biochemically CAM is very similar to C_4 photosynthesis, but whereas in C_4 plants CO_2 fixation by PEP carboxylase and by Rubisco occur simultaneously but separated in space, in different cells, in CAM plants the two enzymes act in the same cells but sequentially, separated in time. CAM metabolism enables plants to conserve water by closing the stomata in the heat of the day, when transpiration would be most rapid. The large vacuoles also serve for water storage. CAM plants are extremely economical on water. But the advantage has its price. To provide a supply of PEP during the night requires both substrate and energy; in CAM plants the ATP requirement is 5.5–6.5 ATP molecules per molecule of CO_2 fixed, and substrate equalling 20% leaf dry weight can be respired nightly. The small surface : volume ratio of succulent organs is moreover unfavourable for gas exchange, and the rate of net photosynthesis of CAM plants is low, as is their growth rate.

The taxonomy, ecology and evolution of C_3, C_4 and CAM plants
The majority of flowering plant species are C_3 species, with C_4 plants accounting for only about 1%, while species exhibiting CAM have been estimated at 5–10%. In assessing their importance, however, their overall biomass as well as species numbers must be considered, and in fact C_4 and CAM plants are of considerable importance ecologically and to some extent also agriculturally.

Both groups of plants are confined to a limited number of families and also have characteristic ecological distributions. C_4 photosynthesis is particularly associated with tropical and semitropical grasses. Sixty per cent of all C_4 species belong to the Poaceae (grasses including cereals); this includes the two very important and highly productive crops, maize (*Zea mays*) and sugarcane (*Saccharum officinarum*). It also includes eight grasses considered at least by some authors to be among the world's most troublesome weeds. Another 32% of C_4 species can be found in five other families: Amaranthaceae, Asteraceae, Chenopodiaceae, Cyperaceae and Euphorbiaceae, with the remainder scattered as odd genera or species amongst another dozen families. No flowering plant family is known to consist exclusively of C_4 species. Almost all C_4 species are natives of warm habitats; few are found in temperate zones and none in arctic regions. C_4 plants are, as stated earlier, adapted to take advantage of high temperatures and high irradiance and to conserve water, which also can run short in warm climates.

CAM plants are found in about 30 flowering plant families, although 80% of the species belong to five families: Cactaceae and Crassulaceae (consisting exclusively of CAM species), Aizoaceae, Bromeliaceae, and Orchidaceae. Some families include examples of both CAM and C_4 plants, e.g. Asteraceae and Euphorbiaceae.

Economically CAM plants do not make a great contribution; the pineapple (*Ananas comosus*), the vanilla orchid (*Vanilla fragrans*) and agave (*Agave americana*) are CAM crops. The significant feature of CAM plants is water conservation, and they are found in arid regions. The succulent flora of a hot desert may consist nearly entirely of CAM species. But CAM plants are found also in water-poor niches in temperate climates; such are, for example, the stonecrops (*Sedum* spp.) which grow on stony ground. Another large group of CAM plants embraces the bromeliads and orchids growing epiphytically on trees in tropical rainforests. One thinks of rainforests as very wet areas; but epiphytes, which have no access to soil water, may be water-limited even in a humid climate and at the top of the canopy, humidity levels during the day can be very low.

There are *facultative* CAM plants (C_3–CAM intermediates), which can switch from C_3 metabolism to CAM in response to water stress; an example is the ice plant (*Mesembryanthemum crystallinum*) from the Mediterranean region. The aquatic plant *Hydrilla verticillata* on the other hand develops C_4 characteristics in response to a low level of CO_2. In some species the induction of CAM is rapidly reversible, in others it is permanent. Under certain environmental conditions, some CAM intermediates do not close stomata fully by day and thus fix CO_2 by day and night.

It is believed that both C_4 photosynthesis and CAM arose several times independently in evolution, since both types of metabolism are found in totally unrelated families, including monocotyledons and dicotyledons, which diverged very long ago; CAM is exhibited also by some spore-bearing plants. One genus can contain representatives of all three photosynthetic types: *Euphorbia corcollata* – C_3; *E. maculata* – C_4; *E. grandidens* – CAM. The selection pressure for CAM has obviously been water stress. It has been suggested that the evolution of C_4 photosynthesis has been stimulated by the decline in atmospheric CO_2 levels which occurred from 120 to 30 million years ago. Evolution is considered further in Section 2.7.1.

2.5 | Limiting factors for photosynthesis

A leaf is photosynthesizing in a garden. What determines its rate of photosynthesis at a given moment? This is not an easy question to answer.

2.5.1 Limitation in the short term

The rate of photosynthesis is affected by the PFD, CO_2 concentration, temperature, wind speed, the plant's water status and the degree of stomatal opening. The levels of some inorganic nutrients, too, may affect the rate; e.g. an adequate phosphate supply is critical for maintaining the levels of phosphorylated intermediates, and for normal opening of stomata. The effect of varying any one of these factors depends on the value of the others so that, in a sense, the rate is determined by *all* these factors acting together. The rate also depends on the intrinsic capacity of the particular system being observed; the leaves of two different species side by side in the garden will probably not have the same photosynthetic rates. It is, however, possible that one particular factor is **limiting** the process at a particular time. If photosynthesis is light-limited, then only an increase in the PFD will achieve an increase in the rate (and of course a decrease in PFD will cause a fall in the rate). But an increase in the irradiance level may bring the plant to a state where the CO_2 concentration becomes limiting, or temperature. At dawn on a warm summer morning, a well-watered plant would be light-limited; by midday the CO_2 concentration is most likely to have become the limiting factor. A final maximal limit to the rate of photosynthesis would be set by the plant's photosynthetic system. In the field, a plant would very seldom be functioning at its theoretical maximal photosynthetic capacity, i.e. with all the external controlling factors at their optimum levels. Photosynthesis is also subject to controls from the rate of utilization of photosynthate: accumulation of end products in leaves is inhibitory, utilization by growing areas ('sinks') is stimulatory.

The effects of temperature on photosynthesis are complex. Temperature has no effect on light absorption and on the primary photochemical reactions, which proceed with a Q_{10} of 1. (The Q_{10} is the ratio of the rate of a process at $[x + 10]$ °C to its rate at x °C.) Except for the primary energizing of the reaction centres, photosynthesis involves thermochemical reactions and the overall process shows a Q_{10} of 2–3 (except at a very low PFD when the photochemical reactions are limiting), up to the temperature optimum of the particular plant. But temperature has indirect effects on photosynthesis, high temperatures promoting water loss, with decrease or even complete closure of stomatal apertures.

2.5.2 Long-term effects; global limitation

The photosynthetic capacity of a plant, or plant organ, depends on the conditions under which it develops. A deficiency of inorganic nutrients will result in poor development of the photosynthetic apparatus. In particular, *nitrogen* is needed in relatively high amounts; Rubisco normally forms 20–25% of total leaf protein and the thylakoid membranes have a high protein content; chlorophylls are nitrogenous compounds. Leaves grown under N-deficient conditions will therefore have lowered rates of photosynthesis. The temperature at which a plant develops affects the amount of photosynthetic apparatus formed per unit leaf area, lower temperatures resulting in higher levels of chloroplast components. The effect of irradiance levels on the development of the photosynthetic apparatus has already been discussed (sun and shade plants, Section 2.3.3). Thus while the rate of photosynthesis of the leaf photosynthesizing in the garden may at a particular moment be limited by one particular factor, it is also limited by the capacity of its photosynthetic system, which during leaf growth will have been subject to various influences.

To take a global and long-term view, the availability of **water** is thought to be the most critical limiting factor for photosynthesis on dry land, and hence for agricultural production; famine caused by drought has scourged humanity down the ages. Not only is water one of the raw materials for photosynthesis, but water stress causes stomatal closure and has deleterious effects on numerous plant processes, which can impinge on photosynthesis. The most productive plant communities are the ones best supplied with water.

So here is the paradox. The basic requirements for photosynthesis are light, CO_2 and water. Of the PAR reaching the earth's surface, perhaps some 0.15% is used in photosynthesis; of the available CO_2, only a fraction is fixed. Nearly four-fifths of the globe is covered by water. Yet shortages of light, CO_2 and water keep limiting photosynthesis. One aspect of the problem is that all the necessities are not always plentiful *in the same place simultaneously*, and combined with a suitable temperature and an adequate nutrient supply. A bright sunny day may be too dry; a rainy day may have a limiting level of irradiance. There is an inexhaustible supply of water in the oceans, but a cactus in the Arizona desert is in a decidedly dry place. Light is bound to become limiting at dawn and dusk. As regards CO_2, although the total global store is very large, its *concentration* in the air (and water in equilibrium with the air) is low, and reaction rates are determined by concentrations.

2.6 | The efficiency of energy conversion in photosynthesis

As discussed at the beginning of this chapter, photosynthesis is essentially a process of energy conversion. The efficiency of this conversion can be considered at a range of levels, from the quantum efficiency (thermodynamic efficiency) of the biochemical energy conversion process, to the efficiency of biomass production on the ecological scale, or even ultimately the efficiency of yield of an agricultural product. These levels are treated in turn.

2.6.1 Quantum efficiency
The overall process of photosynthesis can be expressed by the van Niel equation:

$$CO_2 + 2H_2O \rightarrow [CH_2O] + H_2O + O_2 \tag{2.7}$$

where $[CH_2O]$ represents one-sixth of a carbohydrate molecule. However, the radiant energy is used in the first instance to synthesize ATP and NADPH, of which minimally three and two molecules respectively are needed to reduce one molecule of CO_2; hence the gain in free energy, ΔG, per mole of CO_2 fixed into carbohydrate, must be calculated as follows (Pi = inorganic phosphate):

$$2H_2O + 3ADP + 3Pi + 2NADP^+ \rightarrow 3ATP + 2NADPH + 2H^+$$
$$\Delta G = 565 \text{ kJ}$$
(2.8)

The *positive* value of the free energy, ΔG, indicates an energy gain, or input. For C_4 and CAM plants the extra synthesis of ATP must be taken into account. The energy input of course comes from the light.

Light has a dual nature. Although it travels as a continuous wave motion, it can also be regarded as composed of particles, photons, each associated with one quantum of energy. Quanta are indivisible and as explained in Section 2.3.3, **one** quantum at a time is absorbed by a molecule and energizes **one** photochemical reaction. The energy per quantum is not a constant but it varies inversely with the wavelength, i.e. light with twice the wavelength has half the energy per quantum. Taking red light at 670 nm as an example, it can be calculated that if one quantum at 670 nm is allowed per every molecule of CO_2, the energy provided per one **mole** of CO_2 is 179 kJ. This is well short of the theoretical minimum of 565 kJ needed per mole CO_2; obviously more than one quantum is used for every molecule of CO_2. Many experimental measurements have given values of 8–12 quanta per molecule of CO_2 and the most generally accepted scheme of the light-driven reactions is compatible with a value of 8. This corresponds to an efficiency of energy conversion of about 30%. Some reports claim a higher efficiency with only 5–6 moles quanta required per mole CO_2, which would raise the efficiency of energy conversion to 40–48%. On the other hand, for C_4 and CAM plants the quantum efficiency is lower because of their need for more ATP per mole CO_2.

Quanta at the blue end of the spectrum, with shorter wavelengths, have a higher energy value per quantum. However, the quantum requirement per molecule of CO_2 is the same throughout the PAR range: the higher energy of the shorter-wavelength quanta is lost as heat because, as a first approximation, the energy level of light-excited chlorophyll molecules corresponds to the quantum energy of **red** light. In terms of joules of energy conserved, red light can be said to be most efficient. According to the wavelength, the efficiency of energy conversion (taking the quantum requirement as 8 quanta per 1 molecule of CO_2) varies over the spectrum from about 22 to 35%, whilst the quantum requirement does not vary throughout the PAR. (Some workers prefer to use the *quantum yield*, molecules of CO_2 fixed per quantum, instead of the quantum requirement; e.g. a quantum requirement of 8 corresponds to a quantum yield of $1/8 = 0.125$.)

2.6.2 Efficiency in the field; crop efficiency

The efficiency of energy conversion as calculated above applies only for the quanta *actually utilized by the photosynthetic pigments*. Of the radiation absorbed by a photosynthetic organ, some will also be absorbed by other molecules and will be transformed to kinetic energy, heat, to be ultimately re-emitted as infrared radiation. The quantum efficiency calculated for the total light absorbed by a leaf will therefore be less than the values presented in Section 2.6.1. The efficiency in terms of total *incident* radiation falling on an area of vegetation will be much lower still, for some of the light will be transmitted or reflected and some will fall on barren ground. Thus the proportion of incident light energy utilized for photosynthesis **in the field** is much lower than the 22–35% efficiency of energy conversion calculated in the preceding section. Under the best field conditions, the maximum gross photosynthesis may achieve a utilization of the order of 10% of the incident PAR. The efficiency becomes much lower if averaged over an extended timescale including unfavourable seasons and periods of dormancy. Water stress, limiting availability of CO_2, and disease can further decrease the efficiency. Moreover, a considerable proportion of the photosynthate is respired by the plant. When all these losses are taken into account, even under conditions of intensive agriculture, dry matter production by a crop over a growing season works out at equivalent to a conservation of less than 1% of the solar energy available. The energy conserved in the utilizable harvest product (seeds, tubers) will be less still.

2.6.3   Prospects for improving crop efficiency

We have just seen that less than 1% of the available PAR is realized as harvestable product. Earlier (Section 2.2.1) we also saw that global photosynthesis utilizes only a fraction of the atmospheric and oceanic C stores per annum. The potential for improved photosynthetic efficiency of crops is theoretically present as far as the supplies of light and CO_2 go.

The quantum efficiency of energy conversion, equivalent to a utilization of some 22–35% of the available energy, is much higher than the overall storage of less than 1% of the available incident energy in organic material, over a growing season. The efficiency of energy conversion, once a quantum has been absorbed by a pigment molecule, is thus not the limiting factor. Since an appreciable fraction of the incident PAR fails to be absorbed by plants, one approach might be to try to increase the absorption of light by crops. The ratio of leaf area to ground area is called the **leaf area index** (LAI), and an increased LAI might lead to increased light absorption; on the other hand, an excessive LAI would lead to severe mutual shading of leaves and defeat the object by decreasing light absorption by the lower leaves. The ideal values can be determined by computer modelling. Plants with roughly vertical leaves, such as cereals, are the best candidates for benefiting from a high LAI, since with vertical leaves mutual shading is minimal. Photosynthetic yields have been improved in wheat (*Triticum aestivum*) and maize by breeding for an increased LAI.

C_4 plants can attain, in the field, rates of net photosynthesis averaging twice those of C_3 plants (Table 2.1). Nearly all important crop plants of the world are C_3 species, but the C_4 crops maize and sugarcane are very productive. Ever since the C_4 pathway was discovered, scientists have been mooting the idea of trying to convert C_3 crops to C_4 metabolism, with the hope of boosting net photosynthesis and harvest yield. Techniques of genetic engineering now permit us to transfer genes between species, but C_4 photosynthesis does not depend on a single extra gene or even on a few defined extra genes. The enzymes of the C_4 cycle are in fact not unique to C_4 plants; nearly all are present in all flowering plants, though amounts may be low. What is unique to C_4 plants is a combination of properties: the quantity in which the enzymes are found and the way in which they are partitioned between the mesophyll and bundle sheath cells; the morphology of the leaves; and the transport and permeability properties of the mesophyll/bundle sheath interface. To transfer the entire 'C_4 syndrome' is a pretty tall order. Many genes – including some as yet unidentified – must be involved. Any genetic engineering not only needs to achieve transfer of genes, but must also ensure that particular genes are expressed in the appropriate cells only. Nevertheless, progress in understanding genetic control of the cycle is being made. Even single-gene transfer has potential: the transfer of multiple copies of the maize PEP carboxylase gene into the C_3 cereal rice has resulted in high levels of the enzyme and a reduction in photorespiration (Ku *et al.* 1999). But overexpression of C_4 enzyme genes has also led to metabolic disturbances and stunted growth.

It should also be remembered that C_4 photosynthesis is not a universal prescription for high photosynthetic rates, but confers an advantage only under appropriate environmental conditions, i.e. high temperature and high PFD, when the CO_2 concentration becomes limiting. Hence the value of converting *temperate*-zone crop plants is doubtful, especially since now the atmospheric CO_2 concentration is increasing and the CO_2 limitation of C_3 plants should become less marked (Section 2.7).

The high net rates of photosynthesis of C_4 plants are in the last analysis largely the result of their lack of photorespiration. Perhaps then one can improve the photosynthetic performance of C_3 plants by the simpler expedient of suppressing photorespiration? Since phosphoglycolate cannot be permitted to accumulate, suppression must be achieved at its source: one needs to eliminate the oxygenase activity of Rubisco, or at least to lower it substantially. Attempts in this direction by modifying the protein's structure are, however, being frustrated by the fact that the two gases react at the same catalytic site. To date, modifications to the active site have changed the reactivity of the enzyme more or less equally towards both substrates (e.g. Whitney *et al.* 1999, Spreitzer & Salvucci 2002). There are naturally occurring variations in the carboxylase/oxygenase ratio of the enzyme, but a reduced oxygenase activity is associated with an overall lowering of the catalytic rate. That is undesirable: Rubisco already has a low turnover number, presumably compensated for by its concentration in chloroplasts, which is unusually high for an enzyme. Whether suppression of the oxygenase activity is a viable proposition depends also on whether photorespiration can be eliminated without inducing photooxidation; but this can be tested only on a plant where the suppression has been achieved.

The introduction of C_4 photosynthesis, or changing the molecular structure of Rubisco, would be direct 'improvements' of the photosynthetic apparatus. One can also try to improve yield by means which do not target the photosynthetic apparatus directly. One possibility is the introduction of genes for drought resistance, for photosynthesis is very sensitive to water stress (Chapter 13). Similarly, since pests and diseases account for enormous agricultural losses, yields would be increased by engineering pest-resistant and disease-resistant crop plants. In many situations, such indirect approaches indeed seem more promising, for water stress, disease, and damage by animals, especially insects, frequently limit the harvest yield to a greater extent than the capacity of the photosynthetic apparatus. Some progress has already been made along these lines. The genes for the toxin synthesized by *Bacillus thuringiensis* have been inserted into plant genomes and have conferred protection against insect attack. Another approach has been to breed for an increased harvest index, that is for more partitioning of the photosynthate to the harvestable product – seeds, fruits – compared with the rest of the plant body. Seed weights of up to 50% of the above-ground biomass have been obtained in wheat, rice (*Oryza sativa*), oats (*Avena sativa*) and barley (*Hordeum vulgare*) in this way. Before any attempts to improve productivity, one needs to understand what is limiting the productivity, so that this limitation can be targeted.

2.7 | Photosynthesis and the increase in atmospheric carbon dioxide

The concentration of CO_2 in the earth's atmosphere is increasing rapidly. After remaining essentially steady at 280 μmol mol^{-1} for the last 50 000–60 000 years, as indicated by air bubble analysis from Antarctic ice, it started rising in the early 1800s. By now the value is around 370 μmol mol^{-1} and it is rising at a rate of about 1.8 μmol mol^{-1} per year. The rise is accelerating, so that by the end of the twenty-first century the concentration may be expected to have doubled to about 700 μmol mol^{-1}. This rise is the result of human activity – the burning of coal, oil, gas and wood. The destruction of forests, both tropical and north temperate/boreal, is moreover decreasing the potential for photosynthesis. Fear of the 'greenhouse effect' and rising of global temperatures as a result of the accumulation of CO_2 is spurring the world's governments to *contemplate* the curbing of CO_2 emissions, but action has been minimal and it is obvious that no curbs will make a significant impact before global CO_2 levels are much higher than now. Too many people derive an economic advantage from activities that contribute to rising CO_2 levels.

Atmospheric CO_2 levels have been very far from constant during the history of the earth. The recent level of 280 μmol mol^{-1} was probably the lowest since the dawn of life some 3500 million years ago. Analysis of rocks has suggested a value of around 5600 μmol mol^{-1} for the Upper Ordovician, shortly before the first land plants emerged, and 1500–3000 μmol mol^{-1} for the Mid-Cretaceous period, when the flowering plants began to spread and evolve rapidly. Since then the levels have been falling (with some fluctuations) until the current rise began. Flowering plant evolution has proceeded during a period of (mostly) falling CO_2 concentration and, as indicated previously (Section 2.4.2, p. 30), this drop in CO_2 levels has been suggested as the evolutionary pressure responsible for the emergence of C_4 photosynthesis. The present-day existence of C_3 – C_4 intermediate plants, with partial C_4 characteristics, has been interpreted as evidence that the evolutionary pressure is still acting on plants. One possible response of flowering plants to rising CO_2 levels could be that such evolution is slowed, if not stopped. This, however, is a long-term effect; most interest is concentrating on the direct effect of the rise in CO_2 concentration on the flora as it currently exists.

Experiments have of course been carried out on the effects of exposing plants to raised levels of CO_2, usually up to 700–1000 μmol mol^{-1}. The immediate response of C_3 plants is an increased rate of photosynthesis per unit leaf area. In longer-term studies, the enhancement of photosynthesis is usually accompanied by increased growth and greater accumulation of biomass. But stomatal conductance decreases: CO_2 stimulates (partial) stomatal closure and, in plants grown at elevated CO_2 levels, stomatal density per unit leaf area is reduced. In C_4 plants the stimulatory effect of higher CO_2 concentration is less; one review quotes growth enhancements of 40–44% for C_3 plants and 22–23% for C_4 species. For CAM plants, data are fewer but dry mass increases of 17–51% have been reported.

Numerous earlier studies on plants exposed to high CO_2 concentrations showed what has been termed acclimation: the initial stimulation of rates of photosynthesis and growth failed to be maintained. But this applied to plants grown in containers, where nutrient shortage was possible. Acclimation has not been detected over several years in field experiments where trees growing in a forest have been treated *in situ* by pumping CO_2-enriched air around them (this technique is known as FACE, Free Air CO_2 Enrichment). It has, however, become clear that for increased crop yields under increased CO_2 levels, maintenance of adequate levels of nutrients, especially nitrogen, is vital, otherwise nutrient limitation sets in.

Predictions of the effect of atmospheric CO_2 enrichment are hampered by the complexity of interaction between all the factors that influence photosynthetic rate (Section 2.5). Nevertheless the general consensus of opinion, based on experimental studies complemented with mathematical modelling, is that the rising CO_2 level will result in a rise in the overall global rate of photosynthesis: plants would benefit (Kirschbaum 1994). Photosynthesis being more sensitive to CO_2 concentrations at higher temperatures, the stimulation would be more pronounced in warmer regions. The stimulation would also be particularly important under conditions of water shortage, when stomatal apertures are small; a higher CO_2 concentration increases its rate of diffusion. C_3 plants would gain more than C_4 plants and become more competitive in the warmer and drier regions of the world. Over the long term, species composition of plant communities would change. There is already evidence that in the Great Plains of the USA, C_3 plants have gained in dominance since the end of the nineteenth century.

Plants are not passively at the mercy of ambient CO_2 levels. As elevated CO_2 levels stimulate the rate of photosynthesis, this will *remove* the gas at a faster rate. The fall in atmospheric CO_2 levels through the geological ages has been attributed at least partly to photosynthesis. Possible effects on plant respiration, too, have to be remembered, but the data on this are unfortunately very contradictory. On average, there is some evidence for a lowering of plant respiration rates by increased CO_2 concentrations, of the order of 10%, but doubt is being cast on earlier reports on the grounds of methodology. There are altogether still too many unknowns to predict when or at what concentration of CO_2 an equilibrium might be re-established. Our understanding of the effects of CO_2 concentrations on plants in the long term is still very modest, especially when it comes to mixed natural communities. The effect would depend strongly on temperature. The majority opinion associates the rise in atmospheric CO_2 levels with the 'greenhouse effect', a rise in temperature. An opposing view is that there will be increased cloud cover which will reflect so much of the incoming solar radiation back into space that the net effect will be a cooling – and increased cloudiness will decrease photosynthesis. Finally, the potential for photosynthesis will depend on the amount of plant biomass, and this is being diminished by deforestation, by desertification following agricultural mismanagement, and by urban development.

2.8 | Respiration: the oxidative breakdown of organic compounds

2.8.1 The overall process and respiratory substrates

Earlier in this chapter (Section 2.2.1, Fig 2.1) it was discussed how respiration counterbalances photosynthesis in the biosphere, oxidizing the C fixed in photosynthesis to CO_2 and water again. Most of the respiration is aerobic and utilizes the O_2 produced in photosynthesis, so that the O_2 is recycled as well. The primary function of respiration is to provide for the energy needs of living cells: some of the potential energy of the oxidizable substrates is conserved in ATP (Fig. 2.1). Respiration is sometimes described as being 'the reverse of photosynthesis'. Chemically, the *overall* result of respiration is the reverse of that of photosynthesis; taking carbohydrate as the end product of photosynthesis or the substrate of respiration, one can write

$$[CH_2O]_x + x\,O_2 \xrightarrow[\text{photosynthesis}]{\text{respiration}} x\,CO_2 + x\,H_2O \tag{2.9}$$

With respect to the reaction mechanism, however, respiration as a whole is *not* the exact reverse of photosynthesis, although there are many common intermediates, and *some* reactions do run in the reverse direction in the two processes. Quantitatively, the amount of photosynthesis by a plant must exceed its respiration or a positive mass balance would be impossible. It has been estimated that a flowering plant respires daily 30 to 70% of its photosynthate, not counting any photorespiratory losses that may have occurred (Lambers 1997).

In most plant tissues, **carbohydrate** is the main respiratory substrate, entering the oxidative pathways as hexose sugars. Such sugars are metabolically reactive and are not stored in cells in large amounts. Carbohydrate is stored as polysaccharides of which starch, a glucose polymer, is the most common; it is insoluble and forms starch grains in plastids. Some species store fructosans, soluble polymers of fructose, in vacuoles. The disaccharide sucrose is the main vacuolar store in yet other plants, and sucrose is also the most common form in which carbohydrate is transported in plants. All these more complex carbohydrates have to be hydrolysed to their hexose monomers for respiration. Many seeds store lipids as oil and this can serve as respiratory substrate, although most of the lipid store is converted to carbohydrate before being respired. Protein is not commonly utilized as a respiratory substrate, but C skeletons from amino acids can enter the respiratory pathways. During starvation, or senescence, and during reserve mobilization in seed storage tissues, there is large-scale hydrolysis of proteins and respiration of at least part of the amino acid pool.

Much of the basic biochemistry of respiration is common to organisms from all the living kingdoms, though some details may be specific to plants. With reference to photosynthetic organisms, the respiration which passes through the universal respiratory pathways, as described below, is often termed *dark respiration*, to distinguish it from the unique, photosynthesis-linked photorespiration (Section 2.4.2). The so-called dark respiration still proceeds in photosynthetic tissues in the light.

As in the case of photosynthesis, which proceeds in two stages, one can distinguish two stages in respiration: the stage of breakdown of substrate, which yields CO_2 and reduced coenzymes, NAD(P)H; and the stage of terminal oxidation, which achieves oxidation of the reduced coenzymes. Both of these stages yield ATP, but by far the greater proportion is produced during terminal oxidation.

2.8.2 Pathways of substrate breakdown

Glycolysis

The word glycolysis means 'sugar breakdown'; it brings about the oxidative breakdown of glucose to pyruvate, as illustrated in Fig. 2.12. The glucose is first converted to fructose-1,6-bisphosphate by

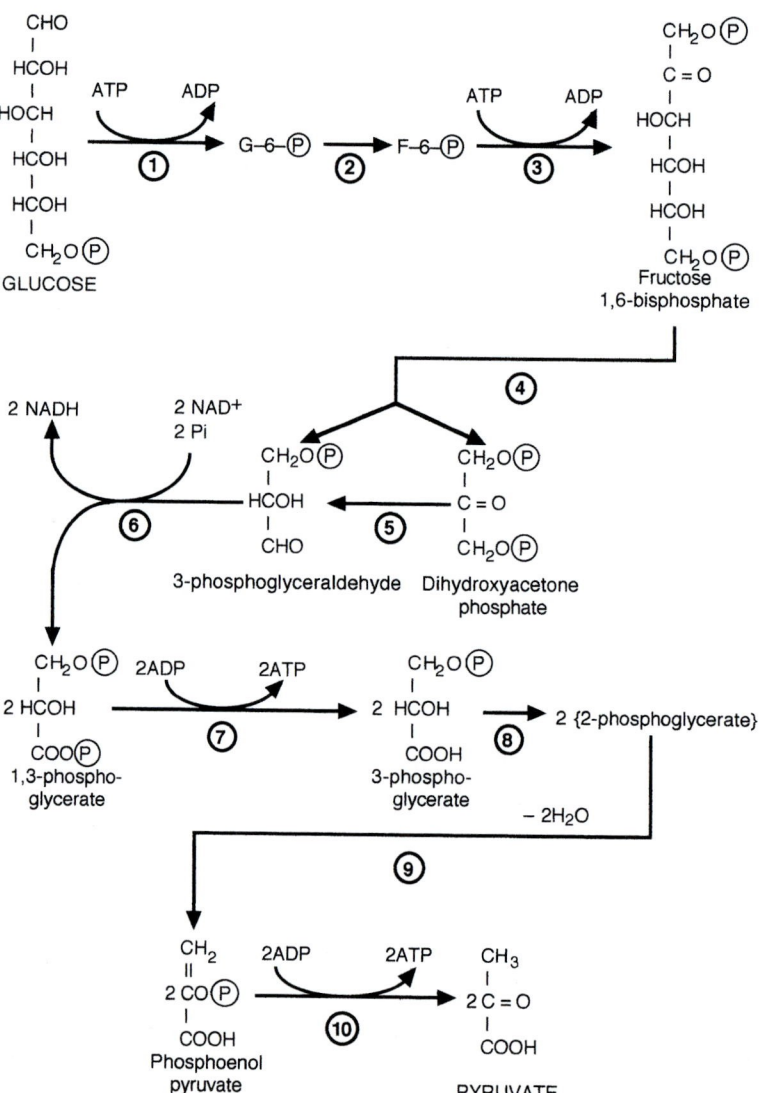

Fig. 2.12 Summary of glycolytic pathway of respiration. The enzymes catalysing the numbered steps are: ① hexokinase; ② glucose phosphate isomerase; ③ 6-phosphofructokinase; ④ aldolase; ⑤ triosephosphate isomerase; ⑥ glyceraldehyde phosphate dehydrogenase; ⑦ phosphoglycerate kinase; ⑧ phosphoglyceromutase; ⑨ enolase; ⑩ pyruvate kinase. P = phosphate group; Pi = inorganic phosphate. Reactions 2 and 5 are isomerization reactions. Note that all reactions from 6 onwards occur *twice* for every molecule of hexose entering the pathway.

phosphorylation at the expense of ATP, catalysed by hexokinase, followed by isomerization and a second phosphorylation by phosphofructokinase. The fructose-1,6-bisphosphate is cleaved by aldolase to the two triose sugar phosphates, 3-phosphoglyceraldehyde and dihydroxyacetone-3-phosphate. Now comes the *oxidative* reaction: the 3-phosphoglyceraldehyde is oxidized by glyceraldehyde-phosphate dehydrogenase to 1,3-phosphoglycerate (PGA) and the coenzyme NAD^+ is reduced. The 1, 3-PGA donates a phosphate group to ADP to synthesize ATP, catalysed by PGA kinase, a process known as *substrate level* phosphorylation. After some molecular rearrangements, phosphoenolpyruvate, PEP, is formed and pyruvate kinase catalyses a second substrate level phosphorylation to give the end product of glycolysis, pyruvate. Since all the reactions from the oxidation step onwards proceed *twice* per molecule of glucose, 4 ATP per 1 glucose can be formed; but 2 ATP are used to prime the system, so that the net gain is 2 ATP per 1 glucose:

$$C_6H_{12}O_6 + 2 \text{ ATP} + 2 \text{ NAD}^+ \longrightarrow 2 \text{ } C_3H_4O_3 + 4 \text{ ATP} + 2 \text{ NADH} + 2 \text{ H}^+$$

$$(2.10)$$

All the C of the glucose is still in organic combination, i.e. no CO_2 has been evolved and most of the potential energy of the glucose is still present in the pyruvate and the NADH; the net gain of 2 ATP represents a very small percentage of the potential total energy. It may also be noted that no O_2 has been used: glycolysis is an anaerobic pathway.

Substrates can enter the glycolytic pathway at several points. Fructose-6-phosphate may come from the hydrolysis of fructans or sucrose, or from the PPP (see below). Triose sugar phosphates are transported out of chloroplasts in the light and may also be derived from the PPP. Starch hydrolysis by the starch phosphorylase enzyme produces glucose-1-phosphate, which is easily isomerized to glucose-6-phosphate. When such phosphorylated sugars enter glycolysis, one or both of the priming reactions with ATP is/are bypassed and the ATP gain is correspondingly greater. Intermediates can also be withdrawn into other metabolic sequences from the glycolytic pathway at various points before pyruvate is produced.

The complete glycolytic sequence is located in the cytosol. In plant cells, however, isozymes of all the glycolytic enzymes are additionally found in *plastids*. (Isozymes are variants of an enzyme, catalysing the same reaction, but differing slightly in structure and properties such as substrate affinity; isozymes may be coded by different genes, or their differences may result from post-transcriptional or post-translational modifications.) The function of glycolysis in plastids appears to be the production of pyruvate for fatty acid biosynthesis, a process confined to plastids in plant cells and particularly active in seed tissues synthesizing oil as nutrient store. Photosynthetic CO_2 metabolism involves several reactions catalysed by glycolytic isozymes, but in the opposite direction to respiratory reactions. For instance, in the C_3 cycle (Fig. 2.7) chloroplast aldolase catalyses the *condensation* of

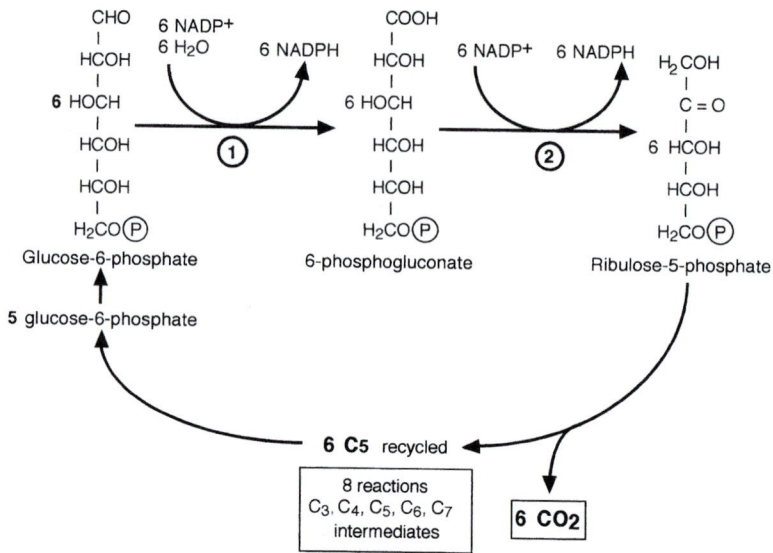

Fig. 2.13 Summary of the pentose phosphate pathway (PPP) of respiration. The enzymes catalysing the numbered steps are: ① glucose-6-phosphate dehydrogenase and ② 6-phosphogluconate dehydrogenase. ℗ = phosphate group. One turn of the cycle effects the oxidation of one-sixth of an individual molecule of glucose; or, as shown, if six molecules of glucose enter the cycle, the equivalent of one molecule of glucose is lost as CO_2, and the equivalent of five molecules is recycled.

triose phosphates to fructose-1,6-bisphosphate; in glycolysis, cytosolic aldolase *cleaves* the sugar.

The pentose phosphate pathway, PPP

This reaction series is also known as the **hexose monophosphate shunt**, or as the **oxidative pentose pathway**, OPP, to distinguish it from an alternative name for the C_3 cycle, which is sometimes termed the **reductive pentose phosphate pathway**. The PPP is located in the cytosol. It is outlined in Fig. 2.13.

The starting substrate for the PPP is glucose-6-phosphate (or glucose followed by the hexokinase reaction). The PPP commences with the oxidation of glucose-6-phosphate to 6-phosphogluconate, followed by an oxidative decarboxylation of the gluconate to ribulose-5-phosphate with the release of CO_2. The respective enzymes for these two reactions are glucose-6-phosphate dehydrogenase and phosphogluconate dehydrogenase; the coenzyme which receives the H equivalents from both reactions is $NADP^+$. The ribulose-5-phosphate is recycled to glucose-6-phosphate:

$$6 \text{ ribulose-5-phosphate} \longrightarrow 5 \text{ glucose-6-phosphate} \qquad (2.11)$$

The recycling reactions here are largely a reversal of the C_3 cycle reactions which regenerate the CO_2 acceptor ribulose-1,5-bisphosphate, but no ATP is expended and the chloroplast and the cytosol each has a distinctive set of isozymes, just as for glycolytic ones.

The balance sheet for the PPP is shown in Equations 2.12 and 2.13; to be able to work with whole molecules, one must start with 6 hexose (C_6) molecules. Phosphates have been omitted for simplicity.

$$6 \text{ C}_6 + 12 \text{ NADP}^+ \longrightarrow 6 \text{ CO}_2 + 5 \text{ C}_6 + 12 \text{ NADPH} + 12 \text{ H}^+$$
$$(2.12)$$

This reduces to:

$$C_6 + 12 \text{ NADP}^+ \longrightarrow 6 \text{ CO}_2 + 12 \text{ NADPH} + 12 \text{ H}^+ \qquad (2.13)$$

The PPP thus does achieve the complete oxidation of glucose to CO_2, but without ATP formation; there is no substrate-level phosphorylation involved. The major function of the PPP is thought to be the provision of NADPH for reductive biosyntheses, e.g. lipid formation, and for the production of metabolic intermediates; the pentose sugars can be utilized for nucleotide synthesis. (Some of the NADPH may be oxidized by mitochondria with ATP formation: see Section 2.9, terminal oxidation.)

The Krebs cycle

The Krebs cycle is the reaction series which achieves the complete oxidation of pyruvate (coming mainly from glycolysis) to CO_2. It is located entirely and exclusively in the **mitochondria**. The reactions are summarized in Fig. 2.14.

The pyruvate (with three C atoms) first loses CO_2 by oxidative decarboxylation catalysed by the enzyme pyruvate dehydrogenase, which also links the remaining 2-C fragment, an acetyl moiety, to coenzyme A producing acetyl-CoA. This condenses with the 4-C acid oxaloacetate to form the 6-C acid citrate. Then, as indicated in Fig. 2.14, there follows a series of molecular rearrangements, oxidation steps and oxidative decarboxylation steps, until the equivalent of the pyruvate has been converted to CO_2 and the oxaloacetate has been regenerated. There is one substrate-level phosphorylation producing ATP, associated with the oxidation of 2-oxoglutarate. At three steps, a molecule of water is added to the reactants. The overall final balance sheet for respiration shows water as a product (Equation 2.9), but water is also a substrate in respiration. (Compare with photosynthesis, where water is not only consumed as indicated by the overall equation, but is also a product.) The 5 pairs of H equivalents removed from the substrates in the Krebs cycle reduce NAD^+, except for the succinate oxidation step, where no coenzyme is involved. The enzymes are present in the mitochondrial matrix, but succinate dehydrogenase is again an exception, being bound to the crista membrane of the mitochondrion.

Between them the Krebs cycle and glycolysis can carry out the complete oxidation of glucose. With 2[H] representing the pairs of H equivalents removed from substrates, one can write

$$\text{Glycolysis} \quad C_6H_{12}O_6 \longrightarrow C_3H_4O_3 + 2 \times 2[H] \qquad (2.14)$$

$$\text{Krebs cycle } 2C_3H_4O_3 + 6H_2O \longrightarrow CO_2 + 10 \times 2[H] \qquad (2.15)$$

$$\text{Sum } C_6H_{12}O_6 + 6H_2O \longrightarrow CO_2 + 12 \times 2[H] \qquad (2.16)$$

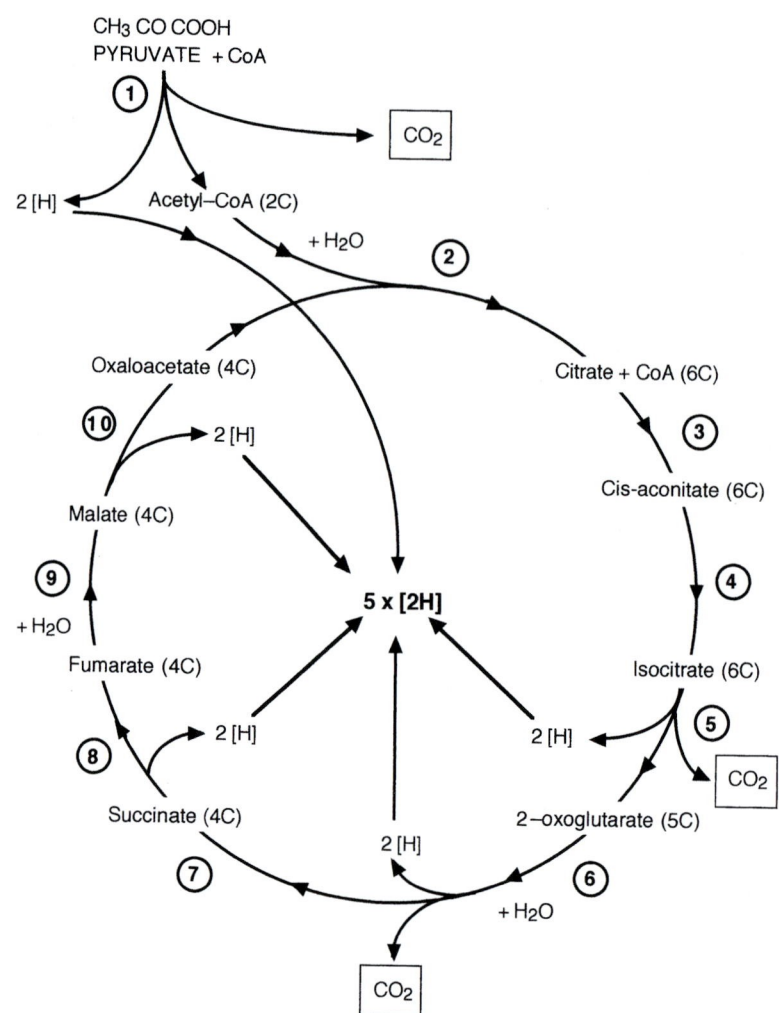

Fig. 2.14 An outline of the Krebs cycle. CoA = coenzyme A. The numbers of C atoms in the intermediates are given as 2C, etc. The enzymes catalysing the numbered steps are: ① pyruvate dehydrogenase; ② citrate synthase; ③ and ④ aconitase (aconitate hydratase); ⑤ isocitrate dehydrogenase; ⑥ 2-oxoglutarate dehydrogenase; ⑦ succinyl CoA ligase; ⑧ succinate dehydrogenase; ⑨ fumarase (fumarate hydratase); ⑩ malate dehydrogenase. All the reducing equivalents, 2[H], are passed to NAD except from the succinate dehydrogenase reaction (which enter the electron transport chain as indicated in Fig. 2.17). For every molecule of glucose oxidized, two molecules of pyruvate enter the cycle. Substrate level phosphorylation is associated with reaction 7. Adapted from Öpik (1980).

The H equivalents can be oxidized in the terminal oxidation reactions using molecular O_2, and producing 12 H_2O per 1 molecule glucose.

Like glycolysis and the PPP, the Krebs cycle can oxidize intermediates fed in at any place in the sequence; its role in this respect is discussed later. The Krebs cycle is also an important source for metabolic intermediates. Malate, oxaloacetate and 2-oxoglutarate are C skeletons for amino acids. Running down of the cycle by removal of intermediates is prevented by the occurrence of *anaplerotic reactions* which replenish it. One such reaction is the carboxylation of PEP by PEP carboxylase to give oxaloacetate; this is of course the initial reaction in C_4 and CAM photosynthesis, but it takes place also in non-photosynthetic cells, at a lower rate. Malate can be produced by malic enzyme from pyruvate, CO_2 and NADPH.

Lipid oxidation

Lipids are frequently stored in seeds as oil bodies (oleosomes, sphaerosomes), where they may account for over 50% of the tissue mass. The storage oils are triglycerides consisting of a glycerol molecule esterified with three long-chain fatty acids, mostly with 16 or 18 C atoms per chain. The first step in lipid oxidation is hydrolysis by lipase enzymes within the oil body to glycerol and fatty acids. Glycerol, a 3-C sugar alcohol, is closely related to triose sugars and is converted to dihydroxyacetone-3-phosphate, which can then enter the glycolytic pathway. The fatty acids are activated by linkage to coenzyme A and are metabolized in the **glyoxysomes** (specialized microbodies of oily seeds) by two metabolic sequences. Firstly, the process of β**-oxidation** results in the cleavage of the fatty acid chain into 2-C fragments, acetyl-CoA. Secondly, the glyoxysomes process the acetyl-CoA via the **glyoxylate cycle** to succinate and this can, through a series of reactions in the cytosol and the mitochondria, finally be synthesized to sucrose. Most of the lipid stored in seeds is converted to sucrose and transported to the growing parts of the seedling. Some of the acetyl-CoA is utilized in the mitochondria of the storage cells as respiratory substrate in the Krebs cycle.

2.8.3 Interactions of pathways

In the cell, none of the pathways of respiratory substrate oxidation functions in isolation. A summary overview of all the pathways, and their interrelationships, is presented in Fig. 2.15. The first stage, 'hydrolysis' in Fig. 2.15, does not involve oxidations and is not specific to respiration: the monomers produced from the polymers serve as substrates not only for respiration but for numerous other metabolic sequences. The free energy change associated with the hydrolytic reactions is minimal, less than 1% of the total available, and cannot support ATP synthesis. The 'incomplete oxidation' stage produces a small number of fairly simple organic acids and the free energy content of the substrate falls by about 33%; some of the energy is conserved in ATP during substrate level phosphorylations, and also in NADH. The organic acids feed into the 'complete oxidation' stage, the Krebs cycle, which in association with the mitochondrial electron transport chain forms the final stage for most aerobic respiratory activity and where most of the ATP synthesis takes place. The PPP does not appear to fit the overall pattern in that it is a direct oxidative pathway; sugars metabolized *exclusively* by the PPP do not pass through the Krebs cycle. However, the PPP and glycolysis share common intermediates, fructose-6-phosphate and triose phosphates. Since both reaction series take place in the cytosol, it is almost inevitable that some exchange of metabolites between the two pathways should occur. If, say, there is a high demand for NADPH, the flow of material through the PPP might be boosted by intermediates from the glycolytic sequence. A large demand for pyruvate, on the other hand, could result in the channelling of PPP intermediates into

Fig. 2.15 Interactions of pathways of substrate breakdown in plant cells. The Krebs cycle and terminal oxidation via the mitochondrial electron transport chain are so closely integrated that they are here considered together, although strictly speaking the electron transport chain is not part of the *substrate* breakdown system.

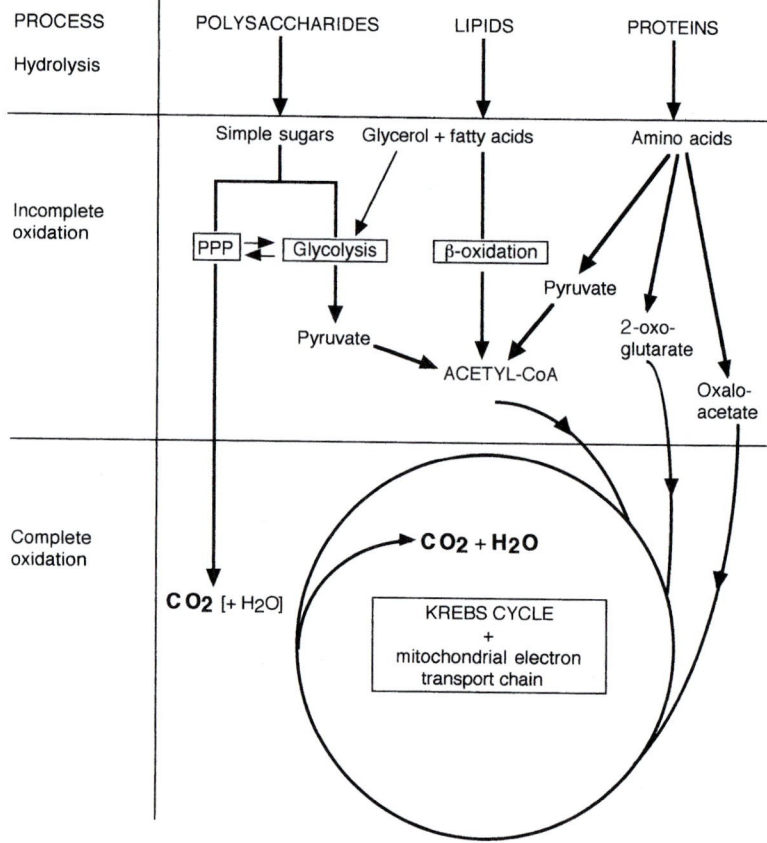

glycolysis. The PPP accordingly should be considered an integral part of a glycolysis – Krebs cycle – PPP network. This network in turn interacts with the rest of cellular metabolism via interchange of metabolites, at many steps. The provision of intermediates and reduced coenzymes is just as much a function of respiration as provision of ATP; respiration, defined as a substrate *breakdown* process, is also the starting point of many biosyntheses.

2.9 | Terminal oxidation and oxidative phosphorylation

2.9.1 The mitochondrial electron transport chain

The pathways of substrate breakdown considered in the preceding section are all concerned with the oxidation of substrate to CO_2; the oxidative reactions throughout these pathways have consisted of the removal of pairs of H atoms from the substrates, producing reduced coenzymes, NADH in glycolysis and the Krebs cycle and NADPH in the PPP. **Terminal oxidation** is the oxidation of the reduced coenzymes to water, with molecular O_2. Most of the respiratory ATP is produced during the terminal oxidation stage, by the process of **oxidative phosphorylation**.

The mitochondria are the main site for terminal oxidation and the sole site for oxidative phosphorylation. The reactions take place in thecrista membranes of the mitochondrion. The cristae, infoldings of the inner envelope membrane, give the mitochondrion a large internal area for the accommodation of the enzyme systems (Fig. 2.16). Electron transfer occurs, in principle similar to the electron transfer occurring in the chloroplast thylakoid membranes during photosynthesis (Section 2.3.2). In photosynthesis, light energy was harnessed to split water and to produce NADPH using the resulting electrons and protons. Now, however, the electrons and protons are **removed** from NADH and combined with O_2 so that water is the **product** of the electron transfer reactions and energy becomes available for ATP synthesis. The mitochondrial electron transport chain is shown in Fig. 2.17. Electrons are fed in at separate points from NADH and succinate respectively and passed to molecular O_2; the enzyme which achieves the reduction of O_2 is **cytochrome oxidase**, the terminal oxidase of the system. The H^+ from the coenzyme and succinate react with O_2 to form water. The flow of electrons and protons drives oxidative phosphorylation as during photophosphorylation. H^+ and positive charge accumulate in the intracristal spaces and move out to the matrix through the proton channel of the mitochondrial ATP synthase complex, driving ATP synthesis. Respiratory electron transport via the cytochrome chain and oxidative phosphorylation are essentially the same in flowering plants and in other organisms including prokaryotes. The ATP is transported out of mitochondria by a translocator in the mitochondrial inner membrane, which exchanges it for ADP produced in the cytosol from ATP utilization.

2.9.2 The alternative oxidase

One peculiarity of higher plant respiration is that it is not fully inhibited by *cyanide*, a very potent inhibitor of cytochrome oxidase (hence the great toxicity of cyanide to animals). Plant mitochondria contain an additional terminal oxidase, known as the **alternative oxidase** or cyanide-resistant oxidase. This siphons off electrons as shown in Fig. 2.17 and passes them to O_2 so that cytochrome oxidase is bypassed. When the cytochrome oxidase pathway is blocked by cyanide, the rate of respiration may be 50% of the original rate, but growth and other metabolic activities are inhibited: terminal oxidation exclusively through the alternative oxidase cannot support normal physiological function. Electron transport through the

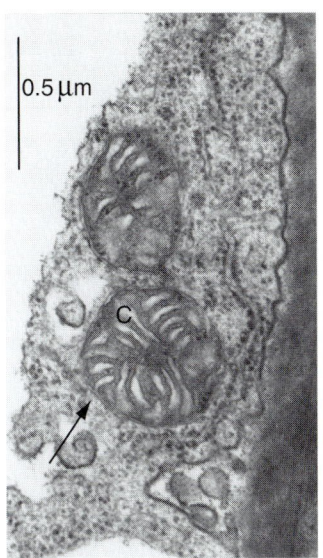

0.5 μm

Fig. 2.16 Electron micrograph of mitochondria from a leaf of mung bean (*Vigna radiata*). The cristae, C, are infoldings from the inner mitochondrial membrane and give a large membrane area. Junctions of the inner membrane with the cristae are visible only when caught at the correct angle by the section (arrow).

Fig. 2.17 The electron transport chain of plant mitochondria. The bold arrows denote the direction of electron flow through the cytochrome chain, terminating in cytochrome oxidase; ATP formation is associated with the activities of Complexes I, III and IV, but the actual synthesis of ATP is mediated by a separate system, Complex V (ATP synthase), not illustrated. The alternative oxidase bypasses the cytochrome system (fine arrows) and is not associated with ATP synthesis.

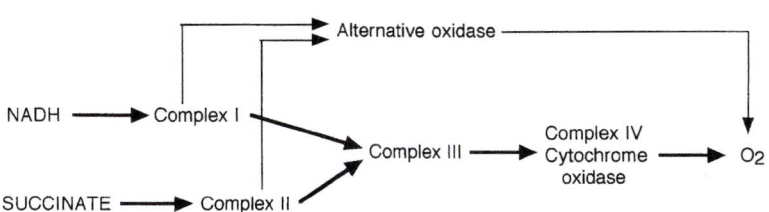

Table 2.2 | Respiration rates of a variety of plant tissues at 25 °C. Where the rate has been measured at some other temperature, the rate for 25 °C has been calculated assuming a Q_{10} of 2 (i.e. a 2-fold change per 10 °C). Rates are given as O_2 uptake per unit dry weight; the *Arum* spadix exhibits thermogenic respiration. From Öpik (1980).

Tissue	Respiration rate ($\mu L\ O_2\ h^{-1}(mg\ dry\ weight)^{-1}$)
Resting dry seeds	≤ 0.0005
Resting tubers, storage roots	0.23–2.6
Leaves	0.80–5.9
Young growing root tips	up to 15
Flower spadix of *Arum maculatum*	600

alternative oxidase is not coupled to ATP synthesis and the energy is liberated as heat. One ATP per NADH can be formed before the electrons enter the alternative pathway but presumably this does not suffice to cover the cell's energy needs. The physiological significance of the alternative pathway is therefore much disputed. There is one situation in which a clear physiological role can be ascribed to the alternative oxidase, namely in the floral organs of certain species of the Araceae, the arum family. In these plants, when the flowers are ready for pollination, a sterile part (spadix) of the inflorescence undergoes extremely rapid **thermogenic respiration** (Table 2.2) which causes the tissue to heat up by 10–20 °C above the ambient. This thermogenic respiration utilizes the alternative oxidase; it can proceed so fast because the alternative oxidase is not subject to the feedback controls associated with ATP-producing electron transport (see below, Section 2.11.2). The heat volatilizes chemicals which attract pollinating flies (although to the human nose the smell is horrible!). In some Araceae, the inflorescence melts its way up through snow. But the aroid inflorescence is a special case; the alternative oxidase is found in many species and in all types of plant organ. Attempts have been made to explain the alternative oxidase as an 'overflow system' enabling plants to dispose of surplus reducing power when a large amount of C skeletons is withdrawn from respiratory pathways while there is a low demand for ATP. The weakness of this argument is that a process requiring large amounts of metabolic intermediates (e.g. growth) is also likely to require a large supply of ATP and/or reductant. Another suggestion is that the alternative oxidase prevents the build-up of damaging superoxide radicals during electron transport via the cytochrome system.

2.9.3 The ATP balance sheet and energy-conversion efficiency of respiration

The *theoretical* balance sheet for respiratory ATP production is easily drawn up. If glucose is completely oxidized via glycolysis and

Table 2.3 The balance sheet for respiratory production of molecules of ATP per 1 molecule of glucose, assuming complete coupling and complete oxidation via glycolysis, Krebs cycle and cytochrome oxidase system.

Substrate level oxidation in glycolysis	4
Substrate level oxidation in Krebs cycle	2
Terminal oxidation, 2 NADH from glycolysis	4
Terminal oxidation, 8 NADH from Krebs cycle	24
Terminal oxidation, 2 succinate	4
Total produced	38
Used up in glycolysis, priming reactions	2
Net gain	36

the Krebs cycle, with terminal oxidation through cytochrome oxidase, one can add up the ATP molecules per 1 molecule glucose (Table 2.3).

The NADH from glycolysis, reacting with mitochondria from the *outside*, is oxidized with the production of only 2 ATP per NADH, whilst mitochondrially produced coenzyme oxidation yields 3 ATP per 1 NADH. The free energy of complete oxidation of glucose is 2880 kJ mol^{-1}. The free energy of hydrolysis of ATP is highly dependent on factors such as ATP concentration and pH, but under cellular conditions is at least 42 kJ mol^{-1}. Assuming this value, a gain of 36 ATP per molecule of glucose is equivalent to an energy conservation of [36 × 42], 1512 kJ mol^{-1}, or 52% of the total available, a very high degree of energy conservation. (The value would be pushed even higher if *phosphorylated* sugars enter the glycolytic pathway, Section 2.8.2.)

The calculation in Table 2.3 is, however, based on an assumption of *complete coupling* of substrate breakdown and terminal oxidation to ATP synthesis. This is unlikely to be the situation in a cell, at least not in all circumstances. Any NADH which is used in reductive reactions does not yield ATP, and terminal oxidation by the alternative oxidase produces only 1 ATP per molecule of NADH oxidized. The 12 NADPH formed per molecule of glucose in the PPP could theoretically be oxidized by the mitochondria with the production of 24 ATP, but, as stated, the PPP probably provides NADPH mainly for reductive reactions rather than for terminal oxidation. It is therefore not possible to say exactly how much ATP is produced per glucose molecule in a particular situation.

2.10 | Anaerobic respiration

2.10.1 Occurrence and endurance of anaerobiosis in plants

Flowering plants are obligate aerobes: no flowering plant can complete its life cycle without O_2. Most flowering plant organs, except dormant seeds, are killed by anoxia (complete lack of O_2) within hours or a few days at most. Nevertheless there are also many situations in which plant tissues regularly survive at least hypoxia, O_2 concentrations too low to support a normal level of aerobic respiration. Anaerobic respiration (fermentation) is therefore not unusual in plants. In hypoxia, both aerobic and anaerobic pathways operate simultaneously.

Germinating seeds often undergo a period of O_2 shortage during imbibition, because the testa can be highly impermeable towards O_2; also before the cells are fully hydrated and expanded there are hardly any air spaces between them. Respiration during this period has a high RQ, *respiratory quotient*, i.e. the ratio CO_2 evolution : O_2 uptake. For aerobic respiration this ratio is unity (Equation 2.9); in anaerobic respiration, CO_2 is evolved without any O_2 uptake, raising the RQ. Analysis shows an accumulation of end products of fermentation, such as ethanol, in the seed tissues. When the emerging radicle splits the testa, the RQ falls. *Complete* anoxia is, however, tolerated by seeds of very few species.

Meristematic tissues carry out partly anaerobic respiration. Cells in meristematic regions are closely packed, without air spaces, making diffusion of O_2 a problem. The vascular cambium lies quite deeply within plant organs and outside it lies the phloem, a living tissue without intercellular air spaces and with a high demand for O_2. In apical meristems the mitochondria are immature, with few cristae per unit volume, so that their capacity for terminal oxidation is low. All these factors favour anaerobic respiration.

Inside bulky organs – large fruits, tubers, thick stems and thick roots – the level of O_2 may be under 2.5% and partially anaerobic respiration would be expected; ethanol accumulation has been detected. However, some bulky organs are well supplied with air spaces and then the problem of O_2 diffusion is much reduced.

In seeds, large organs and meristems, hypoxia is the result of the structure of the plant itself. In aquatic plants hypoxia is imposed by the environment. The solubility of O_2 in water is low. Water at $10\,^\circ C$ in equilibrium with normal air, which is 21% O_2 by volume, contains only about 0.80% O_2 and the solubility falls with rising temperature. Roots and rhizomes growing in mud at the bottom of a body of water, or in boggy ground, will be in an essentially anoxic environment. By daytime, green parts of submerged aquatics can respire aerobically on the O_2 evolved in photosynthesis and O_2 is then also conducted to the roots through the extensive air spaces of these plants. By night the O_2 levels in and around the plants will fall very low, especially for the parts in the muddy substratum. These parts are highly tolerant of anoxia. The marsh plant *Acorus calamus* grows from a rhizome and normally overwinters in a submerged resting state, with a low metabolic rate. Explants from the resting plants, consisting of segments of rhizome with attached roots and small leaves, have survived laboratory incubation under total anoxia, in the dark, for two months.

Seeds of a limited number of aquatic species can germinate in a totally anaerobic environment. Examples are rice, a few species of the grass *Echinochloa* found as a weed in rice fields, and the yellow water lily (*Nuphar luteum*). This is obviously an adaptation to their natural environment, where these seeds germinate at the bottom of the water. Rice and *Echinochloa* seedlings can survive for at least four days anaerobically.

As the above examples demonstrate, varying degrees of anaerobiosis are normal for, and tolerated by, numerous plant species or tissues. On the other hand, many species are intermittently damaged by anaerobiosis owing to flooding. Once the air spaces in the soil are filled with water, the limited amount of O_2 that is dissolved is rapidly used up by the respiration of plant roots and soil microorganisms and the roots then suffer an O_2 shortage; this can have serious consequences for an agricultural crop. An understanding of anaerobic respiration is therefore of considerable practical interest. Knowledge of how e.g. aquatic plants endure anaerobiosis might help the development of flooding-tolerant crop plants.

2.10.2 Respiratory metabolism under anaerobiosis

The respiratory pathway which is functional in anaerobic plant tissues is glycolysis, just as it is in anaerobic microorganisms; this pathway is essentially an anaerobic reaction sequence. On transfer to anoxic or hypoxic conditions, the transcription and translation of most plant genes is inhibited, but there is enhanced transcription and translation of some 20 genes, a major part of which code for glycolytic enzymes: aldolase, glucose-phosphate isomerase, enolase and glyceraldehyde phosphate dehydrogenase, including some isozymes not synthesized aerobically. The pyruvate produced in glycolysis in the absence of O_2 becomes the oxidant which receives the H equivalents from glycolytic NADH and recycles it to NAD^+, which is vital for the continuance of glycolysis; cells contain only very small amounts of the coenzyme. Three pathways of fermentation have been demonstrated in plant tissues:

(1) Lactic fermentation: the enzyme lactate dehydrogenase catalyses the reaction:

$$\text{pyruvate} + \text{NADH} \longrightarrow \text{lactate} + \text{NAD}^+ \tag{2.17}$$

(2) Alcoholic fermentation: firstly pyruvate decarboxylase decarboxylates the pyruvate to acetaldehyde, then this is reduced with NADH by alcohol dehydrogenase (ADH):

$$\text{pyruvate} \longrightarrow \text{CO}_2 + \text{acetaldehyde} \tag{2.18}$$

$$\text{acetaldehyde} + \text{NADH} \longrightarrow \text{ethanol} \tag{2.19}$$

(3) Malic fermentation: malic enzyme carboxylates the pyruvate to malate:

$$\text{pyruvate} + \text{CO}_2 + \text{NADH} \longrightarrow \text{malate} + \text{NAD}^+ \tag{2.20}$$

Increased transcription of mRNA for lactate dehydrogenase, pyruvate decarboxylase and ADH occurs as a response to anaerobiosis.

Why should anaerobiosis be so harmful to plants? There is still no clear answer to this, nor to what confers tolerance on such plants as show it. One serious drawback of anaerobic metabolism is the extremely low yield of ATP per unit of substrate respired. A switch from complete aerobic oxidation to complete anaerobiosis could mean a fall from 36 to 2 ATP produced per molecule of glucose oxidized, a decrease to 5.5%, insufficient to keep a cell alive (Section 2.11). Even if aerobic respiration were producing only half the theoretical maximal ATP, the decrease induced by anaerobiosis would still be drastic. The low ATP yield can be offset to some degree by the increases in the rate of glycolysis which typically occur under anaerobiosis (the Pasteur effect). But increases in the rate of glycolysis induced in plant tissues by anaerobiosis are only 1.5- to 6-fold. With high rates of glycolysis there is a danger of substrate exhaustion and starvation, and also much accumulation of potentially harmful products of fermentation. Lactate and malate are acids and acidification of the cytoplasm has been cited as the main cause of cell death in some instances. Acetaldehyde, the intermediate in alcoholic fermentation, is highly toxic; ethanol is, however, tolerated comparatively well. Maize mutants deficient in an ADH gene succumb rapidly to anaerobiosis, presumably because they cannot remove the acetaldehyde. There is also, strange as it may seem, evidence of oxidative stress, with formation of hydrogen peroxide, a dangerous oxidant, from trace amounts of O_2. Tissue damage and death could result from substrate exhaustion, acidification, lipid peroxidation and toxin accumulation, singly or in combination.

In plant tissues tolerant to anoxia or hypoxia, no basic differences in anaerobic metabolism have been detected, in comparison with sensitive material. Differences seem to be quantitative rather than qualitative; alcoholic fermentation predominates. In submerged plants, at least a sizeable proportion of the ethanol leaches out into the surrounding water, the permeability of cellular membranes towards ethanol being high. This helps to keep down the concentration of alcohol in the tissues. The *Acorus calamus* rhizome which overwinters in mud (Section 2.10.1), lays down an abundant store of starch during the summer. Rice grains, which are very tolerant to anaerobiosis, are able to synthesize α-amylase in anoxic conditions and hence can mobilize their starch reserves. The sensitive wheat grain is unable to synthesize the enzyme and cannot mobilize starch in the absence of O_2, but can be induced to germinate anaerobically by feeding with sucrose or glucose. These examples show that an ability to maintain an adequate supply of respiratory substrate can be a factor in survival under anaerobiosis.

Tolerance towards anaerobiosis may need induction. Rice seedlings (as distinct from the ungerminated grains), which are killed within 24 hours when suddenly exposed to complete anoxia, can survive anoxia for several days when first pretreated at low O_2 levels. During the pretreatment period, there are increases in the activities of ADH and pyruvate decarboxylase, giving the plants a greater potential for alcoholicfermentation. In maize, development of flooding tolerance involves increases in activity of several genes coding for ADH and other enzymes of anaerobic respiration (Sachs *et al.* 1996).

2.11 | Respiration and plant activity

2.11.1 Correlation of respiration rate with physiological activity

Respiration proceeds unceasingly in all active (i.e. non-dormant) living cells. Even when a cell is not performing any *net* metabolic work and is simply subsisting unchanged – say a mature pith parenchyma cell – it still requires repair and resynthesis of protoplasmic components, which are labile and in a constant state of turnover. Membrane potentials are sustained only by continued pumping of ions across membranes, requiring ATP. This aspect of cellular activity has led to the concept of **maintenance respiration**, required for such processes. The proportion of respiration that supports net cellular work is then termed **growth respiration** (synthetic respiration). More precise definitions of the terms have been attempted, while some plant physiologists have disputed the validity of any division of respiration into these components. The general concept of maintenance respiration is, however, useful as a reminder that cells must spend energy for their survival, without cessation. It is not implied that there is a biochemical distinction between growth respiration and maintenance respiration. Measurement of the proportion of maintenance respiration is even more controversial than its definition, but for plant cells, its magnitude has been estimated at up to 50% of the total.

Respiration rates are positively correlated with physiological activity. On a unit mass basis (fresh or dry mass), high rates are found in young, actively growing regions, such as growing apices, or in tissues performing metabolic work at a high rate, such as glands (Table 2.2). To some extent this is the result of the higher ratio of living protoplasm per unit mass in such tissues; mature and metabolically more inert tissues have larger proportions of cell wall and/or vacuoles and storage materials per unit mass, and these tissue compartments do not contribute to respiratory activity. On a unit nitrogen basis, which reflects more truly the 'living' mass, differences between tissues of varying maturity and metabolic activity become less marked. But in the same tissue, respiration rate can be shown to increase with increasing activity. For instance, when roots are washed in distilled water and then transferred to a nutrient solution from which they proceed to take up ions, their respiration rate increases concomitantly with ion absorption. In nectaries, the period of rapid sugar secretion coincides with a period of rapid respiration.

The relationship between respiration and growth has been the subject of much study and speculation. As already noted, growing tissues are characterized by high rates of respiration. The quantitative relationship between growth rate and respiration rate is, however, not a simple one. On the whole, plants with higher growth rates tend to have the higher respiration rates, although there are also reports to the contrary, e.g. of faster-growing genotypes of a grass (*Lolium perenne*) having lower respiration rates. When plant systems are compared, the differences in their rates of respiration are typically *less* than the differences in their growth rates. In a study with nine species of grasses, for a 2–3-fold higher relative growth rate, RGR (for definition of RGR see Section 6.6.1), the increase in respiration was only 1.4–1.7-fold (Scheurwater *et al.* 1998). Explanations must be sought in differences in metabolism, resulting in a greater efficiency of respiratory energy production (or utilization) in the faster-growing species. One such difference might be in the degree to which the alternative oxidase (Section 2.9.2) participates in the respiration of different tissues. In a study of root respiration and growth in four inbred lines of the greater plantain (*Plantago major*) it was found that higher RGR were associated with a higher proportion of respiration passing through the more energy-efficient cytochrome pathway as opposed to the alternative oxidase.

An understanding of relationships between growth and respiration is of great relevance to considerations of plant productivity. Respiration results in a loss of biomass, even while it is necessary to support growth. There have been arguments as to whether it is more favourable to breed a crop with a *low* respiration rate, for minimal loss of biomass, or with a *high* rate, associated with a high RGR. It appears now that it is not just overall respiration rate that is relevant to productivity. To maximize productivity, one would need to maximize the efficiency with which the potential energy of the respiratory substrates is converted to usable form – ATP and reduced coenzymes. We are not yet certain wherein this efficiency lies.

2.11.2 Metabolic control of rates: feedback mechanisms

Respiration rate is geared to the requirements of tissues by numerous mechanisms. One of these is the cellular concentration of ADP, and the ATP : ADP ratio. Respiration produces ATP, growth and maintenance processes consume it and produce ADP, the total cellular amount of [ATP + ADP] remaining constant for prolonged periods. ADP is an essential reactant in respiration. It is directly used in substrate-level phosphorylations, and mitochondrial electron transport through cytochrome oxidase is normally coupled to oxidative phosphorylation, which requires ADP. When there is an increase in an ATP-utilizing cellular activity, more ATP per unit time is converted to ADP. The higher ADP concentration stimulates a faster rate of substrate breakdown and mitochondrial electron transport – with a higher rate of ATP synthesis. The result is a higher rate of respiration and a higher turnover of ATP and ADP. The ratio ATP : ADP seems to be fairly constant in cells.

The effects of ATP and ADP are not only simple concentration effects. The glycolytic enzyme phosphofructokinase is inhibited by ATP, which is one of its substrates; pyruvate kinase is also inhibited by ATP. On the other hand, ADP activates pyruvate kinase. These effects would result in slowing glycolysis by an increase in ATP concentration and a speeding up by a rise in the concentration of ADP.

Accumulation of intermediates of the respiratory pathways causes inhibition of enzymes acting earlier in the pathways – the well-known phenomenon of **feedback inhibition.** Phosphofructo-kinase, one ofthe first enzymes of the glycolytic pathway, is inhibited by phosphoglycerate and by phosphoenolpyruvate. Citrate (from the Krebs cycle) inhibits both phosphofructokinase and pyruvate kinase. A fall in the demand for ATP would slow down the mitochon-drial electron transport; citrate metabolism would slow down corre-spondingly, being dependent on the mitochondrial terminal oxidation. The accumulating citrate would cause slowing down of glycolysis. Similarly, a fall in the demand for biosynthetic intermedi-ates would result in a build-up of these intermediates, and in an inhibition of preceding steps. Any change in requirement for ATP or/and metabolic intermediates would result in an adjustment in respiration rate, until a new equilibrium between supply and demand was established. For the PPP, the ratio of $NADP^+$: NADPH is a regulatory factor.

When metabolic pathways were first elucidated, rate control was postulated to be concentrated at a few key steps catalysed by 'pace-maker' enzymes. For glycolysis, phosphofructokinase and pyruvate kinase were considered to be pacemakers, these being the two gly-colytic enzymes for which the reactions are irreversible under physiological conditions. Several metabolites are moreover known to regulate the activities of these enzymes, as noted in the previous paragraph. But changing the amounts of individual enzymes in cells by genetic manipulation has shown that large changes in the activ-ities of phosphofructokinase and pyruvate kinase have little effect on respiration rate. It now appears that *all* steps in glycolysis and the Krebs cycle contribute to regulation of respiration rate, though not all to the same degree. From the viewpoint of the cell, the larger the number of control points, the more opportunities there are for fine-tuning and interactions between pathways. From the viewpoint of the investigator, it makes the study of rate control exceedingly com-plex. There is nevertheless no doubt regarding the basic principle: respiration rate is integrated with cellular activity so that a depletion of respiratory products (ATP and metabolites) leads to an increase in the rate of respiration. An accumulation of the same leads to a decrease in the respiration rate and in each case a new steady state is achieved.

2.11.3 Plants at work: energy, ATP and heat production

In Table 2.3 it is shown that the respiration of 1 molecule of glucose can result in the synthesis of up to 36 molecules of ATP, equivalent to a conservation of about 52% of the potential energy of the glucose. One can compare this value with the actual amounts of energy conserved as measured experimentally. If E represents the total amount of energy produced by the complete oxidation of glucose (i.e. 2880 kJ mol^{-1}), calculated for a tissue from its measured respiration rate, then some of this energy will be released as heat, H; some will be expended on physical work, W (e.g. movement), and some on synthetic work, S:

$$E = H + W + S \qquad (2.21)$$

The energy used for W and S would be in the first instance conserved in ATP. The value of H can be measured directly by calorimetry. When such measurements are carried out on plant tissues, it is found that 90–99% of the potential energy of the oxidation of glucose is actually released as heat, leaving only 1–10% to cover work, $W + S$. This is a very low value compared with the maximum possible conservation of 52% in ATP. What has happened to all the ATP?

This loss of nearly all the respiratory energy, as heat, has caused some plant scientists to wonder whether most of plant respiration might be a wasteful process. If, however, the way in which ATP is utilized in biosyntheses is considered, it is seen that energy released as heat during metabolism does *not* necessarily represent a waste. The high reactivity of ATP is utilized to make phosphorylated or nucleotide-linked derivatives of metabolites such as sugars or amino acids. This synthesis entails the loss of some energy as heat. The 'activated' derivatives are themselves now highly reactive and able to participate in reactions not possible with the parent compounds – these reactions again releasing energy as heat. When macromolecules are synthesized from monomers, in many instances for each bond made one phosphorylated (or nucleotide-linked) reactant molecule is needed and one ATP molecule is used up while most of the energy of the ATP is released as heat. *It is this loss of heat energy that makes the reactions possible, according to the laws of thermodynamics.* In the case of protein synthesis, there is an expenditure of at least five molecules of ATP for every peptide bond formed, equivalent to a conservation of ATP energy of no more than 1%. Yet the 99% of energy released as heat is not wasted: it is needed for the information transfer, for the placing of each amino acid in its correct location. The electrical energy used in transmitting a telephone call is not conserved in the transmitted message, but it is not wasted. Most of the energy *conserved* in cellular macromolecules is expended in the original synthesis of the monomers, and most of this energy input has already occurred during photosynthesis. The transport of many chemicals across cellular membranes is linked to the hydrolysis of ATP (Chapters 4 and 5); again the energy of hydrolysis is released as heat. The energy output of maintenance respiration must appear entirely as heat, there being no *net* work or synthesis accomplished.

Nevertheless it cannot be assumed that the coupling of respiration to ATP synthesis is 100% efficient, so *some* of the observed heat release can represent a waste in the sense of failure to synthesize ATP. The problem of the alternative oxidase has already been discussed (Section 2.9.2). But respiration which is (partially) uncoupled could still be fulfilling a vital function in producing metabolites.

According to the laws of thermodynamics, any system should assume the state of maximum **entropy**, i.e. maximum randomness, disorganization, and minimum free energy. A growing plant increases in complexity and organization both at the chemical and at the structural level. This is possible only at the expense of entropy increasing somewhere else, in the plant's environment. The high energy quanta from the sun are transformed during photosynthesis to chemical bond energy, but always with some energy loss as heat, low energy quanta. Then, as the products of photosynthesis are incorporated into cells, more heat energy is released during respiration and the subsequent utilization of the respiratory ATP. All this heat loss represents an increase in entropy and dissipation of energy, which compensates for the increased complexity and free energy content of the living plant. In this sense again, the loss, the 'waste' of energy as heat, is necessary to keep living systems functional.

Complementary reading

Ainsworth, E. A., Davey. P. A., Hymus, G. J. *et al.* Is stimulation of leaf photosynthesis by elevated carbon dioxide concentrations maintained in the long term? A test with *Lolium perenne* for 10 years at two nitrogen fertilization levels under Free Air CO_2 Enrichment (FACE). *Plant, Cell and Environment*, **26** (2003), 705–14.

Bowes, G. Facing the inevitable: plants and increasing atmospheric CO_2. *Annual Review of Plant Physiology and Plant Molecular Biology*, **44** (1993), 309–32.

Bryce, J. H. & Hill, S. A. Energy production in plant cells. In *Plant Biochemistry and Molecular Biology*, 2nd edn, ed. P. J. Lea & R. C. Leegood. Chichester: Wiley, 1999, pp. 1–28.

Bunce, J. A. Responses of respiration to increasing atmospheric carbon dioxide concentrations. *Physiologia Plantarum*, **90** (1994), 427–30.

Dennis, D. T., Huang, Y. & Negm, F. B. Glycolysis, the pentose pathway and anaerobic respiration. In *Plant Metabolism*, 2nd edn, ed. D. T. Dennis, D. H. Turpin, D. D. Lefebvre & D. B. Layzell. Harlow: Addison Wesley Longman, 1997, pp. 105–23.

Dodd, A. N., Borland, A. M., Haslam, R. P., Griffiths, H. & Maxwell, K. Crassulacean acid metabolism: plastic, fantastic. *Journal of Experimental Botany*, **53** (2002), 569–80.

Drake, B. G., Gonzàlez-Meler, M. A. & Long, S. P. More efficient plants: a consequence of rising atmospheric CO_2. *Annual Review of Plant Physiology and Plant Molecular Biology*, **48** (1997), 609–39.

Drennan, P. M. & Nobel, P. S. Responses of CAM species to increasing atmospheric CO_2 concentrations. *Plant, Cell and Environment*, **23** (2000), 767–81.

Ehleringer, J. R., Sage, R. F., Flanagan, L. B. & Pearcy, R. W. Climate change and the evolution of C_4 photosynthesis. *Trends in Ecology and Evolution*, **6** (1991), 95–9.

Hall, D. O. & Rao, K. K. *Photosynthesis*, 5th edn. Cambridge: Cambridge University Press, 1994.

Hamilton, J. G., Thomas, R. B. & DeLucia, E. H.. Direct and indirect effects of elevated CO_2 on leaf respiration in a forest ecosystem. *Plant, Cell and Environment*, **24** (2001) 975–82.

Lawlor, D. W. *Photosynthesis*, 2nd edn. Harlow: Longman, 1993.

Leegood, R. C., Sharkey, T. D. & Von Caemmerer, S. *Photosynthesis, Physiology and Metabolism*. Dordrecht: Kluwer, 2000.

Mauseth, J. D. *Plant Anatomy*. Menlo Park, Benjamin/Cummings, 1988.

Mauseth, J. D. *Botany: an Introduction to Plant Biology*, 2nd edn. Sudbury, MA: Jones & Bartlett, 1998.

Nishio, J. N. Why are higher plants green? Evolution of the higher plant photosynthetic pigment complement. *Plant, Cell and Environment*, **23** (2000), 539–48.

Norby, R. J., Kobayashi, K. & Kimball, B. A. Rising CO_2 – future ecosystems. *New Phytologist*, **150** (2001), 215–21.

Perata, P. & Alpi, A. Plant responses to anaerobiosis. *Plant Science*, **93** (1993), 1–17.

Reiskind, J. B., Madsen, T. V., Van Ginkel, L. C. & Bowes, G. Evidence that inducible C_4 photosynthesis is a chloroplastic CO_2-concentrating mechanism in *Hydrilla*, a submerged monocot. *Plant, Cell and Environment*, **20** (1997), 211–20.

Rolletschek, H., Borisjuk, L., Koschurreck, M., Wobus, U. & Weber, H. Legume embryos develop in a hypoxic environment. *Journal of Experimental Botany*, **53** (2002), 1099–107.

References

Bartley, G. E. & Scolnick, P. A. (1995). Plant carotenoids: pigments for photoprotection, visual attraction and human health. *The Plant Cell*, **7**, 1027–38.

Goodwin, T. W. & Mercer, E. I. (1972). *Introduction to Plant Biochemistry*. Oxford: Pergamon Press.

Grahl, A. & Wild, A. (1972). Die Variabilität der Größe der Photosyntheseeinheit bei Licht und Schattenpflanzen. *Zeitschrift für Pflanzenphysiologie*, **67**, 443–53.

Hatch, M. D. (1992). C_4 photosynthesis: an unlikely process full of surprises. *Plant and Cell Physiology*, **33**, 333–42.

Kirschbaum, M. U. F. (1994). The sensitivity of C_3 photosynthesis to increasing CO_2 concentration: a theoretical analysis of its dependence on temperature and background CO_2 concentration. *Plant, Cell and Environment*, **17**, 747–54.

Ku, M. S. B., Agarie, S., Nomura, M. *et al.* (1999). High-level expression of maize phosphoenol pyruvate carboxylase in transgenic rice plants. *Nature Biotechnology*, **17**, 76–80.

Lambers, H. (1997). Oxidation of mitochondrial NADH and the synthesis of ATP. In *Plant Metabolism*, 2nd edn, ed. D. T. Dennis, D. H. Turpin,

D. D. Lefebvre & D.B. Layzell. Harlow: Addison Wesley Longman, pp. 200–19.

McCree, K. J. (1972). The action spectrum, absorptance and quantum yield of photosynthesis in crop plants. *Agricultural Meteorology*, **9**, 191–216.

Öpik, H. (1980). *The Respiration of Higher Plants*. Studies in Biology, 120. London: Edward Arnold.

Sachs, M. M., Subbaiah, C. C. & Saab, I. N. (1996). Anaerobic gene expression and flooding tolerance in maize. *Journal of Experimental Botany*, **47**, 1–15.

Scheurwater, I., Cornelissen, C., Dictus, F., Welschen, R. & Lambers, H. (1998). Why do fast- and slow-growing grass species differ so little in their rate of root respiration, considering the large differences in rate of growth and ion uptake? *Plant, Cell and Environment*, **21**, 995–1005.

Spreitzer, R. J. & Salvucci, M. E. (2002). Rubisco: structure, regulatory interactions and possibilities for a better enzyme. *Annual Review of Plant Biology*, **53**, 449–75.

Whitney, S. M., von Caemmerer, S., Hudson, G. S. & Andrews, T. J. (1999). Directed mutation of the Rubisco large subunit of tobacco influences photorespiration and growth. *Plant Physiology*, **121**, 579–88.

Chapter 3

Water relations

3.1 | Introduction

Liquid water is absolutely necessary for life as we know it. Firstly it is the solvent and reaction medium of all living cells, which contain some 75–90% water by weight; secondly it is a reactant in many metabolic processes; and thirdly, as the hydration water of macromolecules, it forms part of the structure of protoplasm, existing as 'liquid ice' in a labile but ordered structure. The physicochemical properties of water (H_2O) are unique; heavy water (D_2O or DHO), containing deuterium, the heavy isotope of hydrogen, differs sufficiently to be toxic. In multicellular organisms, water provides the transport medium. Additionally, for plants, water is one of the raw materials for photosynthesis and produces the turgor pressure of water-filled vacuoles which gives mechanical rigidity to thin-walled tissues, while some movements of plant organs occur as a result of turgor pressure changes. Plant cell expansion is driven by turgor pressure and hence growth rates depend on hydration levels.

On 'dry' land, the highly hydrated body of a terrestrial plant in many situations tends to lose water to the environment, especially to the atmosphere, in accordance with gradients of free energy of water. There are few habitats where plants do not suffer some water shortage at least intermittently. The necessity for maintaining an adequate internal water content has been a major factor in the evolution of land plants with respect to structure and numerous aspects of physiology. It is not an exaggeration to say that the colonization of land by plants has depended upon the evolution of systems for the absorption and conservation of water. Another difficulty for land plants is the transport of water from the underground supply tapped by the roots to the aerial shoots. For the tallest flowering plants this may mean moving water some 100 m against gravity.

This chapter deals with the forces and factors involved in water uptake and loss in flowering plants, the mechanisms of water movement within the plants, and the controls exerted on water exchange and water transport by the plant and the environment respectively.

3.2 | Water movement and energy: the concept of water potential

Water moves into plants, in the case of terrestrial plants mainly from the soil; and water moves out of plants, mainly into the atmosphere. There is also much movement of water within plants. Movement implies the involvement of *energy*. It was noted at the end of the last chapter that metabolism is driven by changes in free energy. Water movement, too, is driven by energy levels. Water will move from a system or area where it is at a higher free energy, to a system where it is at a lower free energy. If we consider a plant cell and its environment, we can therefore say:

When the free energy of cell water is **lower** than the free energy of external water, net flux will be *into* the cell.

When the free energy of cell water is **higher** than the free energy of external water, net flux will be *out of* the cell.

When the free energy of water is **equal** inside and outside, there will be *no net movement in or out*, and if water movement results in equalization of the energy levels, net movement will cease. In both cases, however, a flux (exchange) of water may still proceed, equal quantities moving in each direction per unit time. (This can be demonstrated by applying radiolabelling to the water.)

In order to predict the direction of movement of water into/out of plants, plant cells or tissues, we therefore need a measure for the free energy of water. This measure is the **water potential**, denoted by the Greek letter Ψ (psi), or Ψ_w. Water moves along gradients of water potential, from higher to lower water potential. Although Ψ_w is basically a measure of free energy, for plant physiology it is most often expressed in pressure units, since hydrostatic pressures and tensions (negative pressures) contribute to water potential and play a very important part in the water relations of plants. The pressure unit is the pascal Pa and its multiples, the pascal being rather small. Throughout this text, the megapascal MPa (10^6 Pa) is used. For the derivation of water potential from basic principles, see note at end of chapter; for the relationship of MPa to the older pressure units, atmospheres and bars, see Appendix.

It should be noted that water potential applies to water in *any* situation, and in *any* form, liquid, ice, or water vapour. Wherever there is water, it has a water potential.

3.3 | Water potentials of plant cells and tissues

3.3.1 Forces determining cellular water potential

The water potential of a plant cell is determined by three kinds of forces which affect the free energy of the cellular water.

(1) In plant cells, the cell wall exerts a hydrostatic pressure, the turgor pressure (wall pressure) on the protoplast; a cell within a compact tissue may also be under pressure from surrounding cells. **Hydrostatic pressure** in excess of atmospheric **increases** the free energy and raises water potential; thus the **pressure potential** Ψ_p is a positive value.

(2) Plant cells contain low-molecular-weight solutes, mainly vacuolar in a vacuolated cell. These exert **osmotic forces**, which **decrease** the free energy and lower the water potential; the **osmotic potential** Ψ_π is therefore a negative value. (Osmotic pressure is numerically equal to osmotic potential, but has a positive sign.)

(3) Plant cells contain high-molecular-weight colloids, in the cytoplasm and the cell wall. **Matric forces** exerted by colloids **decrease** the free energy of water and lower the water potential; their effect is represented by the **matric potential** Ψ_m. Surface tension forces at air/water interfaces in cell wall capillary spaces also contribute to Ψ_m.

The overall water potential of a plant cell is the sum of these three quantities:

$$\Psi = \Psi_p + \Psi_\pi + \Psi_m \tag{3.1}$$

One can think in terms of the wall pressure tending to squeeze the water out, while osmotic and matric forces tend to draw the water in. In vacuolated cells of high water content, the matric potential is thought to make a relatively minor contribution and, for such cells, the water potential is often given simply as

$$\Psi = \Psi_p + \Psi_\pi \tag{3.2}$$

But matric potential is important in controlling water uptake and retention by tissues of low water content, such as 'dry' and partly imbibed seeds and in the soil. In the soil, colloid content also can be high.

Osmosis of water is water movement driven by an osmotic potential gradient through a semipermeable membrane, i.e. a membrane permeable to water but impermeable, or very much less permeable, to solutes that are present. Osmosis drives water movement from a lower to a higher solute concentration: a high solute concentration gives a *low* osmotic potential, hence a *low* Ψ. All membranes of living cells are semipermeable towards nearly all the solutes within the cell, or encountered in the environment, hence osmosis is important in plant water relations. However, as the preceding discussion shows, osmosis is not the *only* process involved in plant water relations. Hydrostatic pressure (or tension) can overrule osmotic forces.

A cell with a positive Ψ_p is said to be **turgid**, though the degree of turgidity varies according to water content. As a cell takes up water, its Ψ rises and its volume expands as the wall is stretched (Fig. 3.1). When a cell is fully water-saturated, i.e. the wall can yield no more, the cell can take up no more water even from pure water. Its Ψ is zero

(= that of pure water, with which it is in equilibrium), the osmotic potential of the cell contents being balanced by the pressure potential:

$$\Psi = 0, \text{ and } -\Psi_\pi = \Psi_p \qquad (3.3)$$

This cell is at full turgidity, at maximal volume, maximal wall pressure and maximal water potential – of 0 MPa! The values of plant cell Ψ usually vary from 0 down, i.e. are negative, and negative values are less than zero.

On the other hand, as water is progressively lost from a cell, its Ψ decreases, both as a result of the reduction of the pressure, as the

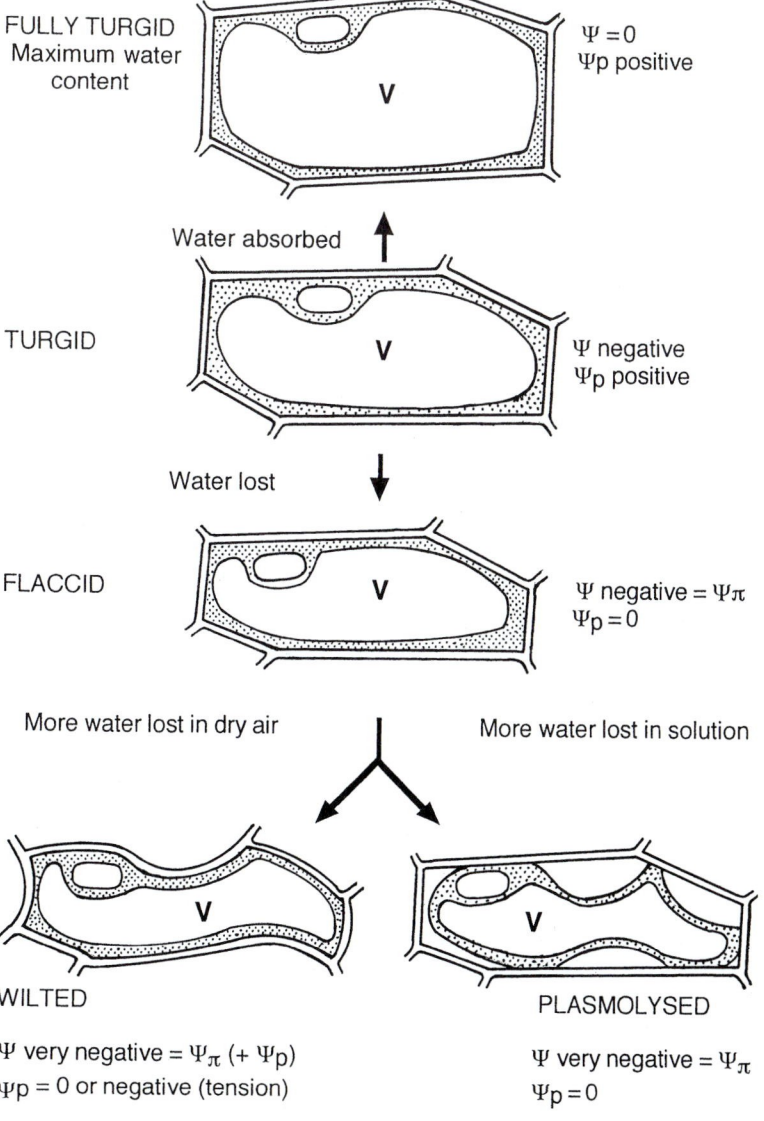

FULLY TURGID
Maximum water content

$\Psi = 0$
Ψp positive

Water absorbed

TURGID

Ψ negative
Ψp positive

Water lost

FLACCID

Ψ negative $= \Psi\pi$
$\Psi p = 0$

More water lost in dry air

More water lost in solution

WILTED

Ψ very negative $= \Psi_\pi (+ \Psi p)$
$\psi p = 0$ or negative (tension)

PLASMOLYSED

Ψ very negative $= \Psi_\pi$
$\Psi p = 0$

Fig. 3.1 Effect of water absorption and water removal on a plant cell. Starting with a turgid cell (second from top), *absorption of water* leads to increase in turgidity and Ψ increases up to a maximum of 0. *Water loss* beyond a certain level in a solution of low Ψ results in plasmolysis, shrinkage of cell contents from the wall. Water loss in dry air results in wilting and the pull of the shrinking contents on the wall can lead to a wall tension instead of a pressure. V = vacuole.

shrinking cell contents press less strongly on the wall, and as a consequence of the solutes being concentrated into a smaller volume and lowering the Ψ_π component. When the stage is reached where the protoplast no longer presses against the wall, the cell is said to be flaccid; now

$$\Psi_p = 0, \text{ and } \Psi = \Psi_\pi \tag{3.4}$$

If still more water is removed, the effect depends on the mode of removal (Fig. 3.1). If the water is removed by evaporation, drying out in air, the cell shrivels in size and the wall caves in or folds as the shrinking protoplast pulls on it. The degree of flexibility of the wall may determine how much water can be removed. The tissue becomes *wilted*. If, however, the water is removed by immersion of the cell in a solution of low Ψ_π, **plasmolysis** results: the protoplast shrinks away from the wall and the external solution fills the space between the plasma membrane and the wall; there is no further decrease in overall cell size. The relationships between Ψ, Ψ_π and Ψ_p are shown graphically in Fig. 3.2. Under field conditions, wilting is more usual than plasmolysis. Loss of water beyond a limit, which varies with the tissue, is fatal; this is discussed in Chapter 13 under desiccation stress.

Actual values of cell Ψ of plants growing in the field in a temperate climate and with a fairly adequate supply of water fall mostly in the range of –0.1 to –2.0 MPa, but may fall well below this in times of water shortage, and in extreme climates or habitats such as deserts or salt marshes; in the latter the apparently generous external water supply is at a low Ψ owing to the osmotic effect of the salt. Plant Ψ values below –10 MPa have been recorded.

Quantitative consideration of plant water potentials needs clear thinking. Because values of Ψ of plants and in the natural environment are negative, one must be careful to keep in mind that e.g. –2 MPa is *lower* than –1 MPa. Sometimes the terms 'higher' and 'lower' are avoided by referring to 'less negative' and 'more negative' values. One must also accustom oneself to thinking of zero as a *high* value, the highest that the Ψ of plant or soil in most cases can attain. (Plant tissues under high pressure, such as squirting glands or squirting fruits, which can eject their contents to considerable distances, may have positive Ψ values. But these play no significant part in overall plant water relations.)

3.3.2 Measurement of water potential and its components in plant cells and tissues

The overall Ψ of a cell or tissue can be measured by exposing replicate samples of the tissue to a graded series of Ψ, either by immersing the samples in solutions of known Ψ, or by enclosing them in atmospheres of known Ψ (vapour pressure). Changes in water content of the samples are detected by weighing the samples before and after

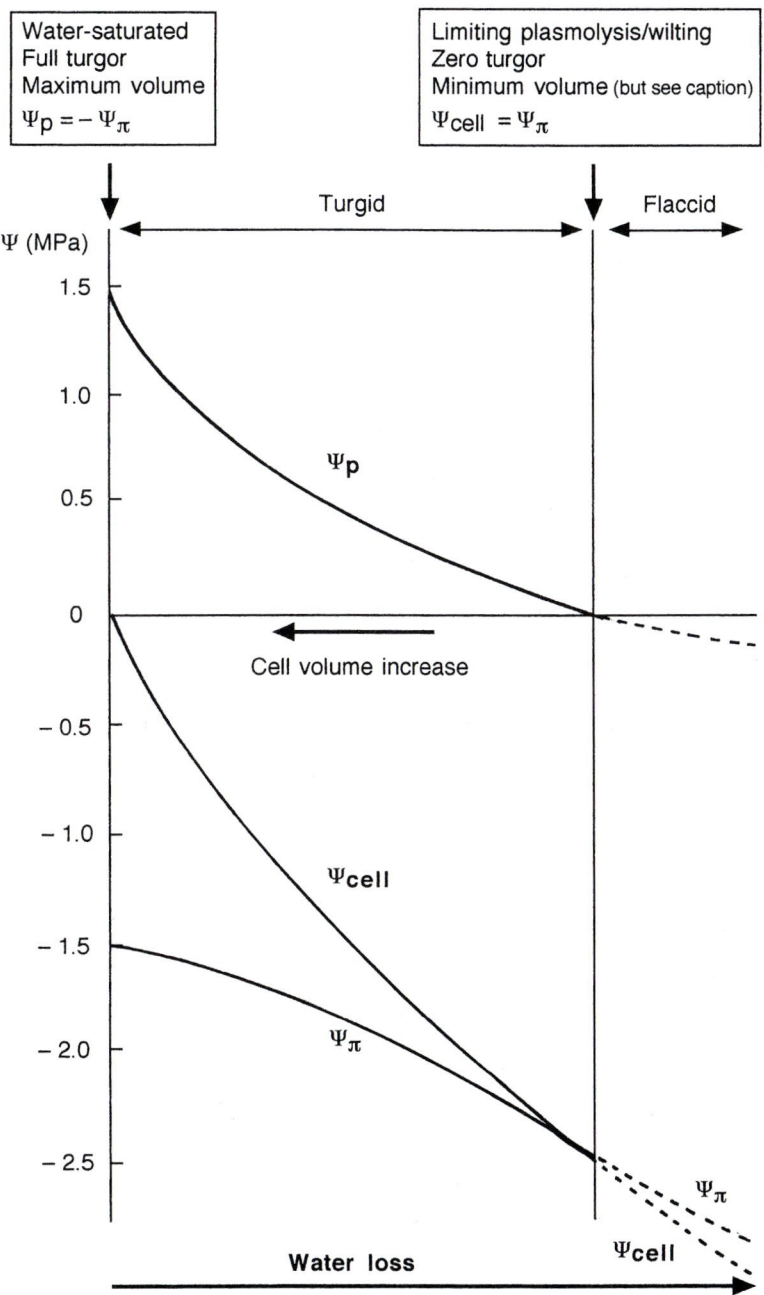

Water-saturated
Full turgor
Maximum volume
$\Psi_p = -\Psi_\pi$

Limiting plasmolysis/wilting
Zero turgor
Minimum volume (but see caption)
$\Psi_{cell} = \Psi_\pi$

Fig. 3.2 Generalized quantitative relationship between overall cell Ψ, Ψ_p, Ψ_π and cell volume. Note that the x-axis scale runs from a maximum value for the volume on the left, at full turgor, to a minimum value at the point of limiting plasmolysis/ wilting. The extent to which the cell volume varies between these points depends greatly on the extensibility of the cell wall; hence no numerical values are given, but if the volume at full turgidity is taken as 100%, the minimum values quoted for different cells vary from about 95% to 70%. The dashed lines indicate what happens if the cell wall caves in under tension after limiting wilting: the volume decreases still further, Ψ_p becomes negative (top section of graph), and the cell Ψ falls below Ψ_π (bottom section of graph). If the wall does not cave in, but the cell wrinkles, Ψ continues to equal Ψ_π with further water loss.

the incubation period; the tissue Ψ equals that of the environment in which it neither gains nor loses water.

Another method for measuring tissue water potential is the **thermocouple psychrometer.** The tissue is allowed to equilibrate with the atmosphere in a small chamber which houses a thermocouple junction and which is incubated at constant temperature. The Ψ of the chamber atmosphere will become equal to that of the tissue.

A small drop of pure water is then introduced on to the thermocouple junction. As this water evaporates, it causes cooling of the thermocouple, causing a current to flow. The current is proportional to the rate of cooling; the rate of cooling depends on the rate of evaporation, which in turn depends on the Ψ of the vapour in the chamber – and this equals the Ψ of the tissue. The instrument is calibrated with material of known Ψ, so that the tissue Ψ can be obtained from the current reading.

The **pressure bomb method** for obtaining water potentials, which is more controversial, is discussed later in connection with xylem transport (Section 3.4.3, p. 80).

It may also be relevant to measure the individual components of Ψ. The *osmotic potential* can be obtained by finding the point of limiting plasmolysis, when the Ψ_p has been just reduced to 0 and the cell's Ψ equals its Ψ_π (Equation 3.4 and Fig. 3.2). This is done by immersing the plant material samples in a series of graded Ψ. Since the precise point at which a cell just becomes flaccid is in practice almost impossible to see, the point is found at which 50% of the cells can be seen to be plasmolysed, i.e. have gone beyond the limiting point, while the remainder still appear turgid. This point is taken as equivalent to an average limiting plasmolysis value for all the cells. The method does have an inherent error. By the time the point of limiting plasmolysis is reached, the cell volume has shrunk somewhat because of the loss of water and cellular solutes have become more concentrated, lowering the value of the Ψ_π. Fortunately the volume change is small and the error is estimated at less than 10%. An alternative method for obtaining the Ψ_π is to express the cell sap and to measure its Ψ_π by physicochemical methods applicable to any solution. Errors can be introduced here through dilution of the sap with water from the cell walls, and by chemical changes resulting from the mixing of the vacuolar contents with the cytoplasm.

The *pressure potential* is more difficult to measure and is often taken as the difference between the overall Ψ and the Ψ_π. However, there is an instrument, the **pressure probe**, which does measure pressures directly. It consists of a very fine hollow glass capillary needle, oil- or water-filled, which is inserted into the cell and at the other end joined to a pressure sensor. The operation is followed under a microscope, and requires skill. It may be difficult to see precisely where the probe tip is positioned and not all types of cell are amenable to this method.

3.3.3 Water permeability of plant membranes

Gradients of Ψ determine the *direction* of water movement between cell and environment, and between cell and cell. The *speed* of water movement into or out of living cells is controlled by **membrane permeability**. In plant cells, two membranes need to be considered: the plasma membrane (plasmalemma) which surrounds the cell; and the tonoplast, the membrane surrounding the vacuole, which contains the bulk of the water in most mature plant cells.

The permeability of biological membranes towards water is generally high. The permeability of plant cell plasma membranes towards water has been estimated to be from several hundred to several thousand times higher than for such small uncharged organic compounds as urea, glycerol or sugars, and for charged ions the difference is greater still. Nevertheless, the resistance of biological membranes towards water is not negligible. The quantitative measure of the permeability of a membrane towards a chemical is the **permeability coefficient**, the rate of diffusion per unit area, unit time and unit concentration gradient. The permeability coefficients of biological membranes towards water are so high that they are difficult to measure; the concentration gradients are small and equilibration between external and internal water is reached rapidly. Very variable values have been reported for the permeability coefficients of plant cell plasma membranes. Because of the practical problems in measuring the permeability coefficient for water, the **hydraulic conductivity** is often measured instead, the rate of diffusion of water per unit area and unit time per unit hydrostatic pressure gradient. This can, however, be used only to compare water permeabilities of various systems, not for comparing the permeabilities of water and other chemicals.

The ease with which water, and other small hydrophilic molecules with masses not above 46 Da, traverse biological membranes was first attributed to minute pores opening up transiently in the bimolecular lipid leaflet of the membrane by the random thermal vibrations of the lipid molecules. Subsequently membrane proteins named **aquaporins**, which make water-specific channels through biological membranes, have been identified in all types of organisms (Box 3.2; Tyerman *et al.* 2002). Plants contain a greater variety of aquaporins than organisms from other kingdoms; over 30 aquaporins are known from maize, for instance. Aquaporins show specificity with respect to organ and membrane, i.e. plasma membrane or tonoplast. Evidence is accumulating that the variability of the reported values of the water permeability coefficients of different plant cells results at least partly from differences in the concentrations (and maybe types) of aquaporins. Development and differentiation are associated with changes in aquaporin density; e.g. during the cell elongation stage which is accompanied by rapid water uptake, aquaporins are particularly abundant. A diurnal rhythm in expression of aquaporin genes has been correlated with leaf movements which depend on turgor changes in their motor cells, mediated by water movements in and out of the cells.

It must be emphasized that all water movement into or out of plant cells is along the Ψ gradient. There is no active pumping of water against its free energy gradient, at the expense of metabolic energy, as occurs with many other metabolites. The permeabilities of the plasma membrane and tonoplast towards water are so high that the water would leak out very fast; the amount of metabolic energy needed to pump it against the leakage would be unrealistic.

Box 3.2

The plasma membrane aquaporins are collectively termed the PIPs (plasma membrane intrinsic proteins) and the tonoplast-located ones, TIPs (tonoplast intrinsic proteins). There exist also aquaglyceroporins, which permit passage of small non-electrolyte molecules as well as water. All the above-mentioned are classed together as MIP, major intrinsic protein family, found in organisms of all kingdoms. Each aquaporin molecule has six transmembrane α-helices embedded in the membrane, and forms one transmembrane water channel; the molecules are commonly grouped in the membrane as tetramers.

3.4 | Water relations of whole plants and organs

The water relations of a whole plant, or even an organ such as a leaf, are much more complex than those of individual cells. The formulae given in the preceding section, relating water potential to pressure and osmotic (and matric) potentials are applicable only at the cellular level. There is no such thing as the Ψ_p of a whole plant; the value will vary between different tissues. The overall Ψ, too, usually varies between different parts of a plant. Most flowering plants are land plants and, in the terrestrial environment, the Ψ of the atmosphere is nearly always much lower than that of plant tissues, often by tens of MPa, and hence there is a great tendency for water loss from the plant. The large surface area necessitated by the photosynthetic mode of life provides a large surface for evaporation of water, **transpiration.** This loss must be made good from the soil, which is generally at a much higher Ψ than the atmosphere or the plant. Hence for most of the time there is a flow of water through the plant, along a Ψ gradient, as below:

$$\text{soil} \longrightarrow \text{root} \longrightarrow \text{stem} \longrightarrow \text{leaf} \longrightarrow \text{air}$$

This flow is frequently termed the *transpiration stream.* Only in times of a water-saturated atmosphere, or in times of extreme drought, may there be equilibrium, more or less, between plants and the environment, with gradients within the plant eliminated and water movement nearly at a standstill. Of the water absorbed by the roots, only a very small fraction is retained by the plant in temperate habitats. For maize, an annual, this fraction has been estimated at less than 1% of the water absorbed during its growing season. During one bright sunny day leaves may transpire several times their own weight of water; a leaf of *Senecio jacobaea* growing on a sand dune can transpire its own weight in water in 45 minutes. The water content of aerial organs of a plant is generally lower in the daytime, when the rate of transpiration is high, than during the night, when the transpiration rate is much lower (owing to the lower temperature and closure of stomata) and the deficit is made good. In the roots, which experience much less temperature change over the diurnal cycle and which have no stomata, the water content fluctuates much less.

3.4.1 Absorption of water by roots

The root systems of plants are often very extensive. Roots of some plants extend much further underground than the shoot rises into the air (Fig. 3.3). The roots of apple tree (*Malus domestica*) may go down to about 10 m, and even in herbaceous plants such depths can be reached, e.g. in alfalfa (*Medicago sativa*). The actual area of a root system is formidable. It has been reported that a rye (*Secale cereale*) plant had a root surface area of over 600 m^2, of which two-thirds was

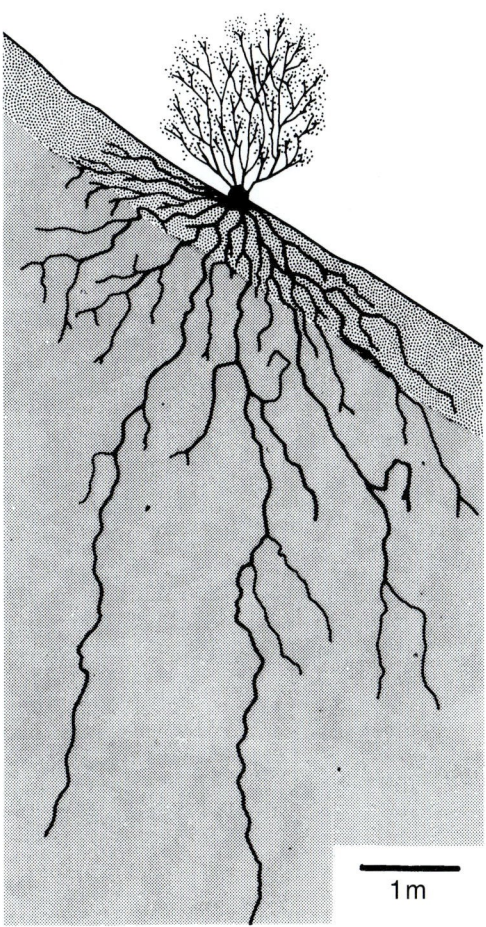

Fig. 3.3 Semi-diagrammatic sketch of the root system of a Californian woody shrub, the chamise (*Adenostoma fasciculatum*). From Hellmers *et al.* (1955).

1 m

root hair area, and the total length of the root system was over 11 000 km, including 10 000 km of root hairs. The total area of the shoot (including areas of cells bordering leaf air spaces) was only 28 m². Most of the water absorption takes place near the root tips, where there is a thin epidermis with root hairs (Fig. 3.4). Not only do the root hairs provide a large area, but they make intimate contact with the soil, bending around soil particles and penetrating into tiny crevices. As the root tissues mature, the epidermis with its hairs is replaced by a more impermeable suberized periderm. For efficient water uptake, root growth must continuously regenerate the absorbing zone behind each growing apex. Continuous growth is also necessary to invade new areas of soil, for there is little lateral movement of water in soil compared with downward drainage directly after water addition. Water will not move to the roots, so roots must grow to the water. Positive hydrotropism (Chapter 12) may direct root growth towards water. Roots can grow rapidly; a rate of 10 mm per day is common for grasses; maize (*Zea mays*) roots can extend by as much as 50–60 mm per day. The average daily increase in length of the total

Fig. 3.4 Scanning electron micrograph of root hairs on a radish (*Raphanus sativus*) root tip. The root hairs appear at a distance of about 1.8 mm from the apex. From Troughton & Donaldson (1972).

1 mm

root system of the rye plant discussed above (hairs excluded) was almost 5 km.

Though the root hair zones provide the most efficient water-absorbing surfaces, uptake in older regions is still appreciable, particularly during conditions of water shortage and at times when root growth is slow, such as in winter. Points of emergence of lateral roots break the suberized layers and enable the entry of water.

3.4.2 The route of water movement through the plant

The xylem as the water-transporting system

The main channel for upward/long-distance movement of water in the plant is the **xylem**, the wood. When the tissues outside the xylem are peeled off over a short length of woody stem where the xylem is central, the conduction of water beyond the stripped region continues unimpeded. The non-living cells of the xylem are filled with a watery sap, at least in young wood, and dyes and Indian ink can be seen to move in the xylem. Toxic solutions have been shown to pass from roots to leaves, indicating that the route does not involve living cells; heat-killed stems, too, can conduct water. Chilling does not stop water movement as long as no freezing occurs. All this points to non-living cells of the xylem as the water-conducting cells. When the lumina of these cells are blocked with mercury or cocoa butter, water movement is inhibited.

The xylem is a complex tissue. In addition to water-conducting cells and lignified fibres which all are dead at maturity, it contains living parenchyma cells and sometimes also living transfer cells (for transfer cells, see Section 5.4.1). Functional xylem is accordingly not a dead tissue, though it contains a large proportion of dead cells. In flowering plants there are two kinds of conducting cells, the **tracheids** and the **vessel elements**. They have lignified secondary walls with the secondary thickening laid down in distinctive patterns leaving areas of primary wall as pits; these facilitate the passage of water from cell to cell (Fig. 3.5). Somewhat confusingly, the thin primary wall across the pit is often called the pit membrane.

The tracheids function as single cells, but vessel elements are joined to make elongate **vessels** by the perforation or partial breakdown of end walls in files of cells; these end walls are then known as perforation plates (Fig. 3.5). Their possession is the distinguishing feature of the vessel element. The diameters of vessel units range from below 10 μm to several hundred μm, even 1000 μm in some lianas. Tracheid diameters overlap with those of the narrower vessel elements. Any one piece of xylem has conducting elements of varied width; this may be of functional importance (Section 3.4.4). The lengths of vessels range from under 1 cm to 10 m or more, and are very variable even within the same plant. In some trees *some* continuous vessels run right from the crown to the roots, but most vessels are shorter than the height of the plant, and even in trees many vessels measure only a few centimetres. The possession of vessels making long continuous channels for water movement is considered to be one of the advanced features of flowering plants; the earliest land plants had only tracheids, and while the evolution of vessels has occurred in several divisions of land plants they are lacking in the conifers. Tracheids offer much more resistance than vessels to water movement.

In woody perennials, new layers of xylem known as **annual rings** are produced each year during the growing season. The older regions of xylem eventually lose their water-conducting capability and become air-filled or blocked by ingrowths (tyloses) from adjacent living parenchyma cells, or by gums, resins and tannins. The water then moves only through the young, outer xylem, the sapwood, which may comprise only the current year's growth or include a few youngest annual rings. The inner non-conducting xylem is known as the heartwood.

Living parenchyma cells make up rays in secondary wood, running radially through the xylem from the pith towards the cortex (Fig. 3.6) and lying also among the conducting cells. They store organic nutrients, and some botanists have assigned a role in water transport to them (Section 3.4.4). Transfer cells are not always present in xylem, but may occur next to conducting cells especially in leaf veins; they have highly involuted cell walls, which gives them a

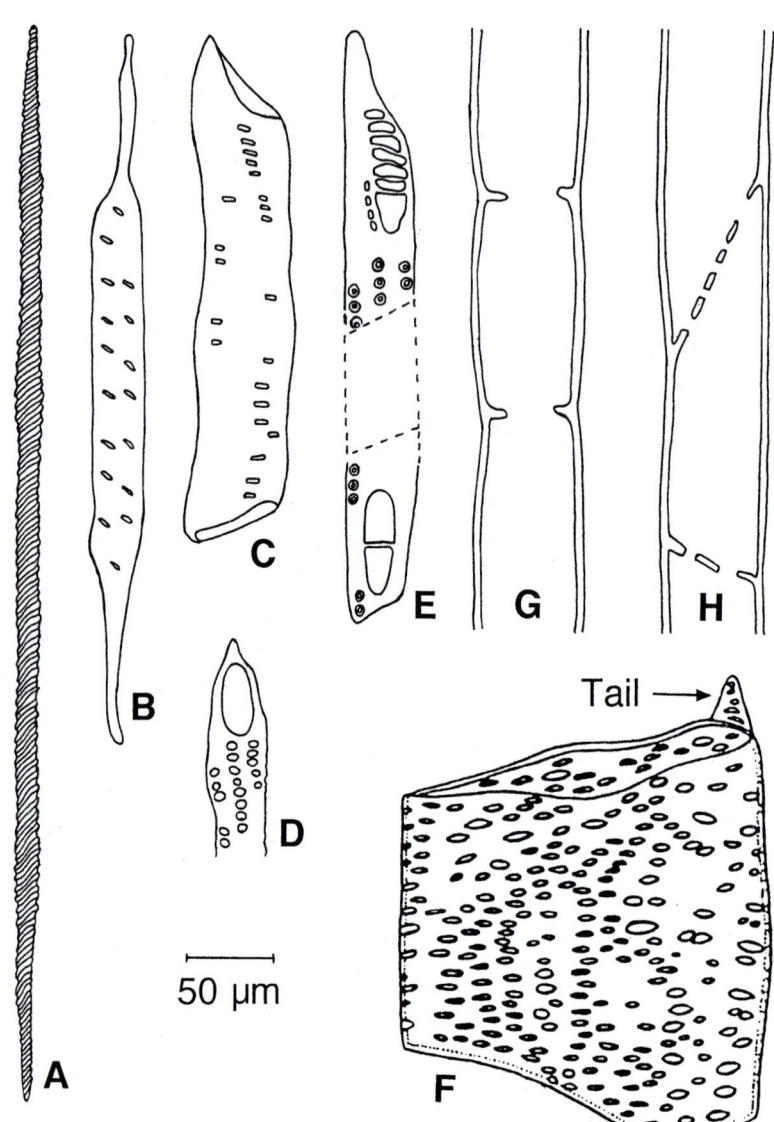

Fig. 3.5 The structure of xylem conducting cells as seen by light microscopy, all to the same scale. (A)Tracheid with helical wall thickening from beech (*Fagus sylvatica*); (B) pitted tracheid from lime (*Tilia* sp.); (C) vessel element, sparsely pitted, with simple perforation plates seen in side view, from oak (*Quercus robur*); (D) vessel tip showing slanted perforation plate in face view, from oak; (E) two ends of vessel element with barred perforation plates, also bordered pits, from oak; (F) very wide pitted vessel element from a vine (liana, *Aristolochia brasiliensis*); (G) and (H) diagrams of longitudinal sections of vessels; in (G) the perforation plates are simple and at right angles to the walls; in (H) the plates are barred and oblique, the oblique orientation giving a larger total plate surface area, which compensates for the loss of open area owing to the barring. (F) from Esau (1965). © John Wiley & Sons. Reprinted by permission.

50 µm

large surface area of plasma membrane, and they function in the transfer of solutes into and out of the conducting cells.

The movement of water into and out of the xylem

In the root, the xylem is the central tissue; to reach it, water must pass radially through the epidermis, cortex, endodermis and pericycle (Fig. 3.7). The precise radial movement pathway is still under discussion. There are three possible routes for the water: the **apoplastic route**, the **symplastic route** and the **transcellular route** (Fig. 3.8). The apoplast is the collective term for all the non-living parts of the plant body: cell walls, intercellular spaces and xylem conducting cells. Apoplastic water movement would occur in the capillary spaces of the root cell walls and perhaps in the intercellular

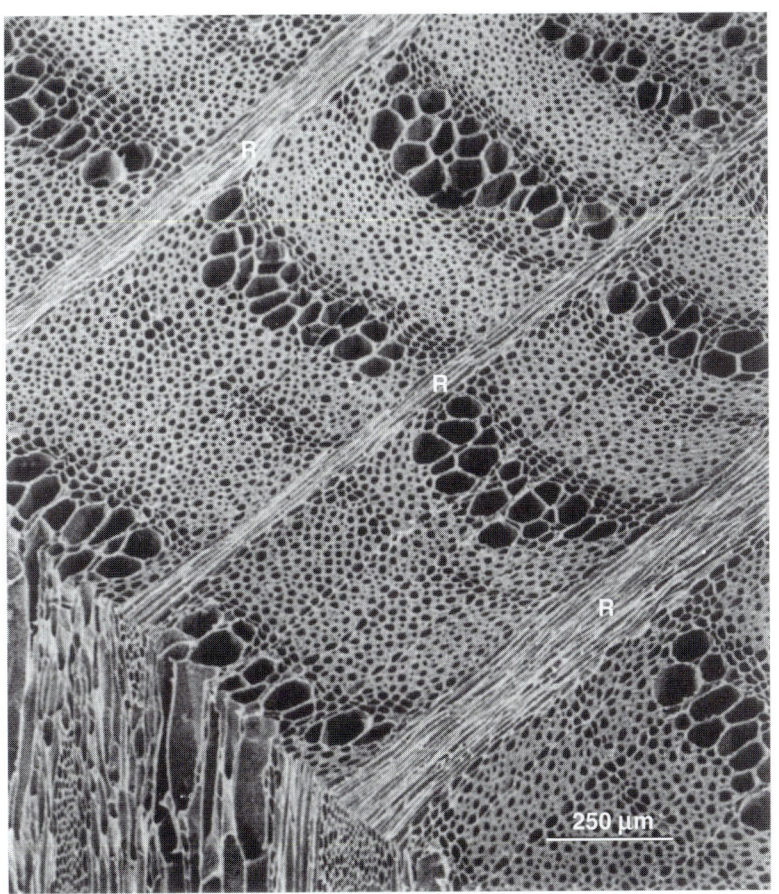

Fig. 3.6 Scanning electron micrograph of wood from *Knightia excelsa* showing mainly a transverse section, with a small area of tangential longitudinal section visible at bottom left. The vessel elements are grouped into bands, with associated bands of parenchyma cells (thin-walled), separated by wide bands of thick-walled fibres. The three bands of cells, R, running diagonally are the parenchymatous rays. From Meylan & Butterfield (1972).

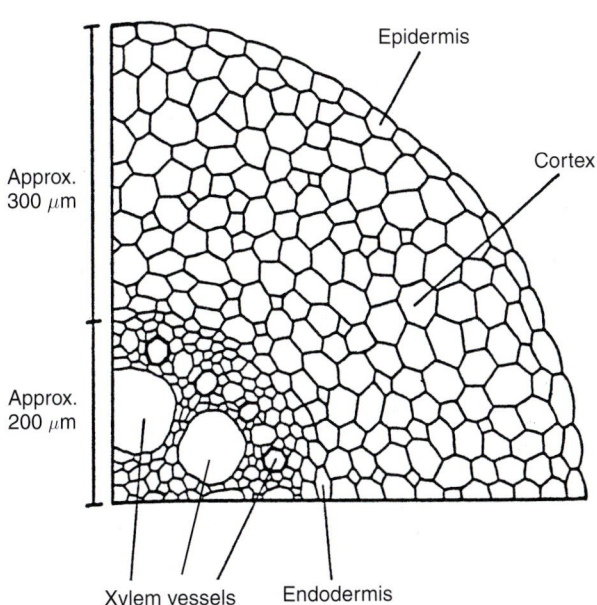

Fig. 3.7 Transverse section of a portion of a young maize root (*Zea mays*), showing the tissues through which water must pass on its way from the soil to the xylem. The epidermis, endodermis and pericycle (to the inside of the endodermis) are single cell layers, but the number of cell layers in the cortex varies widely according to species and type of root. From Frensch & Hsiao (1993). © Springer-Verlag GmbH & Co. KG.

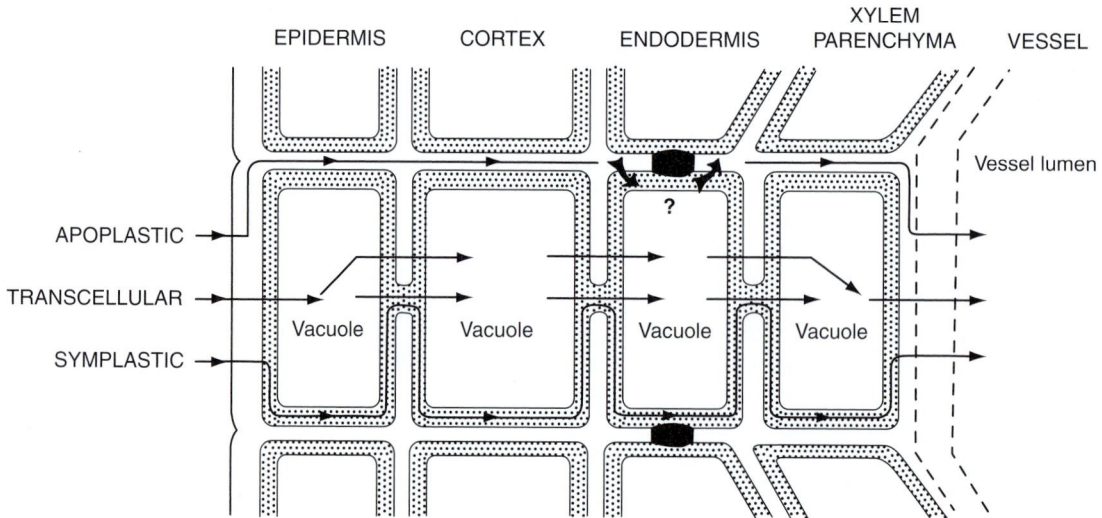

EPIDERMIS   CORTEX   ENDODERMIS   XYLEM PARENCHYMA   VESSEL

Vessel lumen

APOPLASTIC

TRANSCELLULAR

SYMPLASTIC

Vacuole   Vacuole   Vacuole   Vacuole

Fig. 3.8 Diagram of the postulated apoplastic, symplastic and transcellular routes of water movement through the root to the xylem. The arrowed lines represent the pathway of water; the symplast (cytoplasm and plasmodesmata) is stippled; the apoplast is white and the Casparian strips of the endodermal wall are shown in black. The xylem vessel wall and lumen are both parts of the apoplast, but their outlines are shown with dashed lines. For simplicity, all but one cortical cell layer and the pericycle (a parenchymatous layer inside the endodermis) are omitted. The transcellular route is shown in two variations; in the strictest sense it crosses every plasma membrane and tonoplast (upper arrows) but could also combine passage through membranes with passage through plasmodesmata (lower arrows). It is not certain to what extent the water in the apoplastic route has to bypass the Casparian strips as indicated by the curved arrows: see text for discussion.

spaces; there are reports of these spaces in roots containing fluid. The symplast is the collective living part of the plant; nearly all the living cells of the plant body are joined by plasmodesmata, submicroscopic protoplasmic connections of diameters around 50 nm. In the symplastic route, water has to cross a plasma membrane to enter the cytoplasm of an outer root cell; it would then move in the cells within the cytoplasm, around the vacuoles, and from cell to cell through the plasmodesmata without the necessity to cross more membranes, till it exited into a xylem conducting cell. The transcellular route envisages movement 'straight' through the vacuoles, crossing the tonoplasts of each cell; cell-to-cell movement could be via plasmodesmata or crossing the plasma membranes.

The apoplast route has been supported on the grounds that it is the path of least resistance, with minimal traversing of membranes. However, the radial walls of endodermal cells at the level of most active water absorption develop strips of wall thickening, the Casparian strips, chemically resembling the water-impermeable suberin (Fig. 3.9). In older parts of the root, all endodermal walls except the outer tangential ones become heavily thickened. It is therefore frequently suggested that at least at the endodermis water must pass through living protoplasts, and that this layer regulates water movement to the root xylem. There is some evidence to support this idea, but also data to the contrary, probably reflecting

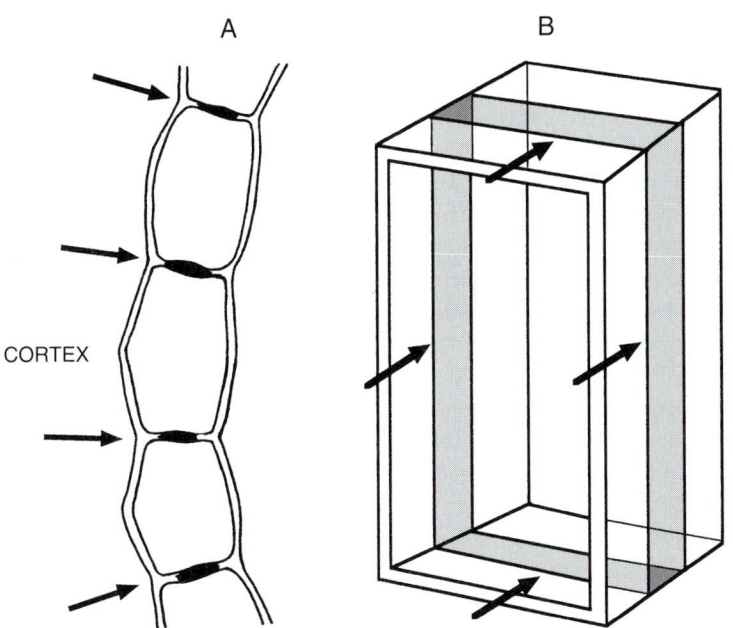

Fig. 3.9 The structure of young endodermal cells of a root, showing the Casparian strips in the radial walls. (A) transverse section, Casparian strip black; (B) 3-dimensional diagram with the cortex-facing wall cut away, Casparian strip shaded. Any water moving in the cell walls from the cortex towards the stele (arrows) will encounter the Casparian strip, which runs like a girdle around all the radial walls. Cell contents not shown. In older endodermal cells, all the walls except the cortex-facing ones may get heavily thickened.

the extent of wall thickening in the material studied. Decreased aquaporin content in plasma membranes of tobacco roots (*Nicotiana tabacum*), achieved by antisense repression of an aquaporin mRNA synthesis, has decreased greatly the roots' hydraulic conductivity. This indicates that there is movement through the plasma membranes and certainly supports the symplast route. But the data do not exclude the transcellular route. This route has been criticized as being the path of greatest resistance in view of the large number of membranes traversed. However, the density of aquaporins in tonoplasts is very high (up to 40% of total tonoplast protein), which may give them a much higher permeability towards water than is shown by the plasma membranes, and crossing the vacuole may offer less resistance than originally supposed. The three routes are not mutually exclusive and it is quite possible that all three contribute to water movement in proportions varying according to circumstances. When the rate of water movement is slow, most of the movement might be along the low-resistance apoplastic route, the higher resistance pathways beginning to contribute when demand increases. But the opposite has also been suggested, with a strong transpiration pull (Section 3.4.3) increasing the apoplastic flow.

In a herbaceous plant, where the vascular strands of the stem have not been joined to a continuous vascular cylinder by secondary growth, any part of the root system normally supplies those parts of the shoot which are directly above it, these being the parts with which it is in direct vascular connection; lateral movement does not occur or is very restricted. But if a part of the root system is deprived of water, lateral movement is activated and the overlying aerial parts receive a water supply from other root sectors. In trees with a continuous cylinder of wood, dye injection has shown that the path of water movement frequently spirals round the stem, following a helical arrangement of the conducting cells around the trunk.

The route of water movement through stem and leaf tissue after its exit from the xylem is also problematic. Fluorescent dyes introduced into the xylem to act as tracers for water movement move from the conducting cells into cell walls or crystallize out in intercellular spaces in the shoot. Such observations have been interpreted as indicating an apoplastic route for water passing out of the xylem. However, careful analysis of the data points to the opposite view: the dye becomes concentrated in the apoplast precisely because the water passes into the living cells near the xylem, leaving behind the dye to which the plasma membranes are impermeable (Canny 1990). It is therefore likely that water in leaves and stems first moves symplastically from the xylem, before it finally passes into cell walls again and evaporates.

3.4.3 The motive forces for water movement: root pressure and transpiration pull

The rate of water movement through plants is very variable; Table 3.1 gives examples of maximum rates attained in a number of types of flowering plants, and conifers for comparison. In any individual plant the rate is of course also highly variable, depending on environmental conditions (Section 3.6.2). The order of increasing maximal speed of water movement is the order of anatomical development towards wider and more numerous vessels in the xylem. Conifers, with the lowest maximal speeds, lack vessels, having only tracheids, while lianas may have very wide vessel elements (Fig. 3.5).

Water moves through the xylem along a Ψ gradient, from roots to all parts of the shoot, most of the flow terminating in the leaves. There must be some motive force, some energy input, which maintains the Ψ gradient and overcomes the frictional and gravitational resistances along the way. There are two possibilities: the xylem sap

Table 3.1 | The midday maximum speeds of the transpiration stream, measured by observation of dye movement or heat conduction. Sclerophylls are plants with tough, woody leaves; diffuse-porous wood has large vessels evenly scattered throughout the secondary xylem; in ring-porous wood the large vessels are concentrated into bands. After Huber (1956). © Springer-Verlag GmbH & Co. KG.

Plant type	Speed (m h^{-1})
Evergreen conifers	1.2
Mediterranean sclerophylls	0.4–1.5
Deciduous diffuse-porous trees	1–6
Deciduous ring-porous trees	4–44
Herbaceous plants	10–60
Lianas	150

could be pumped up under pressure, or it could be pulled up under tension, negative pressure. Capillary rise (a surface tension phenomenon) could not account for a rise of more than 1 m in the finest conducting elements; many plants are much taller than this, trees reaching 90–100 m.

Root pressure

In certain circumstances, the xylem sap is under positive hydrostatic pressure; when the plant is decapitated just above root level, the stump exudes sap, a phenomenon often called 'bleeding'. A manometer fitted over such a bleeding stump registers a pressure known as the **root pressure**, usually in the range of 0.1 to 0.2 MPa, exceptionally reaching 0.5 to 0.6 MPa. The development of this pressure is dependent on the metabolic activity of the roots. No positive root pressure is found when the roots are subjected to treatments inhibiting metabolic activity, such as lack of oxygen, application of respiratory inhibitors, low temperature, or starvation. The mechanism is thought to be osmotic. The exuded sap has a Ψ value *below* that of the soil because of a higher concentration of solutes, mainly inorganic ions, but sometimes including organic solutes, too. It is postulated that the living parenchyma and transfer cells of the xylem secrete the solutes into the conducting cells using respiratory energy. The lowering of Ψ causes water to follow the ions into the xylem, building up a pressure – the classical osmotic pressure – which pushes up the sap. The xylem Ψ does not equilibrate with that of the soil, since the ions are continuously swept away with the water movement, and the root cells continuously secrete more. In favour of this hypothesis is the fact that when the Ψ of the medium around the roots is suddenly lowered, reversing the gradient, exudation rate falls and may even become negative, so that externally applied liquid is sucked in at the cut stump. Root pressure can result in **guttation**, drops of liquid appearing at leaf tips and edges where the xylem sap is forced out through pores overlying vein endings. This liquid is much more dilute than bleeding sap at a stump, solutes having been absorbed by leaf cells. The manifestation of root pressures in temperate climates is most frequent during warm humid weather and the pressure shows a diurnal rhythm with maxima at nights, often dropping to near zero by day. The drops of guttation fluid are easily mistaken for dew in the morning, but an attentive examination shows that the drops are arranged regularly, corresponding to the positions of the pores. In a strawberry (*Fragaria ananassa*) leaf, for instance, there is a neat droplet at the tip of every tooth of the leaf edge. In tropical rainforests, where it is warm and humid all the time, guttation fluid drips from shrubs and small trees, mimicking rain.

Although the development of root pressures is well authenticated, in the majority of cases root pressure cannot account for water movement. Some species apparently never develop root pressure. The observed pressures are too low to raise water to the required level in tall plants. A pressure of 0.2 MPa can raise water maximally to 10 m, so root pressures in the usual range could raise water no higher than 10 m. The maximal values of 0.6 MPa could raise water to 30 m, still far short of the height of many trees. The *quantity* of water that can be moved by root pressure is small: e.g. wheat seedlings (*Triticum aestivum*), which transpired about 3 mL water per hour, exuded only 0.5 mL per hour by root pressure. In many instances maximum bleeding rates are only 1–2% of the water loss occurring by transpiration. Root pressure persists only as long as the water-yielding capacity of the soil is high, but plants can still extract and transport water after root pressure becomes inactivated through a lowering of the soil Ψ. As stated, root pressures in temperate climates are most frequently developed during warm nights; most of water transport occurs during daytime. It is moreover mostly herbaceous species that develop root pressures. In deciduous trees root pressures are demonstrable in the spring before the buds open, but once the leaves have expanded and rapid water movement through the plant begins, root pressures can no longer be detected.

During the periods of rapid water movement associated with rapid transpiration, the vast majority of evidence in fact indicates that the water is not under positive pressure but under **tension**. In such circumstances, a cut in the xylem does not result in sap exudation, but if the cut is made under water, the water is sucked in (as seen if a dye is added to the water). Transpiring twigs can pull water against an artificial resistance more effectively than a vacuum pump; leafy twigs can raise a column of mercury to heights greater than can be supported by atmospheric pressure. On the basis of such evidence it is generally accepted that water movement in plants, particularly in woody species, is the result of water being pulled up to replace that lost by transpiration. We therefore need to explain how water can be pulled up under tension to the topmost leaves of the tallest trees.

The cohesion–tension theory (transpiration–cohesion theory) for the ascent of xylem sap

A theory of the ascent of sap based on the cohesive properties of water was advanced independently in 1894 by Dixon and Joly, and in 1895 by Askenasy. This theory, as described below, is still supported in its essentials by most plant physiologists, although alternatives have also been postulated (Section 3.4.4).

The motive force for root pressure is generated at the root end of the plant. The generation of tension takes place at the *leaf end*. The living cells of the leaf contain solutes and commonly have a Ψ well below 0, say down to –2 MPa under 'average' conditions in a temperate climate. But they are still relatively water-saturated compared with the atmosphere for most of the time in most climates. The Ψ of the atmosphere is usually very much lower than that of the leaf cells, say –10 to –50 MPa. To put it another way, the atmospheric vapour pressure is usually very much lower than would be in equilibrium with the leaf cells. There are of course occasions when the atmosphere is very humid and water may condense out as mist or dew. But generally, there being a tremendous drop in Ψ between leaf and atmosphere, amounting to tens of megapascals, there is a great tendency for water to evaporate, or *transpire* from the leaf. The cells in contact with air lose water, mainly into the intercellular air spaces: it should be remembered that most of the leaf surface is *inside* the leaf, as discussed in the preceding chapter. Moreover the external surfaces are covered by a cuticle which strongly impedes the passage of water vapour. As the cells bordering the air spaces lose water, their Ψ drops and water moves into them by osmosis from the deeper-seated cells with which they are in contact. These in turn replenish their loss from cells still deeper in the tissue, until water is extracted from the xylem conducting cells, especially at the veinlet endings. The 'pull' on the water is transmitted right down the xylem and the water loss is made good by further uptake of water from the soil by the roots. There is thus envisaged an uninterrupted column of water being pulled under tension from the roots to the leaves. The energy for the movement comes from solar heat, which provides the latent heat of evaporation: transpiration is the one plant process which uses solar energy directly without the intervention of photosynthesis. Lignification gives the xylem conducting cells the strength to endure the tensions without collapsing inwards. This briefly summarizes the transpiration–cohesion theory for the movement of water in a qualitative way. The *quantitative* aspects are now considered.

It was stated (p. 77) that the *positive pressure* needed to pump water up the xylem by 10 m is 0.2 MPa; the *negative tension* needed to pull it up by 10 m is numerically equal but opposite in sign, −0.2 MPa. Of this, −0.1 MPa is needed to counteract the force of gravity and a further −0.1 MPa is needed to overcome frictional forces opposing the flow. If 100 m is taken as the height of the tallest tree, the mechanism for xylem transport under tension requires that tensions of at least −2 MPa must be generated by transpiration. That is perfectly feasible. The final evaporation of water takes place from the capillary spaces between cell wall fibrils; these spaces may be as narrow as 5 nm and the tensions generated at the tiny menisci (the curved air–water interfaces) can be calculated to reach −29 MPa (Canny 1995). There is accordingly no problem with the generation of tensions.

A second requirement for this mechanism of water movement is that the water in the xylem must be able to withstand the necessary tensions without the columns breaking. Here the situation is more complicated. The cohesive power of water is great, resulting from hydrogen bonding between the molecules: theoretical calculations of the tensile strength of pure water have given values as low as −1400 MPa. Tensions of −20 to −30 MPa are claimed to have been withstood experimentally. But xylem sap is not pure water: it contains dissolved materials. Taking into account the presence of these, together with the diameters of xylem cells and the adhesive properties of their walls, the expected tensile strength of xylem sap *in situ* has been put at about only –3 MPa. This, however, would still be enough to raise the sap to 150 m, higher than the tallest trees.

The experimental measurement of tensions in the xylem is not easy and most workers have relied on indirect estimates. One technique is known as the pressure bomb method of Scholander, who introduced it in the 1960s. Branches or leaves are cut from transpiring plants, where, if the cohesion theory is correct, the xylem fluid should be under tension. The leaf/branch is enclosed in a pressure chamber with the cut end protruding and nitrogen gas at increasing pressures is applied inside the chamber; eventually a liquid droplet is seen appearing at the cut end. The xylem tension before cutting is taken to equal the pressure at this point, only with a minus sign. The argument is that, when the cut is made, the water column in the xylem snaps and the water retreats deep into the xylem strand; to force it back to its original position, a pressure is needed numerically equal to the original tension. With this technique, tensions of – 0.5 to –8.0 MPa have been recorded, the highest values occurring in halophytes and desert plants, which must extract their water against a very low Ψ in the soil. Twigs of trees have been centrifuged, the centrifugal force exerting a tension on the xylem water until the water columns broke; according to species, tensions from –0.4 to below –3.5 MPa were endured (Pockman *et al. 1995*). Another indirect method for estimating the xylem tension is to measure the water potentials of living leaf cells in contact with the xylem. In wilted plants of tomato (*Lycopersicon esculentum*), privet (*Ligustrum lucidum*) and cotton (*Gossypium barbadense*), leaf Ψ values were found to reach –4.1, –7.0 and –14.3 MPa respectively. It was then assumed (without direct proof) that the xylem sap was under a tension of the same magnitude, since the tension must be balanced by the Ψ of the leaf cells if any water is to move into the leaves. If numerical values for xylem tensions of magnitudes as quoted above are accepted, these support the cohesion theory. There have also been measurements showing *gradients* of tension or leaf water potential, with the values becoming progressively more negative higher up the plant – as is of course required for the flow.

There is nevertheless a serious problem which has led to reservations in accepting the transpiration–cohesion theory. Water columns under tension are liable to **cavitate** (embolize) on mechanical disturbance, breaking up into droplets of water and water vapour. When water is sealed into a glass capillary under tension, the slightest tapping or shaking will bring about cavitation. Plants are buffeted by the wind, sometimes very violently. Any introduction of an air bubble would similarly break up the column. Twigs, even large branches, are broken off by the wind; animals bite away pieces; any such damage might be expected to let air spread through large expanses of xylem, if not the whole plant. Even without external damage, air seeding, i.e. the sucking in of air through the cell walls, is very likely when the xylem water is under high tension. Yet the plants go on conducting water. The actual great stability of the water-conducting system under natural conditions seems at variance with the metastable state of water columns under tension. Twigs which are cut so as to allow air into the vessels will resume water uptake when the severed end is placed in water. In the winter xylem sap may freeze; the dissolved air trapped as bubbles in the ice should remain as bubbles when the sap thaws – yet sap flow is resumed in the spring. On the other hand, cavitation does certainly occur in plants. Each cavitation event in the xylem makes a minute noise that can be amplified by means of suitable electronic equipment to an audible 'click'. These clicks can then be counted and recorded over time. During rapid drying out one can count several hundred clicks per minute in a wilting leaf. When the water supply is restored, cavitation ceases and water uptake is resumed. Another method for detecting cavitation is to freeze plant material very rapidly with liquid nitrogen and then to examine the xylem in the frozen state by cryoscanning electron microscopy. Empty (i.e. embolized) and ice-filled vessels are easily distinguished. Cavitation has turned out to be frequent in the field under natural conditions. These observations are difficult to reconcile with the cohesion theory and some plant physiologists postulate alternative mechanisms.

3.4.4 Validity of the cohesion–tension theory

Alternative hypotheses

All the experimental values for xylem tensions quoted above in support of the cohesion–tension theory have been obtained by *indirect* means. It was considered for many years that the direct measurement of tensions in the xylem was not practicable. Since 1990, however, a group of workers led by U. Zimmermann has published measurements which are claimed to be direct readings of pressures within intact, water-conducting xylem using a pressure probe (Section 3.3.2). It is not feasible to insert this delicate glass needle, 10 µm in diameter, into thick wood, but fine veins such as in leaf midribs were probed and the probe failed to register tensions as strong as obtained by the indirect methods, –0.4 MPa being the usual limit (Zimmermann *et al.* 1994). Occasionally down to –0.6 MPa was registered, but then the xylem vessels cavitated within minutes of probe insertion. Moreover, many of the probe readings failed to register any tensions at all but recorded weak *positive* pressures, up to 0.1 MPa.

On the basis of the pressure probe readings, alternative mechanisms for xylem transport have been put forward. Zimmermann and coworkers suggest that the mass flow of sap in the xylem is aided osmotically (in a manner similar to the building up of root pressures), by the secretion of solutes into the conducting cells, from the living cells of the xylem. There would still occur a transpiration pull, but the magnitude of the tension would be reduced and hence there would be no high tensions and less risk of cavitation. Zimmermann claims that xylem sap is quite rich in solutes, contrary to the generally accepted view that it is extremely dilute (except in early spring, before transpiration commences). Another idea (Canny 1995) is that, whilst transpiration provides the driving force, there is no building up of large tensions because the living turgid cells of the xylem exert a positive pressure on the vessels and tracheids, as does the phloem, where cells are under a high positive pressure.

In defence of the cohesion–tension theory

The majority of plant physiologists nevertheless support the cohesion–tension theory. The new postulates of Zimmermann and Canny rely on tension measurements from one single type of instrument, the pressure probe, whereas evidence for strong tensions has been obtained with several methods, with consistent results. The discrepancy may be due to problems with handling the pressure probe (Milburn 1996). Its insertion could cause cracks in the xylem cell walls, letting air in and causing cavitation. The probe diameter of 10 μm is not negligible compared with the diameters of the xylem vessels probed, 50–90μm; the insertion of the probe could do appreciable damage. With living cells, the plastic cell contents pressing against a pierced wall could seal up cracks, but in the xylem conducting cells there is nothing to protect against air entry. Claims for a high solute content could derive from the probe sampling preferentially the *young* xylem (which is outermost), where immature cells might still contain solutes derived from the breakdown of cell contents (Milburn 1996); such cells might also be under more positive pressures. Later experiments by other workers have succeeded in measuring stronger tensions in the xylem with the probe, down to –1 MPa, and have shown agreement between pressure-bomb and pressure-probe data. The osmotic theory implies a high energy expenditure by the living xylem parenchyma cells for pumping solutes into the xylem, but the low O_2 level in the xylem of woody axes precludes high aerobic respiration rates and the limited volume of living cells would moreover have to control a large volume of dead conducting cells (Richter 1997). There would also have to be a mechanism for recycling the solutes. Mercury is pulled up by a leafy, transpiring twig inserted into the top of a water-filled glass tubing, which has its bottom end in a reservoir of mercury. The twig sucks the mercury into the xylem and right into very narrow tapering cell tips, which would require a tension of –2 MPa, and there is of course no question of osmotic forces acting on the mercury. Strong positive pressures, postulated to be exerted by living xylem parenchyma and by the living phloem, would probably have little effect on the conducting cells with their rigid walls.

At present, the bulk of the evidence seems to be in favour of the cohesion–tension theory; but then one has to account for the fact that airlocks can be introduced into the xylem in nature by cavitation owing to water stress, freezing, mechanical damage – or experimentally, without permanently or even temporarily stopping the overall flow. How is this to be reconciled with the need of continuous water columns for the movement under cohesion?

To account for the resumption of sap flow in the spring after freezing, the suggestion has been made that the air bubbles released from thawing ice might be forced into solution again by root pressures, which are highest in early spring, although root pressures tend to be low in trees. Another mechanism by which cavitated vessels could be refilled at any season is by capillary forces. Perhaps refilling of cavitated cells is a situation where the activities of the *living* cells of the xylem and of the phloem, emphasized by Canny and Zimmermann, do come into play in building up pressures and forcing fluid into the empty vessels. Counts of embolized vessels in sunflower leaves at different times of day, and at various stages of drying out, have revealed far fewer empty vessels than expected from the rate of cavitation. From this it was deduced that the embolized vessels were refilled within minutes by pressure from living cells (Canny 1997). A similar conclusion was reached from scoring maize roots from dawn to dusk for embolized vessels (McCully *et al.* 1998). It is also possible that many airlocked conduits never resume water conduction, since in perennials a new ring of fresh sapwood is produced annually; in some species, only the current year's growth is active in transport. Redundancy of xylem does occur. The total amount of xylem in a plant is generally much in excess over that needed to satisfy the plant's water requirements, as shown when large amounts of wood are removed. The presence of quite large numbers of cavitated vessels can therefore be tolerated without too much ill effect.

There is also evidence that airlocks produced by cavitation do not necessarily spread from their origins over large volumes of xylem. Vessels are formed by perforation of end walls between individual vessel element cells, but very often a perforation plate is crossed by bars of cell wall material (Fig. 3.5 E, H) which break up the opening into numerous fine slits. It is suggested that in such vessels an airlock may remain confined to the single vessel element in which it forms, unable to pass the perforation plate due to water menisci holding firm in the narrow openings because of surface tension forces. Even if a whole vessel cavitates, its length may be limited compared with the height of the plant. In trees, some vessels may extend to over 10 m, but most are much shorter; e.g. in a holly (*Ilex verticillata*) the maximum vessel length was 1.3 m, but 99.5% of the vessels were under 5 cm long. Similar data can be quoted from other species. Cavitations of limited extent can be bypassed by lateral flow through the cell walls (Fig. 3.10), mainly through pits. Water movement can continue even in the presence of two overlapping cuts from two sides of a stem, as long as the distance between the two cuts is great enough to leave some intact vessels within this region. An increase in the resistance to flow can be detected when such cuts are made, as would be expected: cell walls offer more resistance to flow than the lumina of vessel elements. But the cohesive system as a whole is not seriously reduced.

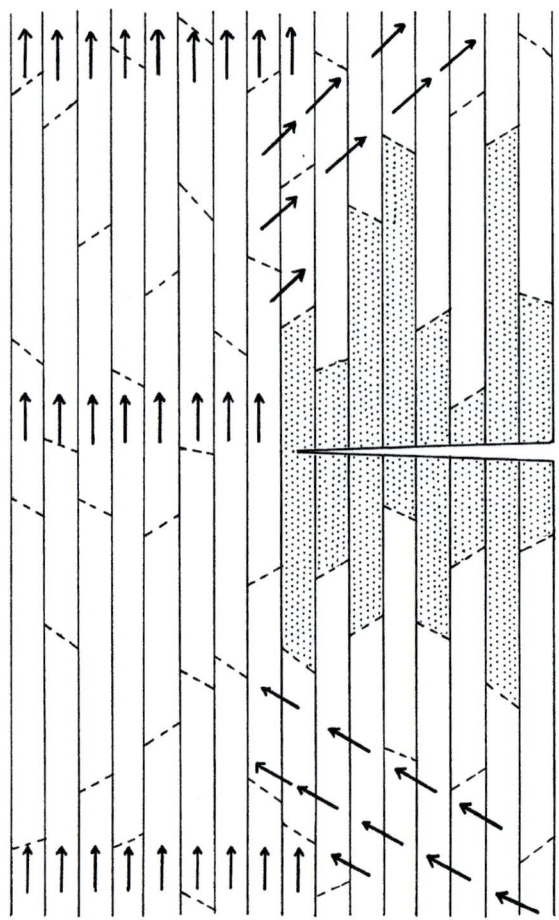

Fig. 3.10 Bypassing an airlock (embolism) in the xylem following a cut at the right. The slanted dashed lines represent vessel end wall perforation plates; the arrows show the pathways of water flow. Air-filled vessel elements are shaded. The necessity to move sideways through cell walls (via pits) increases the resistance to water movement in the damaged area.

The necessity to protect the xylem transport system against the effects of cavitation and airlocks may explain why tracheids have persisted in the flowering plants alongside vessels. The vessels, with no barriers to water flow between cell and cell, offer much less resistance to water flow than tracheids, where water must pass from tracheid to tracheid through the cell wall, mainly through pits. Moreover some vessel elements are extremely narrow, there being normally a range of widths of vessel elements in the same plant organ. Volume flow through a tube is proportional to the *fourth power* of the radius of the tube. Hence a two-fold increase in the radius of a vessel from say, 5 µm to 10 µm would increase volume flow (for the same tension) 16-fold ($2^4 = 16$). Since wide vessels are so much more efficient for a bulk flow of water, why should a vascular bundle contain, in addition to wide vessels, narrow vessels and tracheids? The most probable answer is that this gives the plant the flexibility to react efficiently to varying environmental water status. When the soil Ψ is high the plant does not require very high tensions in the xylem to extract the water; in such a situation, most of the transpiration stream would pass through the widest vessels, which offer the least resistance. But when water stress sets in and xylem tension increases, it will be the widest vessels that are the most vulnerable to cavitation. The narrower vessels and the tracheids then can take over the function of water conduction; the same high tensions which cause vessels to cavitate also overcome the resistance in the narrower conducting cells. In tracheids, any cavitation event is confined to one single cell, and tracheids are also relatively narrow, so they are the least vulnerable to water stress.

The movement of cohesive water columns in the xylem under transpiration pull may thus be buffered against serious disruption from cavitation by excess capacity; by the regular annual replacement of old xylem by new in perennials; by refilling of air-filled vessels by root pressures, by capillarity and by pressure from adjacent living cells; by the bypassing of airlocks in cell walls; and by the presence in xylem of tracheids and narrow vessels, which are less susceptible to cavitation. There is not enough evidence for accepting alternative theories for the ascent of sap which deny the existence of high tensions in the xylem. Nevertheless, data from the pressure-probe measurements, and other apparently anomalous observations, are drawing attention to the possible functions of the *living* cells of the xylem, and of the neighbouring phloem, in water movement.

3.5 | The transport of solutes in the xylem

Xylem transport is often considered primarily in terms of water movement, but the xylem also has an important function in the transport of solutes. Samples of xylem sap can be obtained for analysis by collecting the exudate when it is under pressure (the bleeding sap); or it can be sucked out under vacuum, pressed out or centrifuged out, from pieces of wood. The concentration of dry matter in xylem sap is low, commonly 0.1 to 0.5%, but the total volume of sap moved in the xylem is high, so the amount of solute that is carried in the xylem is significant. The mineral ions absorbed by the roots are distributed around the plant in the xylem; this can be shown by tracing the pathways of the ions with radioactive labelling, as well as from sap analysis. From two-thirds to three-quarters of the solids are, however, organic, including amino acids, amides and carboxylic acids, giving the sap a pH of about 5. The xylem sap composition in any individual plant varies with the environmental conditions; during rapid transpiration, when large amounts of water are passing through the xylem, the concentration of solids may fall very low. There are also more regular seasonal variations correlated with plant development. In woody perennials, the mineral content of xylem sap is highest in the spring, when active growth is resumed.

The carbohydrate content of xylem sap is usually below 0.05% and may be undetectable. But in woody species in the early spring, before leaf expansion and the onset of transpiration, there may be a period of high sugar content in the xylem sap, up to 8%. This sugar is derived from reserve starch, stored in the woody stem over the winter. During this period of sugar mobilization, the xylem sap acquires a positive pressure. The sugary sap flows out in quantity from cuts made into the wood, and the tapping of birch (*Betula* spp.) and sugar maple (*Acer saccharum*) in the spring has been a tradition in northern regions of Europe and America probably for millennia. Maple syrup is commercially produced from sugar maple sap.

Plant hormones can also move in the xylem; e.g. the sensing of water stress in roots stimulates the synthesis of abscisic acid, which is transported in the transpiration stream to leaves, where it induces stomatal closure. Other hormones transported up in the xylem are important in growth correlations between root and shoot.

3.6 | Water uptake and loss: control by environmental and plant factors

The rates of water absorption and water loss, and consequently of water movement through the plant, are determined by an interaction between plant and environmental factors. The environmental factors can be classified as soil (edaphic) and atmospheric. With regard to the soil, important considerations are the amount and availability of soil water, soil temperature and soil aeration. Above ground, the relevant factors are atmospheric humidity, temperature, wind speed and light. The plant factors are the area and water permeability of the absorbing surface in the roots; the area and water permeability of the evaporating surfaces of the shoot; the frequency of stomata and the degree of their opening.

3.6.1 Soil water and uptake by the roots

The soil is a complex system. Physically, it consists of particles with sizes ranging from large stones to submicroscopic colloidal material, and it contains pores of varied dimensions. Chemically the particles are of various composition, organic and inorganic, and there are many solutes in the soil water. The high colloidal content of most soils (coarse sand is an exception) gives it a significant matric potential; solutes such as mineral ions give it an osmotic potential. The pressure potential is represented by tensions (negative), i.e. the surface-tension forces at water menisci in small pores. Electrostatic forces around particles and capillary forces in pores also decrease free energy and help to retain water in the soil.

When a soil is saturated with water, all the pores are filled, but a well-drained soil does not remain water-saturated for long. Water drains away quickly under gravity from the larger spaces, but some is retained in the smaller pores by the colloidal, surface-tension and capillary forces, and as adsorbed surface films around soil particles.

When a soil contains as much water as it can hold against gravity, it is said to be at **field capacity**. The amount of water present at field capacity depends on the soil type. Soils with fine particles have many small pores and much total particle surface area, and can hold more water than coarse soils (Fig. 3.11). A clay soil at field capacity may hold 55% water on a dry-weight basis (i.e. 55 g water per 100 g dry soil), while a coarse sand may hold only 17%. Once the water content has fallen to field capacity, there is almost no movement of liquid water in the soil, though water evaporates to the atmosphere.

The Ψ of a soil at field capacity is very high, just below zero (unless the soil is highly saline) and uptake by plants can proceed freely. As the water content of a soil falls, its Ψ decreases progressively. The concentration of solutes rises and the Ψ_π falls and the smaller volumes of water between soil particles have more curved menisci; this increases the surface tension forces and lowers the Ψ. Also, as the outer layers of water are removed from the surface films, the inner layers are held more strongly by electric charges and van der Waals forces. At first the lowering of the soil Ψ is matched by lowering of the Ψ in the plant and water absorption continues. Eventually a stage is reached, however, when the soil Ψ falls so low that the plant can no longer obtain enough water to compensate for transpirational losses, and it wilts. At first this may be temporary, the plant wilting by day but recovering at night, when the transpiration is low and water uptake catches up with water loss. Eventually there comes the **permanent wilting point** (PWP), defined as the stage when the plant will not recover from wilting unless more water is added to the soil. Numerically the PWP is expressed as the percentage of water left in

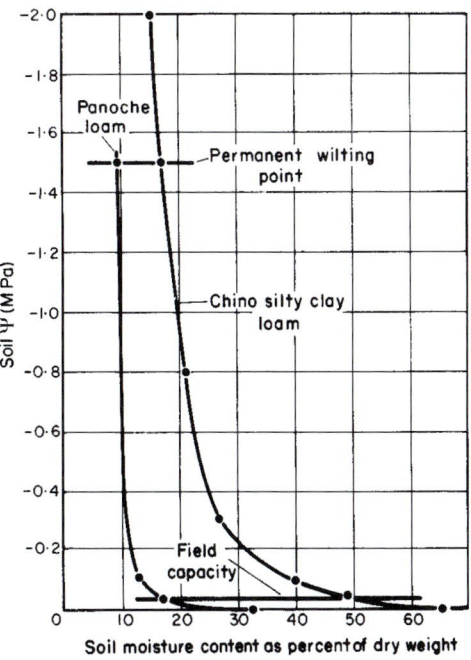

Fig. 3.11 The relation between soil moisture content and soil Ψ in a sandy soil (Panoche loam) and a clay soil (Chino silty clay loam). The permanent wilting point (PWP) varies between plants, but many species reach PWP at about -1.5 MPa, as indicated in the graph. Adapted from Kramer (1949).

the soil. The value of the PWP depends on the species as well as on the soil, different species reaching the PWP at different values of soil Ψ. Water uptake does not totally cease at the PWP but leaf turgor pressure remains at zero. Once the PWP has been reached, removals of very small amounts of soil water cause very large decreases in the soil Ψ: the relationship between soil water content and soil Ψ is not a linear one (Fig. 3.11). Whether the plant can survive wilting depends on the species, on the degree of water loss and on the length of time in the wilted state.

Most of the water uptake by a plant takes place when the soil moisture is between field capacity and PWP. As the soil dries out, the forces opposing plant water uptake increase; the rate of water uptake, plant hydration and plant growth rate can be impaired by water stress even if wilting is not reached. This reduction of growth rate can aggravate the effect of water shortage, for the slowing down of root growth decreases the rate at which new areas of the soil are tapped by the roots.

Soil aeration affects water uptake. Field capacity is the ideal state of soil for plants since it has a high Ψ but also has air-filled spaces. A fully water-saturated soil has an even higher Ψ, but it is water-logged, without air spaces. An adequate O_2 supply is necessary for root growth; a lack of O_2 and a high concentration of CO_2 (accumulating from anaerobic respiration of roots and soil microorganisms) are moreover reported to decrease the permeability of roots to water. Soil temperature affects root growth and root permeability, both being decreased at low temperatures. The viscosity of water on the other hand increases as the temperature falls; low soil temperature may thus considerably reduce the uptake of water by roots under transpiration pull, leading to 'physiological drought' when there is water available in the soil, yet the plants suffer water stress. Root pressures develop only in warm, well-aerated soils of favourable moisture content.

3.6.2 The atmosphere and transpiration

The daily course of water absorption by plants closely follows, with a time lag, the course of transpiration (Fig. 3.12). Thus the atmospheric factors which determine the rate of transpiration also largely determine the rate of water uptake.

The rate of transpiration, the outward diffusion of water vapour from plants, is subject to the same physical laws as the inward diffusion of CO_2, as discussed in Chapter 2. Transpiration is directly proportional to the water potential gradient $\Delta\Psi$ between the leaf and the air. Or, since atmospheric water status is often expressed in terms of vapour pressure e, which is proportional to Ψ (Equation 3.7, page 97), we can substitute Δe, the water vapour pressure gradient, for $\Delta\Psi$. Transpiration, on the other hand, is inversely proportional to R_a (boundary layer resistance) and R_s (stomatal resistance). Denoting the rate of transpiration by T, we have:

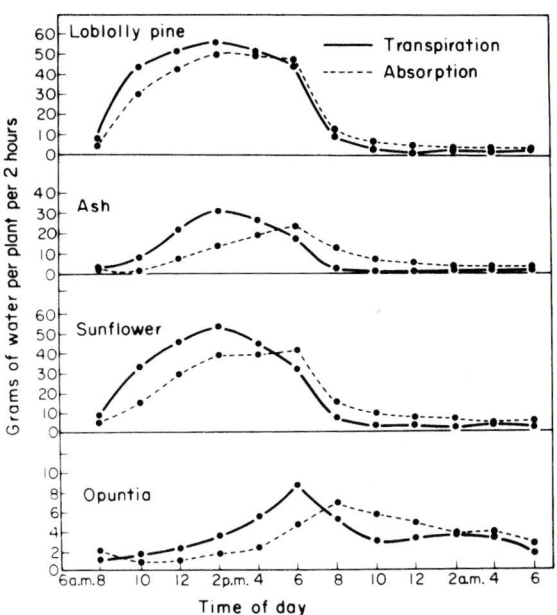

$$T = \frac{\Delta\Psi}{R_a + R_s} = \frac{\Delta e}{R_a + R_s} \tag{3.5}$$

There is no term in this equation to correspond with the mesophyll resistance which is involved in CO_2 diffusion (Section 2.4.1), for water vapour does not pass through cells on its way to the outside.

Transpiration rate increases with increasing temperature, a rise in temperature resulting in a steeper concentration gradient of water vapour out of the leaf, i.e. Δe increases. The air spaces within the leaf are normally at near saturation vapour pressure, c. 100% RH (relative humidity, the ratio of actual vapour pressure to saturation vapour pressure as a percentage). The absolute concentration of water vapour at a given RH increases with increasing temperature, i.e. air at 100% RH at 20 °C will contain more water vapour than air at 100% RH at 10 °C. A rise in leaf temperature therefore increases the vapour pressure in the leaf without a corresponding rise in the external air. The gradient is approximately doubled for a 10 °C rise in temperature. Transpiration decreases with increasing atmospheric humidity, for this decreases the Δe. Wind stimulates transpiration, decreasing R_a as it sweeps away the water vapour accumulating in the boundary layer at the leaf surface. By causing bending of the leaf it may cause mass flow of air into and out of the leaf, thereby enhancing water loss. Light has no direct effect on water loss, but it does have a profound effect on transpiration indirectly: it warms up the tissues, increasing transpiration, and it promotes stomatal opening (Section 2.4.1). The combined effects of light and temperature result in the diurnal changes in the rates of transpiration (and hence of water uptake) illustrated in Fig. 3.12.

3.6.3 Stomatal control of transpiration

The features of the plant which control the rate of passage of water through it act largely through controlling the rate of transpiration. If comparison is made between plants of different species, or different individuals of the same species (which can show variation according to their growing conditions), the following shoot characters are seen to favour rapid transpiration: a thin cuticle; lack of hairs; a high stomatal frequency per unit area; a large surface area and a large ratio of internal to external surface area. The water uptake capacity of the roots can also limit the rate; when half the leaves are removed from a plant, the remainder may transpire more rapidly per unit area of leaf, being now able to draw on the whole root system rather than on half of it.

However, when an individual plant is considered over a short term so that developmental changes in, say, the extent of the root system do not come into play, the degree of opening of the stomata is often the most important single plant factor directly controlling the rate of transpiration. By far the greater proportion of the water lost from the leaf comes from the leaf air spaces via the stomata, although the stomatal pore area may be only 1–2% of the total leaf surface. The daily course of transpiration rate follows closely the course of stomatal opening (Fig. 3.13). Where there is a midday closure of stomata, this is always accompanied by a reduction of transpiration rate, as shown in Fig. 3.13. In plants with a thick cuticle, such as the bay laurel (*Laurus nobilis*) water loss through the epidermal surface

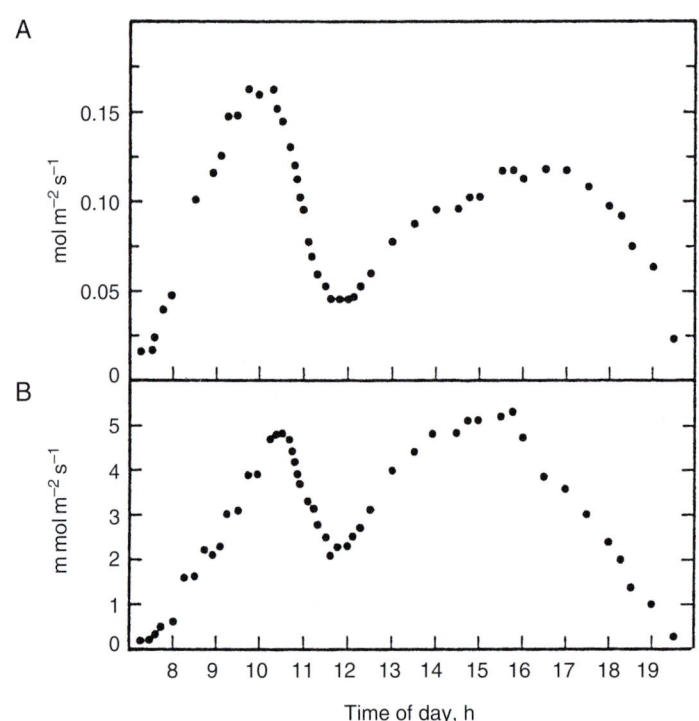

Fig. 3.13 The relationship between stomatal aperture measured by stomatal conductance (A) and the rate of evaporation (B) in an attached leaf of the strawberry tree (*Arbutus unedo*) under conditions typical of a Mediterranean day. Note the marked midday decrease in stomatal aperture and transpiration. Adapted from Raschke & Resemann (1986). © Springer-Verlag GmbH & Co. KG.

excluding stomata, the 'cuticular transpiration', may be as low as 2% of the total transpiration. When the cuticle is thinner, cuticular transpiration can constitute up to about 50% of the total; the cuticular transpiration of the 'average' mesophyte is 10–25% of the total. Closing of the stomata will therefore, according to species, reduce transpiration to some 2–50% of that occurring with fully open stomata.

Transpiration is slowest when stomata are completely closed and increases with increasing stomatal opening. If the atmospheric conditions favour rapid transpiration, the increase continues right up to maximum opening, the stomatal aperture being the limiting factor. But if the atmospheric conditions do not favour rapid transpiration, maximal transpiration rate may be reached when the stomata are only partly open, the external conditions becoming limiting (Fig. 3.14). In the pathway of water movement from soil to atmosphere, the stomata are situated between the leaf air spaces and the atmosphere:

soil ⟶ root ⟶ stem ⟶ leaf ⟶ stomata ⟶ air

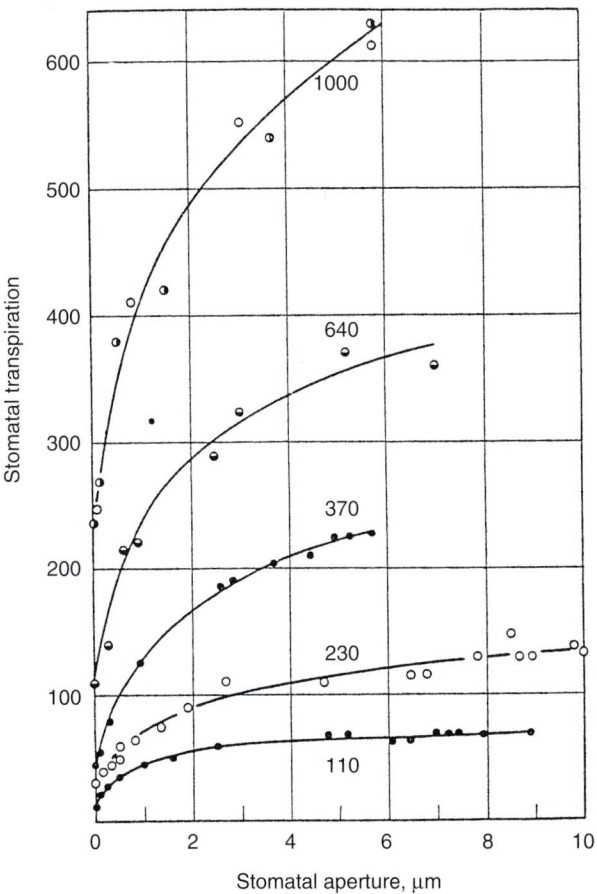

Fig. 3.14 The relationship between transpiration rate and stomatal aperture in birch (*Betula pubescens*) under different conditions of evaporation. The number by each curve equals the rate of evaporation of water, in mg per hour, from 25 cm^2 wet blotting paper surface under the same conditions. At low rates of evaporation (and of transpiration), full transpiration rate is reached at a stomatal aperture of only 2 μm since the atmospheric conditions are limiting. At high evaporation rates, maximal transpiration rate is scarcely reached with a stomatal aperture of 8 μm because now the stomata are limiting. From Stålfelt (1956). © Springer-Verlag GmbH & Co. KG.

That is the point where the drop in Ψ is the greatest, and therefore the stomata can exert very effective control over water movement when, in moving air, the boundary layer resistance is low. In still air, however, the boundary layer resistance may become the limiting factor and be more important than the stomatal resistance. Partial stomatal closure cuts down the rate of transpiration more than the rate of CO_2 diffusion; there being no mesophyll resistance for water vapour movement, the stomatal resistance assumes proportionally more importance.

As already mentioned in Chapter 2, stomata are highly sensitive to water stress and some species react by (partial) closure at quite low levels of water deficit, protecting the plant against further water loss. In wilted plants the stomata are shut. The early stages of wilting may, however, be accompanied by a widening of the stomatal aperture, the guard cells being pulled apart by the shrinking of surrounding epidermal cells, which lose water more rapidly. In extreme wilting, too, the protective mechanism may break down as the epidermal cells shrink and again pull the pores open.

3.6.4 Waterproofing the surface: cuticle and wax

One of the key features that arose during the evolution of terrestrial plants is the xylem, which makes possible the transport of water to parts of the organism not in contact with a water supply. Another key feature was the evolution of waterproofing chemicals which cut down the rate of evaporation from plant surfaces. The control of water loss by stomata can be significant only if the rest of the surface is waterproofed to some extent. This is achieved by the presence of the **cuticle** and **wax**.

The whole outer surface of all land plants (even the Bryophyta) is covered by a cuticle. In the flowering plants, the cuticle covers not just the external surface, but the walls of cells lining internal air spaces, although here it is extremely thin; there is a very thin cuticle on the root epidermis, too. A cuticular ridge commonly overarches stomata (Fig. 3.15). The cuticle is a continuous skin over an organ, not separate for each individual cell; it can be detached by chemical treatment in one piece from e.g. a leaf. The cuticle is made up of several layers which differ in chemical composition, but the main component in all layers is **cutin**, a complex hydrophobic polymer of mainly hydroxy fatty acids. Some wax (see below) may be present in the cuticle, and secondary compounds such as tannins. The innermost layer of the cuticle is rich in pectins; when the pectins are hydrolysed, the cuticle is detached. The thickness of the cuticle varies from a fraction of a μm to over 10 μm, being thick in plants of dry habitats.

On top of the cuticle aerial organs have a layer of epicuticular wax. This is attached only loosely and can be wiped or rubbed off. The wax

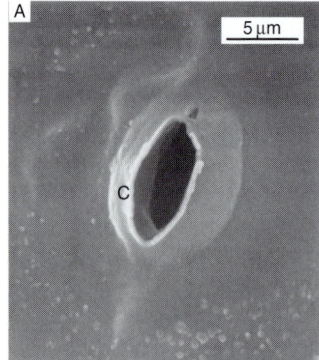

A   5 μm

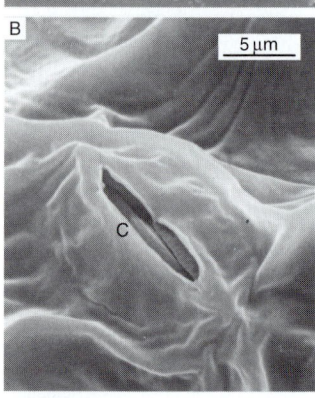

B   5 μm

Fig. 3.15 Scanning electron micrographs showing the cuticle ridge C arching over stomata. (A) Open stoma in a leaf of mung bean (*Vigna radiata*); wax granules are visible on nearby cells. (B) Closed stoma in wilting leaf of bean (*Phaseolus vulgaris*); the edges of the guard cells can be seen closely adpressed. Photo (B) courtesy of Dr. J. M. Milton.

gives the 'bloom' to glaucous leaves and fruits; some dark plums and so-called black grapes look almost sky-blue when untouched, owing to the wax, but much of that gets rubbed off when the fruit is handled. The sheen and texture of many floral parts result from their wax layer. The wax is not a pure chemical, but a mixture of long-chain hydro-carbons, long-chain fatty acids, long-chain hydroxy acids, esters, alco-hols, aldehydes and terpenoids. Each species produces its own mixture; about 50 different chemicals have been detected in apple wax. Even as for the cuticle, the thickness varies from a fraction of a μm upwards. The wax palm (*Klopstockia cerifera*) native to the Andes, has 5 mm of wax on its leaves; carnauba wax, from the leaves of *Copernicia cerifera*, is harvested commercially in Brazil.

The thickness and chemical composition of the cuticle and wax determine the extent of transpiration with closed stomata, the 'cuti-cular transpiration' discussed in Section 3.6.3. The more imperme-able these layers, the smaller is the cuticular transpiration (and the more complete is the control exerted by the stomata).

The cuticle and wax also prevent the *entry* of water. It would not be a healthy situation if rain penetrated plants freely, flooded the air spaces, soaked the cell walls and leached out solutes from the apo-plast. Raindrops largely roll off owing to the hydrophobic nature of the surface. The wettability is determined mainly by the epicuticular wax, its thickness and even more by its structure. To the eye a waxy leaf or fruit looks smooth. Microscopy, especially scanning electron microscopy, reveals that the wax is present as plates, rods, granules or tufts (Fig. 3.15, 3.16). These formations increase the hydrophobi-city; even smooth wax is water-repellent, but a surface bristling with small wax projections is almost unwettable. Agricultural and horti-cultural sprays are mixed with detergent to lower the surface tension and enable the fluid to wet plant surfaces. The cuticle and wax also give some protection against the penetration of pathogens, and against ultraviolet radiation.

The mode of production of these extracellular layers by the plant is something of a puzzle. The components are generally assumed to be secreted as liquid precursors by the epidermal cells and to solidify on the outside, otherwise continuous layers cannot be produced, nor can solid wax rods be transported through a cell wall. The most difficult problem is the wax, which must move not only through the cell wall, but through the cuticle as well. It might be suggested that the wax layer is formed first and then the cuticle under it; but when the wax is wiped off, leaving the cuticle intact, a new layer of wax is formed by young leaves (and in some species by mature leaves also). Searches for channels in the epidermal cell walls have not given clear-cut results. There have been reports of channel-like structures in cell walls seen by light microscopy, but these have not been confirmed by electron microscopy. Another idea is that the wax molecules are carried along with water vapour molecules during cuticular transpiration, a process that has been compared to steam distillation.

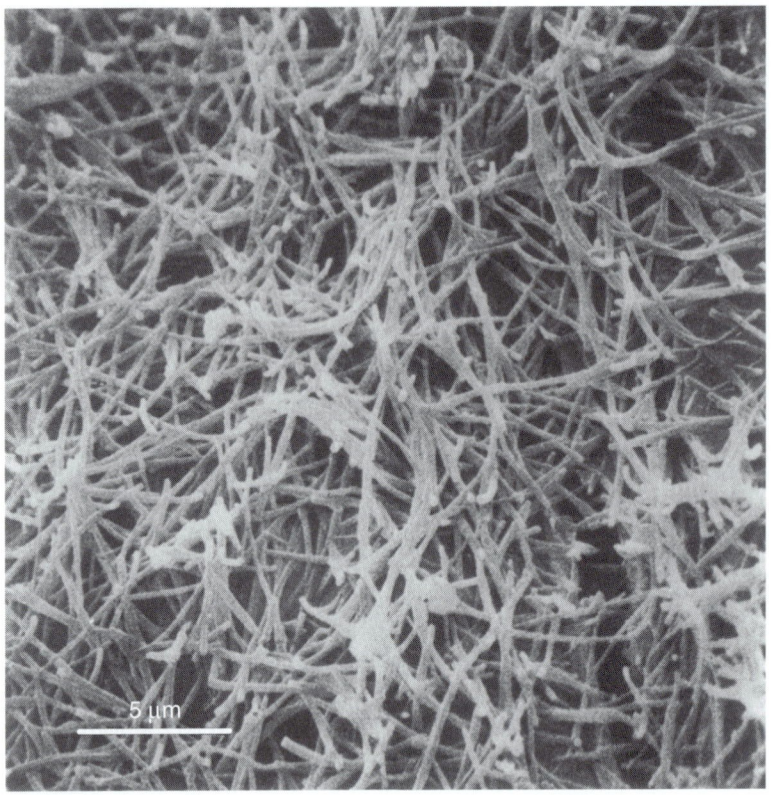

Fig. 3.16 Scanning electron micrograph of the leaf surface of wheat (*Triticum aestivum*) showing a meshwork of wax rodlets. From Troughton & Sampson (1973).

The *patterns* of the wax can be produced physically, with no control from the living cells. Waxes from various species dissolved in organic solvents and allowed to crystallize out as the solvent evaporates have been found to crystallize into the pattern originally exhibited on the epidermis. However, in some cases the pattern has differed from the original.

3.6.5 Is transpiration really necessary?

To put it another way, could water be moved to the tops of plants in sufficient quantities by any other mechanism, root pressure having been shown to be insufficient?

Continuous columns of water under tension could in fact be maintained as water is used up in photosynthesis, and in any other chemical reactions where it is a reactant. As long as something removes water at the leaf end, the water would be pulled up; it does not have to be removed by evaporation. Osmotic forces in the leaf cells could also maintain Ψ gradients between the leaf and the root (and soil). But the amount of water moved in such circumstances would be no more than is used up in growth, in chemical reactions, and in maintaining the hydration levels of the leaf cells. Xylem transport is important also in distributing mineral ions, organic nitrogenous compounds (Chapter 4) and hormones. The critical question is whether the much slower water movement that would be maintained in the absence of transpiration would suffice for the distribution of these solutes. Information on this is limited. One investigation with sunflower plants (Tanner & Beevers 2001) indicated that the suppression of transpiration had no adverse effect on mineral ion uptake and distribution over 30 days. But one wonders whether *rapid* xylem transport might be advantageous for the transmission of hormonal stimuli in situations such as the sensing of drought by the roots.

The *cooling effect* of transpiration is also very important in keeping down the temperatures of plant organs, especially in hot weather. Throughout the day, a leaf absorbs radiant energy. The amount of energy stored in photosynthate is about 1% of that absorbed, maybe up to 4% in some C_4 species. The rest of the absorbed energy is converted to heat. If the leaf is not to heat up steadily, this heat must be dissipated, and it has been calculated that transpirational cooling accounts for about half of this heat dissipation. (The remainder is lost through convection and conduction.) Hence transpiration is generally regarded as a 'necessary evil'.

3.7 | Water conservation: xerophytes and xeromorphic characters

Plants growing in extremely dry habitats ('xeric habitats') are termed **xerophytes**. They exhibit a number of structural features which are regarded as potentially conserving water and are called **xeromorphic characters**. Some of these features may also be found in plants of habitats which are not particularly deficient in water. It is common to find xeromorphic characters in evergreen plants of temperate and cold climates, where the plants may suffer water shortage during cold seasons owing to an inability to absorb water adequately by chilled roots, or even because of the soil water freezing. The character of succulence is typical of plants of saline habitats, where water is abundant, but the plants must absorb it from a medium of low Ψ. Xeromorphic characters include:

(1) Deep and/or extensive root systems. These enable the plant to reach water at considerable depths. In the Mediterranean region, tree roots grow right into the porous limestone rock. In some species, the ratio of root weight to shoot weight is very high, as is root length to shoot height.

(2) Water storage organs and tissues: succulence. Any part of the plant may store water – root, stem or leaf. The storage organs are succulent; they contain large, highly vacuolate cells which swell up when water is available and gradually release water to growing regions when there is no external supply, shrinking in the process. Examples of stem succulents are cacti, whilst *Aloe* species and stonecrops (*Sedum* spp.) have succulent leaves.

(3) Low surface : volume ratio of shoot organs. Leaves usually are the organs with the highest surface : volume ratios. Xeromorphic plants tend to have small leaves, or succulent leaves, succulence leading to a decrease in the ratio and hence a low water loss by transpiration. Sometimes the leaves lose photosynthetic activity altogether and are replaced by scales or spines – the cacti are the classic example of the latter. The stem then becomes the photosynthetic organ. The possession of spines discourages animals from using the plants as their supply of water and food in an area where both are scarce.

(4) Hidden stomata. The stomata may lie in deep depressions or grooves, which trap a volume of still air and increase the boundary layer resistance. In dry conditions, the leaves may roll up and protect the stomata. Sometimes the surface cuticle and wax form a dome over the stoma, with a small aperture, again enclosing a region of still air.

(5) Thick cuticle and wax. These clearly serve to conserve water by preventing evaporation through the outer epidermal surface.

(6) Hairiness. The hairs trap air and effectively increase the boundary layer. The hairs also help to keep the temperature down since they are usually colourless or weakly pigmented, resulting in a pale surface which reflects light and decreases heat absorption.

(7) Lignified leaves. When the tissues do become dehydrated, lignification prevents collapse.

Several of these features can be present simultaneously. It may be noted that the xeromorphic features are incompatible with fast growth rates. A high root-to-shoot ratio puts a burden on the limited amount of photosynthetic tissues. A low surface : volume ratio as in cacti (and other succulents) results in slow diffusion of CO_2 into the photosynthetic organs and only the outermost cell layers are well illuminated, so that photosynthesis is confined to these layers. In succulent leaves and stems, the ratio of photosynthetic cells to total mass is low. The net assimilation rate is consequently low and growth is slow; cacti are notorious for their low growth rates. The sinking or overarching of stomata, and hairy surfaces, increasing the boundary layer, slow down the inward diffusion of CO_2 even as these features protect against excessive outward diffusion of water. That is the price the plants pay for survival in xeric habitats.

Box 3.2 Derivation of water potential and its relationship to vapour pressure

Water potential is derived from basic principles as follows. The amount of free energy contributed per mole water in a system is called its chemical potential, μ_w. This quantity cannot be measured directly, but the chemical potentials of water in different systems can be compared against μ_w^0, the chemical potential of pure water at atmospheric pressure and the same temperature as the system being studied:

$$\Psi = \frac{(\mu_w - \mu_w^0)}{\bar{V}_w} \qquad (3.6)$$

The term $\bar{V}_w$ is the partial molal volume of water, i.e. the volume of 1 mole water, 18 mL. By dividing through by this value, the water potential is obtained in convenient units:

$$\Psi = (energy\ per\ mole\ water)/(volume\ per\ mole\ water)$$
$$= energy\ per\ unit\ volume\ of\ water$$

The basic units of Ψ are thus energy per unit volume, e.g. $J\,m^{-3}$. But energy per unit volume can easily be converted to an equivalent force per unit area, i.e. pressure, in pascals Pa:

$$1\ J\,m^{-3} = 1\,Pa,\ or\ 10^6\ J\,m^{-3} = 1\ MPa$$

Water potential of the atmosphere is of great importance in plant water relations and can be calculated from its water vapour pressure according to the formula:

$$\Psi = \frac{RT\ \ln(e/e^0)}{\bar{V}_w} \qquad (3.7)$$

where R is the gas constant, T the absolute temperature, e the vapour pressure of water in the air and e^0 the vapour pressure of pure water at the same temperature.

The water potential of pure distilled water at STP (standard temperature and pressure, i.e. $0\,°C$ and 1 atmosphere) is by definition zero. See Equation 3.6, above: for pure water, the right-hand side becomes $(\mu_w^0 - \mu_w^0)/\bar{V}_w$, which $= 0$.

Complementary reading

Balling, A. & Zimmermann, U. Comparative measurements of the xylem pressure of *Nicotiana* plants by means of the pressure bomb and pressure probe. *Planta,* **182** (1990), 325–38.

Clearwater, M. J. & Clark, C. J. *In vivo* magnetic resonance imaging of xylem vessel contents in woody lianas. *Plant, Cell and Environment,* **26** (2003), 1205–14.

Cochard, H., Forestier, S. & Améglio, T. A new validation of the Scholander pressure chamber technique based on stem diameter variations. *Journal of Experimental Botany,* **52** (2001), 1361–5.

Cochard, H., Lemoine, D. & Dreyer, E. The effects of acclimation to sunlight on the xylem vulnerability to embolism in *Fagus sylvatica* L. *Plant, Cell and Environment,* **22** (1999), 101–8.

Kramer, P. J. *Plant and Soil Water Relationships: a Modern Synthesis.* New York, NY: McGraw-Hill, 1969.

Meylan, B. A. & Butterfield, B. G. *Three-Dimensional Structure of Wood*. London: Chapman & Hall, 1972.

Milburn, J. A. *Water Flow in Plants*. London: Longman, 1979.

Milburn, J. A. Sap ascent in vascular plants: challengers to the cohesion theory ignore the significance of immature xylem and the recycling of Münch water. *Annals of Botany*, **78** (1996), 399–407.

Neinhuis, C., Koch, K. & Barthlott, W. Movement and regeneration of epicuticular waxes through plant cuticles. *Planta*, **213** (2001), 427–34.

Richter, H. Water relations of plants in the field: some comments on the measurement of selected parameters. *Journal of Experimental Botany*, **48** (1997), 1–7.

Salleo, S., Lo Gullo, M. A., Raimondo, F. & Nardini, A. Vulnerability to cavitation of leaf minor veins: any impact on leaf gas exchange? *Plant, Cell and Environment*, **24** (2001), 851–9.

Siefritz, F., Tyree, M. T., Lovisolo, C., Schubert, A. & Kaldenhoff, R. PIP1 plasma membrane aquaporins in tobacco: from cellular effects to function in plants. *The Plant Cell*, **14** (2002), 869–76.

Steudle, E. The cohesion–tension mechanism and the acquisition of water by plant roots. *Annual Review of Plant Physiology and Plant Molecular Biology*, **52** (2001), 847–75.

Stiller, V. & Sperry, J. S. Cavitation fatigue and its reversal in sunflower (*Helianthus annuus* L). *Journal of Experimental Botany*, **53** (2002), 1155–61.

Tyree, M. T., Salleo, S., Nardini, A., Lo Gullo, M. A. & Mosca, R. Refilling of embolized vessels in young stems of laurel. Do we need a new paradigm? *Plant Physiology*, **120** (1999), 11–21.

Tyree, M. T., Cochard, H. & Cruiziat, P. The water-filled versus air-filled status of vessels cut open in air: the 'Scholander assumption' revisited. *Plant, Cell and Environment*, **26**, (2003). 613–21.

Xia, Y., Sarafis, V., Campbell, E. O. & Callaghan, P. T. Non-invasive imaging of water flow in plants by NMR microscopy. *Protoplasma*, **173** (1993), 170–6.

Zimmermann, M. H. & Jeje, A. A. (1981). Vessel-length distribution in stems of some American woody plants. *Canadian Journal of Botany*, **59** (1981), 1882–92.

Zimmermann, U., Meinzer, F. C., Benkert, R. *et al*. Xylem water transport: is the evidence consistent with the cohesion theory? *Plant, Cell and Environment*, **17** (1994), 1169–81.

References

Canny, M. J. (1990). What becomes of the transpiration stream? *New Phytologist*, **114**, 314–68.

Canny, M. J. (1995). A new theory for the ascent of sap: cohesion supported by tissue pressure. *Annals of Botany*, **75**, 343–57.

Canny, M. J. (1997). Vessel contents during transpiration: embolisms and refilling. *American Journal of Botany*, **84**, 1223–30.

Esau, K. (1965). *Plant Anatomy*, 2nd edn. New York and London: Wiley.

Frensch, J. & Hsiao, T. C. (1993). Hydraulic propagation of pressure along immature and mature xylem vessels of *Zea mays* measured by pressure probe techniques. *Planta*, **190**, 263–70.

Hellmers, H., Horton, J. S., Juhren, G. & O'Keefe, J. (1955). Root systems of some chaparral plants in southern California. *Ecology*, **36**, 667–78.

Huber, B. (1956). Die Gefäßleitung. In *Encyclopedia of Plant Physiology*, vol. 3, ed. W. Ruhland. Berlin: Springer. pp. 541-82.

Kramer, P. J. (1937). The relation between rate of transpiration and rate of absorption of water in plants. *American Journal of Botany*, **24**, 10–15.

Kramer, P. J. (1949). *Plant and Soil Water Relationships*. New York: McGraw-Hill.

McCully, M. E., Huang, C. X. & Ling, L. E. C. (1998). Daily embolism and refilling of xylem vessels in the roots of field-grown maize. *New Phytologist*, **138**, 327–42.

Meylan, B. A. & Butterfield, B. G. (1972). *Three-Dimensional Structure of Wood*. London: Chapman & Hall.

Pockman, W. T., Sperry, J. S. & O'Leary, J. W. (1995). Sustained and significant negative water pressure in xylem. *Nature*, **378**, 715–16.

Raschke, K. & Resemann, A. (1986). The midday depression of CO_2 assimilation in leaves of *Arbutus unedo* L.: diurnal changes in photosynthetic capacity related to changes in temperature and humidity. *Planta*, **168**, 546–58.

Stålfelt, M. G. (1956). Die stomatäre Transpiration und die Physiologie der Spaltöffnungen. In *Encyclopedia of Plant Physiology*, vol. 3, ed. W. Ruhland. Berlin: Springer.

Tanner, W. & Beevers, W. (2001). Transpiration, a prerequisite for long-distance transport of minerals in plants? *Proceedings of the National Academy of Sciences (USA)*, **98**, 9943–7.

Troughton, J. H. & Donaldson, L. A. (1972). *Probing Plant Structure*. London: Chapman & Hall.

Troughton, J. H. & Sampson, F. B. (1973). *Plants: a Scanning Electron Microscope Survey*. Sydney: Wiley.

Tyerman, S. D., Niemietz, C. M. & Bramley, H. (2002). Plant aquaporins: multifunctional water and solute channels with expanding roles. *Plant, Cell and Environment*, **25**, 173–94.

Chapter 4

Mineral nutrition

4.1 | Introduction

Of the naturally occurring 92 elements of the periodic table, about a quarter are essential to plants. Water and CO_2 provide the plant with the elements C, H and O; the remaining necessary elements are obtained by flowering plants as inorganic mineral ions, mostly from the soil solution. Water uptake and ion uptake are to some extent linked, e.g. water uptake mediated by root pressure depends on ion uptake, and the rate of ion uptake tends to increase with increasing rate of transpiration. But the uptake of mineral ions differs greatly from water uptake in that it proceeds against the free energy gradient of the ions and is dependent on metabolic energy. The transport of ions through cellular membranes is mediated by numerous membrane-bound transport proteins which enable the plant to exert considerable control and selectivity over the process. This is vital if the nutritional needs of the plant are to be satisfied. Heterotrophic organisms obtain nearly all their essential elements via plants and the element composition of plants is accordingly of major interest and importance also for human nutrition.

4.2 | Essential elements

4.2.1 Definition: macronutrients and micronutrients
An element is classed as **essential** to a plant if the plant cannot complete its life cycle without it and no other element can substitute for it. The effect of the element must also be direct, i.e. it should not act by promoting the uptake of another essential element, or by retarding the absorption of a toxic one. To test for the essentiality of an element, the test plants must be placed in an environment totally free from that element. In practice this means growing the plants in a liquid culture medium of precisely known composition. Some elements are required in such minute amounts that they are very difficult to eliminate from solutions to levels below those required by plants. Even distilled water, glass of containers, gaseous

pollution of the atmosphere and dust particles may provide enough of certain elements to support plant life. Specially purified water, spectroscopically pure chemicals and a filtered air supply must be used. As techniques have been refined, elements have been added gradually to the list of those needed. Whilst the macronutrient elements (see below) were known by the second half of the nineteenth century, the essentiality of chlorine was not established until 1954. Nickel was added to the essential list even later, in 1987. There is the further problem of the minerals already present in the propagule used to start the culture; the supply in this may suffice for a considerable period of growth. More elements may possibly be added to the current list, which stands at 18–21 (Table 4.1). The precise number depends on the species studied, some elements being apparently essential for certain species but not for others; it also depends on how strictly the criteria for essentiality are applied. It may well be that some elements, hitherto known to be required by only a few plants, may eventually be found to be universally essential.

As shown in Table 4.1, the essential elements are classified as **macronutrients** and **micronutrients**. The macronutrients are required in large amounts relative to the micronutrients; in culture solutions, macronutrients are supplied at 10^{-3} to 10^{-2} mol L^{-1}, whilst the micronutrient concentrations may be as low as 10^{-7} mol L^{-1}. Most of the micronutrients become toxic at quite moderate concentrations, say above 10^{-4} mol L^{-1}.

In addition to essential elements there are **beneficial elements**, as indicated in Table 4.1, which are not absolutely necessary for survival but promote the growth and vigour of plants. Non-essential elements are also taken up by plants; any element present in the environment will be absorbed at least in small amounts. For plants grown in the soil, large amounts of Al and Na are frequently present as these are common in soils. Though inessential, such elements are far from being inert. They influence the ionic balance and osmotic potential of the cells and may affect the uptake of essential ions. Many non-essential elements are toxic in quite low amounts and their uptake is detrimental to the plants and to the animals which feed on them.

4.2.2 The physiological functions of the elements in plants

The roles of the essential elements in plants are at least partly known; the majority of the essential elements are indeed universal for all living organisms and many of their functions are the same in plants, animals, fungi and prokaryotes. In discussing the functions of the elements, most emphasis is here put on those functions that are characteristic of flowering plants. The macronutrient elements are constituents of cellular macromolecules, including all the major building blocks of protoplasm, whereas many micronutrients are enzyme cofactors or occur as parts of prosthetic groups of enzymes.

Carbon and **hydrogen** are of course constituents of all organic molecules and the majority of organic molecules of living cells

Table 4.1 | Macronutrients, micronutrients and beneficial elements for flowering plants, and the main forms in which they are obtained. An asterisk* marks elements so far found to be essential only in some species; of these, silicon, chlorine and sodium are *beneficial* in numerous other species. Additional beneficial species are listed at the end of the table. Nitrate is the main N source. Of the three types of phosphate, the most important P source for plants is $H_2PO_4^-$; all three phosphates can be referred to as Pi, inorganic phosphate, especially when it is uncertain which ion(s) is/are involved.

Element	Symbol	Form absorbed
Essential macronutrients		
Carbon	C	CO_2, CO_3^{2-} (carbonate), HCO_3^- (bicarbonate)
Hydrogen	H	H_2O
Oxygen	O	O_2, H_2O, CO_2
Nitrogen	N	NO_3^- (nitrate), NH_4^+ (ammonium)
Sulphur	S	SO_4^{2-} (sulphate)
Phosphorus	P	$H_2PO_4^-$, HPO_4^{2-}, PO_4^{3-} (phosphates)
Calcium	Ca	Ca^{2+}
Potassium	K	K^+
*Silicon	Si	H_4SiO_4 (silicic acid)
Essential micronutrients		
Iron	Fe	Fe^{2+} (ferrous)
Magnesium	Mg	Mg^{2+}
Manganese	Mn	Mn^{2+}
Copper	Cu	Cu^{2+} (cupric)
Zinc	Zn	Zn^{2+}
Boron	B	H_3BO_3 (boric acid)
Nickel	Ni	Ni^{2+}
*Cobalt	Co	Co^{2+}
*Molybdenum	Mo	MoO_4^{2-} (molybdate)
*Chlorine	Cl	Cl^- (chloride)
*Sodium	Na	Na^+
Beneficial elements		
Selenium	Se	SeO_4^{2-} (selenate)
Rubidium	Rb	Rb^+
Strontium	Sr	Sr^{2+}
Aluminium	Al	Al^{3+}

contain **oxygen** as well; these three elements are present in the greatest amounts.

Nitrogen, too, is a constituent of many cellular molecules, in particular proteins and nucleic acids, the key macromolecules of life as we know it. There are many lower molecular weight nitrogenous organic compounds vital to cell metabolism – vitamins, cofactors, hormones, the chlorophyll pigments and the phytochrome photoreceptors. Flowering plants additionally contain an extraordinary variety of nitrogenous *secondary compounds* not involved in basic metabolism. These include alkaloids, among which are compounds used as drugs, e.g. morphine, nicotine and quinine. Plants also contain numerous non-protein amino acids, which are not incorporated into normal proteins. There has been much dispute about the possible physiological functions of the secondary chemicals. Both the alkaloids and the non-protein amino acids are toxic and often bitter tasting; one possible function is protection against herbivores. In seeds, non-protein amino acids, with a high proportion of N by weight, can act as N storage compounds. Some non-photosynthetic pigments contain N, e.g. betacyanin, the red pigment of beetroot (*Beta vulgaris*).

Sulphur performs an important structural role in proteins where the disulphide bridges –S–S– stabilize tertiary protein structures. Sulphydryl groups, –SH, are found in the active sites of many enzymes. There are also –SH-containing coenzymes, e.g. coenzyme A, whilst glutathione, again with a –SH group, is an important anti-oxidant. Several iron–sulphur proteins, e.g. ferredoxins, occur in the electron transfer systems of chloroplasts and mitochondria; these proteins contain clusters of linked S and Fe atoms at their reactive sites. Membrane sulpholipids are structural molecules which contain a sulphate group, found in chloroplast thylakoid membranes. Numerous flowering plants contain pungent secondary S-containing compounds appreciated as flavours; these are very common in the Brassicaceae (cabbage family) which includes mustard (*Sinapis alba*). Onions (*Allium cepa*), garlic (*Allium sativum*) and related species are also flavoured with S-containing chemicals. The presence of such compounds may deter some herbivores.

Phosphorus is contained in nucleic acids and also in membrane phospholipids which make up the bimolecular lipid leaflet of biological membranes. As a component of the adenosine phosphates (ATP, ADP and AMP) and related nucleotides, the phosphate group is involved in 'energy metabolism' and intermediary metabolism involves many phosphorylated intermediates. In metabolically active cells there is a continuous turnover of phosphate from organic combination to Pi (inorganic phosphate) and back again.

Calcium, as in cells of other kingdoms, contributes to membrane stability in plant cells by its association with membrane phospholipids, and it is necessary for the maintenance of the normal permeability of the plasmalemma. In plants it also contributes to cell wall structure as calcium pectate; this is a major component of the middle lamella which cements adjacent cell walls together. The Ca^{2+} ion is extremely important in stimulus perception; one of the first effects in the chain of reactions set off by a stimulus, environmental or hormonal, is very often a change in the cellular concentration of Ca^{2+} which is termed a 'second messenger' (see Section 7.4.4). Ca^{2+} further acts as activator to some enzymes – amylases, ATPases and phospholipases.

Potassium is something of a mystery element. It is present in cells as the free K^+ ion; it does not enter into organic combination. It is known to be the activator of some enzymes, but other elements which act as enzyme activators are required only in micronutrient amounts. The affinity of proteins for K is, however, low and it may be that fairly high concentrations are needed to make enzyme-potassium complexes. It is the chief cation of protoplasm and as such it balances the charges on cytoplasmic anions, organic cations being few. Chloroplasts have a high content of K^+; the movement of H^+ from the stroma into the thylakoid lumen during photosynthesis (Section 2.3.2) is electrically balanced by a movement of K^+ into the stroma from the cytosol and a shortage of the element leads to a low rate of photosynthesis. The K^+ ion is very important in controlling Ψ (the water potential) and hence the water content of plant cells. Cell expansion is associated with the accumulation of K^+ in vacuoles, which induces water uptake into the vacuoles and an increase in size. In plant cells which function in movements involving turgor changes, K^+ ions are concerned in turgor control; such cells are stomatal guard cells and the pulvinar cells (hinge cells) of leaves and petioles. In these cells, increases and decreases of turgor are achieved by K^+ moving in or out, water following according to the resulting Ψ gradient. Since so many of the effects of K^+ are physical effects on Ψ or electric potential, the question is why K^+ should be so specifically required and why the very similar Na^+ can replace it to only a limited extent.

Silicon in the form of silica gel, a hydrated oxide of Si, gives the cell walls of grasses, including cereals, their characteristic rigidity; this is very conspicuous in the dried-out straw. Si is not known to take part in any biochemical reactions within cells and Si-requiring plants can be nursed to maturity in culture in the absence of the element. Lack of Si does, however, result in some wilting, withering and necrosis, and under natural conditions such Si-deficient plants would have little chance of survival; hence it is reasonable to consider Si as an essential element for species which normally have highly silicified cell walls, e.g. wetland grasses. Many other species contain smaller amounts of Si in their walls and Si can be regarded generally as a beneficial element. It can ameliorate toxic effects of Al and Mn and can increase resistance to fungal disease.

Many of the micronutrients have been identified as enzyme activators or as parts of the prosthetic groups of enzymes. **Iron** and **copper** are present in the respiratory and photosynthetic electron transfer chain cytochromes. They are also needed for other oxidative enzymes: Fe for catalase and peroxidase, Cu for ascorbic acid oxidase and polyphenol oxidase; Fe is present in iron–sulphur proteins, as mentioned in connection with S (p. 103) and Fe is necessary for chlorophyll synthesis.

Magnesium and **manganese** activate many dehydrogenases and phosphate transfer enzymes and are also important in photosynthesis, a Mg atom being part of the chlorophyll molecule whereas Mn is present in the O_2-evolving complex. All the three elements Fe, Cu and Mn are transition metals, able to change valency as they lose or gain an electron; hence their association with oxidoreduction activities, which involve the transfer of electrons between reactants.

Zinc is again an activator for many enzymes. Particularly important in plants are alcohol dehydrogenase, superoxide dismutase (which degrades the highly reactive and dangerous superoxide radicals formed during certain oxidative and photosynthetic reactions), and carbonic anhydrase. The last-named catalyses the reaction

$$CO_2 + H_2O \; \underset{\longleftarrow}{\longrightarrow} \; H_2CO_3 \text{(carbonic acid)} \tag{4.1}$$

In aquatic plants, where carbonate is the main C source for photosynthesis, this reaction produces CO_2 as substrate for Rubisco. In C_4 plants the reverse reaction occurs, resulting in the formation of carbonic acid which dissociates to form bicarbonate, the substrate for PEP carboxylase.

Nickel is a constituent of the enzyme urease, which hydrolyses urea; the enzyme is needed for N metabolism in plants.

Molybdenum is present in the enzyme nitrate reductase, which is needed to utilize nitrate, the major source of inorganic N for most plants, and it is needed for symbiotic N_2 fixation. It is also part of the cofactor (Moco) for aldehyde oxidase, an enzyme involved in the synthesis of ABA, and in a few other oxidases (Mendel & Hänsch 2002). The amount required is extremely small.

Cobalt also is needed in minute quantities only, and is known to be needed for symbiotic N_2 fixation, which involves the Co-containing vitamin B_{12}. Since plants normally associated with symbiotic N_2 fixers can survive without the symbionts, Co might be argued to be beneficial rather than essential. However, symbiotic N_2 fixation is very important not only for the species in which it occurs, but in the overall ecological context. Hence it seems appropriate to include Co in the essential list.

Chlorine is required for the O_2-evolving system of photosynthesis. For this it is needed in only micronutrient amounts. However, the element is taken up by cells in large quantities and the chloride ion Cl^- is the chief inorganic anion in cells, often accompanying K^+, e.g. during K^+ fluxes in stomatal guard cells, so that it is beneficial in much larger amounts than required to fulfil its essential biochemical role.

Sodium is required for C_4 photosynthesis in some C_4 species where it seems to be involved with the conversion of pyruvate to PEP. It is present in cells as the free Na^+ ion and like Cl^- is tolerated in relatively high concentrations. Chemically it is very similar to K and to some extent it can interchange with that element; e.g. in *Commelina benghalensis* Na can replace K in the control of turgor of the stomatal guard cells. In succulent halophytes, plants which live in saline habitats, Na^+ acts as an osmoregulatory ion, with Cl^-.

Boron is the element for which the physiological role has proved most difficult to elucidate. Much of the B in the plant is associated with cell walls where it cross-links cell wall polymers, such as pectins.

It also is needed for normal membrane function. In B-deficient roots, ion uptake capacity deteriorates but when such roots are supplied with B, recovery is considerable by 20 minutes and complete within an hour. Such fast action suggests a primary action at the membrane level, B either affecting membrane permeability or acting on membrane-bound enzymes. There is also some evidence for B affecting enzymes of auxin and ascorbate metabolism. The greatest demand for B is during the reproductive phase; in the absence of B, pollen grain formation fails, and pollen tubes germinating in the absence of B swell and burst.

Regarding the **beneficial elements**, individual species differ in their requirements. **Sodium** benefits many species, being able to substitute to some extent for K. **Rubidium** and **strontium** also probably owe their beneficial effect to an ability to replace some of a plant's requirements for K and Ca respectively. Rubidium enhances growth most markedly in K-deficient media. **Aluminium** is more limited in its beneficial effects; it is beneficial to tea (*Camellia sinensis*), *Fallopia sachalinensis*, and a number of grasses. **Selenium** is accumulated in large amounts by some 'accumulator species' e.g. certain species of *Astragalus* growing in Se-rich soils, also by *Lupinus albus* and *Phleum pratense*; it affords some protection against insect attack and protects against toxicity from excess Pi. The beneficial effects of **silicon** have already been discussed in connection with its essentiality for some species.

Complex interactions take place between mineral elements and metabolism over and above the primary roles of the minerals. Several examples of interactions between mineral supply and growth hormone metabolism have been reported. In sunflower plants, deficiency of N, P or K in the rooting medium has been found to decrease the flow of cytokinin hormones from the roots to the shoots. Macronutrient deficiency can thus act on plant development not only through direct shortage of elements, but via the hormone supply. The biosynthesis of the gaseous hormone ethylene is promoted by Ca^{2+} ions. Ethylene acts antagonistically to the hormone auxin in a number of effects. Thus Ca^{2+} antagonizes auxin by promoting ethylene biosynthesis. Cobalt inhibits the biosynthesis of ethylene and hence Co salts are used to prolong the life of cut flowers, ethylene being normally produced by the flowers and promoting senescence. More such interactions have been reported and undoubtedly more still remain to be noted.

4.3 | Ion uptake and transport in the plant

4.3.1 | Ions in the soil

With the exception of C, H and O, which are derived from water and CO_2 and are incorporated by photosynthesis as described in Chapter 2, plants acquire all other elements as inorganic ions. Even C can be obtained as the carbonate or bicarbonate ion. Organic nitrogenous compounds can act as the N source, but normally are available to plants in limited amounts if at all. The ions which serve as sources of the essential elements for flowering plants are listed in Table 4.1. For terrestrial flowering plants the chief source of mineral ions is the soil. The mineral rock particles of the soil yield ions by weathering which gradually brings them into solution; ions are also released by the action of microorganisms on dead organic material. The ion concentration of the soil solution rises as the water content of the soil falls, but except under very dry conditions the solution is very dilute. It has been shown that a solution corresponding in ionic composition to that of a soil solution will support good growth of crop plants provided it is frequently renewed or applied as a flowing solution so that it is not depleted. In a natural soil the ions in the solution are constantly being replenished. The laws of physicochemical equilibrium ensure that, as ions derived from soil solids are removed from the solution, more ions dissolve from the rock particles. Soil Pi concentration is always low, 1 mg L^{-1} or less; it has been calculated that the Pi of the soil solution may need to be renewed 10 or more times per day to meet the P demands of a growing crop. Nitrate also needs rapid replenishment; it is absorbed rapidly by plants, N being needed in larger quantities than most other elements, and being extremely soluble, nitrate is easily washed downwards in rainwater.

Not all the ions in the soil are totally free in the soil solution. The colloidal matter in the soil, both inorganic clay particles and organic particles, 'humus', which help to retain water in the soil (Chapter 3) also serve to retain ions by adsorption. The colloidal constituents of the soil usually carry a net negative charge; cations, being positively charged, are adsorbed to the negatively charged groups on the clay and on the organic particles. These ions are held at the surface of the soil particles by electrostatic attraction only loosely and can be exchanged for other cations; by washing a soil with a concentrated solution of a salt such as NH_4NO_3 the soil cations can be displaced into the solution in exchange for the ammonium ions (NH_4^+). For most of the anions, there is little adsorption because of the lack of positive charges on the soil colloids. There may be some adsorption of phosphate ions, especially of the trivalent PO_4^{3-} which has a high electric charge density, and phosphate ions can also replace hydroxyl and silicate anions in clays.

4.3.2 Ion uptake by the root

Adsorption, absorption and accumulation

The region of most active ion uptake by roots is the same as for water uptake, i.e, the young region of the root behind the apical meristem, the root hair region. The uptake of cations begins with the **adsorption** of the cations to the cell walls where polysaccharides carry negative charges, which attract H$^+$ ions moving out of cells by the action of proton pumps (Section 4.3.4, p. 117). These H$^+$ are then accessible to be exchanged for soil solution cations, which thus become electrostatically bound to the cell walls. The binding sites lie not just on outer surfaces of walls, but inside the capillary spaces within the cell walls; a living cell's wall may be thought of as a hydrated sponge, with much 'wet space' and much adsorptive surface inside it. There is also the possibility of some cation exchange, termed contact exchange, between H^+ ions adsorbed on the root surface and the cations adsorbed onto solid colloidal soil particles. Anions have few adsorption sites in cell walls; they must simply diffuse into the water-filled capillary spaces of the cell walls before passing through the plasmalemma. To some extent, they 'follow' the cations by electrostatic attraction.

The adsorption of ions may then be succeeded by **absorption**, passage through the plasma membrane. There is an important difference in meaning between these very similar words, **ad**sorption for attraction outside the plasma membrane, **ab**sorption for actual entry into the cell, to the inside of the plasma membrane. When the ions are taken up to a greater concentration inside the plasma membrane than outside, this is termed **accumulation** (Fig. 4.1).

Uptake of ions is by no means confined to root cells. Roots take up the ions in the first instance, but these ions are distributed around the whole plant in the xylem, where they are in the apoplast, outside living cells, and cells in various parts of the plant have to take up ions from the supply carried up in the xylem sap. All living cells of the plant are capable of ion uptake from their environment; in the

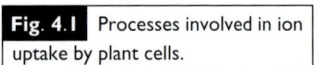

Fig. 4.1 Processes involved in ion uptake by plant cells.

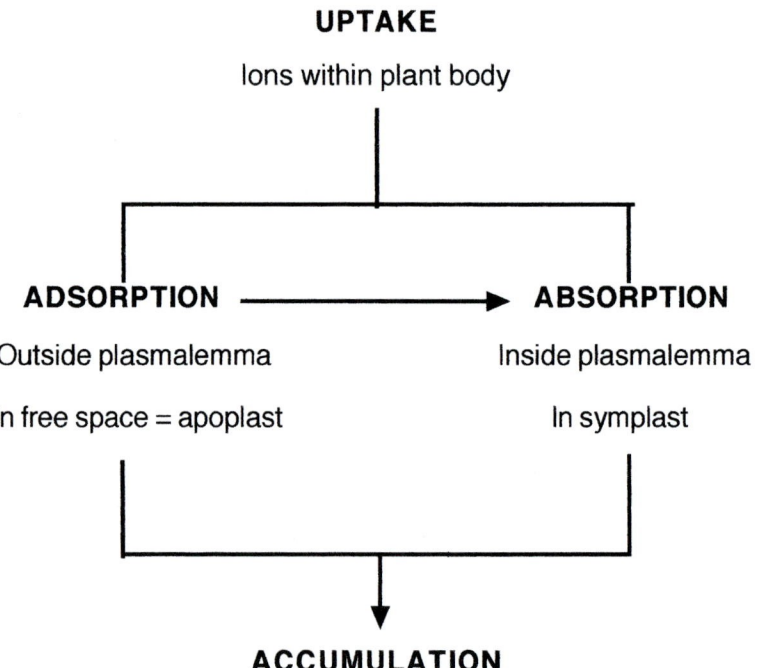

UPTAKE

Ions within plant body

ADSORPTION ⟶ **ABSORPTION**

Outside plasmalemma   Inside plasmalemma

In free space = apoplast   In symplast

ACCUMULATION

Internal concentration > external concentration

natural state, for cells other than the outer root cells, the environment for ion uptake is the apoplast.

Compartmentation of the plant cell and the concept of free space

Throughout discussions of ion uptake reference is made to 'inside' and 'outside'. Plant cells are, however, more complicated than a sac surrounded by a membrane. When plant tissue is thoroughly prewashed and then immersed in a solution containing mineral ions, there is first a very rapid uptake, followed by a much slower rate (Fig. 4.2). When the immersion takes place at low temperature, around 4 °C, or under anaerobic conditions, only the initial rapid uptake occurs, suggesting that the initial uptake is a physicochemical process not requiring metabolic energy. When the tissue is washed in distilled water after a period of uptake, some of the ions taken up are quickly washed out. These ions are said to occupy the 'water free space' which seems to correspond to the cell walls of the tissue. Some more cations can be removed from the tissue by immersion in a solution of cations, e.g. K^+ might be washed out in exchange for excess NH_4^+. The exchangeable cations are again present in the cell walls, but they are associated with the negative charges on the cell wall polysaccharides and at the plasma membrane surface; these ions are described as being in the 'Donnan free space', named after Donnan, an investigator in this field. A certain fraction of ions is, however, retained firmly; these are the ions which have passed

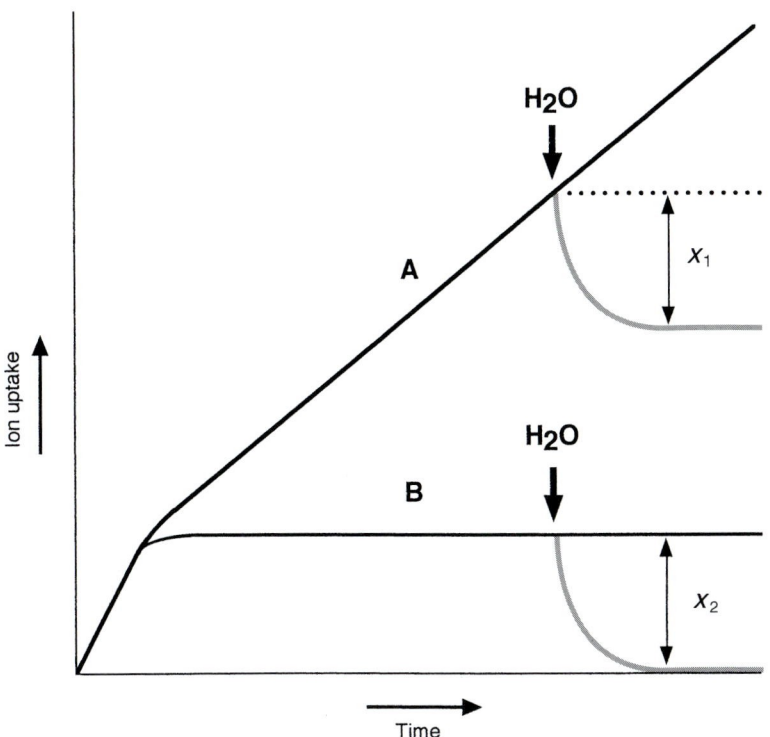

Fig. 4.2 Diagrammatic representation of a typical time course of ion uptake by plant tissue. Curve (A) under aerobic conditions and at a relatively high temperature, say 15–25 °C; curve (B) under anaerobiosis, or in the presence of respiratory inhibitors, or at a low temperature, say 0–5 °C. The grey curves indicate what happens when the tissue is placed in distilled water at the time indicated by the arrow. The initial rapid uptake lasts 10–20 minutes and represents uptake into the free space; in water, the free-space ions are washed out and $x_1 = x_2$.

through the plasma membrane. Quantitatively, the fraction that is washed or exchanged out with ease corresponds to the fraction acquired during the rapid uptake phase; the firmly held fraction corresponds to what was taken up in the slow, energy-requiring process. Diffusion and adsorption into the free space or apoplast, being physicochemical processes, do not use metabolic energy and hence proceed even at low temperature or under anaerobic conditions, but passage through the plasma membrane, into the symplast, requires ATP. Only what has passed through the plasma membrane is truly inside the living cell, although free space ions are inside the plant or tissue as a whole.

The cell wall is not a highly selective barrier to movement of solutes except in so far as, owing to negative charges on the wall polysaccharides, it tends to attract cations. The larger pore spaces between the wall macromolecules reach dimensions up to about 5 nm; the diameters of the ions, including water of hydration, are below 1 nm. The wall does slow down the diffusion of chemicals with masses above some 200–300 Da, but usually transmits molecules up to several thousand Da.

The plasma membrane is generally regarded as the primary barrier to ion movement. There are, however, further barriers within plant cells. In mature cells the vacuole forms a compartment occupying 80–90% of the protoplast volume and the tonoplast (vacuolar membrane) is an important barrier to ion movement, for a large proportion of the free inorganic ions in a plant cell passes into the vacuole. The organelle membranes constitute further barriers limiting the free diffusion of ions.

4.3.3 The transport of ions within the plant

The long-distance transport of ions takes place in the xylem concurrently with water transport (Chapter 3). Just as for water, it has been questioned whether the ions move through the outer root tissues by an apoplastic, symplastic or transcellular route (Fig. 3.8).

Apoplastic ion movement could be partly by diffusion, partly along with the flow of water moving into the transpiration stream. The endodermal Casparian strips are believed to form a barrier to ion movement, as for water. Indeed for ions there is more direct evidence for this being so. Solutions of salts of the heavy elements lanthanum (La), lead (Pb) and uranium (U) have been used as tracers for ion movement. These elements can be located by electron microscopy because of their high atomic masses (La = 139, Pb = 207, U = 238) and can be seen in cell walls of the root cortex and in the endodermal cell walls as far as the Casparian strips, but not beyond; La and U are apparently unable to cross the plasmalemma and are excluded from the stele; Pb, however, enters the endodermal cell cytoplasm and also the stele. The above observations strongly support the suppositions that (1) ions move in the apoplast; (2) the Casparian strip is an effective barrier to ion movement; and (3) the endodermis can be crossed by ions only if they can enter the symplast. The weakness of the argument lies in extrapolating from data obtained with these ions of high atomic weight to the behaviour of ions of essential elements, with lower masses. There is, however, indirect evidence for similar behaviour by the nutrient ions. In barley (*Hordeum vulgare*) roots, Pi ions and K^+, which readily enter the symplast, are translocated to shoots even from the older parts of the roots, where the endodermal Casparian strips are fully developed. The ion Ca^{2+}, which penetrates the symplast with difficulty, is translocated to the shoots mainly from apical regions where the endodermal cell walls are still permeable.

When the cortex was stripped from a segment of barley roots (still attached to the shoots) so as to break the endodermal cells across at the Casparian strips, the concentrations of ^{32}Pi and $^{85}Sr^{2+}$ in the transpiration stream equalized with the external medium, which was not the case when the endodermis was intact. The endodermis can thus regulate the entry of ions into the stele. The endodermis also acts as a barrier against the outward leakage of ions from the stele. Experiments with $^{45}Ca^{2+}$ in barley roots have shown that over 60% of the radioactivity was exchanged out from the cortex for excess unlabelled Ca^{2+} ions in 10 minutes; but from the pericycle (the cell layer just inside the endodermis), only 19% was lost.

In summary, the mineral ions absorbed by roots can travel to the xylem in both the apoplast and the symplast. In the extreme apices of the roots, the apoplastic pathway is continuous, but once the Casparian strips have fully developed (this may occur within a few millimetres of the tip) the endodermis can be crossed only via the symplastic route. In regions of the root where secondary rootlets emerge, the endodermis may be interrupted, giving freer access to the stele again. Control of what passes into the xylem is possible by the selective transfer of ions to the symplast anywhere between the epidermis and the endodermis, by the retention of ions by living root cells, and perhaps by a filtering action in the endodermis. The sap composition is further modified during long-distance transport by the absorption of ions from the flowing stream by living xylem parenchyma cells adjacent to the conducting cells.

When the rate of water transport from the soil to the xylem is increased, particularly by increased transpiration, ion uptake and transport to the shoots are also increased. It may be that the additional ions entering the xylem represent a passive mass flow of ions carried along in the water current through the apoplast. Alternatively, it has been suggested that the dilution of the xylem sap by an increased rate of water uptake into the conducting cells stimulates a higher rate of ion secretion into the sap from the symplast of living xylem cells. A still further possibility is that the tension pull created by a high transpiration rate lowers the resistance to ion (and water) movement through membranes and thereby increases the rate of secretion or of passive leakage from the symplast. There is evidence that tensions developed in the xylem vessels are transmitted across the root diameter.

Movement in the xylem is one-way traffic. In leafy plants, most of the solution in the xylem eventually arrives in the leaves, though some solutes and water are of course absorbed on the way by older root segments, stems and petioles. For the ions moving away from the conducting cells, there are again the two possible pathways: via diffusion through the apoplast, or by absorption into adjacent living cells with subsequent distribution through the symplast.

Some of the elements are retained in the tissues to which they were first transported; these are known as immobile elements. One such example is Ca. Crystals of calcium carbonate or calcium oxalate are often precipitated in vacuoles of older cells of leaves and stems. Silicon is deposited as the insoluble silica and is immobile. In many species B also is not relocated; a common symptom of B deficiency is the death of meristems: once the external supply fails, the newly formed tissues immediately run short, while the older tissues still have the B originally deposited in them. There are, however, species which can translocate and redistribute B in the phloem; in these species the phloem sap contains sugar alcohols, with which B can form complexes. Other ions, the mobile ones, can be transported out of old, senescing tissues and relocated to young, growing regions; K^+ is a mobile ion. This transport occurs in the phloem (Chapter 5). Some elements can be moved around the plant in organic combination: N can move as amides or amino acids, S as S-containing amino acids. Proteins in senescing tissues are largely hydrolysed and the soluble amino acids transported to regions of growth and storage, so that most of the N is conserved.

The constant supply of minerals in the xylem stream leads to the accumulation of ions in the leaves. It has been suggested that the shedding of leaves is equivalent to excretion, ridding the plant of waste minerals (and organic waste, too). There occurs also some leaching of minerals by rain. Around trees one can sometimes distinguish a drip-zone flora, which thrives on the leachings from the tree.

4.3.4   Ion transport across cellular membranes

Methods of study of ion concentrations, and ion fluxes

In studies of ion uptake, it is often necessary to measure the cellular concentration of an ion in order to determine concentration gradients, or to follow the increase in concentration, during an uptake period. But the various cellular compartments all have their own specific ion concentrations, and an average concentration from whole tissue analysis may not be sufficient or meaningful for a particular study. Obtaining separate analyses for individual subcellular compartments is on the other hand difficult and may not be feasible for all types of plant material. Much of the work on ion uptake by 'plants' has been carried out on algae with giant cells allowing the separate sampling of cytosol, vacuole and sometimes also chloroplasts. These algae include *Nitella translucens*, *Hydrodictyon* sp. and *Valonia* sp. One *Nitella* cell, 1 mm wide and up to 100 mm long, yields 40–60 µL of vacuolar sap, ample for analysis, and the cytoplasm, too, can be sampled separately. By contrast, the vacuolar volume of a barley root cortex cell is 7×10^{-3} µL.

X-ray microanalysis under the electron microscope has been used to estimate the ion concentrations in individual cells and sub-cellular compartments. The electron beam which produces the image also causes the emission of X-rays from the atoms of the specimen; each element emits X-rays at specific wavelengths. By collecting the X-rays the elements present can be identified and quantitative analysis can be carried out by measuring the intensity of X-ray emission. To avoid the loss of the water-soluble ions during specimen preparation, they may be immobilized by precipitation, or, better still, by very rapid freezing. The frozen but still fully hydrated cells can then be analysed in the deep-frozen state.

Radioisotopes have proved invaluable in studies of ion movements. On the whole-plant scale, they can be used to trace the pathways of movement. At the cell or tissue level, radioisotopes are the routine means for measuring rates of ion flux, in or out. For short-term experiments (minutes to hours) it would be impossible to detect reliably the changes in concentration of an ion within a tissue by chemical analysis, because the amounts would be too small. But radioactivity can be detected accurately when the total amount of chemical is minute. Radioisotopes with a suitable half-life are not available for all elements of interest; thus K has no 'good' isotope. In such cases, it may be possible to substitute the radioisotope of another element with similar physicochemical properties as a marker. Plants may transport such markers by the same mechanisms as the ion of interest, even though the marker cannot substitute metabolically for the essential element.

For detailed studies on cellular physiology and molecular biology, the whole plant cell, with its numerous compartments and several barrier membranes, is too complex. Intact vacuoles can be isolated from cells to investigate transport properties of the tonoplast. The plasma membrane breaks up during cell fractionation, but the vesicles formed from it can be isolated and purified for experimentation. Small areas of plasma membrane or tonoplast can be sealed across the tip of a glass microcapillary for study, a technique known as patch-clamping, enabling the detection of movement of as few as 60 ions (Tester 1997).

Reference to the literature on ion transport at the cellular level shows that much of the work has been carried out on the three ions K^+, Na^+ and Cl^-. There are two reasons for this. Firstly, these ions are taken up by cells in large amounts, and are tolerated in the external medium in relatively high concentrations. This makes them convenient to handle. Secondly, they remain in the cells as free ions. The ions NO_3^-, NH_4^+, SO_4^{2-} and phosphates are also taken up by plants in large amounts, but are rapidly incorporated into organic combination. Analytical methods such as use of radioisotopes or X-ray analysis detect the *element* only, with no information as to the form in which it is present. If cellular Cl^- concentration is estimated by X-ray analysis, one can be confident that a count of the Cl wavelength X-rays gives a measure of ion concentration. But X-rays at the P wavelength would come from organic compounds as well as from Pi and would be unreliable as an estimate of the concentration of phosphate ions.

Active accumulation: the electrochemical potential gradient

Ion uptake usually is a process of ion accumulation, plant cells acquiring a total ion content greater than that in their environment, and this applies for many individual ions as well. For some ions, the internal concentration can be very much higher than the external; several hundred-fold accumulation is common. Phosphate, which is present in very low concentrations in natural soils, can be accumulated several thousand-fold. This immediately suggests movement against a free energy gradient, for higher concentration of a solute means higher free energy. However, there is another component to the free energy gradient of ions, namely the electric field. For electrically charged particles, an electric potential gradient is a free energy gradient. A cation, being positively charged, will move towards a more electronegative region. An anion, being negatively charged, will move towards an electropositive region. For ions therefore the free energy gradient is the combined **electrochemical potential gradient** to which both the concentration of the ion (determining its chemical potential) and the electric potential contribute. This is highly relevant to ion uptake by plant cells, because there does exist an electric charge difference across the membranes of cells. As mentioned earlier, positively charged protons are pumped to the outside of plant cells, into the walls; this results in the inside of the plasma membrane being left electronegative with respect to the outside. The potential difference is in the range of 100–250 millivolts (mV). Similarly, pumping of protons from the cytoplasm across the tonoplast into the vacuole makes the cytosol side of the tonoplast electronegative with respect to the vacuolar side.

The cytoplasm is accordingly more electronegative than the apoplast outside it. Therefore, when a cation, say K^+, is found to be in a higher concentration within a cell than outside it, the question arises: has it been moving *against* the free energy gradient (as suggested by the concentration gradient) or *along* the free energy gradient (as suggested by the electric potential gradient)? When the overall electrochemical potential gradient is taken into account, it is found that the accumulation of cations, even allowing for the electronegativity of the cytoplasm, is very often, though not exclusively, against the electrochemical potential gradient. For anions, any accumulation into the cytoplasm must be against the free energy gradient, for anions are negatively charged particles moving into a more electronegative area as well as against the concentration gradient. Movement against the free energy gradient is often termed active and requires an input of energy by the cell. Where an ion is accumulated along the free energy gradient, the cell is still not getting something for nothing, since maintenance of the potential difference across membranes requires energy. As knowledge of the mechanisms of ion uptake has developed, the distinction between active and passive accumulation has become rather blurred. It is perhaps best to regard accumulation generally as an active process, and then to consider, for individual cases, precisely how the energy input is achieved, directly or indirectly: see below on mechanisms of membrane transport.

When concentrations of ions in the cytoplasm and vacuole are analysed separately, and the electric charge difference across the tonoplast is measured, it is found that the movement of ions from the cytoplasm to the vacuole is also an active process. In mature vacuolated plant cells most of the ions of the cell are in fact transferred to the vacuole, which occupies much of the volume inside the cell wall. The vacuolar store is tapped as required. It is also found that ions may be actively transported out of cells. This is commonly the situation with Na^+. Although the cytoplasmic concentration of Na^+ is generally higher than the external, it is usually lower than would be in electrochemical equilibrium with the external medium. Indeed the active transport of ions into the vacuole can also be regarded as being out of the cytoplasm into a non-living space. Some cells are specialized for the outward transport of ions: these include the cells of salt glands of halophytes, which secrete excess Na^+ and Cl^- ions from the plants; and xylem parenchyma and transfer cells, which secrete ions into the xylem conducting cells. Directions of active flux of some nutrient ions across the plasmalemma and the tonoplast of plant cells are shown in Fig. 4.3.

The maintenance of physiological concentrations of ions in plant cells is thus an active process requiring an energy supply in the form of ATP. It is estimated that up to half of the energy from root respiration is expended on membrane transport of ions. In photosynthetic tissues, ATP from the photochemical reactions can be utilized to power ion transport.

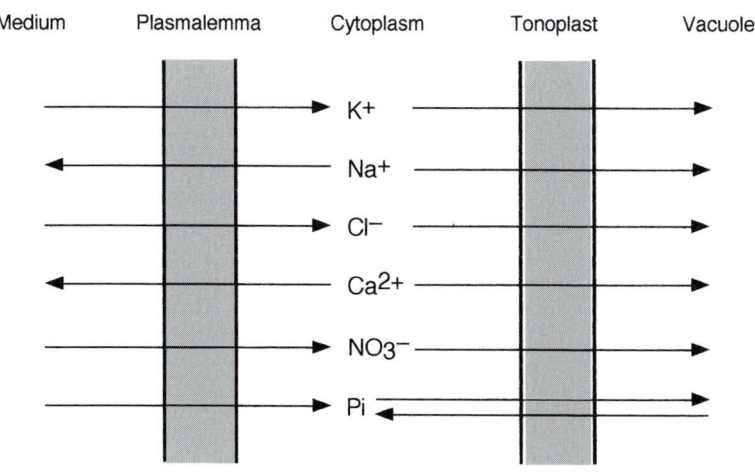

Medium   Plasmalemma   Cytoplasm   Tonoplast   Vacuole

K^+

Na^+

Cl^-

Ca^{2+}

NO_3^-

Pi

Fig. 4.3 Directions of active flux of some important ions across the plasmalemma and the tonoplast of plant cells. Phosphate is recorded as Pi since several ionic species may be involved and the (main) species may differ according to the membrane. The direction of movement of Pi between cytosol and vacuole depends on the Pi concentrations in the two compartments: the vacuole acts as a reservoir, releasing Pi to the cytosol when the cytosolic concentration falls low.

Mechanisms of membrane transport

The two main characteristics that determine the ease with which a particle can diffuse through a biological membrane are the lipid-solubility of the particle and its molecular size. The more lipid-soluble (lipophilic) and the smaller the particle, the easier the penetration. The mineral ions are very hydrophilic, so they do not dissolve in the lipid bilayer. The atomic weights of some of the nutrient ions are quite low. However, the electric charge on the ions attracts hydration shells of water molecules; e.g. K^+ (mass 39 Da) carries 4 molecules of water whereas the divalent Ca^{2+} (mass 40 Da) has about 12 associated water molecules. These hydration shells increase the effective size of the ion considerably. The permeability of biological membranes towards ions is therefore very low. The flux of ions across membranes is enabled by specific **transport proteins** in the membranes which facilitate the movement of ions and not only provide the physical means of passage for the ions, but utilize the energy of ATP to transport the ions against their electrochemical potential gradients. The total number of ion transport proteins in plant membranes is much greater than the number of nutrient ions, there often being more than one transport protein for the same ion. In *Arabidopsis*, 16 genes have been identified coding for proteins involved with nitrate uptake, and the same number for phosphate (Vance 2001). The specificity of transport proteins varies; some are highly specific to single ions, but often they can transport several related ions, i.e. ions of similar physicochemical properties such as valency and size. The rubidium ion, for instance, is transported by a number of cellular systems for K^+ transport, and the radioactive ^{86}Rb is often used in experiments as a substitute for K^+, there being no convenient K radioisotope available.

The ion transport proteins can be divided into the pumps, the porters (carriers) and the channels; their main characteristics are summarized in Table 4.2 and Fig. 4.4. The term 'carrier' was originally used for all membrane transport proteins, before different types were distinguished.

(1) Pumps. These are transport proteins which hydrolyse ATP and simultaneously transfer an ion across the membrane. The energy is derived from the ATP hydrolysis, directly. Pumps are vectorial, i.e. a

Table 4.2 Plant ion transport systems and their basic features. Porters and channels exist also for organic chemicals. All cellular membranes contain transport proteins, mostly specific for a particular membrane.

	Pump	Porter	Channel
Energetics	Uses ATP directly	Uses H^+ gradient from H^+ pump	Movement along free energy gradient
Ions moved s^{-1}	up to 5×10^2	5×10^2 to 10^4	10^6 to 10^8
Examples of ions transported	H^+, Ca^{2+}	Most if not all	Ca^{2+}, K^+, Cl^-, H^+

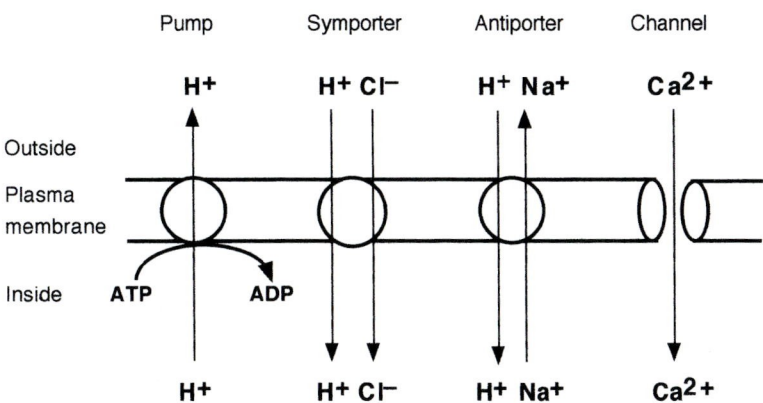

Fig. 4.4 Ion movement through a membrane via pumps, porters or channels: highly diagrammatic. The plasma membrane is used as the example, but similar systems exist in all cellular membranes. The circles represent the transporters, which are integral transmembrane proteins and may have multiple subunits. Pumps split ATP directly and undergo a conformational change which moves an ion, e.g. H^+, through the membrane. The porter proteins must simultaneously bind a H^+ ion and their specific substrate ion, to move the substrate ion across the membrane, either inwards with the H^+ ion (symport) or outwards in exchange for the H^+ ion (antiport). Some porters require several H^+ per substrate ion.

particular pump can move an ion only in one direction. The hydrolysis of the ATP results in a conformational change in the pump protein, which causes the transmembrane passage of the ion. Two of the most important ion pumps of plant cells are noted in Table 4.2. Ca^{2+} pumps are situated in both the plasma membrane and the tonoplast and they pump Ca^{2+} ions out of the cytosol (Fig. 4.3), keeping the cytosolic concentration very low, 1–5×10^{-7} mol L^{-1} (compare with $c.\ 2.5 \times 10^{-2}$ mol L^{-1} for K^+), much lower than in the vacuole and the apoplast. Another supremely important ion pump is the proton pump, also known as the proton ATPase. One might not tend to think of H^+ as an important metabolite. In fact the proton pumps constitute the metabolic machinery driving the porters and, in some instances, controlling the channels. Most of the energy utilized in membrane transport in plant cells is via the proton pumps. The previously mentioned movement of protons out of cytoplasm across the plasmalemma and the tonoplast, building up the potential differences across these membranes, is achieved by proton pumps. Their activity of course builds up also a proton gradient. In combination the electric gradient and proton concentration gradient make up an electrochemical potential gradient for protons, favouring the inward movement of the protons. This free energy gradient is harnessed by the porters.

(2) **Porters** are mostly transport proteins which couple the transport of an ion with the inward movement of a proton or protons. The transport can be **symport** (cotransport), the proton and the ion moving in the same direction, or **antiport** (countertransport), the two moving in opposite directions (Fig. 4.4). The carrier in the plasma membrane for Cl^-, for instance, is a symporter, transporting one Cl^- in with $2\,H^+$. The outward pumping of Na^+ is achieved by an antiporter, which exchanges one proton going in for one Na^+ moved out. There is also a $H^+/\,K^+$ symporter for the entry of K^+ ions into the cytoplasm, amongst the numerous transport systems existing for this ion. The energy input occurs during the activity of the proton pumps which build up the proton gradients.

(3) Channels. As the name suggests, channels are formed by proteins with several subunits enclosing an aqueous pore. An open channel is an open hole and movement through a channel is *along the free energy gradient*; it is very much faster than movement mediated by pumps and porters (Table 4.2); presumably because of that, channel proteins are present in low numbers per cell (except for aquaporins, Section 3.3.3). The channels nevertheless show specificity; some are extremely specific, for one single ion; others will permit passage according to size. Some interaction is believed to occur between the ion and the channel protein as the ion moves through: 'Passage of an ion through a channel may be likened more to a python swallowing its prey than to a ball rolling through a drainpipe' (Tester 1990). Passage through a channel has been termed **uniport** since only one chemical moves.

A very important feature of ion channels is that they open only transiently, in response to some stimulus. Any one channel is estimated to be open over only a few per cent of the lifetime of a cell. A permanently open ion channel would be fatal, the cell losing control of its ion concentration. Movement through a channel is along a free energy gradient, so active accumulation would become impossible. (Aquaporins are permanently open and active water accumulation does not occur.) The antibiotic gramicidin acts by inserting into the plasma membrane and forming permanently open channels for K^+ which leak out of the cells. Over 20 ion channels are known from plants. The Ca^{2+} pumps keep the cytosolic Ca^{2+} concentration below the apoplastic and vacuolar levels. Numerous stimuli cause a transient opening of Ca^{2+} channels and flooding of Ca^{2+} ions from the apoplast or internal compartments into the cytosol, where a reaction chain is started. The Ca^{2+} pumps then restore the cytosolic concentration to its previous level. K^+ channels are very important in turgor control of stomatal guard cells (Section 2.4.1). The rapid changes in turgor of motor cells also depend on K^+ movements through K^+ channels.

4.3.5 Control of ion uptake by plant and environment interaction

The ion content and the elemental composition of a plant reflect an interaction between the plant and its environment. Plants show great *selectivity*: ions are not taken up in the proportions in which they are present in the surroundings (Table 4.3). For example, most flowering plants show a strong preference for K^+ over Na^+ and maintain a higher internal K^+ concentration irrespective of the external proportions of the two ions. There are accumulator species which concentrate some element to a particularly high degree; the selenium accumulators mentioned earlier (Section 4.2.2, p. 106) accumulate Se to 200 times higher levels than non-accumulators in the same habitat. When several species are grown with an identical external ion supply, each shows a different internal content of mineral elements

Table 4.3 The differences between ion contents of plant tissues and the external medium, and between ion contents of different species. Bean and maize plants were placed into a culture medium of identical initial composition for four days; the plant ion content was measured on sap pressed from roots. The points to note are (1) the ions have been accumulated in the plants to much higher concentrations than present in the original solution; (2) the ions have been accumulated in proportions very different from those in the original solution; (3) the two species have very different ionic contents. Data from Marschner (1995). © Reprinted by permission from Elsevier Science.

Ion	Initial concentration in medium (mM)	Concentration in root press sap (mM)	
		maize	bean
K^+	2.00	160	84
Ca^{2+}	1.00	3	10
Na^+	0.32	0.6	6
Pi^a	0.25	6	12
NO_3^-	2.00	38	35
SO_4^{2-}	0.67	14	6

^{a}Pi = inorganic phosphate, which has several ionic forms – see Table 4.1.

(Table 4.3). Sometimes a specific preference can be interpreted in terms of function; in a pasture, the grasses contain much higher levels of Si than other herbs, and this is correlated with the presence of silica in the grass cell walls, on which the grasses depend for rigidity. But the physiological significance of, say, maize sap having double the K^+ content of bean sap in the experiment illustrated in Table 4.3 is unknown.

The rate of uptake of an ion is dependent on the physiological requirements of the plant. The rate of nitrate uptake by a grass has been found to vary with the diurnal growth rhythm of the plants, the highest uptake rates coinciding with the maxima of growth rate. There are reports of nitrate uptake remaining approximately constant over a wide range of external concentrations, and of some tendency to maintain more or less constant internal concentrations of several ions – e.g. K^+, Cl^-, phosphate and nitrate. There is a negative feedback between the tissue content of an ion and the rate of its uptake: other things being equal, plants with a high content of an ion take it up at lower rates than plants with a low internal concentration (Table 4.4). Higher external concentrations do usually promote higher ion uptake rates, and result in higher internal concentrations, but not in direct proportion to the external. In one experiment, as the medium Pi concentration was raised from 0.03 to 30 μmol L^{-1}, a thousand-fold increase, the shoot P content rose only four-fold from 0.23 to 0.96% of dry weight.

Physiological control of ion uptake is thus well documented, but the selectivity of plants is far from absolute and the environment also

Table 4.4 | The effect of the internal K^+ concentration on the rate of K^+ uptake by barley (*Hordeum vulgare*) roots. The roots were allowed to accumulate K^+ to the levels shown, then excised and transferred to ^{86}Rb-labelled KCl solution for measurement of the rate of uptake. The values are averages from three replications. With increasing internal K^+ concentration, the rate of K^+ uptake falls. Data from Glass & Dunlop (1979). © Springer-Verlag GmbH & Co. KG.

Root K^+ concentration (μmol (g fresh weight)$^{-1}$)	K^+ uptake (μmol (g fresh weight)$^{-1}$ h^{-1})
20.9	3.05
32.1	2.72
47.9	2.16
57.8	1.61

Table 4.5 | Induction of an increased sulphate uptake capacity in the roots of a tropical legume, *Macroptilium atropurpureum*, by sulphate deprivation. Plants were preincubated in a medium containing 0.25 mM sulphate or without sulphate for times as shown, and then the capacity of the roots to take up sulphate from a 0.25 mM solution was measured. After six days, the roots' uptake capacity is more than 10 times higher in the sulphate-deprived plants. Data from Bell *et al*. 1995. By permission of Society for Experimental Biology.

Days incubated	Sulphate uptake (μmol h^{-1} (g fresh weight)$^{-1}$)	
	with sulphate	without sulphate
0	0.306	0.306
1	0.196	0.811
6	0.254	2.78

exerts a very considerable influence on a plant's elemental composition. It was noted above that representatives of different species in the same environment differ in their ion contents; it is equally true that specimens of the same species in different ionic environments acquire distinctive elemental compositions. Inessential ions and toxic ions are absorbed. Not only do these have direct effects, but their uptake may compete with that of essential elements; e.g. selenate competes with phosphate for transport proteins, and arsenate competes with sulphate. Any elements present in the environment will be found in plant tissues, even the artificially produced transuranium elements such as plutonium.

The uptake capacity of a plant adapts to the current environmental concentration of that ion. Plants grown at low concentrations of an ion develop an enhanced capacity for absorbing that ion compared to plants grown with an abundant supply (Table 4.5). For any one ion there may be several transport systems available, with different affinities for the ion. At low external concentrations,

high-affinity systems are activated, which are efficient under these conditions. For K^+, the high-affinity system predominates at external concentrations below 0.5 mmol L^{-1}. The mobilization of high-affinity transport systems involves transcription of genes for the high-affinity transport proteins. These can be numerous; in *Arabidopsis*, 6 genes for high-affinity phosphate transporters have been identified. Other responses also enhance uptake in a nutrient-deficient environment, including increased root : shoot ratios, increases in number and length of root hairs and increased development of mycorrhiza.

4.3.6 Mycorrhiza

The word **mycorrhiza** means 'fungus-root'. It is the name given to a symbiotic association between a plant root and a fungus, which in most cases enhances the mineral nutrient supply of the plant, whilst the fungus benefits from a supply of organic C from the plant. Mycorrhizal associations in some species have been known for many years; gradually it has become apparent that, far from being an odd exception, the formation of mycorrhizal associations occurs in most species of flowering plants in the field and it is highly beneficial. Fossil evidence indicates the presence of mycorrhiza-like associations already in the primitive terrestrial plants of the Devonian era.

In a mycorrhizal association, part of the fungal mycelium is free in the soil, part is closely associated with roots. In **ectomycorrhiza** there is a thick sheath of fungal mycelium around the outside of young roots, and from this sheath hyphae grow into the intercellular spaces of the root cortex. In **endomycorrhiza**, all the plant-associated part of the mycelium is inside the root. The most common type of endomycorrhiza is the vesicular–arbuscular (VA) type. The hyphae grow into the cortex cells, penetrating the walls and branching greatly to form the **arbuscules** ('little trees', Fig. 4.5), but they do not penetrate the plasma membrane and the cells remain alive. The plasma membrane grows to surround the arbuscules and there is a very large area of surface contact between the plant cell and the fungal arbuscule. Some of the fungal hyphae swell into vesicles, hence the name of vesicular–arbuscular. It is estimated that about 80% of field-grown plants have VA mycorrhiza. There is also ectendomycorrhiza, with limited cell penetration by the fungus. Special associations occur in some plant groups, e.g. orchids. Exchange of nutrients takes place over the large area of contact between the fungus and the root cells. The surface area of fungus exposed to the soil is also very large, enabling efficient mineral absorption.

Mushrooms (agarics) beneath trees are often the fruiting bodies of ectomycorrhizal fungi. Some fungal species are associated with particular genera of trees: the brown birch bolete (*Leccinum scabrum*) is confined to birch (*Betula* spp.) whereas others have a broad host range; e.g. the famous red-and-white patterned fly agaric mushroom

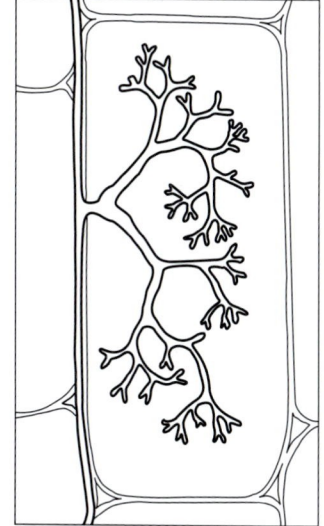

Fig. 4.5 Part of an arbuscule in a cell; the entire arbuscule has hundreds of branches and spreads out in three dimensions. The plasmalemma of the host cell is invaginated around it so that the fungus is still *physiologically* outside the cell.

(*Amanita muscaria*) is associated with birch, oak (*Quercus* spp.), pine (*Pinus* spp.) and other trees.

Most flowering plants can survive without mycorrhiza, and in laboratory culture with abundant nutrients they grow vigorously, although even in well-nourished cultures the presence of mycorrhiza may enhance growth. But in the field, especially if the soil is deficient in mineral nutrients, growth is very much poorer without mycorrhiza and the plants are less tolerant of stress. The VA association is particularly important in transferring Pi to the plant, whereas ectomycorrhiza is known to provide increased access to both N and P.

The occurrence of mycorrhiza demonstrates how closely organisms in an ecosystem interact. Flowering plants are autotrophic organisms and their basic needs are just light, CO_2, water and some 20 mineral ions. But as parts of natural communities, the majority of flowering plants are provided with an appreciable proportion of their mineral requirements by mycorrhizal fungi.

4.4 | Nitrogen assimilation, fixation and cycling

4.4.1 Nitrogen assimilation

C, H and O enter organic combination by photosynthesis, on dry land in greatest amounts by the activity of flowering plants, and these elements are returned to inorganic form again through the universal process of respiration; in Chapter 2 it was shown how these elements, especially C, are cycled globally. Next to C, H and O, nitrogen is the most abundant element in living organisms and N, too, is brought into organic combination by flowering plants as well as by algae and certain prokaryotes. The following paragraphs explain how N is incorporated into organic molecules in flowering plants, and how N is globally recycled.

The main store of N on this planet is in the atmosphere, where N makes up about 79% as nitrogen gas (N_2), but this gas is very unreactive (inert). It is totally unavailable to flowering plants, which obtain their N from the soil as the nitrate anion NO_3^- or the ammonium cation NH_4^+; one or other predominates at a particular site, but in most soils nitrate is the main source or almost the sole one. Nitrate is a highly oxidized compound, but in all organic compounds of living cells N is found in a highly reduced form and the nitrate must be reduced in plant cells to the ammonium level before it can be incorporated into organic molecules. This process of **nitrate reduction** is carried out by two enzymes acting sequentially, **nitrate reductase** followed by **nitrite reductase**. Nitrate reductase contains molybdenum. It reduces nitrate (NO_3^-) to nitrite (NO_2^-); this then acts as the substrate for the second enzyme, nitrite reductase, which produces ammonium as its end product. In summary the reactions can be represented as follows:

$$NO_3^- + 2H^+ + 2e^- \xrightarrow{\text{nitrate reductase}} NO_2^- + H_2O \qquad (4.2)$$

$$NO_2^- + 8H^+ + 6e^- \xrightarrow{\text{nitrite reductase}} NH_4^+ + 2H_2O \qquad (4.3)$$

$$\text{Sum } NO_3^- + 10H^+ + 8e^- \longrightarrow NH_4^+ + 3H_2O \qquad (4.4)$$

The reducing power (as protons and electrons) is supplied for nitrate reductase by NADH whilst reduced ferredoxin is the reductant for nitrite reductase. As the equations (4.2) to (4.4) show, a total of 10 reducing equivalents is needed for every molecule of NO_3^- fully reduced to the ammonium level – a considerable energy input. The reactions can take place in the roots so that organic N is transported to the shoot, or in the leaves. In roots, or in the dark, the reduction of nitrate uses reduced coenzyme from respiratory reactions and nitrate competes with O_2 and acts as the terminal H and electron acceptor, so that the RQ (respiratory quotient) rises when nitrate reduction is stimulated. In photosynthetic tissues in the light, photosynthesis supplies the reductant and the reduction of nitrate competes with CO_2 reduction so that the assimilatory quotient of CO_2/O_2 falls. Nitrate reductase is an inducible enzyme; its levels rapidly increase in response to an increase in the nitrate supply whereas in plants having ammonia as the N source activity is very low or undetectable.

The ammonium is incorporated into organic molecules mainly by reacting with the amino acid glutamate to produce the amide glutamine, a molecule with two amino groups; the glutamine then donates one amino group to the acid 2-oxoglutarate, resulting in two molecules of glutamate:

$$\text{2-oxoglutarate} + \text{glutamine} \longrightarrow \text{glutamate} + \text{glutamate} \qquad (4.5)$$

Other amino acids can be produced from glutamate by aminotransferase reactions, and the amino acids form starting points for syntheses of all other nitrogenous compounds of the cells.

4.4.2 Symbiotic nitrogen fixation

Whilst the atmospheric N_2 is inert with respect to flowering plants, there are numerous prokaryotes that can utilize gaseous N_2 and fix it into the form of ammonia. In a limited number of flowering plant species symbiotic relationships with N_2-fixing prokaryotes have evolved. These symbioses mainly involve heterotrophic bacteria, but there are a few symbioses also with the photosynthetic cyanobacteria. The most widespread and most fully studied symbioses occur between legumes (i.e. plants of the family Fabaceae, the pea and bean family, formerly called the Leguminosae) and bacteria of the family Rhizobiaceae, especially the genera *Rhizobium* and *Bradyrhizobium*. The bacteria reside in nodules on the plant roots but are also able to live saprophytically in the soil. When living symbiotically the bacterium receives organic nutrients from the plant while the plant gains ammonium ions from the symbiont.

The symbiotic association begins with the attraction of the bacteria to young roots of the plant by flavonoids, iso-flavonoids and betaines diffusing from the root; these activate bacterial genes and the bacteria in turn produce chemicals which induce curling of root hairs and stimulate nodule formation (Fig. 4.6). The bacteria attach to the root hairs, and invade the hair cells. Breakage of the hair cell's wall is the combined result of wall loosening due to auxin secretion by the bacteria and the action of wall-digesting enzymes of the plant. The bacteria become enclosed within an infection thread formed by the host plant. The thread grows into the root and discharges the

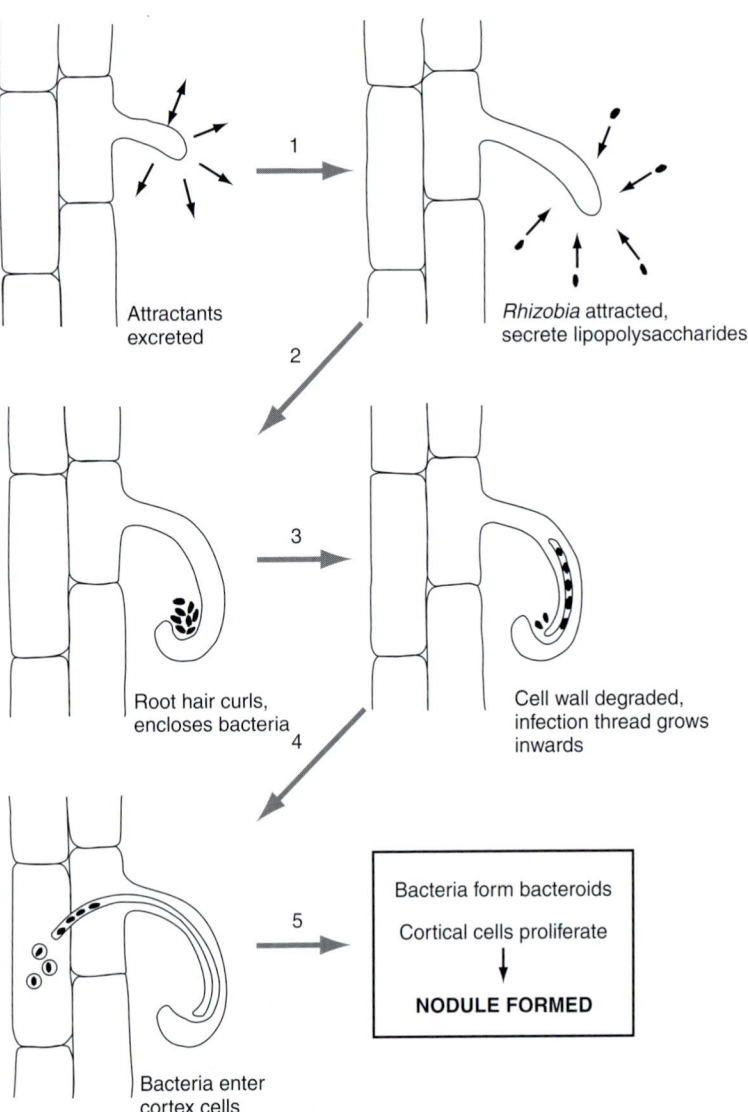

Fig. 4.6 Infection of a legume root by *Rhizobium*. Each stage leads to the next by a complex interaction between host plant and bacteria. Chemicals from the plant attract the bacteria and activate genes in the bacterium, which responds (1) (among other stimulated activities) by synthesizing chemicals that induce further gene activity in the plant, leading to (2) curling of the root hair (3) forming of infection thread (4) entry of bacteria into cortex, and eventually (5) nodule formation. The infection thread is an ingrowth of plasma membrane and cell wall material of the plant cell.

Attractants excreted

Rhizobia attracted, secrete lipopolysaccharides

Root hair curls, encloses bacteria

Cell wall degraded, infection thread grows inwards

Bacteria form bacteroids

Cortical cells proliferate

NODULE FORMED

Bacteria enter cortex cells

bacteria into a cell of the root cortex. There the bacteria grow and multiply, taking on a form known as **bacteroids**, much larger than the free-living bacteria (Fig. 4.7). Each bacteroid or group of a few bacteroids forms a **symbiosome,** surrounded by a peribacteroid membrane produced by the plant cell, which separates the bacteria from the host cell. As the bacteroids multiply, the cortical cells also

A

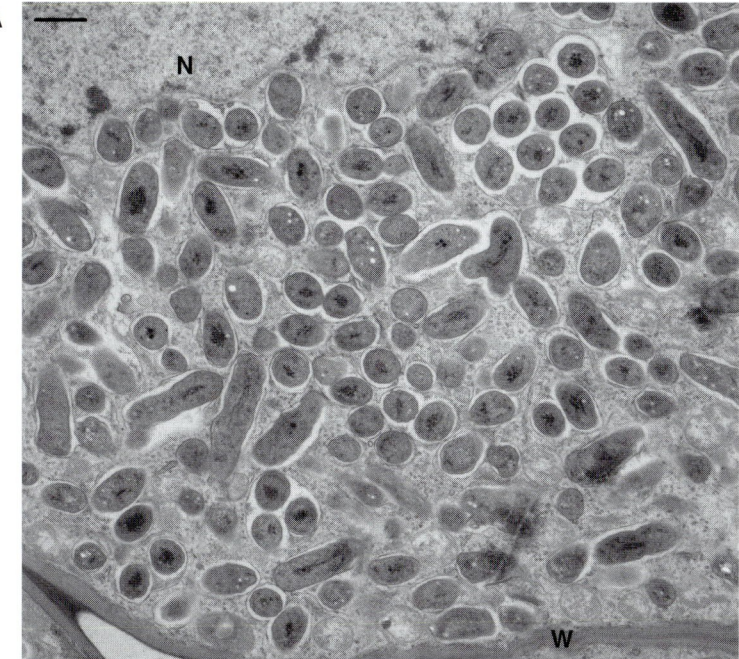

B

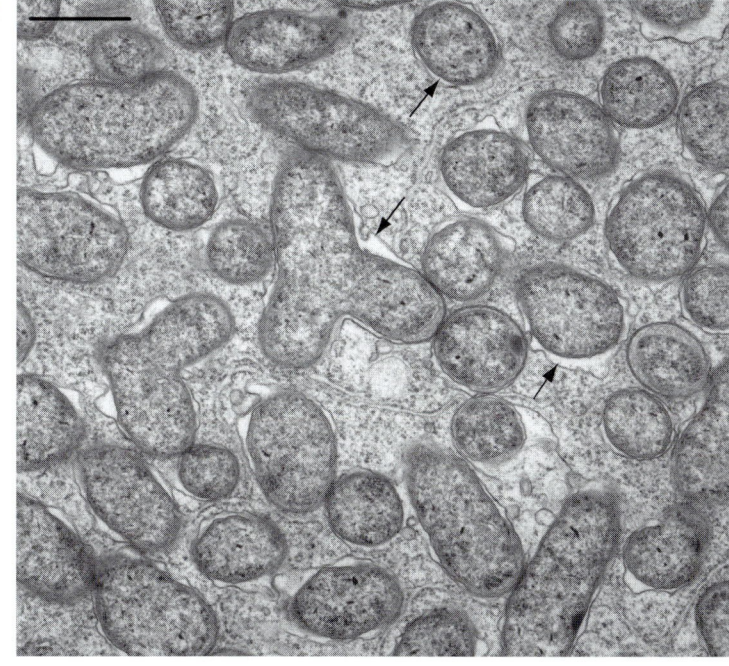

Fig. 4.7 (A) Low-power transmission electron micrograph of bacteroids in a root nodule cell of cowpea (*Vigna unguiculata*); N = host cell nucleus; W = cell wall. (B) Higher–power view of a nodule cell of pea (*Pisum sativum*); the peribacteroid membrane can now be clearly seen (arrows). Scale line = 1 μm in each case. Micrographs kindly provided by Dr Euan James, CHIPs, University of Dundee, Scotland.

divide and form the nodule, shape and size depending on the legume species. The infected cells fill the central part of the nodule and are characteristically tetraploid.

The fixation of N_2 is catalysed in the bacteroids by the enzyme **nitrogenase**, an enzyme present only in prokaryotes. There are several forms of this enzyme; the type found in the symbionts contains Mo, like nitrate reductase, and also Fe. It carries out a multistep process reducing the N_2 to ammonium:

$$N_2 + 10H^+ + 8e^- \longrightarrow 2NH_4^+ + H_2 \tag{4.6}$$

Sixteen ATP are also used per molecule of N_2 fixed. Thus, like nitrate reduction, the process needs a large supply of reducing power (H^+ and electrons delivered by coenzymes) and a large supply of ATP. Some bacterial strains are able to oxidize the H_2 produced as a by-product (Equation 4.6), yielding some ATP, but even then most of the ATP and all the reductant are produced by aerobic respiration. Nitrogenase is, however, irreversibly denatured by O_2. The nodule must fulfil two contradictory requirements: the enzyme nitrogenase must be protected from O_2 while the high demand for ATP and reduced coenzymes requires a high rate of aerobic respiration and by implication a high O_2 flux into the nodule. The nodules contain **leghaemoglobin**, a nodule-specific protein resembling vertebrate haemoglobin and with a very high affinity for O_2. The leghaemoglobin is found in the host cell cytosol where it combines with O_2 coming from the outside so that the concentration of free O_2 in the infected cells is kept low but O_2 is delivered to the bacteroid at the peribacteroid membrane in response to demand from respiration. The leghaemoglobin gives nodules a pink colour. The protein of the leghaemoglobin is coded for by the host cells, but the gene is expressed only in the nodules and the haem part of the molecule is synthesized by the bacterium. The outer layer of nodules is also such as to limit the diffusion of gaseous O_2.

The ammonium ions which diffuse from the bacterium to the host cells are used to synthesize glutamine as described above (Section 4.4.1). In some species this reacts to transaminate aspartate to its amide asparagine, which is transported from the nodules. In other species there is a more complex biochemical pathway producing ureides such as allantoin for transport. These compounds have high N : C atomic ratios, 1 : 2 for asparagine and 1 : 1 for allantoin; hence they are efficient means of transporting organic N.

The legume–*Rhizobium* symbiosis is a highly developed system. Strains of *Rhizobium* are specific to legume species or even to cultivars of a species. The genetics of the symbiosis has been studied in detail. In the bacterium there are 20 *nif* genes involved in the N_2 fixation process which are also present in free-living N_2 fixers. In addition there are *nod* genes essential for infection and nodule formation and *fix* genes needed for fixation specifically by the symbiotic fixers. In flowering plant hosts about 60 genes involved with the symbiosis have been identified. The two partners of the symbiosis control genetic activity in each other, inducing the appropriate genes by chemical signals. For instance, plant mutants are found in which non-functional nodules are formed, the bacteria failing to synthesize nitrogenase: this indicates that a plant signal is needed for the synthesis of the bacterial enzyme.

While the legume–*Rhizobium* symbioses are the most intensely studied, they are not the only N_2-fixing symbioses known for flowering plants. The filamentous actinobacterium *Frankia* sp. forms N_2-fixing nodules in the roots of numerous woody dicotyledons, e.g. alders (*Alnus* spp.) and the bog myrtle (*Myrica gale*).

4.4.3 The nitrogen cycle

The cycling of N in the biosphere is shown in Fig. 4.8. For flowering plants it is the supply of available N in the soil that is critical.

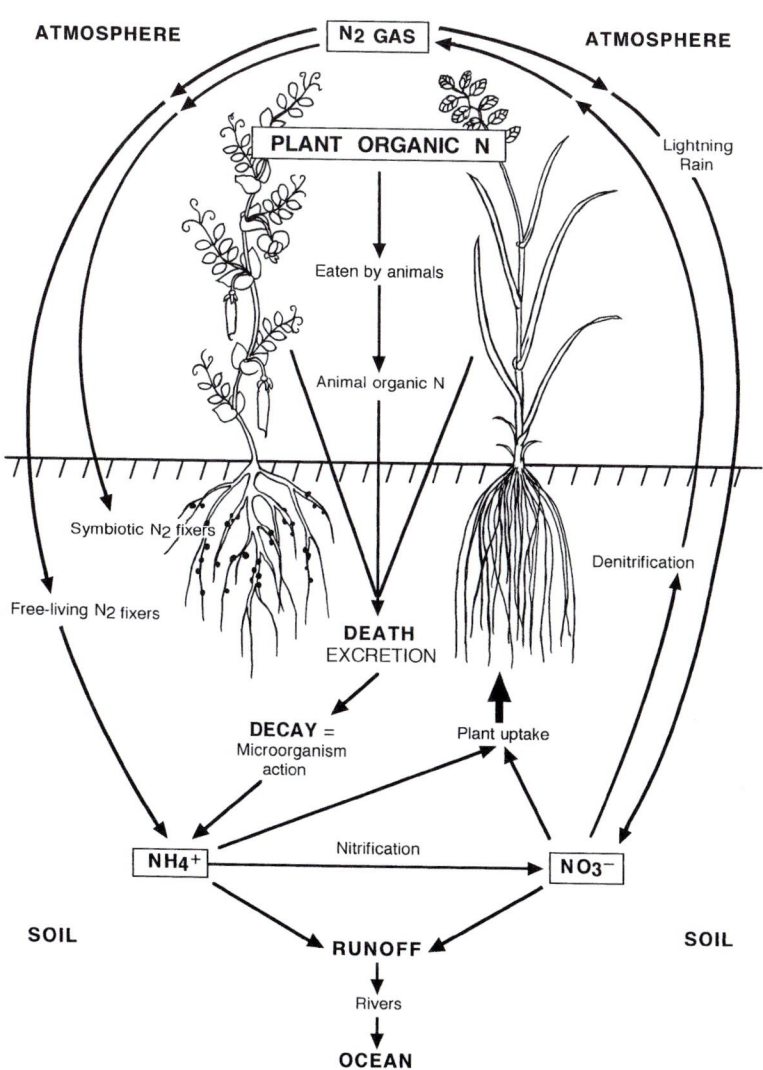

Fig. 4.8 The nitrogen cycle. The main N source for flowering plants is NO_3^-. The soil NO_3^- is **depleted** by plant uptake, by denitrification and by runoff. It is **replenished** by regeneration from decaying organic matter via ammonium, by N fixation (by symbiotic and free-living microorganisms) and by the action of lightning on atmospheric N_2. Processes omitted from the diagram are (1) addition of oxides of N to the atmosphere by volcanic activity and by industrial pollution, and (2) addition to the soil of nitrogen fertilizer derived from industrial N fixation from atmospheric N_2.

Death and activity of microorganisms return the N of plants and other organisms to the soil as ammonia, which dissolving in soil water yields ammonium ions. Some of the ammonia is converted to nitrate by nitrifying bacteria, e.g. *Nitrosomonas* and *Nitrobacter*, in the soil. But nitrate and ammonium ions are very soluble and easily leached from the soil by rain into bodies of water; ammonia is volatile and escapes to the atmosphere. Moreover the soil contains also denitrifying bacteria, e.g. *Thiobacillus denitrificans*, which convert nitrate to N_2 gas or to nitrous oxides. The N supply in the soil, on which flowering plants depend, and hence on which most other terrestrial organisms also depend, needs an input over and above that which comes from the direct recycling of organic material. There are no nitrogenous minerals in the soil. The supply comes from the atmospheric N_2 via N_2-fixing microorganisms, of which the legume–*Rhizobium* symbioses account for 25–50%. The rest comes from the activities of symbioses involving non-leguminous plants, and from free-living N_2-fixing bacteria, e.g. *Azotobacter* and *Clostridium*. A non-biotic fixation is achieved by lightning: the electric discharges convert N_2 to oxides of N which dissolve in rainwater to give nitrite and nitrate ions.

4.5 | Problems with mineral elements: deficiency and toxicity

4.5.1 Nutrient deficiencies

When the supply of any of the essential elements falls below a minimal level, plants usually show visible deficiency symptoms and the yield of cultivated plants can be severely reduced; in extreme cases the plants die. Many economically important and symptom-characteristic plant diseases are recognized as resulting from mineral element deficiencies. These have been given names descriptive of the visible symptoms of the deficiency. Examples are 'tea-yellows'– S deficiency of tea; 'grey-speck' of oats (*Avena sativa*) – Mn deficiency; 'sickle leaf' of cocoa (*Theobroma cacao*) – Zn deficiency; 'brown heart' of swede (*Brassica napus*) – B deficiency; 'scald' disease of beans – Mo deficiency. Texts are available with such symptoms described and photographed for a diagnosis to be made. Unfortunately, by the time the symptoms are fully developed a crop may be beyond recovery. Earlier diagnosis can be achieved by leaf analysis, or by checking for curative effects of foliar sprays containing various elements, applied to crops suspected of incipient mineral deficiency. Plant species which rapidly develop characteristic deficiency symptoms for particular elements are used as **indicator species** for testing suspect soils.

In natural habitats, the elements are recycled and returned to the soil by decomposition of dead organisms and by excretion from animals. Losses by leaching downwards with water drainage are made good by fresh solution from mineral particles. Large-scale harvesting of crops, however, removes minerals wholesale and necessitates the application of fertilizers to avoid the development of deficiencies. Even when deficiencies are not severe enough to produce disease symptoms, mineral supply may still limit growth and productivity. It is in the interests of producers to maximize the yield, and agriculture in the developed world makes extensive use of fertilizers to make good nutrient limitation as far as is feasible. The addition of fertilizers, however, creates its own problems. For maximal yield, fertilizer is added in excess of that absorbed by the plants; the surplus is leached by rainwater and is liable to find its way into rivers and lakes. The consequence can be **eutrophication**, giving excessive growth of algae and other water plants. The dense mat of plant material can physically choke up shallow water. When microorganisms in turn grow on and decompose the organic material produced, they deplete the water of O_2 and aquatic animals may die. Nitrate which is added in large amounts as a fertilizer is extremely soluble and, being an anion, is not retained by attraction to soil colloids so that it is particularly easily washed into bodies of water and has become a serious source of pollution directly endangering human health. Nitrate levels in the drinking water of some areas have become so high that the water is not recommended for drinking by babies.

The molecular biology of mineral uptake is being elucidated, and the genes for numerous transport proteins have been identified. Can this knowledge be utilized to decrease the dependence of cultivated crops on fertilizer? In addition to the deleterious effects of large-scale application of fertilizer, the supplies of minerals used for fertilizer are not inexhaustible. One promising line of investigation for genetic modification is to encourage the expression of genes coding for the high-affinity transport proteins, which would enable lower amounts of fertilizer to be used. The overexpression of single genes for NO_3^- and Pi uptake has been found to increase the uptake of N and P respectively, with an increase in plant biomass (Vance 2001). The idea of engineering more plants able to support N_2 fixation, symbiotically or even independently, is also attractive, but the complexity of the process and the large number of genes involved present major problems. Not only must all the biochemical reactions be coded, but nitrogenase must be protected from O_2. Some tropical C4 grasses have a loose association with a N_2-fixing genus of soil bacteria, *Azospirillum*. The bacteria are attracted to the vicinity of the roots by excreted organic compounds; nitrogenous compounds released by the bacteria are absorbed by the plant. This type of association might be easier to engineer. Some of these problems are discussed in a review by Vance (2001).

4.5.2 Mineral element toxicity

Most of the micronutrient elements can become toxic to plants in excess, i.e. when soils contain more than micronutrient amounts – B, Co, Cu, Mo, Ni, Zn – and many non-essential elements are toxic, especially heavy metals – Ag (silver), As (arsenic), Cd (cadmium), Cr (chromium), Hg (mercury) and Pb (lead). At the cellular level, toxic metals cause oxidation of protein –SH groups to –S--S– bridges, oxidation of lipids and damage to DNA. The same elements are also toxic to organisms from other kingdoms, so that plants growing on metal-rich areas can cause poisoning when eaten. Metal-contaminated areas occur naturally; e.g. serpentine soils are characterized by high levels of Co, Cr, Fe, Mg, Mn and Ni whilst calamine soils are enriched with Cd, Zn and often Pb. Others are the result of mining, ore extraction, use of agricultural sprays, and contamination with lead from petrol. The main elements involved are Cu, Zn, Ni, and Pb. Species vary in their sensitivity to heavy metals; areas of high heavy metal content exhibit a characteristic flora of tolerant species. Some species are naturally confined to sites of unusually high heavy-metal content and may be resistant to more than one toxic element, e.g. *Viola calaminaria*, *Thlaspi caerulescens* and *Minuartia verna*. Such species are at a high competitive advantage in such sites, but the metals are not directly beneficial to them. Many species of flowering plants have metal-tolerant ecotypes growing in areas of high metal content. The evolution of such ecotypes can be rapid, over a few generations, and it has accordingly been possible to breed and select metal-tolerant races of numerous species for the revegetation of many areas of industrial dereliction. The tolerance depends to a large extent on the exclusion of metals from the cytosol by binding of the metals in the cell walls, which possess binding sites specific to individual metals, and by vacuolar storage of the metal ions chelated with various organic compounds. Since the metals are absorbed by the plants, the contaminated areas can be landscaped, but they are not suitable for production of crops, even if tolerant strains could be developed, for these would contain the elements at unacceptable concentrations. However, there is the potential for using the plants to 'mine' the metals out of contaminated soil, a process known as **phytoextraction**. Some of the tolerant species are **hyperaccumulators**, taking up large amounts: 20% of Ni by dry weight has been found in the latex of a tree, *Sebertia acuminata*! This of course leaves the problem of disposing of the accumulator plants at the end of the growing season. It may be possible to recover and use the metal from ashed plant material.

Complementary reading

Beevers, L. *Nitrogen Metabolism in Plants*. New York, NY: Elsevier, 1976.
Blevins, D. G. & Lukaszevski, K. M. Boron in plant structure and function. *Annual Review of Plant Physiology and Plant Molecular Biology*, **49** (1999), 481–500.

Brown, P. H., Bellaloui, N., Hu, H. & Dandekar, A. Transgenically enhanced sorbitol synthesis facilitates boron transport and increases tolerance of tobacco to boron deficiency. *Plant Physiology*, **119** (1999), 17–20.

Crawford, N. M. Nitrate: nutrient and signal. *The Plant Cell*, **7** (1995), 859–868.

Dupont, F. M. & Leonard, R. T. The use of lanthanum to study the functional development of the Casparian strip in corn roots. *Protoplasma*, **91** (1977), 315–23.

Epstein, E. The anomaly of silicon in plant biology. *Proceedings of the National Academy of Sciences (USA)*, **91** (1994), 11–17.

Flowers, T. J. & Yeo, A. R. *Solute Transport in Plants*. Glasgow: Blackie, 1992.

Imsande, J. & Touraine, B. N demand and the regulation of nitrate uptake. *Plant Physiology*, **105** (1994), 3–7.

Krämer, U. (2000). Cadmium for all meals – plants with an unusual appetite. *New Phytologist*, **145** (2000), 1–5.

Raghothama, K. G. Phosphate acquisition. *Annual Review of Plant Physiology and Plant Molecular Biology*, **50** (1999), 665–93.

Raskin, I. & Ensley, B. D. *Phytoremediation of Toxic Metals: Using Plants to Clean Up the Environment*. New York, NY: Wiley, 2000.

Read, D. J. & Perez-Moreno, J. Mycorrhizas and nutrient cycling in ecosystems – a journey towards relevance? *New Phytologist*, **157** (2003), 475–92.

Smith, R. J., Lea, P. G. & Gallon, J. R. Nitrogen fixation. In *Plant Biochemistry and Molecular Biology*, 2nd edn, ed. P. J. Lea & R. C. Leegood. Chichester: Wiley, 1999, pp. 137–62.

Smith, S. E. & Read, D. J. *Mycorrhizal Symbiosis*, 2nd edn. London: Academic Press, 1997.

Terry, N., Zayed, A. M., de Souza, M. P. & Tarun, A. S. Selenium in higher plants. *Annual Review of Plant Physiology and Plant Molecular Biology*, **51** (2000), 401–32.

van der Heijden, M. G. A., Klironomos, J. N., Ursic, M. *et al.* Mycorrhizal fungal diversity determines plant biodiversity, ecosystem variability and productivity. *Nature*, **396** (1998), 69–72.

Vance, C. P. The molecular biology of N metabolism. In *Plant Metabolism*, 2nd edn, ed. D. T. Dennis, D. H. Turpin, D. D. Lefebvre & D. B. Layzell. Harlow: Addison Wesley Longman, 1997, pp. 449–77.

Zhao, F. J., Lombi, E., Breedon, T. & McGrath, S. P. Zinc hyperaccumulation and cellular distribution in *Arabidopsis halleri*. *Plant, Cell and Environment*, **23** (2000), 507–14.

References

Bell, C. I., Clarkson, D. T. & Cram, W. J. (1995). Sulphate supply and its regulation of transport in roots of a tropical legume *Macroptilium atropurpureum* cv. Siratro. *Journal of Experimental Botany*, **46**, 65–71.

Glass, A. D. M. & Dunlop, J. (1979). The regulation of K^+ influx in excised barley roots. *Planta*, **145**, 395–7.

Marschner, H. (1995). *Mineral Nutrition of Higher Plants*, 2nd edn. London: Academic Press.

Mendel, R. R. & Hänsch, R. (2002). Molybdoenzymes and molybdenum cofactors in plants. *Journal of Experimental Botany*, **53**, 1689–98.

Tester, M. (1990). Plant ion channels: whole-cell and single-channel studies. *New Phytologist*, **114**, 305–40.

Tester, M. (1997). Techniques for studying ion channels: an introduction. *Journal of Experimental Botany*, **48**, 353–9.

Vance, C. P. (2001). Symbiotic nitrogen fixation and phosphorus acquisition: plant nutrition in a world of declining renewable resources. *Plant Physiology*, **127**, 390–7.

Chapter 5

Translocation of organic compounds

5.1 Introduction

Flowering plants are described as being **autotrophic**, 'self-feeding', capable of synthesizing all their organic material via photosynthesis. But a flowering plant is a complex organism with cells and organs specialized for diverse functions, and only the green photosynthetic cells are truly autotrophic; they must accordingly supply all the non-photosynthetic parts with organic carbon. Over small distances, i.e. between individual cells and within small groups of cells, chemicals can move by diffusion through plasmodesmata, or across plasma membranes by diffusion and by active transport. But organic materials must move for long distances; the growing tips of the roots of a tree are many metres away from the nearest photosynthetic leaves and even in a herbaceous plant diffusion would be too slow for the distances involved. We have already seen (Chapter 3) how water moves in plants over long distances in a specialized transport tissue, the xylem. The subject of this chapter is the long-distance, multi-directional movement or **translocation** of organic compounds which takes place in the phloem.

5.2 Phloem as the channel for organic translocation

5.2.1 Evidence for translocation in the phloem

In flowering plants, the xylem is regularly associated with the **phloem**, the two together making up the vascular tissues. In young organs the two tissues are in contact; when secondary growth occurs they become separated by the vascular cambium, the meristem which then adds xylem to one side and phloem to the other.

In woody stems, where the vascular tissues form complete cylinders, it is fairly easy to cut through the outer stem tissues down to the vascular cambium and to remove the 'bark', which includes the phloem, leaving the central xylem intact. Regions of defoliated stem separated from all leaves by such 'bark rings' become deficient in carbohydrates, but the transport of water and minerals continues, showing that xylem transport is still functional. This suggests that the transport of carbohydrates occurs in the phloem. Admittedly the cortex (if present) and periderm are also removed in the bark ring; but these tissues do not contain cells which are structurally suited for long-distance transport. A variation on debarking is to kill the living tissues in a length of woody stem by the application of hot steam. The heat-killed ring stops the translocation of organic substances whilst still permitting xylem transport. This type of experiment shows that organic material is translocated through *living* cells, as are found in the phloem, in contrast to xylem, which functions with dead conducting cells. The high concentration of organic compounds in phloem sap also strongly supports the postulate of its function in the transport of organic nutrients.

The most direct evidence for phloem as the channel of organic translocation comes from the use of tracers. Various fluorescent dyes, such as fluorescein and its derivatives, can be directly observed under the microscope to move in the phloem. At first this observation was regarded with caution, since these are artificial compounds and might move along paths different from those of natural metabolites. Final confirmation has come from radioactive labelling. When radioactive CO_2 is supplied to photosynthesizing leaves, the radioactivity soon appears in the phloem of the petiole and the stem, as radioactive sugars. Here there has been no introduction of any foreign substance, nor any interference with the plant's normal activity, and the data prove that the products of photosynthesis move from their sites of production in the phloem. Radiolabelling has shown that the phloem is also the pathway of translocation out of non-photosynthetic storage organs. Since it has been established that naturally transported metabolites and numerous fluorescent dyes move along the same pathway, the dyes are now frequently used as tracers for phloem transport.

5.2.2 The structure of phloem

The phloem of flowering plants consists of several types of cell. Tracer experiments have shown that, at the cellular level, translocation proceeds through the **sieve tubes**, built of longitudinal files of individual *sieve tube elements* (sieve tube cells). The sieve tube diameter usually lies between 10 and 50 μm and the length of individual cells is 150–1000 μm, but in palms diameters of 400 μm and lengths of 5000 μm have been reported; in minor veins, however, sieve tubes can be very narrow, below 2 μm. The transverse or oblique end walls between the individual sieve tube cells, some 0.5–2 μm thick, are pierced by pores giving them a sieve-like appearance and are known as the **sieve plates**, hence the cells' name (Fig. 5.1). The diameter of the sieve plate pores is extremely variable between species and in different parts of a plant. The narrowest are only 0.1 μm wide; 0.5–1.0 μm might be considered an 'average' value. But in the Cucurbitaceae (marrow family) pore diameters up to 10 μm are found, and 14 μm has been reported for *Ailanthus altissima* (tree of

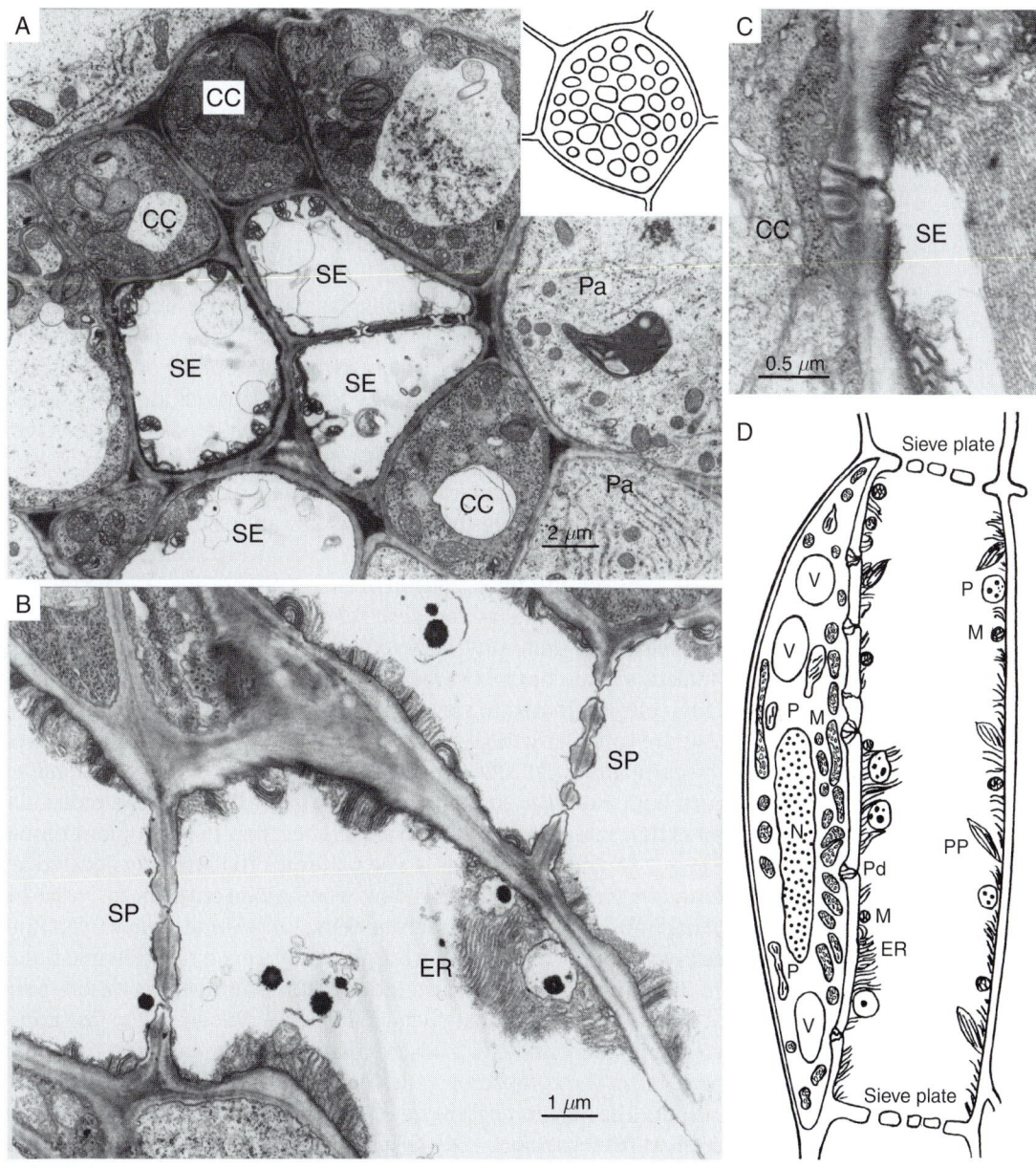

Fig. 5.1 The fine structure of phloem. (A), (B) and (C) are transmission electron micrographs. SE = sieve tube element; CC = companion cell; Pa = parenchyma; SP = sieve plate. *Loading* phloem is illustrated, hence the narrow diameter of the SE. (A) Transverse section of part of a vascular bundle in the cotyledon of a French bean seedling (*Phaseolus vulgaris*). All the SE are almost devoid of contents except for some mitochondria and indistinct peripheral material. The CC are full of cytoplasm containing mitochondria, plastids and a high density of ribosomes. Inset: diagram of a wide-pored cucurbit sieve plate as seen in surface view. (B) Longitudinal section of sieve tubes in a leaf of mung bean (*Vigna radiata*); two sieve plates, with open pores, are visible. At this magnification, peripheral material is recognizable as endoplasmic reticulum, ER, and P-protein fibrils; there are a few mitochondria and a number of plastids with black-stained starch grains; some plastids have burst.
(C) Branched plasmodesma between a SE and its CC. Three distinct branches and the suggestion of two more can be seen on the CC side, but the thinness of the section means that some branches may have been missed.
(D) The SE–CC complex: composite diagram based on electron microscopy and confocal laser microscopy. M = mitochondrion; N = nucleus; P = plastid; Pd = plasmodesma; PP = P-protein; V = vacuole in CC.

heaven). Associated with the sieve tube elements there often are **companion cells**, one or more to each sieve tube cell, lined up longitudinally beside the conducting cells. The sieve tube elements have a very unusual structure (Fig. 5.1 A, B, D). In the mature state they lack nuclei and the only organelles identifiable are small plastids, sparse mitochondria and endoplasmic reticulum, but these organelles do not occupy much of the cellular volume, lying next to the cell walls. Most of the lumen is filled with a kind of sap but there is no tonoplast and no demarcation of a vacuole. The cells are still, however, bounded by a functional plasmalemma which also lines the sieve plate pores. The sap contains P-proteins (P for phloem), sometimes visible by light microscopy as protein bodies or longitudinal strands, though at least some of the reported strands are now thought to be damage artefacts. The companion cells on the other hand have a full complement of organelles (Fig. 5.1 A, D). They are rich in cytoplasm, with small vacuoles and very numerous, highly cristate mitochondria, which gives them a high potential for metabolic energy production. The sieve tube cell and its adjacent companion cell are derived by longitudinal division of the same mother cell and the pair is often referred to as the sieve element–companion cell (SE–CC) complex. It is generally believed that the companion cell with its nucleus exerts control over the enucleate sieve element. The companion cells are joined to sieve tube cells by abundant plasmodesmata that are unusually large on the sieve element side and branched on the companion cell side (Fig. 5.1 C). These plasmodesmata also have a very high *molecular exclusion limit*, i.e. a high limit for the molecular mass that can pass through, much higher than in other plant tissues (see p. 138 and Section 5.4.2). The materials that are translocated are found in the companion cells, too; when radioactive photosynthate is being translocated, the companion cells also become radioactive. Other cell types found in the phloem are phloem parenchyma cells and fibres.

There is some difference in the structure of phloem in various parts of the plant. Photosynthate is loaded into the phloem in the fine minor veins of leaves, and this *loading (collecting) phloem* has sieve tubes narrower than the companion cells. In the petioles, stems and older parts of roots, the *transport phloem* sieve tubes are wider than the companion cells. In the *unloading (release) phloem*, where the solutes leave the transport system, the companion cells are very small and may be absent altogether. There are moreover two basic mechanisms of phloem loading, and phloem structure varies accordingly (Section 5.4).

In most perennials with secondary growth, the sieve tubes and their companion cells die after one growing season and are replaced by the cambium in the next season. In perennial monocotyledons, such as the palms, which lack secondary growth, sieve tubes persist many years; in palms 50-year-old sieve tubes have been identified. In a few dicotyledons, too, e.g. lime (*Tilia* sp.) and grapevine (*Vitis vinifera*), the phloem conducting cells survive for several seasons. In such species, the sieve plate pores may be blocked during the winter by a deposit of dormancy callose, a polysaccharide of glucose subunits.

5.2.3   The composition of phloem sieve tube sap

Obtaining uncontaminated samples of phloem sieve tube sap in quantities adequate for analysis presents problems. Species differ greatly in the ease with which they yield phloem sap. Sometimes exudate is obtained by cutting into the phloem, but this poses the risk of contamination with the contents of other cells. Very often sap fails to exude at all from cut phloem, although exudation may be elicited by mechanical stimulation prior to cutting. The most reliable method of obtaining sieve tube sap from numerous species is to use phloem-feeding aphids to tap the sieve tubes. These insects feed by inserting their stylets into sieve tubes from the outside. When the insertion is accomplished, the insect is cut away under anaesthesia. Sap will then continue to drip from the stylets for up to several days, at a rate of 1–2 µL per hour. This liquid has been assumed to represent more-or-less unadulterated sieve tube sap. There is evidence that the saliva initially released by the aphid exerts no digestive function on the phloem and the composition of the sap certainly remains unchanged over many hours of collection. There still does remain the possibility of some continuous seepage of liquid from surrounding tissues into the tapped sieve tube units.

Sap analyses vary markedly from species to species. The sap is quite viscous, reflecting its high content of organic solutes; sugars generally make up 90% of the solids and may be present at concentrations of 2–25% w/v; the water potential is correspondingly low, from –0.6 to –3.4 MPa. In the majority of species, **sucrose** is the main sugar, often present at 0.4–0.5 mol L^{-1}, with traces of the oligosaccharides raffinose (a trisaccharide), stachyose or verbascose (two tetrasaccharides); but in some species one of these oligosaccharides predominates. In yet other species, the predominant carbohydrate is a polyol (sugar alcohol) such as sorbitol, dulcitol or mannitol. Glucose on the other hand is found only in very low concentrations and may be undetectable.

Amino acids and amides are regularly present in phloem sap, amounting to 0.2–12% of the transported solutes, with glutamate, glutamine, aspartate and asparagine being the most abundant. In the Cucurbitaceae soluble nitrogenous compounds make up a high percentage of the solids, and in perennials nitrogenous compounds are abundant at certain times (see below, seasonal patterns). Phloem which translocates materials out of seed storage tissues can have high levels of nitrogenous compounds; in *Ricinus communis* seedlings, a total amino acid/amide content of 0.16 mol L^{-1} has been found, more than half the sucrose content of 0.27 mol L^{-1}. In species where nitrate reduction (Chapter 4) takes place mainly in the leaves, roots are dependent for their N supply on amino acids and amides translocated down in the phloem. Protein is detectable in the sap in variable amounts and will be discussed separately. Sulphate reduction is located in leaves, roots receiving S-containing amino acids via the phloem. In some plants, different regions of the phloem transport different materials; in cucurbits, phloem within minor leaf veins transports mainly carbohydrate, whereas phloem strands outside the veins carry mainly amino acids.

Some mineral elements are found in the phloem sap; K^+ is the predominant ion, reaching concentrations of 0.03–0.5 mol L^{-1}. ATP is a regular constituent. In small quantities, many other compounds have been detected including organic acids, hormones and secondary products; numerous plant viruses, too, spread in the phloem. The pH is usually alkaline at 7.5–8.6, although in perennials it may be faintly acid in the spring. This contrasts with xylem sap and vacuolar saps, which are typically acid.

The carbohydrate concentration of phloem sap derived from photosynthesizing leaves is strongly dependent on the rate of photosynthesis, and hence on weather. It also exhibits regular diurnal variation. In the cotton plant (*Gossypium barbadense*), the highest concentration in the stem is recorded in the latter part of the day. In a number of trees, the highest concentration occurs at night and close to the leaves, a concentration wave moving down the tree.

In perennials there are seasonal patterns in phloem exudation and sap composition. In several tree species, abundant exudate is obtained in late summer but no flow occurs before about mid-June. Marked seasonal changes are found with respect to amino acid content; this is high in spring, drawing on N stored in woody tissues over winter, falls in the summer, and rises to a second peak in the autumn, when leaf proteins break down prior to abscission. The N translocated out of leaves at this time is deposited as organic nitrogenous compounds in the stems, where it remains stored during dormancy. Carbohydrate, too, may be transported to woody stems for storage.

Proteins and RNA in phloem sap

Phloem sap contains proteins, mostly not more than 0.1 mg mL^{-1}, but in the Cucurbitaceae the values reach 10–60 mg mL^{-1}. Gel chromatographic methods have revealed a large number of sap proteins, with molecular masses from 9000 to 200 000 Da; over 200 protein species have been detected in wheat (*Triticum aestivum*) phloem sap, and individual species exhibit specific patterns. RNA including mRNA can also be detected. Since the sieve tube elements lack a system for transcription or translation, the proteins and RNA must be synthesized in the companion cells and enter through the connecting plasmodesmata. Such movement is vividly demonstrated in transgenic plants programmed to synthesize green fluorescent protein (GFP) in leaf companion cells; the fluorescence appears in the sieve tubes. A very interesting finding is that some proteins in phloem sap have the property of *increasing* the exclusion limits of plasmodesmata. Proteins from *Cucurbita maxima* phloem sap were microinjected into individual cells in *Cucurbita* cotyledon mesophyll, along with fluorescently labelled dextrans, i.e. high molecular mass carbohydrates (Balachandaran *et al.* 1997). These dextrans showed no cell-to-cell movement if injected by themselves. But in the presence of the phloem sap proteins, the spread of fluorescence indicated that dextrans with masses of at least 20 000 Da moved through up to 20 cells from the site of injection within 1–2 minutes. The accompanying phloem proteins, with masses of up to 100 000 Da, must have moved, too. The exclusion limit for numerous 'ordinary' plant tissues is 800–1000 Da. Phloem proteins from *Cucurbita* and castor bean (*Ricinus communis*) also increased the exclusion limits of plasmodesmata in the leaf mesophyll of other species, e.g. *Nicotiana tabacum*. The presence of such proteins in the SE–CC complex presumably maintains the high exclusion limits of the plasmodesmata. Additionally, some proteins in the SE–CC complex may act as chaperonins and unfold large proteins for passage.

Some of the proteins in phloem sap have been identified as enzymes, or as components of the P-protein filaments, and together with others as yet unidentified are presumably necessary for the maintenance of the life functions of the sieve tube cells. The question then arises how such macromolecules avoid being swept with the translocation stream into the sinks: maybe by adsorption on the P-protein fibrils, or to the peripheral ER (endoplasmic reticulum)? But proteins do move in the phloem, even through graft unions. Evidence is mounting that some sieve tube sap proteins and RNA are information molecules destined for transport to the sinks, where they are unloaded (Oparka & Santa Cruz 2000). These information molecules include, for instance, signals (believed to be small RNA molecules) which can suppress the activity of specific genes in the sinks. The phloem is emerging as an important carrier of macromolecular information as well as hormonal signals.

5.3 | The rate and direction of translocation

5.3.1 The rate of translocation: velocity and mass transfer

One measure of the rate of translocation is **velocity**, the distance moved by the translocated material per unit time, expressed in, say, cm h^{-1}. This seems a simple value, but it is difficult to measure. Most estimates to date have been made by introducing a marker into the phloem at a specified point and noting the time of its arrival at another specified point. Fluorescent dyes have been timed, but there is a time delay of unknown length between the external application of the dye and its entry into the sieve tubes, and the dye, as a foreign substance, might conceivably adversely affect the phloem and change the velocity of translocation. The timing of the movement of radioactivity from ^{14}C-labelled photosynthate avoids the introduction of foreign material, but is not straightforward; among other problems, ^{14}C is a weak β-emitter and to detect its presence, plant segments have to be extracted for assay; it is not feasible to hold a detector to the plant. More recently the measurement of nuclear magnetic resonance (NMR) has been utilized; NMR depends on the magnetic properties of atomic nuclei. This technique is used on living plants and causes no damage. The specimen is placed within a detector which gives information about the presence, concentration and movement of selected atoms in known locations in the plant – in this case, the H atoms of the water molecules in the phloem. Actual values of the velocity of translocation, measured in different specimens and using a variety of methods, range from below 10 to 660 cm h^{-1}, with woody plants coming in the lower part of the range, at 20–30 cm h^{-1}. NMR measurements have yielded a value of 200 cm h^{-1} for castor bean (*Ricinus communis*) seedlings and 90 cm h^{-1} for the adult plants.

A somewhat different measure of the rate of translocation is **mass transfer**, the total mass of translocate moved per unit time. If traffic is used as an analogy, the mass transfer would be equivalent to counting the total number of cars passing an observation point in a given time interval; the velocity is equivalent to the speed at which the cars pass the observation point. The units for mass transfer are g h^{-1} (cf. velocity, cm h^{-1}). If the mass transfer is calculated per unit area of phloem, it is termed **specific mass transfer** (**SMT**) and expressed in g cm^{-2} h^{-1}. The (specific) mass transfer is measured by increases in dry weight in growing organs. It also is not easy to measure accurately, for a correction must be applied for the amount of translocate respired away in the organ, and for SMT precise measurements of transporting cell cross-sectional areas are required.

5.3.2 The direction of translocation

In contrast to movement in the xylem, which is a strictly one-way traffic, the direction of translocation in the phloem is variable. Physiologically it is described as passing '**from source to sink**', i.e. from organs of synthesis (or organs of storage) to organs of utilization. In a more-or-less mature leafy plant, the mature leaves are sources by virtue of their photosynthetic activity. The main sinks are the **growing regions**, which are numerous (Fig. 5.2). The roots and underground perennating organs such as tubers, corms and rhizomes require downward translocation through the shoot axis. In a young vegetative plant, most of the total translocate goes to support the rapidly growing root system. The growing shoot tips, on the other hand, require upward translocation, as do young leaves before they have developed their full photosynthetic capacity. In the reproductive stage, flowers and fruits need upward translocation, fruits in their most rapid growth stage receiving nearly all the translocated nutrients. In the appropriate season, underground storage organs mobilize their reserves and translocate the soluble products upwards to growing shoot organs. Mature axial tissues in stems, petioles and roots of course also need organic nutrients for their maintenance. They draw on the translocation flow as it passes through the axis and are termed **axial sinks**, to distinguish them from the **terminal sinks** in the growing regions. Generally, a sink is supplied by the nearest source; thus upper leaves tend to feed the

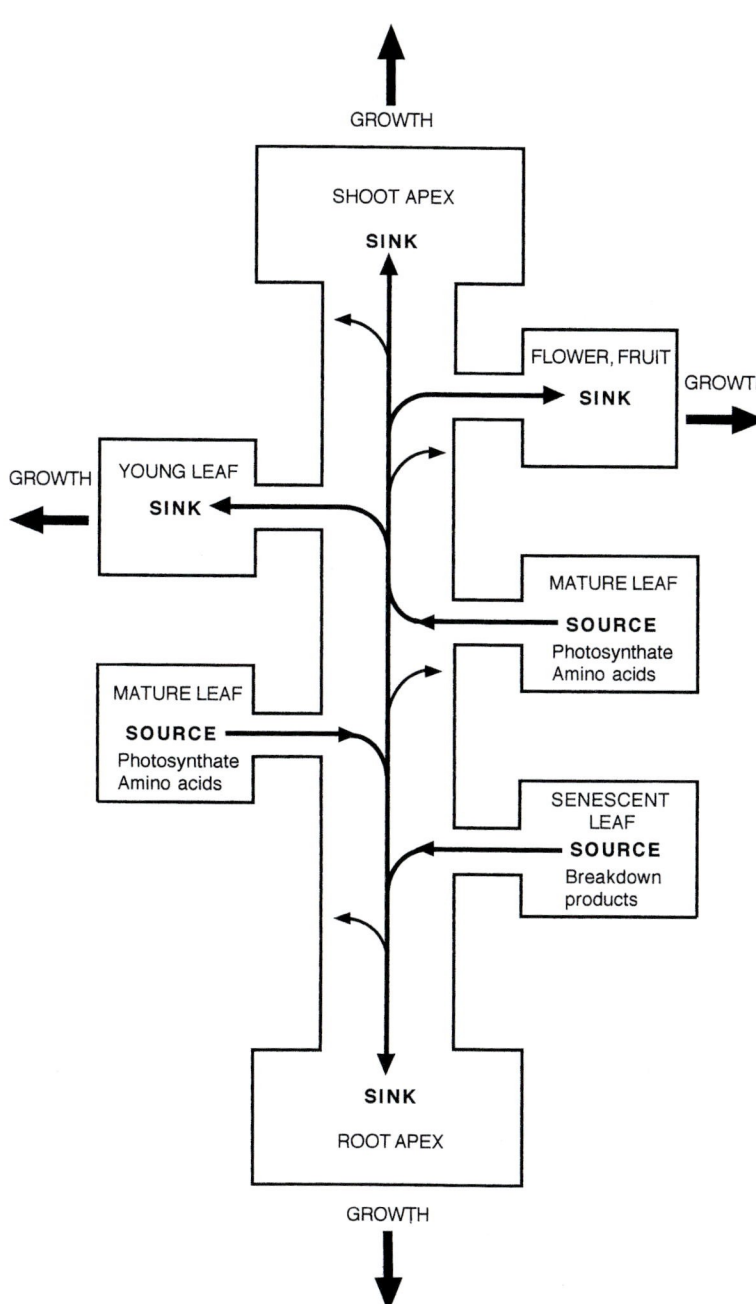

Fig. 5.2 Phloem translocation 'from source to sink' in a flowering plant. The small curved arrows indicate translocate moving into the axial sinks. Lower mature leaves feed (mainly) the roots, the higher mature leaves feed (mainly) the young leaves and the shoot apex. There are situations when the root, or some other underground storage organ, becomes the main source and the direction of transport is upwards from it.

shoot growing points, whilst the roots are fed mainly by the lower leaves. If, however, leaves close to fruits are removed, the fruits start to receive translocate from further away. Tracer experiments have shown that simultaneous upward and downward translocation can occur in the same part of an axis.

Since a source = organ of production where solutes are passed into the phloem, and a sink = organ of utilization where solutes are removed from the phloem, it might be postulated that the source-to-sink translocation represents movement along a concentration gradient. There is typically indeed a concentration gradient of translocated material, decreasing from source to sink. However, the situation is more complicated. If a mature leaf is shaded so that it falls below its light compensation point (Section 2.3.3) it might be expected to act as a sink, but in fact shaded mature leaves starve, senesce and die. Sugar transport towards a sink is sometimes faster than its utilization in the storage or reproductive organ to which it is moving and temporary accumulation of sugar occurs in petioles and fruit stalks without preventing or reversing the flow. It is obvious that there are metabolic controls for the direction of flow over and above simple concentration gradients. There must also be control of the amount of mass transfer to various sinks.

The overall direction of phloem translocation is under metabolic control; the precise pattern of movement is determined by the arrangement of the vascular tissue. Leaves generally export photosynthetic products to young leaves or floral organs directly above them, and to parts of the root system directly below them. This corresponds to the pattern of vascular connections. Lateral transport from one vascular strand to another hardly ever occurs in either root or stem, and usually the removal of leaves from one side of a plant results in asymmetric, lop-sided growth as a consequence of one-sided transport. Lateral diversion is possible in species with a suitably anastomosing vascular system, as occurs in e.g. the beetroot (*Beta vulgaris*).

5.4 | Phloem loading and unloading

5.4.1 Phloem loading

Phloem loading is the transfer of material into the phloem at the source; **unloading** is the removal of this material from the phloem in the sink. Of the various organs that can act as sources, the minor veins of leaves, which collect the photosynthate from the mesophyll cells, have been studied in most detail. Flowering plant species can be broadly divided into two groups according to their method of phloem loading in the minor veins: the **apoplastic loaders** and the **symplastic loaders**.

In the apoplastic loaders, sucrose is discharged from the mesophyll cells into the apoplast (the cell walls), possibly via carriers. Subsequent uptake into the SE–CC complexes is mediated by sucrose-proton symporters in the plasma membranes (Fig. 5.3); the loading process is thus driven by the proton pumps which pump protons into the apoplast at the expense of ATP hydrolysis (Section 4.3.4, p. 117). The apoplastic loaders can be anatomically recognized by having very few plasmodesmatal connections between the companion cells and phloem parenchyma or mesophyll cells; their SE–CC complex is a highly isolated unit, relying on membrane transport to move sucrose against a concentration gradient into the phloem. The companion cells of the

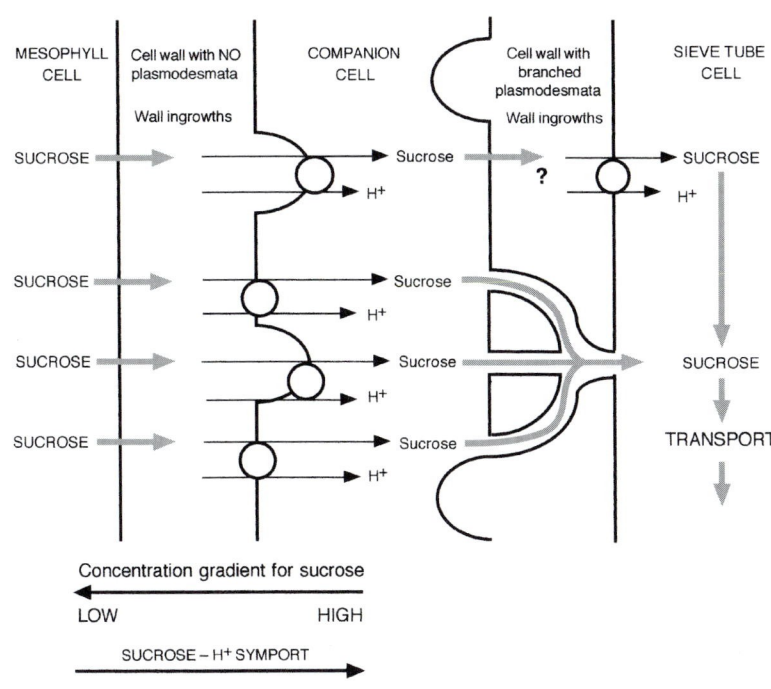

MESOPHYLL CELL | Cell wall with NO plasmodesmata | COMPANION CELL | Cell wall with branched plasmodesmata | SIEVE TUBE CELL

Wall ingrowths — Wall ingrowths

Concentration gradient for sucrose

LOW — HIGH

SUCROSE – H⁺ SYMPORT

Fig. 5.3 Apoplastic phloem loading. Sucrose moves (grey arrows) from a mesophyll cell into the cell wall (apoplast) adjoining a companion cell, with wall ingrowths, and is transported by a sucrose-proton symporter (circles) into the companion cell. The sucrose can then diffuse into the sieve tube cell via plasmodesmata, and there are also sucrose-proton symporters in the sieve tube cell plasma membrane which can achieve sucrose transfer. There is an energy input via ATP-splitting proton pumps which pump the protons into the apoplast, and wall ingrowths increase the cell surface area available for transport.

apoplastic loaders are typically of transfer cell structure, with wall infoldings and hence a large surface (i.e. plasma membrane) area. A number of sucrose transport proteins and their genes has been identified. For example, in rice seedlings (*Oryza sativa*), the sucrose transporter SUT1 is confined to the companion cells, implying that (most of) the sucrose is taken up first by the companion cells and then passes into the sieve tube elements via plasmodesmata (Matsukura *et al.* 2000). In the potato (*Solanum tuberosum*), tomato (*Lycopersicon esculentum*) and tobacco (*Nicotiana tabacum*), however, SUT1 is distributed throughout the SE–CC complex, so that the sieve elements can take up sugar directly. The expression of the sucrose porter genes occurs when leaves mature and begin to export photosynthate. The repression of sucrose porter gene expression by transforming potato plants with antisense DNA leads to inhibition of sucrose transport, a great accumulation of starch and lipid in the leaves, and severe reduction of root and tuber growth.

The loading of amino acids has been studied less intensely than that of sucrose, but a number of amino acid/H⁺ symporters are found in mature leaves. They are not specific for single amino acids.

In symplastic loaders (Fig. 5.4) the photosynthate moves to the phloem through plasmodesmata, and there are abundant plasmodesmata between companion cells and phloem parenchyma, and between companion cells and mesophyll cells if in contact; the sieve elements must receive their supply via plasmodesmata from the companion cells since they do not have direct plasmodesmatal connections to other cells. The companion cells lack transfer cell

Box 5.1

Sugars are among the most hydrophilic chemicals known, due to the large number of OH groups in the molecules, which H-bond with the OH groups of water. The permeability coefficients of cellular membranes towards sugars are therefore extremely low and their movement through membranes depends on transport proteins, mainly of the porter type. For sucrose, SUT1 (Sucrose Transporter No.1) is only one of several types of transporter known from plants, some located in the plasmalemma, some in the tonoplast. Amino acid transporters are also important; it is mainly the hydrophilic amino acids that are transported in the phloem (and xylem), especially glutamate, aspartate and their amides glutamine and asparagine.

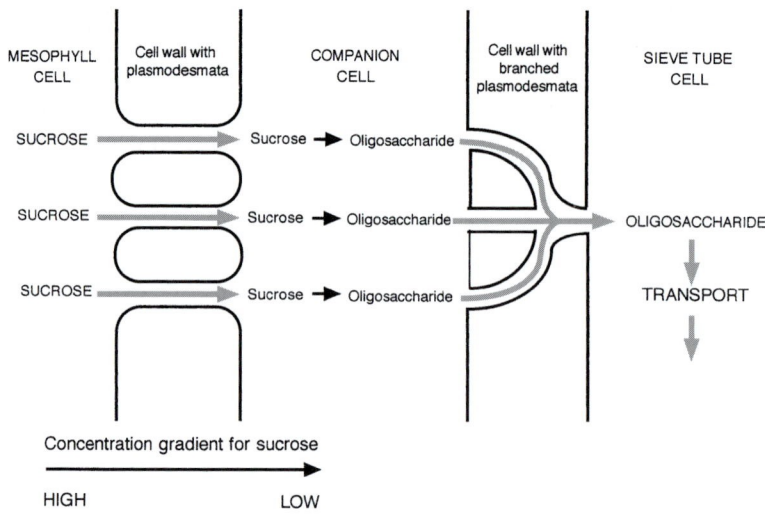

Fig. 5.4 Symplastic phloem loading by the polymer trap mechanism. Sucrose moves from the mesophyll cell to the companion cell (lacking wall ingrowths) through plasmodesmata by diffusion and possibly mass flow. In the companion cell the sucrose is converted to oligosaccharide, which diffuses through branched plasmodesmata into the sieve tube, but the plasmodesmata on the mesophyll side do not permit back-flow.

characteristics. The precise mode of movement of the solutes through the plasmodesmata is not certain. It has been noted that symplastic loaders translocate mainly galactose-containing oligosaccharides rather than sucrose, predominantly raffinose and stachyose. Chemically, these are equivalent to sucrose with respectively one and two galactose residues added. It has therefore been suggested that, in symplastic loaders, sucrose diffusing into companion cells is there converted to oligosaccharides, and these are transferred to the sieve element (Fig. 5.4). This would keep up a concentration gradient for *sucrose* from mesophyll to the companion cells. This 'polymer trap' hypothesis implies that the oligosaccharides cannot back-flow into the mesophyll through the plasmodesmata, which in the relevant cell walls must have rather low exclusion limits, permitting the passage of sucrose (molecular mass 342 Da) but not of raffinose (504 Da) or stachyose (666 Da).

In some symplastic loaders sucrose is the main carbohydrate translocated. There are not many data on such species, but in two cases at least, willow (*Salix babylonica*) and poplar (*Populus deltoides*) there is evidence that the mesophyll cytosol has a high sucrose concentration and there is a diffusion gradient from mesophyll cytosol to the SE–CC complex.

Apoplastic loading is found in many herbaceous plants and in families generally regarded as more highly evolved: Asteraceae (= Compositae), Brassicaceae (= Cruciferae), Chenopodiaceae, Fabaceae (= Leguminosae) and Solanaceae. Apoplastic loading is considered more efficient. It achieves higher loading rates, is relatively insensitive to low temperatures and is found in temperate-zone species. *Symplastic* loading is characteristic of the more primitive families, including tropical and subtropical families with many tree species; it is cold-sensitive. Apoplastic loaders, transporting sucrose, can build up higher sugar concentrations in the phloem sap since sucrose is highly soluble; concentrations of 0.5 mol L^{-1} are not rare in phloem sap and values up to

0.9 mol L^{-1} have been reported, but raffinose saturates at 0.25 mol L^{-1}. Osmosis depends on the number of molecules per unit volume of solution, so that for the same weight, sucrose, with the smaller molecular mass, gives a lower osmotic potential and therefore a higher turgor pressure, which is postulated to drive translocation in the phloem. Table 5.1 summarizes the features of apoplastic and symplastic loaders. The two types of loading are not mutually fully exclusive, a certain extent of apoplastic transfer occurring in symplastic loaders, and vice versa.

5.4.2 Phloem unloading and post-phloem transport in the sinks

Unloading of translocate from the SE–CC complex can also proceed either symplastically or apoplastically, but in this respect no taxonomic division of species according to unloading type has become apparent; rather, the type of unloading depends on the plant organ, and sometimes also on physiological conditions. In apical meristems and in fruits, unloading is mainly symplastic but with some apoplastic contribution. Symplastic movement has been traced in root apices with fluorescent and radioactive labelling. In unloading phloem the companion cells are small and there is some direct contact between sieve tube elements and other cells. Plasmodesmata pass from sieve tube cells in the root apex protophloem to meristematic cells. In mature stems, unloading from the transport phloem (feeding the axial tissues) tends to be apoplastic. The movement of sugar out of the SE–CC complex into the apoplast is a simple leakage process, diffusion-driven by the steep concentration gradient. This leakage

Table 5.1 Characteristics of plants with predominantly apoplastic or symplastic modes of phloem loading. CC = companion cells.

	Apoplastic	Symplastic
Plasmodesmata between CC and phloem parenchyma or mesophyll	Few	Many
CC of transfer cell type	Yes	No
Main carbohydrates translocated	Sucrose	Oligosaccharides, polyols
Loading rates	Higher	Lower
Loading sensitivity to low temperature	Moderate	Marked
Geographical distribution	Mainly temperate	Many tropical, subtropical
Type of plant	Many herbaceous	Many trees
Taxonomic distribution	Mostly more highly evolved families	Mostly more primitive families

would in fact deplete the translocation stream, if it were not counteracted by a continuous (partial) retrieval of the sugar by the sucrose-proton symport system. In the terminal sinks, the retrieval capacity of the phloem is low, favouring exit of solutes from the phloem.

After unloading from the transporting cells has taken place, the nutrients must still move from cell to cell, for distances ranging from a few hundred micrometres (e.g. apical meristems) to several centimetres (e.g. some fruits). This post-phloem movement is mainly symplastic. In immature sinks, the exclusion limit of plasmodesmata is very high, up to about 50 000 Da in young leaf tissue (cf. about 1000 Da for mature leaf mesophyll) and 27 000 in developing wheat grains. This enables the distribution of protein and small RNA molecules from the sap (Section 5.2.3) through the sink; GFP can be seen to move from the sieve tubes into all tissues of immature leaves. There may be bulk flow of solution through the plasmodesmata, especially in apical meristematic regions, where there is as yet no functional xylem and the phloem, which differentiates much closer to the apex, must supply water as well as nutrients. In root apices, the first sieve tubes differentiate within 250–$750\,\mu m$ of the tip of the meristem, but the first protoxylem matures at 400–$8500\,\mu m$, or even further away. The possibility of bulk flow in the apoplast, too, has been suggested. The driving force for solute movement would be the diffusion gradient, from the high concentration in the sieve tubes to the sink cells, where the concentration is kept low by utilization in growth and metabolism. Bulk flow could be maintained by the utilization of the water.

5.5 | Partitioning of translocate between sinks: integration at the whole-plant level

In addition to control of the *direction* of translocation, there must be control of how *much* passes to different sinks at any particular time – i.e. how nutrients are *partitioned*. Balanced, integrated growth of a plant would otherwise be impossible. The photosynthetic potential of a plant depends on its leaf area; translocation of organic C to the leaf primordia enables leaf growth and increases the photosynthetic potential of the plant. But the growth and subsequent maintenance of the leaves need mineral and water supplies from the roots: during leaf growth there must be translocation of enough photosynthate to the root meristems to support a concomitant increase in the mass (and maintenance cost) of the root system. The organic nutrients available at any particular moment must be partitioned in appropriate proportion between the shoot and root sinks. In fact the growth rates of various parts of a plant can keep in proportion with mathematical precision (see allometric growth, Chapter 6). This implies that partitioning of organic nutrients, too, is precisely controlled. There must be signalling between the sinks and the sources.

Partitioning control is a complex process. Farrar (1996) lists the steps involved in C flow to a sink: '*phloem loading in a source leaf, phloem transport into the sink, unloading and short distance transport within the sink and then metabolism and storage in the sink.*' Just about everything that affects the physiology of a plant has the potential to affect partitioning, and control of partitioning is far from being fully understood. One important signalling molecule is sucrose itself (Fig. 5.5). Sucrose acts in growing regions as a promoter of the expression of genes involved in growth and respiration (which provides the energy for growth); in the photosynthetic leaves, sucrose represses the expression of genes involved in photosynthesis, including the gene coding for the small subunit of Rubisco. Leaf growth leads to an increased production of sucrose and its translocation to the roots, where the sugar would stimulate the activity of genes which promote growth of more root biomass – corresponding to the growth of the leaves. A fall in the sink demand has been shown to decrease the rate of loading at the source, although it is not clear how the message is transmitted. Turgor has been suggested, a change in turgor at the sink being transmitted to the source and suppressing loading activity. The consequent increase in sugar concentration in the mesophyll

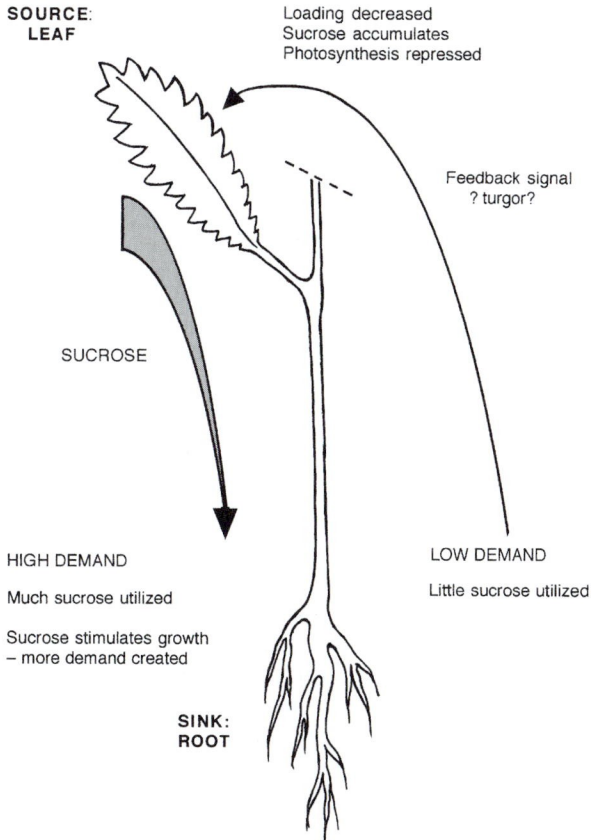

SOURCE:
LEAF

Loading decreased
Sucrose accumulates
Photosynthesis repressed

Feedback signal
? turgor?

SUCROSE

HIGH DEMAND

Much sucrose utilized

Sucrose stimulates growth
– more demand created

LOW DEMAND

Little sucrose utilized

SINK:
ROOT

Fig. 5.5 Sucrose as an information molecule integrating activity between source and sink.

would eventually repress the rate of photosynthesis. Some of the proteins mobile in the phloem may also act as signals.

The axial sinks can act as buffers between the sources and the terminal sinks: when demand by terminal sinks falls, more photosynthate is unloaded laterally from the transport phloem into the surrounding tissues.

The status of various nutrient elements affects the partitioning of photosynthate. When plant growth is limited by a low supply of the elements N, P, S or Fe, the ratio of root to shoot mass rises, more organic C being delivered to the roots, whereas limitation of the supply of Mg, Mn or K leads to a decrease of the root : shoot mass ratio. Water stress increases the root : shoot ratio, root growth being stimulated while shoot growth is reduced.

Hormones are molecules that transmit various signals from one part of the plant to another. Translocates can be attracted to a cut stem stump by applying the hormone auxin to it, or diverted away from a growing wheat ear by the application of auxin to a point remote from the ear. The hormone, however, appears to act by stimulating growth and/or metabolic activity at its point of application, creating a new sink. Cytokinins, too, can induce sink activity. There seems to be no evidence of a hormonal signal specifically and directly controlling loading or unloading.

5.6 | The mechanism of phloem translocation

For the movement of any substance in solution, there are two possibilities. Either it is carried along by **mass flow** (bulk flow) of the solution, or else only the solute moves, e.g. by diffusion, while the solvent remains stationary. The rates of phloem translocation are of the order of 10^5 times faster than diffusion, so this mechanism can be discounted: there must be either mass flow, or some special, rapid transport of the solutes. Movement in the xylem is undisputedly a mass flow of the whole solution, and the motive force is either the tension pull of transpiration, or root pressure. For phloem translocation, too, the current consensus is that mass flow occurs in the sieve tubes, driven by an osmotic gradient. But the structure of sieve tube cells is less obviously suited for mass flow than that of the xylem conducting elements. The hypothesis of mass flow in the phloem needs to be examined critically.

5.6.1 The Münch hypothesis of mass flow driven by an osmotic gradient

One of the earliest theories for the mechanism of phloem translocation was Münch's 1930 hypothesis, which postulated a mass flow along a turgor pressure gradient, driven by a physiologically maintained gradient of osmotic potential (Fig. 5.6). At the source end,

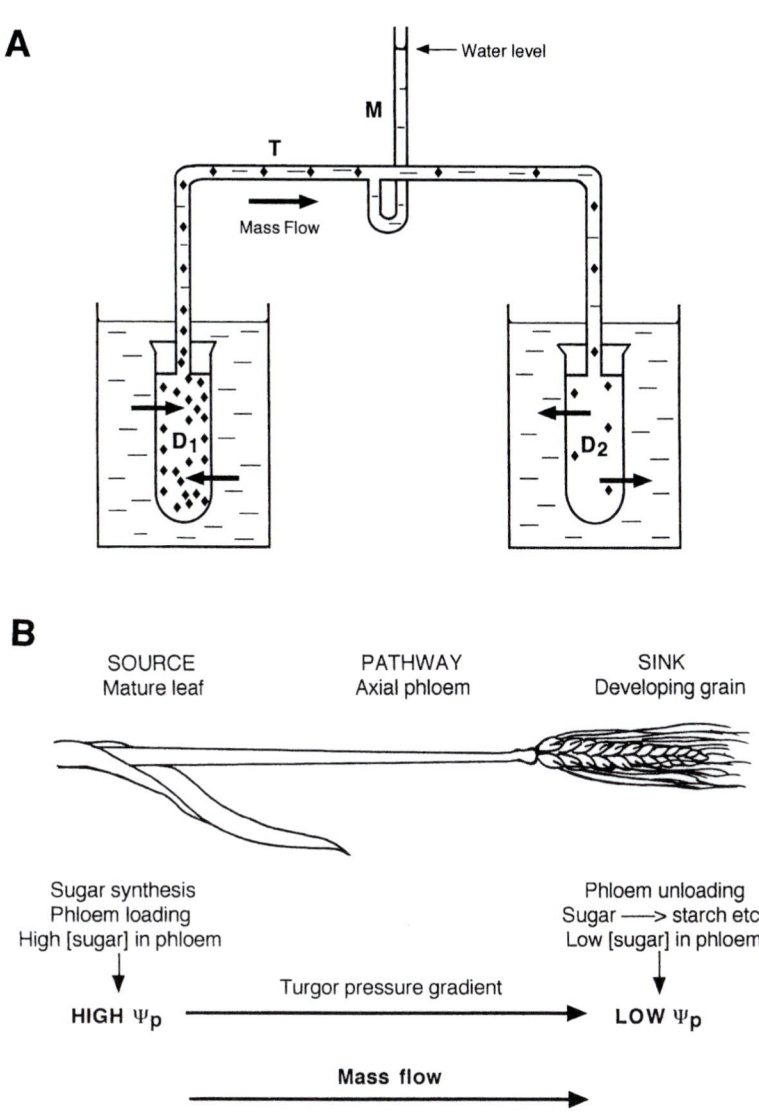

A

Water level

M

T

Mass Flow

D₁ D₂

Fig. 5.6 The Münch mass-flow hypothesis. (A) Principle demonstrated by a simple laboratory experiment. Two dialysis sacs permeable to water but not to sucrose are filled respectively with a concentrated sucrose solution (D_1) and a dilute sucrose solution (D_2), then connected with glass tubing T and immersed in beakers of distilled water. The greater Ψ gradient between water and D_1 results in a greater rate of water uptake into this sac; a 'turgor' pressure builds up, solution can be seen to pass along tube T into D_2 and pressure is registered by manometer M. The flow stops when the sucrose concentrations in the sacs have equalized. (B) Operation in the plant. A continuous loading of the phloem at the source and unloading at the sink can keep up the turgor gradient and the flow indefinitely.

B

SOURCE	PATHWAY	SINK
Mature leaf	Axial phloem	Developing grain

Sugar synthesis
Phloem loading
High [sugar] in phloem

Phloem unloading
Sugar ——> starch etc.
Low [sugar] in phloem

HIGH Ψ_p —— Turgor pressure gradient ——> LOW Ψ_p

Mass flow ——>

sugars (or other solutes) are loaded into the sieve tubes; this lowers their water potential and induces an inflow of water. The turgor pressure in the sieve tubes rises and the solution is pushed along the phloem by the pressure. At the sink end the solutes are unloaded and water moves out with them, according to the water potential gradient, so that the turgor pressure here is kept low. A lateral movement of solutes along the way into surrounding tissues of stem and root, which all need to be supplied, would also work to maintain a solute concentration gradient (and hence pressure gradient) from source to the final sink. Water would be absorbed into the phloem from the xylem at the source and released at the sink, returning into the xylem, or the water might be utilized at the sink end, as noted earlier.

Evidence in favour of the mass-flow hypothesis

Many observations can be quoted in favour of a mass flow in the phloem. When one single sieve tube unit is pierced by an aphid stylet, the volume of sap exuded in an hour is around 1 μL, equivalent to the volume of some 2500 individual sieve tube cells; the flow can continue for days at this rate. On a macroscopic scale, the phloem sap of sugar palms and agaves is tapped for sugar production or fermentation into an alcoholic drink. One tree can exude for several months and yield several thousand litres of sap during this period. This is manifestly a mass flow of liquid! Moreover, the sugar concentration remains steady during the period of exudation in both the aphid stylet exudate and in the palm sap, so that the flow of liquid cannot be attributed to a leakage of water into punctured sieve tube cells from the nearby xylem. If mass flow is rejected, all the above-mentioned exudations of sap must be ascribed to an injury reaction unrelated to normal translocation in the phloem. NMR techniques (Section 5.3.1) have demonstrated water flow in the phloem in intact plants.

The Münch hypothesis requires that sieve tube contents should be under a positive turgor pressure, and that there should be a gradient of turgor pressure (and of osmotically active solutes), decreasing from source to sink. The sieve tubes are certainly under positive turgor pressure; this is demonstrated even by the simple fact that sap can exude from cuts. Aphid stylets contain only a very narrow channel, offering a considerable resistance to the flow of the viscous sap; a pressure of 1–3 MPa is required to force the sap through the stylets. Direct measurements of the turgor of sieve tubes have been achieved, either by insertion of sensitive pressure probes, or by the external application of pressure cuffs (analogous to the apparatus used for measuring blood pressure). Such measurements are in practice fraught with great difficulty; nevertheless, not only have positive pressures been recorded, but some workers have succeeded in demonstrating gradients of decreasing turgor passing down trees, away from source leaves. Gradients of sugar concentration have been observed. In soybean (*Glycine max(soja)*) stalks, for example, the sugar concentration in the sieve tubes of the leaflet stalk has been found to be 10.5–12.5%, when the sieve tubes of the root contained only 4.4–6.3%.

All this is good evidence for the Münch mass-flow hypothesis. But the hypothesis cannot be accepted until it has been shown to be feasible quantitatively as well as qualitatively: until it is shown that mass flow *at the experimentally measured velocities* can be driven *by the actually existing pressure gradients*, through channels *of the dimensions present in the sieve tubes*. It is difficult to prove this with complete certainty, the main problem being the interpretation of the fine structure of the sieve plate area.

The dimensions of transport channels in the sieve tube cells

Mass flow requires continuous open channels, the wider the better, for the more narrow the diameter, the greater is the frictional resistance to the flow, and the greater is the force that is required to drive the liquid along. The diameter of the sieve tube cells at 10–50 μm would pose no problems. But at each sieve plate the solution must pass through the sieve plate pores. The crucial question for mass flow is: are these pores open channels of sufficient cross-section to permit transport at the observed rates?

In mature sieve tubes the pore diameters, as stated, range from 0.1 to over 10 μm. These are the diameters of the holes in the wall. If it is assumed that the entire hole is open to flow, a hydrostatic pressure gradient of 0.06–0.10 MPa m^{-1} has been regarded as adequate to drive mass flow at the measured rates even for narrow pores. (That means, for every metre length along the phloem towards the sink, the pressure in the phloem must *decrease* by the above value.) Some workers have put the value higher, up to 0.5 MPa m^{-1}. Experimentally measured pressure gradients vary from 0.02 to 0.55 MPa m^{-1}, which lie within the required range. But if there is any cytoplasmic structure within the pores, much higher pressure gradients become necessary, of tens of MPa m^{-1}. It is clear that turgor-driven mass flow is possible only if the pores are more or less fully unobstructed, at least for species with narrow pores.

Sieve plates and their pores are usually seen lined with a distinctive polysaccharide, callose (a glucose polymer), which may narrow the pore diameters to 0.1 μm even when the aperture in the wall proper is over 1 μm, but it is arguable how much of the callose seen in any preparation was there in the living state, for injury, e.g. resulting from excision, has been shown to induce callose deposition on sieve plates within seconds. Because of this, it can be practically impossible in many instances to know the precise diameter of the pores in a functional sieve plate.

It is even more difficult to decide on the content of a sieve plate pore. The pore diameters are often too close to the limit of resolution of the light microscope for a satisfactory *in vivo* examination. Sieve tubes cannot be dissected out singly undamaged, and when whole vascular bundles are viewed by conventional light microscopy, clarity is lost. *Confocal laser microscopy*, however, enables the observer to focus on a narrow plane even in a thick specimen without interference from surrounding material. This method has been used to study living phloem in a leaf still attached to the plant and actively translocating. As far as could be seen the pores were unobstructed, but submicroscopic contents cannot be ruled out.

Electron microscopy overcomes the problems of resolution and interference by surrounding structures, but the results remain disputable. The sieve tubes are extremely delicate, being highly hydrated and under high pressure; this makes them very susceptible to fixation damage. In particular, as the pressure is released by cutting, the contents surge towards the cut. According to the method of preparation and the species studied, the appearance of the pores varies. At one extreme pores appear quite empty, or contain some sparse P-protein filaments; at the other extreme,

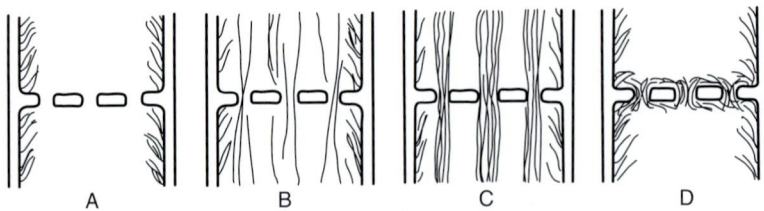

Fig. 5.7 Diagrams of interpretations of sieve plate pore contents based on electron micrographs of phloem from a number of plant species and organs, and prepared with a range of methods. The lines represent P-protein filaments. (A) Pores empty, P-protein peripheral; (B) Sparse P-protein filaments pass through pores; (C) Compact P-protein strands pass through pores; (D) Dense plugging by P-protein, sometimes combined with ER. Mass flow would be feasible with models (A) and (B), doubtful with (C) and not feasible with (D). Models with membrane-bounded strands or tubules traversing the pores are not illustrated, being based on limited observations not confirmed by subsequent work.

the pores are completely packed with plugs of P-protein and sometimes also with endoplasmic reticulum (Fig. 5.7). Proponents of the mass-flow theory can regard the empty pores as the natural state and dismiss the densely filled pores as artefacts, containing material swept in as turgor was released on cutting, and/or material coagulated on fixation. Opponents of the mass-flow theory are, however, equally free to argue that empty-looking pores have had their natural contents destroyed by the fixative! Freeze-fracture electron microscopy, where the tissue is frozen extremely rapidly by being plunged into a liquid-nitrogen-cooled bath at about -196 °C, avoids the artefacts caused by chemical fixatives. But in sieve tubes, ice crystal formation has obscured the cell contents and, where structure could be discerned, it has been very variable. In the same sieve plate, some pores may look empty, others are traversed by various amounts of filaments, and some have filaments lying across them. Even 'instantaneous' freezing may not be fast enough to show the true structure for sieve tubes pores through which material is moving rapidly. A translocation velocity of 100 cm h^{-1} is slow on a macroscopic scale; a snail can crawl faster. But relative to cellular dimensions, translocation is very fast (see Sjölund 1997). A particle would traverse a 1-μm-thick sieve plate in 0.0036 seconds (indeed probably less: the measured velocity is an average for the whole sieve tubes and the rate should be even faster in the narrow pores). A fraction of a second would permit structural disturbances to occur in the pores between the start of freezing and final total solidification.

Current view inclines to accept the pores as more or less unoccluded. When damage has been identified with certainty, it has resulted in formation of P-protein strands and *deposition* of P-protein material on sieve plates rather than loss of discrete structures. This was observed by confocal microscopy when sieve tubes were deliberately wounded (Knoblauch & van Bel 1998). The blocking of pores on wounding confines leakage of sap to the damaged cell.

Metabolic activity in the phloem

According to the original mass-flow hypothesis, the sieve tubes act as passive conducting channels, with no necessity for the expenditure of metabolic energy on the way, for the energy input would take place at the source during loading and possibly also at the sink during unloading. Normally the transport phloem has a high rate of respiration; sieve tube sap contains ATP at an average concentration of 0.4 mmol L^{-1} and the turnover rate of ATP is high, suggesting an expenditure of ATP all along the pathway. This has been quoted as an argument against the mass-flow hypothesis. It has become clear, however, that all along the transport route there is leakage of translocate, and whilst some of this goes to nourish the axial tissues, a major proportion is retrieved by energy-requiring membrane pumps. A high rate of turnover is claimed for phloem proteins and protein synthesis requires an energy supply. Perhaps the unusual state of the sieve tube cells, lacking nuclei, and traversed by a rapid flow of sap, requires a higher than usual rate of maintenance respiration. The high respiratory rate in the phloem therefore cannot be taken as a contraindication to mass flow driven by an osmotic gradient built up locally in the loading region.

The function of the sieve plates

For mass flow, sieve plates appear to be obstructions and their presence has been interpreted as being against the mass-flow hypothesis. They do nevertheless perform an important function in preventing sap loss from damaged phloem: on injury, the deposition of callose narrows the pore diameters, P-protein surges towards the cut because of release of turgor and then piles up against the sieve plates forming slime plugs over the already narrowed pores, making them impassable. Plastids burst, releasing starch grains which help to plug the pores, and possibly they also release chemicals which promote P-protein coagulation. As noted (Section 5.2.3), from most species there is little exudation from cut phloem in spite of the fact that the sieve tube contents are under considerable hydrostatic pressure. The system can be desensitized by repeated rubbing or even beating in the case of robust woody specimens so that cutting no longer induces blockage, and this effect is utilized when sugar palms are tapped. The sieve plates may also give some mechanical support to the sieve tubes. Different functions for the sieve plates have been suggested in connection with alternative views on the mechanism of phloem translocation.

Alternatives to the mass-flow hypothesis

Alternatives to mass flow have from time to time been proposed. The concept of *spreading along interfaces* visualizes the translocated molecules forming a monomolecular film on surfaces (P-protein fibrils?), like oil films at an air–water interface. This film would then remain intact, with molecules added at the source end and removed at the sink. The hypothesis does not account for the flow of water, which does occur in the phloem. It is also difficult to conceive of surfaces able to form films with all the variety of compounds that can be carried in the phloem – sugars, amino acids, mineral ions, ATP, fluorescent dyes, hormones, etc. Some sieve tubes moreover contain very little P-protein. *Protoplasmic streaming* has been claimed to proceed in the sieve tubes, with transcellular strands running from cell to cell through the sieve plate pores, but after the original observations were published in the 1960s and 1970s, other workers have been unable to confirm them. The apparent streaming rates of 3.5 cm h^{-1} are also far too slow compared with the velocity of translocation which reaches several m h^{-1}. Submicroscopic tubules passing through the sieve plate pores, and pumping the tubule contents along by contractions (as in peristalsis) have been proposed without any solid evidence; the same goes for the idea of lashing protein filaments, like miniature cilia, anchored to the outsides of the hypothetical tubules and to sieve tube walls, wafting along a flow in the bulk of the cell. The theory of *electroosmosis* regards the sieve plates as pumping stations for ions: an electric potential difference (PD) is built up across the plate, by pumping of H^+ ions from companion cells into the sieve tube upstream of the plate, and out of the sieve tube into the companion cells downstream of the plate. This would result in the movement of the positively charged K^+ ions in the phloem sap across the plate, towards the negatively charged downstream side and the K^+ ion flow would carry with it water and other solutes. The attractiveness of the electroosmotic hypothesis is that it ascribes both a specific function to the high K^+ content of the phloem sap, and a physiological function to the sieve plates. But the idea has no experimental data to support it. The energy expenditure for the pumping would be high; there would have to be localized differentiation of sieve elements and companion cells with the direction of H^+ ion movement between the cells reversed over a micrometre or two near sieve plates, and all other reported cases of H^+ pumping involve pumping of the ions *out* of cells. Quantitatively, too, with sieve tube cells often over 100 μm long, it seems hardly credible that so much force could be built up by sieve plates so far apart. In summary, no viable alternative to turgor-driven mass flow in the phloem has been proposed. But it is conceivable that in species with very narrow sieve plate pores there is facilitation of mass flow through the pores from some mechanism like electroosmosis. With sieve pore diameters in the flowering plants covering two orders of magnitude, from 0.1 μm to over 10 μm, the forces required to drive mass flow through the narrowest pores must be very much larger than for the widest ones.

5.6.2 The translocating system in the plant kingdom and macroalgae

All vascular terrestrial plants possess a phloem of elongated living cells joined by some kind of 'sieve areas', i.e. fields of pores on lateral walls, on end walls or on both. In the Psilotaceae, probably the most primitive group of living vascular plants, the sieve cells of the phloem are described as just elongated parenchyma cells with lateral sieve areas. True sieve *tube* cells, with transverse sieve plates and pores reaching 10-μm dimensions, are found only in the flowering plants. Highly differentiated companion cells are also confined to the flowering plants, although the sieve cells of conifers have associated cells known as albuminous cells. It is reasonable to suppose that the mechanism of organic translocation is essentially the same throughout the kingdom Plantae.

The brown algae (Phaeophyceae) include species with differentiated bodies measurable in metres, and these macroalgae translocate organic compounds through sieve elements, elongate cells joined by pores. In most of the algal species these cells have nuclei and are filled with cytoplasm containing small vacuoles, small plastids and many mitochondria; the sieve plate pore diameters range from 0.03 to 0.1μm. But in the giant kelp (*Macrocystis pyrifera*), which can grow to a length of 50 m, the cells are enucleate, with cytoplasm confined to a peripheral layer, and the pores are much larger, 2.4–6.0 μm wide; the pores are lined with callose and callose deposits block them on injury – features strikingly similar to those exhibited by flowering-plant sieve tube cells. The compounds translocated are the algal photosynthetic products – mannitol (and amino acids) in *Macrocystis*. The direction of translocation is 'source to sink', from mature photosynthetic regions to meristems, and the algal translocation velocities fall within the range measured in flowering plants, 10 cm h^{-1} in *Laminaria*, 70 cm h^{-1} in *Macrocystis*. There seems to have been a high degree of parallel evolution of the translocating systems in terrestrial plants and these macroalgae. Both groups share the photosynthetic mode of life, and have a common fundamental structural feature in the possession of a *cell wall*. When walled cells form a multicellular body, the only possibility of direct intercellular communication is via pores in the cell wall. Once the basic structure of cells joined by communicating pores was established, it is not surprising that these pores should have evolved in several evolutionary lines into a means of large-scale transport of nutrients, when transport was necessitated by size increase accompanied by functional differentiation of the body of the green terrestrial plant or alga.

Complementary reading

Esau, K. & Cheadle, V. I. Size of pores and their contents in sieve elements of dicotyledons. *Proceedings of the National Academy of Sciences (USA)*, **45** (1959), 156–62.

Fisher, D. B. The estimation of sugar concentration in individual sieve-tube elements by negative staining. *Planta*, **139** (1978), 19–24.

Fisher, D. B. & Cash-Clark, C. E. Sieve tube unloading and post-phloem transport of fluorescent tracers and proteins injected into sieve tubes via severed aphid stylets. *Plant Physiology*, **123** (2000), 125–37.

Flowers, T. J. & Yeo, A. R. *Solute Transport in Plants.* London: Blackie/Chapman & Hall, 1992.

Golecki, B., Schulz, A., Carstens-Behrens, U. & Kollmann, R. Evidence of graft transmission of structural phloem proteins or their precursors in hetero-grafts of Cucurbitaceae. *Planta*, **206** (1998), 630–40.

Haupt, S., Duncan, G. H., Holzberg, S. & Oparka, K. J. Evidence for symplastic unloading in sink leaves of barley. *Plant Physiology*, **125** (2001), 209–18.

Köckenberger, W., Pope, J. M., Xia, Y., Jeffrey, K. R., Komor, E. & Callaghan, P. T. A non-invasive measurement of phloem and xylem water flow in castor bean seedlings by nuclear magnetic resonance microimaging. *Planta*, **201** (1997), 53–63.

Mauseth, J. D. *Plant Anatomy*. Menlo Park, CA: Benjamin/Cummings, 1988.

Patrick, J. W. Phloem unloading: sieve element unloading and post-sieve element transport. *Annual Review of Plant Physiology and Plant Molecular Biology*, **48** (1997), 191–222.

Pickard, B. G. & Beachy, R. N. Intercellular connections are developmentally controlled to help move molecules through the plant. *Cell*, **98** (1999), 5–8.

Sasaki, T., Chino, M., Hayashi, H. & Fujiwara, T. Detection of several mRNA species in rice phloem sap. *Plant and Cell Physiology*, **39** (1998), 895–7.

Sideman, E. J. & Scheirer, D. C. Some fine structural observations on devel-oping and mature sieve elements in the brown alga *Laminaria saccharina*. *American Journal of Botany*, **64** (1977), 649–57.

Sweetlove, L. J. & Hill, S. A. Source metabolism dominates the control of source to sink carbon flux in tuberizing potato plants throughout the diurnal cycle and under a range of environmental conditions. *Plant, Cell and Environment*, **23** (2000), 523–9.

van Bel, A. J. E. The phloem, a miracle of ingenuity. *Plant, Cell and Environment*, **26** (2003), 125–49.

van Bel, A. J. E. & Gamalei, Y. V. Ecophysiology of phloem loading in source leaves. *Plant, Cell and Environment*, **15** (1992), 265–70.

Versch, J., Kalusche, B., Köhler, J. *et al.* The kinetics of sucrose concentration in the phloem of individual vascular bundles of the *Ricinus communis* seed-ling measured by nuclear magnetic resonance. *Planta*, **205** (1998), 132–9.

Wark, M. C. Fine structure of the phloem of *Pisum sativum*. II. The companion cell and phloem parenchyma. *Australian Journal of Botany*, **13** (1965), 185–93.

Wark, M. C. & Chambers, T. C. Fine structure of the phloem of *Pisum sativum*. I. The sieve element ontogeny. *Australian Journal of Botany*, **13** (1965), 171–83.

References

Balachandaran, S., Xiang, Y., Schobert, C., Thompson, G. A. & Lucas, W. J. (1997). Phloem sap proteins from *Cucurbita maxima* and *Ricinus communis* have the capacity to traffic cell to cell through plasmodesmata. *Proceedings of the National Academy of Sciences (USA)*, **94**, 14150–5.

Farrar, J. F. (1996). Sinks: integral parts of a whole plant. *Journal of Experimental Botany*, **47**, 1273–9.

Knoblauch, M. & van Bel, A. J. E. (1998). Sieve tubes in action. *The Plant Cell*, **10**, 35–50.

Matsukura, C., Saitoh, T., Hirose, T., Ohsugi, R., Perata, P. & Yamaguchi, J. (2000). Sugar uptake and transport in rice embryo. Expression of companion cell-specific sucrose transporter (*OsSUT1*) induced by sugar and light. *Plant Physiology*, **124**, 85–93.

Oparka, K. J. & Santa Cruz, S. (2000). The great escape: phloem transport and unloading of macromolecules. *Annual Review of Plant Physiology and Plant Molecular Biology*, **51**, 323–47.

Sjölund, R. D. (1997). The phloem sieve element: a river runs through it. *The Plant Cell*, **9**, 1137–46.

Part II

Growth and development

Chapter 6

Growth as a quantitative process

6.1 | Introduction

Growth is one of the most fundamental and conspicuous characteristics of living organisms, being the consequence of increase in the amount of living protoplasm. Externally this is manifested by the growing system getting bigger, and growth is therefore often defined as an irreversible increase in the mass, weight or volume of a living system. The size increase must be permanent; the swelling of a cell in water is not growth, being easily reversed by returning the cell to a solution of lower Ψ. It is, however, possible to consider as growth developmental changes not immediately involving an increase in size. An amphibian embryo, or a *Selaginella* female gametophyte, for a long time utilizes the nutrient store with which it was released from the parent, to produce many new cells without any increase in overall size, yet growing in the sense that living protoplasm is increasing at the expense of stored nutrients. Again, if dry mass is measured, a flowering plant seedling loses dry mass while utilizing reserves and growing.

Growth is an exceedingly complex process. Every reaction associated with the synthesis and maintenance of living protoplasm is associated with it, which makes it complicated enough at the cellular level. At the organismal level, it means the coordinated multiplication, size increase and specialization of millions of cells, all arranged in precise positions. Growth processes are also synchronized with seasonal changes, plants responding to appropriate environmental stimuli to achieve this synchronization. This chapter is an introduction to growth and development of flowering plants, discussing the overall process, methods of measurement, elementary quantitative analysis of growth patterns, and growth rhythms. This is followed in the next chapters by a more detailed discussion of events at the cellular level and then by an analysis – in so far as it is feasible – of the controlling and integrating factors, hormonal, genetic and environmental, which lead to the visible patterns of development at tissue and higher organizational levels.

6.2 | The measurement of plant growth

Growth can be measured in a variety of ways. Since increase in the amount of protoplasm, the fundamental growth process, is difficult to measure directly, generally some quantity is measured which is more or less proportional to the increase in protoplasm. In higher plants or their organs the four most commonly employed measures (parameters) are:

(1) Fresh weight. This measurement is technically easy. However, a plant organ must be detached from the plant to be weighed, killing the organ in the process, and an entire plant which has been removed from its growth environment can rarely be replaced undisturbed. Thus growth measurements by weighing almost always necessitate the taking of successive samples from a series of plants.

(2) Dry weight. This is sometimes considered more meaningful than fresh weight, because the increase in fresh weight may be largely the result of water uptake, and fluctuations in fresh weight may occur as a result of chance fluctuations in a plant's water content. In a germinating seedling, living on seed reserves, the dry weight decreases until the seedling builds up an adequate photosynthetic capacity, and in such a case the fresh weight increase is a better indication of growth.

(3) Linear dimensions. For an organ growing predominantly in one direction, such as a root tip or a pollen tube, length is a suitable measure. Indeed linear elongation can be used to assess the growth of an entire shoot. An increase in width (diameter) may be relevant in other cases, e.g. an expanding fruit or a thickening axis.

(4) Area. This can be used to assess growth in a system extending mainly in two dimensions, such as an expanding leaf.

Measurements of length, width and area have the attraction that they can be carried out on the same plant or organ over a period of time without destroying it. Growth in length can be measured with highly sophisticated *auxanometers* which detect or magnify size increments far too small to be measured directly, and auxanometers record the data automatically. Auxanometers can be sensitive enough to record growth increments of less than 1 μm and hence can detect changes in growth rate within minutes of applying, say, a hormone to an organ with an overall growth rate of no more than 1–2 mm h^{-1}. At the same time the apparatus can be robust enough to be used in the field. The drawback of auxanometers is that they involve some kind of attachment of the plant to the apparatus, although the force thereby exerted on the plant is generally considered to be too small to affect the growth rate.

Refined *optical methods* for measurement of length, width and area are available. Beginning with simple time-lapse photography, where growth is estimated from successive photographs with a camera electronically controlled to take exposures at set time intervals, the techniques have progressed to the use of digital cameras interfaced with computers to analyse the images. Great sensitivity can be attained both spatially and temporally, detecting small increments over short time periods (Schmundt *et al.* 1998). Photography involves minimal interference with the growing system although some physical fastening of the plant (organ) to a support may be necessary, for vibrations might be recorded as 'growth movements' by the analyser. The system is illuminated with near-infrared radiation, since much of the visible spectrum has the potential to affect plant growth rates.

For assessing the growth of plant cell cultures, the methods applied are those used for estimating the growth of microorganisms. Volume of a cell culture can be determined by centrifuging the cells into a graduated centrifuge tube; for single cells, the volume can be calculated from measurements made under the microscope, often utilizing confocal microscopy. Increase in the *turbidity* of a cell suspension is proportional to the increase in cell number and this can also be obtained by direct counting in a special counting chamber on a slide, a *haemocytometer*. Automatic cell counters are available. Other criteria for estimating the mass or number of cells in culture are analyses of protein nitrogen or rate of respiration.

The units in which growth rate is expressed are as diverse as the methods of measurement and depend also on the overall size of the system under study. The length increase of a pollen tube might be measured in μm min^{-1}; the length (height) growth of a tree might be expressed in m year^{-1}.

6.3 | Growth, development and differentiation

Growth is always accompanied by a change in form and in physiological activity, by **differentiation**. The identical cells produced by cell division in a meristem (region of cell division) enlarge, i.e. grow, and at the same time become different from the meristematic cells, and from each other, forming for instance pith parenchyma, or xylem vessels, or companion cells. In their mature form these cells are of very different structure and function. Nevertheless, certain growth processes are common to all plant cells and, during the initial stages of a cell's development, these common processes predominate. In the meristem, the newly formed cells first grow by *plasmatic growth*, a synthesis of protoplasm including multiplication of organelles. In cells destined to remain meristematic, a doubling of cell mass is followed by division. In cells destined to undergo further growth, there follows a phase of *expansion* (elongation) growth, characterized by a rapid volume increase, extensive water uptake accompanied by vacuolation, and by cell wall synthesis. Then, as expansion slows down, divergences in cell development become dominant. While basic protoplasmic components such as proteins and nucleic acids increase in quantity in all cells during growth, the proportional increases in particular proteins differ so that cells of varied structure and metabolism are formed. Cell wall growth occurs in all cells, but as differentiation proceeds cell-type-specific differences in the chemistry and structural arrangement of the cell wall components become apparent. These processes are considered in more detail in subsequent chapters. The completion of differentiation leads to the formation of the mature tissue cells. Finally, mature cells ultimately age and die. Such a sequence of changes constitutes the overall sequence of development:

$$\textbf{development} = \textbf{growth} + \textbf{differentiation}$$

Although the growth and differentiation stages overlap in time at least partly, it is possible to separate growth from differentiation conceptually and experimentally. Thus it is possible to suppress differentiation by inhibitory chemicals, while permitting growth to continue. Further, these two aspects of development seem in many developing systems to be mutually competitive: conditions which favour rapid growth often suppress differentiation, and vice versa. For instance, plants whose growth is retarded by a deficient water supply may show an enhanced degree of cellular differentiation.

In the light of the above discussion it is possible to advance a definition of growth as follows:

Growth is a synthesis of protoplasm, usually accompanied by a change in form and an increase in mass of the growing system. The total mass increase may be many times that of the increase in the mass of the protoplasmic components proper.

Levels of growth

Growth can be studied at several levels of structure. In order of increasing complexity, one can consider the physiology of growth at levels of

cell ⟶ tissue ⟶ organ ⟶ organism

The growth of a system above the cell level is brought about by a combination of *cell multiplication* and *cell growth*. The extent to which these two processes contribute to the growth of a tissue or organ depends on the particular system under study, and on its developmental stage: it is common to find during the growth of a tissue or organ a stage of cell division followed by cell expansion (Section 8.4).

6.4 | Localization of growth in space and time

Flowering plants continue growth throughout their life history by virtue of persistent localized growth regions. Cells are formed by cell division in **meristems** in these growth regions, and size increase and organ formation result from the activities of these meristems; the associated cell expansion, differentiation and maturation take place in close proximity to the meristems. Whilst still in the seedling stage a flowering plant already contains cells at all stages of development – meristematic, expanding, differentiating, mature and dead. The age of a plant and the age of its cells are consequently two quite different things. Trees live for hundreds and even thousands of years; ages of ancient bristlecone pines (*Pinus longaeva*), growing at high altitudes in California, have been estimated to exceed 5000 years. But few living cells in a tree are more than a few years old. Leaves typically survive for one growth season and even the so-called evergreen leaves are shed after a few years, or a few decades at the most. Tree pith cells may live to 100 years. This continuous replacement of the living tissues of a perennial is probably one of the secrets of tree longevity. Another characteristic of plant growth is its *indeterminacy*: the total size and the size and number of organs are not precisely fixed. The plasticity that plants can exhibit in response to environmental conditions is quite remarkable. A common garden weed, the groundsel (*Senecio vulgaris*), growing in a well-fertilized vegetable bed will grow to a bushy plant some 30 cm high and bear over 50 flower heads. A groundsel established in a crack between paving stones produces a tiny plantlet, a few centimetres tall with a single flower head, scarcely recognizable as the same species. Continued growth enables a plant to respond to environmental stimuli with growth reactions to the end of its life. Individual organs – leaves, fruits, floral parts – on the other hand typically have *determinate* growth which ceases when the mature size is reached, and their shapes, too, are fixed.

Localization of growth in particular areas and along particular directions achieves the shaping of plant organs, the process of morphogenesis, which is the subject of later chapters.

For flowering plants, which are immobile, growth takes the place of movement in response to certain environmental stimuli, particularly directional ones. A motile unicellular alga might swim towards a source of light (phototaxis); a flowering plant shoot will grow towards the light (phototropism). Such responses are known as growth movements and are the subject of Chapter 12.

6.5 | Conditions necessary for growth

To grow, a plant tissue must be in a potentially growing state. A mature tissue is no longer capable of growth except in response to special stimuli such as wounding or the application of growth hormones; such stimuli restore it to a potentially growing state. The 'potentially growing state' is an internal, physiological condition that must be satisfied. Additionally, environmental conditions and other factors within the plant must be favourable. Some of the requirements for plant growth have already been mentioned in preceding chapters. The environment

Table 6.1 Temperature limits for the growth of some flowering plants. Data from Pfeffer (1903) and Stiles (1950).

Species	Minimum (°C)	Maximum (°C)
Pisum sativum (garden pea) roots	−2	44.5
Sinapis alba (white mustard)	0	>37
Triticum vulgare (= *aestivum*) (wheat)	0–5	42
Lepidium sativum (cress)	2	28
Acer platanoides (Norway maple)	7–8	26
Zea mays (sweetcorn, maize)	9	46
Cucurbita pepo (pumpkin)	14	46
Cucumis sativus (cucumber)	15–18	44–50

must provide a supply of water, O_2 and inorganic nutrients. Plants can grow only when they are turgid and in nature, as stated, water supply often limits growth. Most flowering plants require O_2 for growth, although some aquatic plants can pass through the stages of seed germination and early seedling growth under more-or-less anaerobic conditions (see Chapter 2, Section 2.10.1). For some types of growth a specific environmental stimulus may be required. The dividing and growing cells must also be nurtured by more mature parts of the plant, which supply organic nutrients and growth hormones or their precursors.

The temperature range compatible with plant growth varies from species to species (Table 6.1). Within this range there is an optimum temperature whose value will depend also on factors such as previous growth conditions and irradiance levels. Plants native to warm habitats require higher temperatures for growth than those of cooler regions. The optimum growth temperature for winter wheat (*Triticum aestivum*), a cereal of temperate climate, is 20–25 °C; for maize (*Zea mays*), a cereal from a warm climate, it is 30–35 °C.

The effect of temperature on plant growth is complex. An alternation of lower temperature by night and higher temperature by day is frequently better for growth than any one constant temperature. The optimum growth temperature varies not only between species but between organs of a plant, and changes with age. What emerges as the optimum temperature also depends on whether one measures growth over a short time interval or over a prolonged period. The temperature limits for growth are generally narrower than the temperature limits for individual physiological processes, or for survival. Respiration for instance continues at temperatures above which growth is inhibited (Fig. 6.1). Growth requires a harmonious interaction of all physiological processes. Individual chemical reactions in the cell do not all have quite the same values of Q_{10} (temperature coefficient, measure of change of rate per 10 °C temperature difference). Hence the balance of reactions needed for growth can be disturbed at temperatures not inhibitory to individual reactions or reaction series.

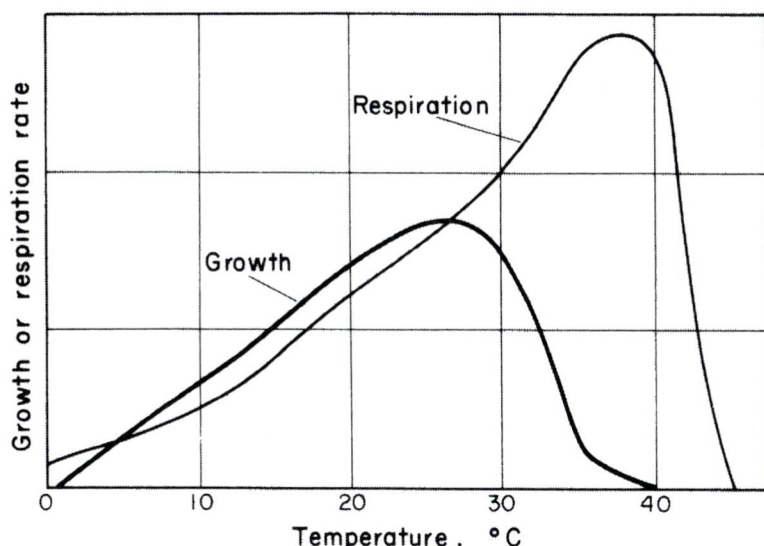

Fig. 6.1 The effect of temperature on growth (elongation) and respiration rate in the bean *Phaseolus coccineus*, measured over a 2-hour period. At extreme temperatures, growth ceases before respiration stops. From Bünning (1953). © Springer-Verlag GmbH & Co. KG.

Light is not an absolute requirement for growth. Flowering plants are normally dependent on light for photosynthesis and will starve in its absence, but they can grow in the dark if a supply of organic nutrient is available, and may be able to complete their life history in darkness. Maize has been grown from seed to seed-setting in the dark on an artificial sugar supply and *Arisaema triphyllum* (Jack-in-the-pulpit, family Araceae) has been raised in the dark from one large original corm for four successive growth seasons. But light does profoundly affect growth and differentiation; it is indeed supremely important as a morphogenetic factor, and plants grown in the dark are highly abnormal (Chapter 10). Light actually suppresses elongation growth while promoting differentiation. In the field plants often grow more in height during the night than during the day provided that the night temperature is not too low. The growth of the date palm (*Phoenix dactylifera*) is stopped completely by direct sunlight. However, the controlling factor in the day/night growth cycle may be water supply rather than light; this is the case with bamboos (*Bambusa* spp.) in various tropical regions which can grow in total height about two to three times more during the night than during the day. Many studies indicate that growth is greater in the day if temperature is the limiting factor and in the night if water is limiting. The highest growth rates often occur in the early morning (temperature rising and water content still high) and early evening (temperature still high and water content increasing).

6.6 | Growth rates

6.6.1 Comparing growth rates

Quantitative comparisons between the growth rates of living systems can be made from two viewpoints. One can measure and compare their **absolute growth rates**, i.e. the total growth per unit time; or their **relative growth rates** (RGR), the growth of each per unit time expressed on a common basis, e.g. per unit mass. To estimate plant yield the absolute amount of growth may be appropriate, but for comparing the growth activities of two different systems values of RGR are more meaningful. If two leaves with respective leaf areas of 5 cm^2 and 50 cm^2 both expand by a further 2 cm^2 in a day, the absolute growth rates are the same, but the smaller leaf has a ten times higher RGR and generally would be considered as growing faster.

The relative rates of linear elongation growth of a number of flowering plant organs, and two fungi, are compared in Table 6.2. The table is compiled from results obtained under varied experimental conditions, but the differences in growth rates are far greater than can be accounted for by differences in external conditions. Under no conditions will the growing zone of a *Vicia* root double its length per minute, as does that of the fungus *Botrytis cinerea*. Pollen tubes and staminal filaments show rates equalling or approaching those of fungal growth, but these rates are short-lived; staminal filaments expand over a period of minutes, whereas the fungal growth is maintained steadily over long periods.

Bacteria have the highest growth rates of all terrestrial living organisms; bacterial cells can double their mass and divide to form daughter cells in 20–30 minutes. The duration of a cell cycle in plants is much longer than this; in pea root tip meristems, mitosis takes one hour at 30 °C and the interval between divisions is much longer (see Chapter 8). The high growth rates of microorganisms reflect their

Table 6.2 Relative growth rates (RGR), using linear elongation as the growth measure, of some higher plant organs and, for comparison, hyphae of two species of fungi. Sources of data as for Table 6.1.

Organ	Species	RGR[a]
Pollen tubes	*Impatiens hawkeri* (New Guinea balsam)	220
	Impatiens balsamina (garden balsam)	100
Staminal filaments	*Triticum* sp. (wheat); *Secale* sp. (rye)	37.5
Shoot growing zones	*Bambusa* sp. (bamboo)	1.27
	Bryonia sp. (white bryony)	0.58
Root, fastest growing zone	*Vicia faba* (broad bean)	0.36
Fungal hyphae	*Botrytis cinerea*	83–200
	Mucor (= *Rhizopus*) *stolonifer*	118

[a] RGR is expresed as percentage increase per unit length of growing zone per minute, i.e.

$$\frac{\text{length increase per min (mm)}}{\text{length of the growing zone (mm)}} \times 100$$

generally high rates of physiological activity and are believed to be related to their small cell size, which allows rapid diffusion of metabolites and gases into and out of the cells.

Table 6.2 presents the RGR, growth per unit length of growing zone. The absolute growth rates depend on these values and on the actual lengths of the growing zones. In the hyphae of *Botrytis* the growing zone is only 0.018 mm long, and even with a 200% increase in length per minute the total extension growth made in 24 hours, the daily absolute growth rate, is about 5 cm. In the bamboo shoot, with a RGR of only 1.27% per minute, the growing zone is 5 cm long and hence the daily absolute growth rate is 90 cm.

In the above examples, length has been used as the growth measure. Other bases can be used for calculating relative growth rates, e.g. mass increase per unit mass is very widely utilized. Once the basis of calculating the RGR has been stated, the growth measure unit may be omitted in quoting the RGR numerically, since it cancels out in the ratio:

$$\text{RGR} = 0.2\,\text{mm per mm per day} = \frac{0.2\,\text{mm}}{} \times \frac{1}{\text{mm day}} = 0.2\,\text{d}^{-1} \quad (6.1)$$

The extent of total growth achieved during a given time period varies enormously between plant species. A marrow plant (*Cucurbita* sp.) or a hop (*Humulus lupulus*) grows from seed to a length of 12 m in a summer; an oak seedling may grow 12 cm in the same time, and since mass varies as [length]^3, the differences in mass are greater still. Marrow and hop are plants with large growing zones and produce much soft, thin-walled tissue; a large proportion of the plant mass is green photosynthetic tissue. The oak, a perennial, has a small growing zone and is woody; it can be regarded as showing a higher degree of differentiation, and growth and differentiation are to some degree mutually antagonistic. Generally, herbaceous species attain higher RGR at comparable developmental stages and under comparable external conditions (Table 6.3). Herbaceous plants of open, disturbed habitats, including species regarded as weeds, have the highest RGR. Such differences are genetically programmed. Even individuals of the same species can show genetically determined variations in growth rates. This has been found for example in a study of the RGR in 48

Table 6.3 | Relative growth rates (RGR), as dry weight increase per day in herbaceous and woody plants measured over 14 days (herbaceous) or 21 days (woody). Young but photosynthetic seedlings were studied, grown from 43 herbaceous and 16 woody species derived from a wide range of natural habitats. Data from Hunt & Cornelissen (1997).

	RGR d^{-1}	
	Herbaceous	Woody
Average	0.20	0.09
Range	0.10–0.31	0.01–0.14

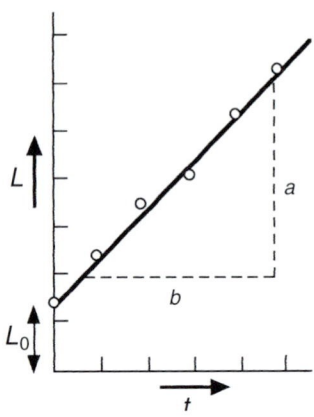

Fig.6.2 Constant linear growth in length (diagrammatic). A plot of total length L against time t gives a straight line. See text for further explanation.

clones of white clover (*Trifolium repens*) sampled from the same field: the RGR varied almost two-fold between individual clones.

6.6.2 Mathematical analysis of growth

Growth often follows regular patterns which lend themselves to mathematical analysis, and formulae can be used to express various types of growth. The simplest case is that of **constant linear** or **arithmetic growth**, as shown for example by a root elongating at a constant rate. If one denotes the length at zero time by L_0, the length at time t by L_t and the growth rate (elongation per unit time) by r, then

$$L_t = L_0 + rt \tag{6.2}$$

If L is plotted against time t, the graph is a straight line (Fig. 6.2), the intercept of the graph is L_0 and the slope a/b gives the growth rate r. Once r is known, the length L at any time can be calculated from Formula 6.2.

Constant linear growth is observed only in relatively few growing systems. More frequently, the linear growth rate does not remain constant. In an entire root system, any one tip may be elongating at a steady linear rate, but more and more tips are continuously formed and hence the linear rate of elongation of the root system as a whole keeps increasing. The relative growth rate per unit of growing mass may still remain constant, and in that case there is **constant exponential** (or **logarithmic**) growth. If the relative growth rate is denoted by r', and the other symbols are as for Equation 6.2, then

$$L_t = L_0 \, e^{r't} \tag{6.3}$$

where e is the base of natural logarithms. This is the formula for continuous compound interest. In this case a plot of L against time will give a curve as shown in Fig. 6.3A. To obtain a straight line the logarithm of L must be plotted against t (Fig. 6.3B). From Equation 6.3, taking logs to base e,

$$\ln L_t = \ln L_0 + r't \tag{6.4}$$

or, converting to logs to base 10,

$$\log L_t = \log L_0 + \log e \, r't \tag{6.5}$$

Then in the graph, the intercept gives $\log L_0$ and the slope a/b gives $\log e \, r' = 0.43 \, r'$.

A period of such exponential growth is found in the development of many living systems but, since it implies growth at an ever-increasing rate, it obviously can proceed only for a limited period in the development of an organ or organism. If the mass of a plant organ (or a whole plant) is plotted against time from the start to the cessation of growth, the graph is an S-shaped or *sigmoid* curve as illustrated in Fig. 6.4 for a single cucumber leaf (*Cucumis sativus*). This curve shows that the growth is slow initially but then speeds up and for a time approximates to exponential, later slowing down and finally stopping. With the onset of

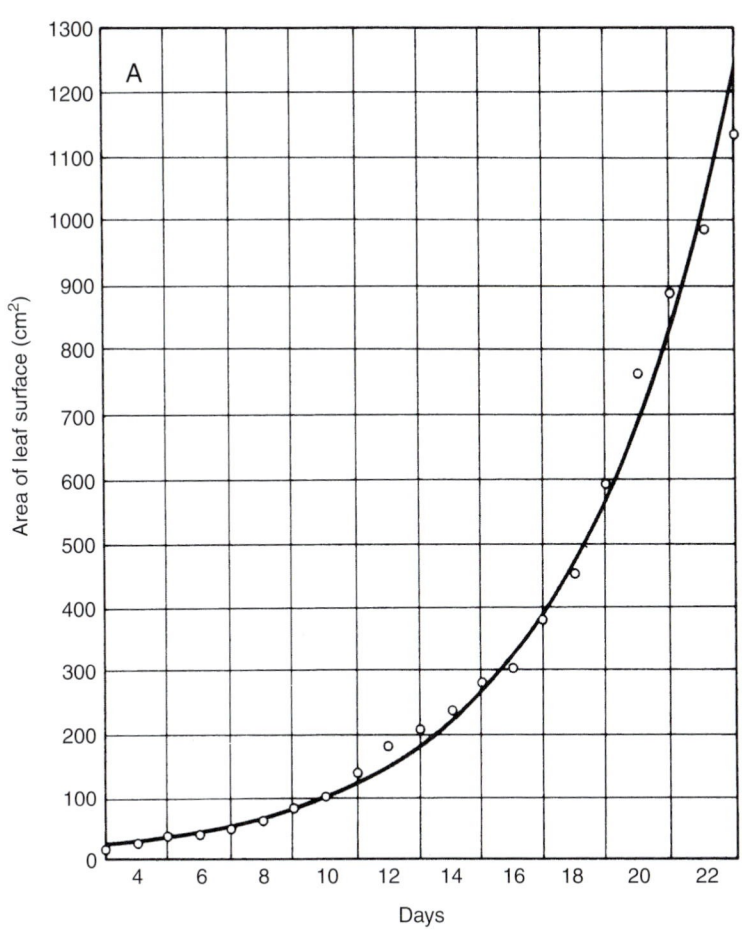

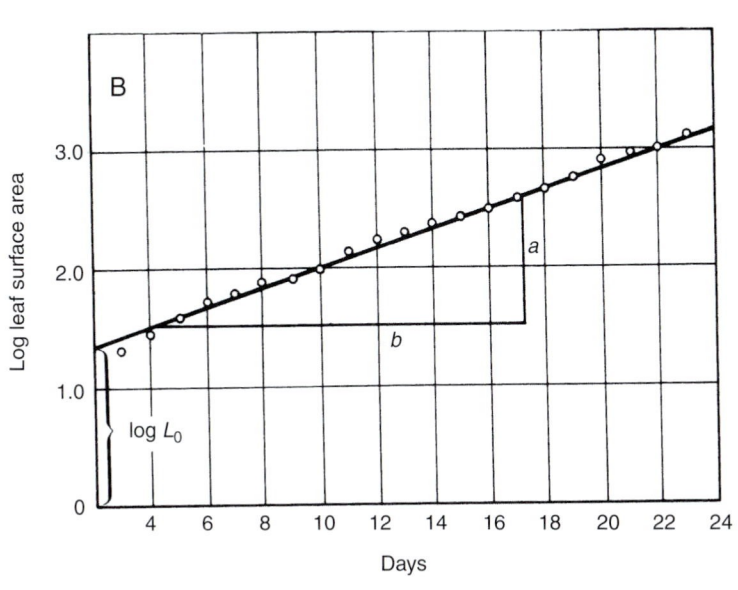

Fig.6.3 Constant exponential growth as illustrated by the increase in *total* leaf area of a cucumber plant (*Cucumis sativus*). (A) A plot of the leaf area (equivalent to L in Equation 6.3) against time gives a curve of ever-increasing steepness. (B) A plot of the logarithm of the leaf area against time gives a straight line. From Gregory (1921).

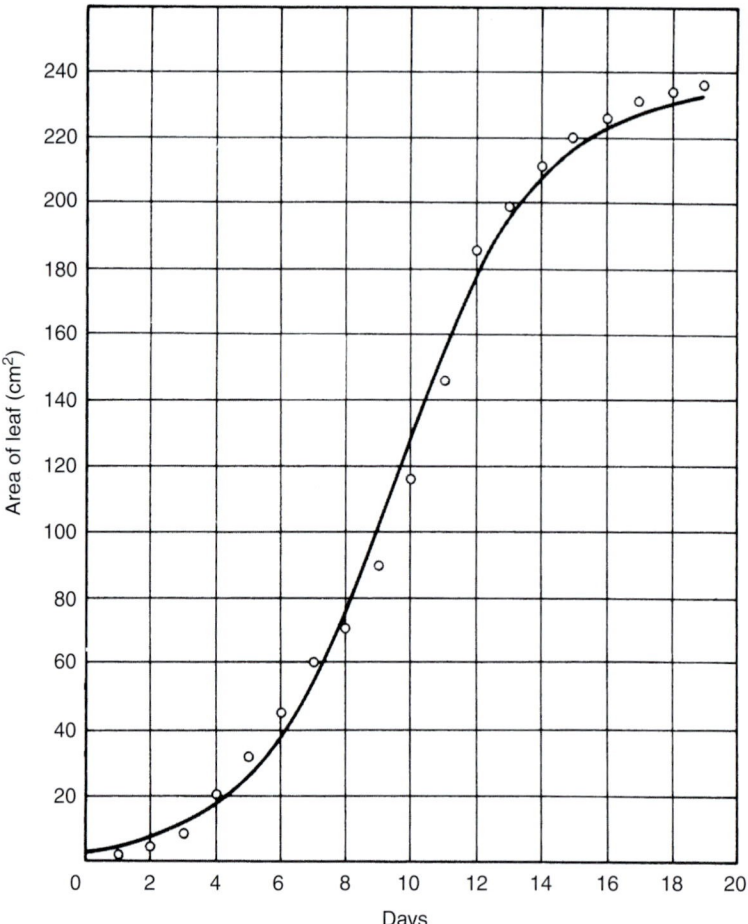

Fig. 6.4 The grand period of growth of a single cucumber leaf (*Cucumis sativus.*) Curve for the growth of a whole plant would be similar. From Gregory (1921).

senescence there may occur a loss of mass. This sequence covers the so-called **grand period of growth** of an organ or organism. It is not difficult to see how the period of exponential growth rate is limited. In the very young plant, mass increase means an increase in growing points, in photosynthetic area and in absorptive young roots. The growth potential increases and, as growth rate increases, the growth potential rises even more. However, not all the new tissue remains growing nor does it necessarily add to the synthetic and absorptive capacity of the plant. The proportion of non-growing and non-photosynthetic tissue soon increases and growth rate declines. During senescence there is finally a loss of mass resulting from an excess of respiration over photosynthesis and from abscission of organs. A phase of vegetative growth is eventually followed by one of reproductive growth, marked by the shoot apex changing quite abruptly from the production of vegetative organs to the production of flowers. This may be reflected in the growth curve of the plant as a new surge of growth after the grand period curve of vegetative growth has levelled off, while floral organs and fruits pass through their individual grand periods. Reproductive growth is discussed in Chapter 11.

The smooth growth curves shown in Fig. 6.3 and 6.4 were obtained with plant material grown under more-or-less uniform external conditions. In the field, environmental conditions which influence growth fluctuate considerably, especially temperature and water supply. These fluctuations are reflected in plant growth rates, but the underlying pattern of the grand period can still be recognized.

The growth analysis formulae and curves so far discussed are applicable to living organisms in general, not just to flowering plants. There is a special measure of RGR which is designed for leafy plants. Since the growth potential of a plant depends very much on its rate of photosynthesis, the growth rate may be calculated per unit *leaf area*, i.e. per unit of photosynthetic area. This gives the **net assimilation rate** (NAR), also called unit leaf rate, or specific leaf rate:

$$\text{NAR} = \frac{\text{dry weight increase of whole plant}}{\text{unit time}} \times \frac{1}{\text{leaf area}} \qquad (6.6)$$

This way of calculating the RGR gives an estimate of the efficiency of the photosynthetic tissues in supporting growth.

Growth ratios: allometric growth

Where a constant ratio is maintained between the growth rates of different parts of the plant these parts are said to show **allometric growth** (heterogonic growth). This relationship can be expressed by the allometry formula. If x and y represent the sizes of two plant parts growing allometrically, and k equals the ratio of their growth rates, then

$$y = bx^k \qquad (6.7)$$

where b is a constant (b = the value of y when x is taken as unity). Taking logarithms,

$$\log y = \log b + k \log x \qquad (6.8)$$

A plot of $\log x$ against $\log y$ gives a straight line and the slope of the graph gives k. If both parts are growing at equal rates, the slope of the graph is $45°$ and $k = 1$. Figure 6.5 shows an allometric relationship exhibited by the growth of roots and shoots of some plant species. An allometric relationship has also been observed in many cases between the growth rates of different parts of leaves, and between the growth rates of the various organs in developing flowers. The functional significance of allometric growth is obvious: since all parts of a plant are physiologically interdependent, they need to grow in certain proportions so that e.g. the root system is of the 'right' mass to supply the leaves with water and minerals. This implies signalling between the allometrically growing parts; both hormones and metabolites such as sucrose (Section 5.5) may be involved. The value of k changes in response to external conditions; e.g. when plants are placed in conditions of low mineral nutrient supply, the ratio of root growth to shoot growth (on a dry-mass basis) increases.

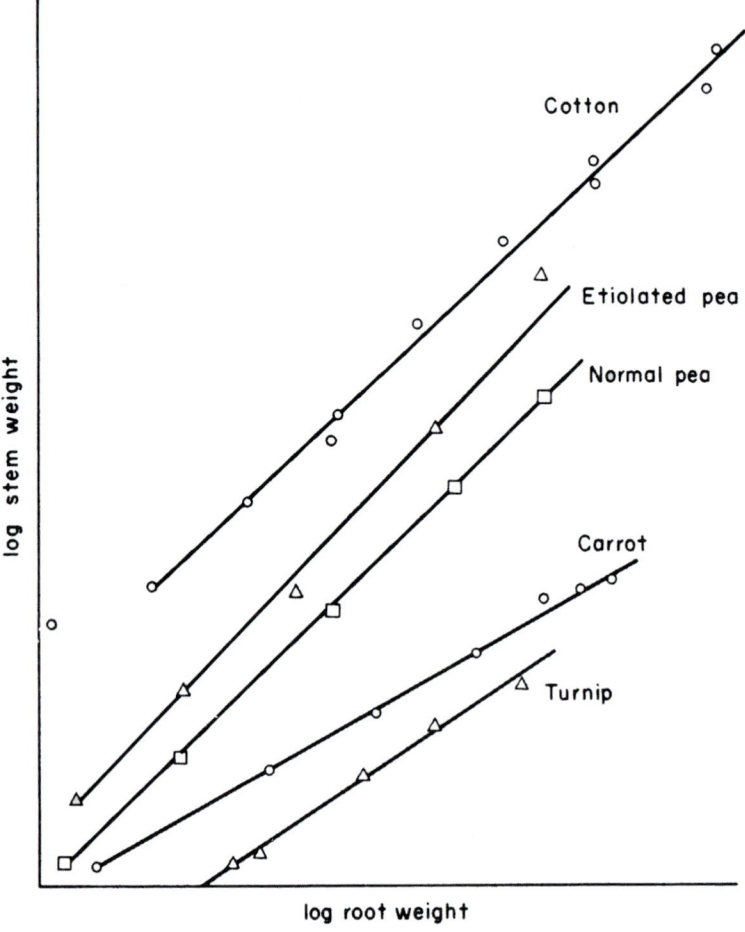

Fig. 6.5 Allometric growth in weight shown between stems and roots of a number of species. The vertical scale is reduced to one-half in the case of etiolated peas. From Pearsall (1927).

6.6.3 Growth rhythms

Diurnal rhythms

If measurements of growth are made at short time intervals, say hourly, then rhythmic changes in growth rate become apparent. Growth has a **diurnal rhythm**, with maxima and minima occurring at definite times of day. This rhythm conforms to the general pattern of endogenous diurnal rhythms controlled by the 'biological clock' mechanism (in conjunction with the regular diurnal alternation of light and darkness) and exhibited by all types of living organisms. A plant grown from seed in darkness and under completely constant conditions does not show a diurnal growth rhythm; but once a rhythm has been initiated – a single period of illumination may suffice for this – the rhythm persists with an approximately 24-hour periodicity in the darkness for several days. Plants may grow better with an alternation of light and darkness than in continuous light, the alternating regime harmonizing with their natural rhythmicity. The alternation of light : dark must fit an approximately

24-hour cycle. The growth of tomato plants (*Lycopersicon esculentum*) is inhibited if the plants are subjected to light : dark cycles of 6 : 6 or 24 : 24 hours; the endogenous rhythm of the plants apparently cannot adjust to cycles so far removed from the natural. A diurnal growth cycle may show more than one maximum in 24 hours.

Short-term rhythms

Growth rhythms not related to the 24-hour period are also known. Growing shoot tips bend so as to rotate in a circle as viewed from above. This movement is known as **circumnutation** or **nutation** and is most pronounced in climbers, in which it aids the plant to find and to twine round a support. Tendrils also nutate till they contact a support. Nutation involves a wave of cell extension growth moving round the axis tip; at any instant growth is most intense in one limited region of the tip and the wave of activity completes a cycle round the tip in anything from 1.25 to 24 hours according to species and environmental conditions. Changes in cell turgor, however, also contribute to the bending. The nutational rhythm is not dependent on any rhythm in the external conditions, though the speed of rotation is affected by environmental factors such as temperature. The direction of nutation is fixed and in most species is anticlockwise as viewed from the top.

Annual and other long-term rhythms

The grand period of growth represents the entire life history of ephemerals and annuals; in perennials it corresponds to the course of growth during one growing season. In most temperate-zone perennials growth has a regular annual rhythm, with shoot growth completed within a short period in the spring and early summer; some trees for example complete 90% of a year's growth in a 30-day period, starting 7–14 days after the commencement of growth. The first days represent the 'slow phase' of the grand period (Fig. 6.4). Such an early cessation of growth during prime climatic conditions must be the result of an internal control mechanism. Sometimes a second flush of growth occurs later in the season. Root growth continues longer into the summer: plants must tap fresh areas of the soil for water and minerals even when the shoots are not growing.

The annual growth cycle of plants living in a climate with distinct seasonal changes is synchronized with the climatic cycle so that the growth period coincides with the favourable season, environmental signals such as changes in daylength and temperature serving as synchronizing agents (Chapter 11). Nevertheless the endogenous rhythm also plays a part. Trees of the same species and age, growing side by side, may show slight differences in their times of e.g. leaf fall, which are consistent from year to year, with the same individuals being the earliest and the latest in each season.

In tropical habitats, growth of a plant community as a whole continues all the year round with almost equal intensity, but in individual plants periods of high and low growth activity alternate, each lasting some months. Perennials seem to have an innate tendency for an alternation of growth and rest periods. In tropical climates where the environment does not exercise a synchronizing influence, each plant – sometimes even a branch of a plant – grows according to its own internal rhythm.

Complementary reading

Bell, A. D. *Plant Form: an Illustrated Guide to Flowering Plant Morphology*. Oxford: Oxford University Press, 1991.

Burdon, J. J. & Harper, J. L. Relative growth rates of individual members of a plant population. *Journal of Ecology*, **68** (1980), 953–7.

Caré, A,-F., Nefed'ev, L., Bonnet, B., Millet, B. & Badot, P.-M. Cell elongation and revolving movement in *Phaseolus vulgaris* L. twining shoots. *Plant and Cell Physiology*, **39** (1998), 914–21.

Groeneveld, H. W., Bergkotte, M. & Lambers, H. Leaf growth in the fast-growing *Holcus lanatus* and the slow-growing *Deschampsia flexuosa*: tissue maturation. *Journal of Experimental Botany*, **49** (1998), 1509–17.

Hunt, R. *Basic Growth Analysis*. London: Unwin Hyman, 1990

Hunt, R. & Lloyd, P. S. Growth and partitioning. *New Phytologist*, **106** (1987), Suppl. 1, 235–49.

Morey, P. R. *How Trees Grow*. Studies in Biology, 39. London: Edward Arnold, 1973.

Scheurwater, I., Cornelissen, C., Dictus, F., Welschen, R. & Lambers, H. Why do fast- and slow-growing grass species differ so little in their rate of root respiration, considering the large differences in rate of growth and ion uptake? *Plant, Cell and Environment*, **21** (1998), 995–1005.

Thomas, H., Thomas, H. M. & Ougham, H. Annuality, perenniality and cell death. *Journal of Experimental Botany*, **51** (2000), 1781–8.

Whaley, W. G. Growth as a general process. In *Encyclopaedia of Plant Physiology*, vol. **14**, ed. W. Ruhland. Berlin: Springer-Verlag, 1961, pp. 71–112.

References

Bünning, E. (1953). *Entwicklungs- und Bewegungsphysiologie der Pflanze*. Berlin: Springer.

Gregory, F. G. (1921). Studies in the energy relations of plants. I. The increase in area of leaves and leaf surface of *Cucumis sativus*. *Annals of Botany*, **35**, 93–123.

Hunt, R. & Cornelissen, J. H. C. (1997). Components of relative growth rate and their interrelations in 59 plant species. *New Phytologist*, **135**, 395–417.

Pearsall, W. H. (1927). Growth studies VI. On the relative sizes of growing plant organs. *Annals of Botany*, **41**, 549–56.

Pfeffer, W. (1903). *The Physiology of Plants*, Vol. 2. Oxford: Clarendon Press.

Schmundt, D., Stitt, M., Jähne, B. & Schurr, U. (1998). Quantitative analysis of the local rates of growth of dicot leaves at a high temporal and spatial resolution, using image sequence analysis. *The Plant Journal*, **16**, 505–14.

Stiles, W. (1950). *An Introduction to the Principles of Plant Physiology*. London: Methuen.

Chapter 7

Plant growth hormones

7.1 | Introduction

A constant theme underlying the study of plant physiology is that plant growth and development are controlled by the environment. Plants being sessile organisms, it is not surprising that their development is exquisitely sensitive to a wide range of environmental factors and is extremely plastic, i.e. very flexible. There are underlying basic patterns in plant development, but there is considerable regulation by environmental signals of how and when these patterns are expressed.

In addition, there are internal signals within the plant. One of the most important factors influencing the development of a cell is its position within the plant. A plant cell develops depending on its location in relation to neighbouring cells, and this in turn will determine its response to environmental signals. For example, the response to drought of a cell within the leaf will differ in many ways from that of a cell within the root. The key question arises of how a complex set of environmental factors can interact with cells to elicit an appropriate response within a given cell type: what are the internal signals that communicate between cells, and mediate between environmental factors and the plant tissues?

It has been known for decades (if not centuries) that plants contain a range of compounds which have profound effects on many aspects of growth and developmental physiology, and act as a means of communication within the plant. These **plant growth hormones**, sometimes referred to as plant growth regulators, are still being discovered. They are of supreme importance in controlling cell division, growth and differentiation, which in turn determine the morphology and ultimately the physiology of the whole plant. The concentrations of the hormones are greatly influenced by environmental factors, and the movement of hormones in the plant can also be under environmental control. Although the mechanisms by which they act cannot be considered to be well understood, we are beginning to unravel their modes of action and the ways in which they mutually interact to regulate plant functions. An overview of the essential features of plant hormones is given here; specific roles and functions are discussed in later chapters.

7.2 | Plant growth hormones

7.2.1 Concepts and definitions

In any area of biology, attempting to formulate definitions is difficult and this is especially true of plant growth hormones. One reasonable definition might be:

> *A plant growth hormone is a substance produced within the plant in very low concentrations and transported to another part of the plant where it causes a response.*

There are a number of key concepts in this definition. Firstly, plant growth hormones are naturally occurring compounds synthesized within the plant. However, similar if not identical compounds can be synthesized by other organisms. For example, many plant pathogens produce substances similar to plant hormones which are important in the progression of the disease. Man-made plant hormones are commercially extremely important as they have such a diverse range of effects on plant development from acting as weedkillers to stimulating fruit development.

The second key concept is that of transport. Plant growth hormones are synthesized throughout the plant but apical meristems (Section 9.2) and young, developing tissues are rich sources of these compounds. They are transported, perhaps through only a few cells, but often throughout the plant, giving rise to a response in target tissues. It is possible, and has been proposed, that volatile chemical signals might be exchanged *between* plants although the importance of this in field conditions remains controversial.

Thirdly, these compounds cause a response. Different cells will respond in different ways (or not at all) to a particular hormone, giving rise to the concept of '**competence to respond**' – that is whether a particular cell type will respond to a given concentration of a plant growth regulator or not.

Fourthly, the hormones are effective at low concentrations. This would rule out compounds such as sucrose which is produced within the leaves by photosynthesis, is transported throughout the plant in the phloem, and has a profound effect on plant development, but is found at high concentrations. Sucrose has been considered by some to be a plant growth hormone. However, in this chapter the use of the term is limited to compounds which have traditionally been classed as plant growth hormones. These have the following characteristics which can be added to the definition above.

In general :

- Plant growth hormones are small, relatively simple compounds.
- Specific receptors exist which bind these compounds.
- Often the presence of one plant growth hormone will affect the synthesis or action of other plant growth hormones.

The concentration of a plant growth hormone at a particular site will depend upon many different factors including the rate of synthesis, degradation and transport to and from the target cell. In addition, plant growth hormones are often chemically modified, which may inactivate them, although – as this process is often reversible – it can also increase the effective concentration of a plant growth hormone in a cell. Finally, as the activity of plant hormones is thought to require binding to specific receptors, transport in and out of subcellular compartments also controls the concentration perceived by the cell. As all of these processes have the potential to be regulated by the environment, it can be seen that plant growth hormones act as a means of integrating environmental signals and distributing them around the plant.

Classically there are five major groups of plant growth hormones. These are the auxins, gibberellins, cytokinins, abscisins and ethylene. However, many more compounds exist in plants which have all of the characteristics of the more established plant growth hormones, and their role in regulating plant development is becoming clearer. These compounds include the jasmonates, salicylates, brassinosteroids and peptide hormones. Recently, small RNA molecules have been identified which regulate gene expression in different parts of the plants; these too might be considered to fulfil many of the criteria of being a plant growth hormone.

7.2.2 Auxins

The first plant growth hormone to be discovered was **auxin** – the name is derived from the Greek *auxein* meaning 'to increase'. At low concentrations (10^{-12} to 10^{-3} M) auxin promotes the elongation of certain plant organs, principally as a result of cell expansion. Figure 7.1 shows the development of a typical grass seedling and the structure of the coleoptile which protects the first true leaf. In many grasses the coleoptile increases in length from under 10 mm to over 80 mm as the cells within it expand. If the tip of the coleoptile is excised, growth is much reduced, but this can be restored if it is replaced. A key breakthrough came when it was found that this restoration of growth would occur even if a small agar block was placed between the tip and the lower part of the coleoptile. This demonstrated that a diffusible substance (auxin) was passing from the tip, through the block, and then stimulating growth lower in the coleoptile. If a series of excised tips is placed on the block, the diffusible substance accumulates within it and the block alone can then stimulate growth. Placing the block (or tip) asymmetrically on the coleoptile causes directional growth and auxin-stimulated growth is a central feature of many directional growth responses or tropisms (see Chapter 12).

Eventually the chemical structure of the active substance was determined as indole-3-acetic acid (IAA) (Fig. 7.2). The coleoptile

Fig. 7.1 Development and structure of a grass seedling. (A, B, C) Stages in germination; stage (C) shows emergence of the first leaf from the apex of the coleoptile. (D) Section taken at stage A; a = coleoptile; b = first leaf; c = apical meristem; d = node; the region below the node is referred to as the *mesocotyl*. (E) Transverse section through the coleoptile; e = coleoptile; f = first leaf; vascular strands in the coleoptile and first leaf are shown as filled circles. Redrawn from James (1943).

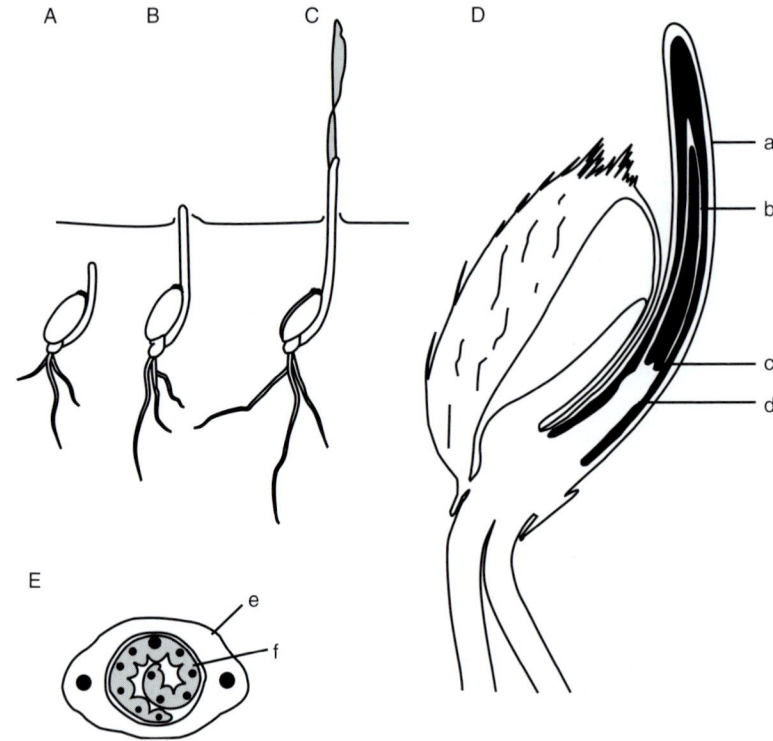

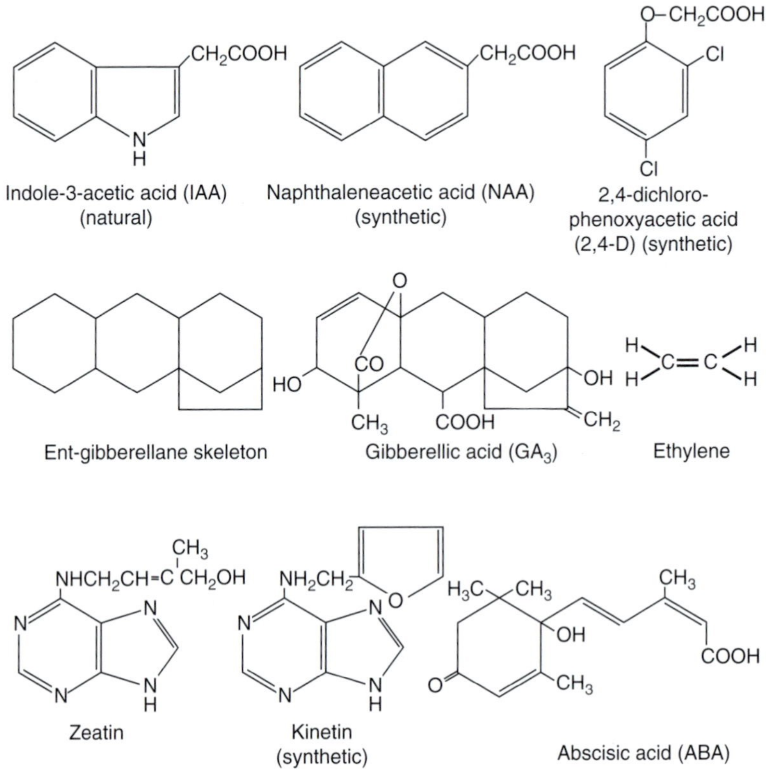

Fig. 7.2 The structures of some natural and synthetic plant hormones. Indole-3-acetic-acid (IAA) is the most common natural auxin found in plants. The synthetic auxins naphthaleneacetic acid (NAA) and 2,4-dichlorophenoxyacetic acid (2,4-D) are widely used in agriculture. All of the gibberellins are based upon the ent-gibberellane skeleton. The structure of a common, naturally occurring gibberellic acid, GA_3, is shown. Ethylene (ethene) has a much simpler structure than other plant hormones. The natural cytokinin, zeatin, was first identified in maize (*Zea mays*), although synthetic homologues, such as kinetin, are more widely used for plant tissue culture. The best known representative of the abscisins is abscisic acid (ABA).

system provided a quantitative and sensitive *bioassay* for auxin which was used for many years although it has now been superseded by modern analytical techniques. If agar blocks are placed asymmetrically on decapitated oat (*Avena sativa*) coleoptiles, the degree of curvature under standardized conditions is related to the auxin concentration. Similarly if coleoptile or hypocotyl segments are floated on different concentrations of auxin solution, the growth rate is logarithmically related to the auxin concentration. Such a system illustrates well the essential features of a plant growth hormone – auxin is produced in the tip of the coleoptile and transported down the coleoptile stimulating a response, cell elongation. The growth response which is observed depends upon both the concentration of the hormone and the target tissue. Figure 7.3 illustrates the growth responses of roots and shoots to different concentrations of auxin. Note that the concentration ranges which stimulate growth are quite different, and in both roots and shoots either promotion or inhibition of growth can result from exposure to auxin depending upon the concentration.

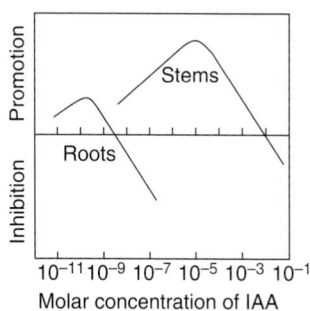

Fig. 7.3 Growth responses of roots and shoots to different concentrations of auxin. Indole-3-acetic acid (IAA) can promote or inhibit the linear growth of roots and shoots depending upon concentration. Note that the auxin concentration is shown on a logarithmic axis. After Thimann (1937).

The structure of IAA is relatively simple and it is the most commonly found natural auxin although many other related compounds with auxin-like actions can be found within plants. A general feature of these compounds is that they have a carboxylic acid group linked to an aromatic ring. Many synthetic auxins have been produced such as naphthaleneacetic acid (NAA) and 2,4-dichlorophenoxyacetic acid (2,4-D) (Fig. 7.2), and these are widely used in agriculture. As well as these simple, free auxins plants also contain large amounts of **conjugated** auxins in which the auxin is covalently bound to another molecule (ranging from simple sugars to proteins). These conjugated auxins are inactive and are transported via the phloem to other parts of the plant where they may be deconjugated, and hence activated. However, the growth responses of coleoptiles described above result from the basipetal (basewards) transport of free auxin from cell to cell (discussed in more detail in Chapter 12).

As well as stimulating growth, auxins have many other effects on plant development. For example, auxin plays a key role in apical dominance. As seen in later chapters, auxin is also important in regulating the formation of lateral roots (Section 9.6.3), the development of the vascular system (Section 9.5.3), parthenocarpy (Section 11.11.3) and in senescence. This diversity of function led K. V. Thimann to state 'The trouble with auxin is that its actions are so numerous and apparently unrelated' (in Palme & Galweiler 1999). Although its actions are certainly numerous, it is clear that its functions can be understood only when interactions with other plant growth hormones are considered. This interaction between plant growth hormones is a theme which will continually arise, and it is fundamental to the way in which plants respond to interacting, and at times conflicting, environmental signals.

Auxin and apical dominance

It has been recognized for many years that auxin, produced in the shoot apex, can repress the development of axillary buds in a process known as **apical dominance**. If the shoot apex is removed, this inhibition is released and the axillary buds develop as side shoots. In many species, application of auxin to the decapitated stump can substitute for the presence of the shoot apex, preventing axillary bud growth. However, auxin does not inhibit axillary bud development directly – it is not transported into the buds and direct application does not inhibit their growth. It has been proposed that the plant growth hormone cytokinin (Section 7.2.4) plays a role in the relief of apical dominance, as application of cytokinin to axillary buds will stimulate their development and cytokinin export from the root increases following shoot apex decapitation. However, recent grafting experiments have indicated that other mobile signals are also involved (Beveridge *et al.* 2003, Sorefan *et al.* 2003). Mutants of pea and *Arabidopsis* have been identified which exhibit increased shoot branching. If the shoots of these plants are grafted onto an unmutated rootstock, normal branching patterns are restored. It is proposed that auxin travels down the stem and generates a mobile, branch-inhibiting signal which is then transported upwards (presumably in the xylem) inhibiting axillary bud development. The mobile signal has yet to be identified, but current evidence points towards a compound derived from carotenoids.

7.2.3 Gibberellins

Gibberellins, like auxin, are also growth-promoting compounds which are active at low concentrations. They were first isolated from the rice fungal pathogen *Gibberella fujikuroi*, which causes infected plants to grow very tall, and it was only later that related compounds were isolated from healthy plants. Plants contain many different gibberellins although their chemical structure is more clearly defined than that of auxin, all being based upon a 5-ring gibberellane skeleton (Fig. 7.2) with different side groups. Over 110 different gibberellins are known but relatively few (GA_1, GA_3, GA_4 and GA_7) are biologically active – the rest are thought to be either precursors or breakdown products (Swain & Olszewski 1996).

Although the chemical structure of gibberellic acid was not determined until the 1950s, plants with alterations in GA biosynthesis or sensitivity to GA have been known for many centuries as these plants are dwarf. One of the traits examined in the famous experiments on heredity in peas by Gregor Mendel was dwarf vs. tall plants (controlled by the *Le* alleles – *length*) and it is now known that dwarf plants have a lesion in GA biosynthesis (Martin *et al.* 1997). Application of different concentrations of GA to these plants causes internode elongation, and this was used as a bioassay for these compounds for many years. In agreement with the idea that gibberellins are growth-promoting substances, mutant peas which have a greater content of biologically active GA are elongated.

The introduction of dwarfing genes into plant breeding schemes was an essential component of the 'Green Revolution' – the programme of crop improvement which increased grain yields massively in the 1960s and 1970s. Dwarf cereal varieties have a short and stocky stem which is strong enough to support the heavy head of grain produced as a result of breeding programmes and the application of nitrogenous fertilizers. Non-dwarf varieties are susceptible to *lodging* – the stems collapsing under the weight of the developing grain. The short stature of a semi-dwarf variety of rice (*Oryza sativa*) known as IR8, one of the first high-yielding varieties, is controlled by the *sd1* allele. This encodes a GA_{20} oxidase, a key enzyme in the biosynthesis of GA_1 (Sasaki *et al.* 2002). In contrast the *Rht* (reduced height) alleles, which cause dwarfing in wheat, have lesions in GA perception but the end result, a dwarf plant, is the same (Peng *et al.* 1999). Figure 7.4 shows wheat (*Triticum aestivum*) varieties with different *Rht* alleles. The difference in stature between the dwarf and non-dwarf plants is marked. If similar alleles, isolated from GA-insensitive mutants of the small crucifer *Arabidopsis thaliana*, are transferred into other cereals such as rice, they too become dwarf. As well as dwarfing genes, chemical inhibitors of GA biosynthesis can be used to similar effect.

The effects of GAs are not limited to stimulating growth. They have been demonstrated to have many roles including the promotion of parthenocarpy in some species, delaying leaf senescence and abscission, dormancy breaking in seeds (and the mobilization of seed reserves) and the stimulation of flowering in long day plants (Section 11.3.6). Again it is possible to make sense of this multiplicity

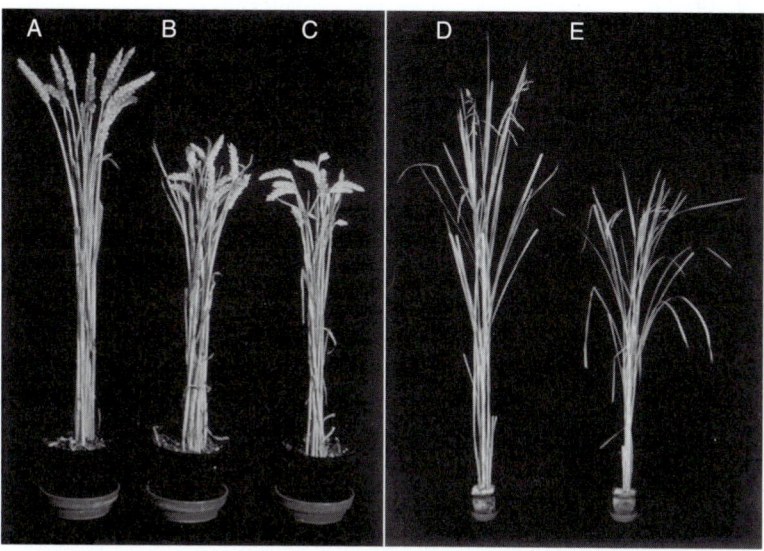

Fig. 7.4 Gibberellin-insensitive wheat *Triticum aestivum* and rice *Oryza sativa* plants are reduced in stature. The Mercia variety of wheat (A) is tall whilst plants containing the *Reduced height-1* alleles *Rht-B1b* (B) or *Rht-D1b* (C) are semi-dwarf. If the orthologous gene from *Arabidopsis* (*gai*) is introduced into Basmati rice plants stature is reduced. (D) Basmati rice; (E) semi-dwarf rice plant containing the *Arabidopsis gai* coding region. Reproduced from Peng *et al.* (1999). © Nature Publishing Group.

of effects only when the interactions of these compounds with other plant growth hormones are considered.

7.2.4 Cytokinins

The cytokinins were discovered after many years of effort in attempting to find compounds which control cell division in plants (*cytokinesis –* cell movement – an important feature of cell division: Section 8.2.2). Plants contain many different cytokinins which are based around 6-substituted adenine derivatives (Fig. 7.2) and these may be conjugated to a variety of other compounds. Cytokinins are important in cell culture where, at low concentrations, they stimulate cell division. Synthetic forms, such as kinetin, are widely used both commercially and in laboratory studies for plant tissue culture.

The list of actions of cytokinins is very long and in many ways overlaps that of GAs and auxins, including roles in cell division and elongation, the promotion of parthenocarpy and flowering, breaking of dormancy, the control of apical dominance and the delay of senescence. Again, this overlap is most probably a result of interactions with other plant growth hormones rather than a duplication of role. Perhaps one of the best-studied examples of such an interaction is the relationship between auxin and cytokinin concentration in determining the development of plant tissue in culture. Skoog and Miller (1957) demonstrated that it is the *ratio* of auxin to cytokinin, rather than the absolute concentrations of these plant growth hormones, which determines the tissues which will be produced.

7.2.5 Plant tissue culture

Modern plant breeding makes extensive use of tissue culture techniques. If plant tissue is placed in a nutritive medium, and supplied with precise amounts of auxin and cytokinin, it will develop as a mass of undifferentiated cells known as **callus**. If these cells are then transferred to a medium in which the concentration of cytokinin has been increased relative to that of auxin, the callus is stimulated to differentiate into shoots. After a period of growth, the shoots can be transferred into a medium with a high auxin content, stimulating root development. In this way an entire plant can be regenerated from a single cell, or many genetically-identical clones can be generated from a single plant.

The ability to manipulate plant growth and development underpins many procedures used in plant breeding. In the 1950s, cells isolated from the pith of tobacco (*Nicotiana tabacum*) stems were cultured and then induced to form new plants – this process is referred to as **somatic embryogenesis**, as embryonic material is generated from vegetative (somatic) cells. It is even possible to generate haploid plants (i.e. plants with only a single copy of each chromosome) by culturing pollen grains, which can then be induced to double their chromosome number by exposure to drugs such as colchicine. This technique allows a true-breeding, homozygous line of plants to be produced in a few months, rather than the years required by traditional methods. Another method allows the production of hybrids between plants which are not sexually compatible – hybrids can be generated by removing the cellulose cell wall by enzymatic digestion and fusing the released protoplasts. With much skill, and patience, the fused cells can be induced to re-synthesize the cell wall and grow and differentiate in culture. These techniques are of great interest to plant breeders as they allow desirable traits to be identified rapidly or transferred between otherwise incompatible species. In fact the process of tissue culture often generates genetic variation resulting from mutation. Whilst this may be a useful way of producing plants with new traits, in other instances it may prove problematical. Plant material maintained in culture for extended periods often loses its ability to regenerate into a viable, entire plant (Section 8.2.1).

On the whole, the general public welcomed these advances in agriculture which formed an important component of the 'Green Revolution'. However, more recently, public perception has shifted as concerns over the genetic modification of crops have been raised. The underlying principles of genetic modification are discussed below, and the reader is referred to Chrispeels and Sadava (2003) for more in-depth information.

7.2.6 The genetic modification of plants

Although many have argued that we have been genetically modifying plants for millennia, modern genetic modification differs fundamentally in that species barriers are completely abolished. Artificial genes can be made, using the **recombinant DNA technology**, which include DNA sequences from different species (not restricted to plants) and even synthesized chemically within the laboratory. These genes can be combined to produce useful traits (e.g. herbicide resistance, alterations in fruit ripening or even the synthesis of biodegradable plastic) and then introduced into a plant cell. As these genes come from another organism they are known as transgenes (*trans* = across) and organisms which contain them are referred to as **transgenic.** It is this ability to generate entirely new traits which makes this technology so exciting, but at the same time raises many concerns.

The transgenes are produced in large amounts by incorporating them into small circles of DNA, known as **plasmids**, which can replicate within bacteria – normally *Escherichia coli*. The plasmid will contain the genes which confer the new trait on the plant and, in addition, one or more genes allowing the genetically modified plant cells to be selected. This is necessary as all methods of introducing transgenes into plants (transformation) are inefficient.

The DNA can be introduced into plant cells in a number of ways. Plasmid DNA can be isolated from *E. coli* and introduced into plant protoplasts using chemical treatments or short high-voltage pulses of electricity (electroporation). Alternatively, the DNA can be used to coat tiny (1–2 μm) gold spheres which are then accelerated to high speeds using a blast of high-pressure helium. The particles hit the plant cells and a few become lodged in the cytoplasm where they release the DNA which becomes incorporated into the plant genome. This procedure is known as **particle bombardment** or, more evocatively, **biolistics**. A third method uses the natural DNA transfer mechanism of a soil bacterium, *Agrobacterium tumefaciens*, to introduce the transgene into the plant. This process of *Agrobacterium*-mediated gene transfer was the first to be developed and is still widely used for producing transgenic dicotyledonous plants (the other methods are generally used for the production of transgenic monocotyledonous plants). This method will be elaborated on as it illustrates how plant growth hormones can be exploited by plant pathogens, or scientists, to modify plant growth and development.

Agrobacterium-mediated gene transfer

In nature, *A. tumefaciens* infects wounds on plants and transfers a short piece of DNA (the **transfer** or T-DNA) from a plasmid within the bacterial cell into the plant genome. The T-DNA encodes a number of genes which direct the plant to synthesize the auxin and cytokinin as well as unusual sugars (e.g. mannopine or octopine). The alterations in plant growth hormone content lead to the unregulated growth of infected plant material and the formation of a gall; *A. tumefaciens* is the causative agent of crown gall disease and is a particular problem in vineyards and orchards where pruning generates wounded plants routinely. The unusual sugars are metabolized by the bacteria, fuelling their growth. Essentially *A. tumefaciens* hijacks the plant cell.

The process of T-DNA transfer requires the action of many bacterial genes (de la Riva *et al.* 1998) but, fortunately, all of the genes on the T-DNA can be replaced without affecting the transfer process – only short DNA sequences at the left and right borders of the T-DNA are required. The plant biologist can therefore replace the *A. tumefaciens* hormone and sugar biosynthetic genes with the transgenes to be introduced into the plant. Plant tissue (typically discs cut from leaves or pieces of roots) is incubated with the modified *A. tumefaciens*, allowing T-DNA transfer to occur. After an hour or so, the plant material is transferred to a tissue culture medium containing auxin and cytokinin to stimulate callus formation, sugar to fuel plant growth, and an antibiotic which will kill *A. tumefaciens*.

Selection and regeneration

Regardless of the procedure used, only a few plant cells will contain the transgenes, and these would quickly be overgrown in culture by the neighbouring, untransformed cells. Therefore the transformation process normally includes a **selectable marker** – typically a gene which will confer resistance to an antibiotic or herbicide. When these are incorporated into the tissue culture medium, only the transformed cells will grow. Different relative concentrations of plant growth hormones can then be used to regenerate entire plants. Much of the concern over the use of genetically modified plants arises from the choice of selectable markers. If herbicide-resistance genes are used these might make the crop difficult to control in the field or they may be transferred to weed species. Similarly, antibiotic-resistance genes might be taken up by gut bacteria causing problems for human or animal health. There is considerable interest in developing techniques by which selectable markers can be removed from transgenic plants once they have fulfilled their function, and in utilizing new selectable markers which will have little impact on human health or the environment (see Scutt *et al.* 2002 for a review).

7.2.7 Abscisic acid

Whereas auxins, cytokinins and gibberellins are best known as promoters of growth, abscisins generally inhibit growth. The best known member of this group is abscisic acid (ABA, Fig. 7.2) which, in spite of its name, is now thought to have little role in controlling abscission. ABA is important in inducing dormancy in seeds (and preventing germination) and buds as well as increasing the stress resistance of these organs. A clear inter-action with GA is apparent in seed formation and germination (Section 11.14). A developing seed exhibits high rates of cell division and expansion and characteristically GA content is high whilst ABA content is low. As the seed matures, cell division ceases, and GA levels fall while ABA levels rise. This increase in ABA concentration is associated with the development of dehydration tolerance and is important in promoting dormancy. Plants with lesions in ABA biosynthesis exhibit vivipary, in which the seeds germinate whilst still attached to the plant. A more appropriate name for ABA (and one of its original names) might be dormin, although the term abscisic acid is too well established in the literature to be changed.

Another important role of ABA is in the control of stomatal aperture. Under water-stress conditions ABA is released from the roots or chloroplasts causing stomata to close, hence reducing transpirational water loss. Long-term exposure to high concentrations of ABA causes a repression of the photosynthetic apparatus, a loss of chlorophyll from leaves, and expression of genes which result in improved drought tolerance.

7.2.8 Ethylene

Ethylene (Fig. 7.2) is the last of the 'traditional' plant growth hormones and, unlike the others discussed so far, it is a gas. Although the impact of ethylene on plants was recognized a hundred years ago – old gas lamps produced ethylene and had a profound effect on seedling development – it was only with the introduction of modern analytical techniques that its role as a plant growth hormone was accepted. Ethylene is considered by many to be a 'stress' hormone like ABA.

Box 7.1

With *Arabidopsis thaliana*, a model organism widely used for studying plant biology, entire flowering plants can be dipped into solutions of *A. tumefaciens*. A few of the seeds (typically 1 in 1000) from plants treated in this way will be genetically modified. This procedure is rapid as it avoids the labour-intensive tissue culture procedure, but unfortunately has been developed only for this one species at present.

Exposure of seedlings to ethylene in the dark produces a characteristic 'triple response'. Plants grown in normal air in darkness are etiolated with long, thin hypocotyls and a small apical hook. Exposure to ethylene results in the production of a short, radially thickened hypocotyl with an exaggerated apical hook (Fig. 7.5). This response is thought to be important in allowing a seedling to push its way through soil. Young seedlings are potent producers of ethylene, which will be trapped in air pockets in the immediate vicinity, eliciting the response. Mutant plants which do not exhibit the triple response cannot grow through even a thin layer of soil.

Exposure of mature plants to ethylene promotes **epinasty** – a downward movement of leaves – which is thought to protect stressed plants from damage by sunlight and protect rosette plants from trampling damage. Prolonged exposure leads to leaf senescence and abscission. Ethylene also has an important role in the response of plants to wounding and pathogen attack. It is important as well in fruit ripening. Ethylene action inhibitors are widely used in fruit transportation (Section 11.13) and this is a major area of interest in the production of genetically modified crops.

7.2.9 Other potential plant hormones

As it is so difficult to write a comprehensive definition of a plant hormone it is not surprising that many other plant compounds might fall within this group. In recent years many signalling

Fig. 7.5 *Arabidopsis* seedlings showing the triple response and the *etr1* mutant. *Arabidopsis* seedlings were grown in darkness in an atmosphere containing 5 μmol mol^{-1} ethylene. Most show a characteristic triple response to elevated ethylene with short, thick hypocotyls and an exaggerated apical hook. An ethylene-resistant mutant, *etr1*, shows etiolated growth. Reprinted from *Science* **241** (4869) 1988, cover page. © American Association for the Advancement of Science 1988. Photo by Kurt Stepnitz, Michigan State University.

molecules have been discovered which may act as hormones. For some molecules the evidence is good, with production, transport, a receptor and responses having been demonstrated. For other substances the evidence is less conclusive, although rapid progress is being made.

Peptides

In animals numerous peptide hormones have been discovered, and it is surprising that so little is known about equivalent molecules in plants. However, there is good evidence that an 18-amino-acid peptide, systemin (Fig. 7.6), acts as a hormone in wounded leaves of tomato (*Lycopersicon esculentum*) (McGurl *et al.* 1992). This peptide is released from wounded cells and is transported through the phloem to unwounded leaves where it induces proteinase inhibitor (*pin*) gene expression. The proteinase inhibitors impair digestion in herbivores and hence have an antifeedant role. Systemin is active at very low (femtomolar) concentrations and a receptor has been isolated. Peptides closely related to systemin have been discovered in other members of the Solanaceae.

Another peptide, sulfokin, has been implicated in the control of cell division in asparagus (*Asparagus officinalis*) and rice cell cultures. This peptide is even smaller than systemin, being only 4–5 amino acids long, two of which are sulphated (Fig. 7.6). Other biologically active peptides have been described (for reviews, see texts by Lindsey in Complementary reading).

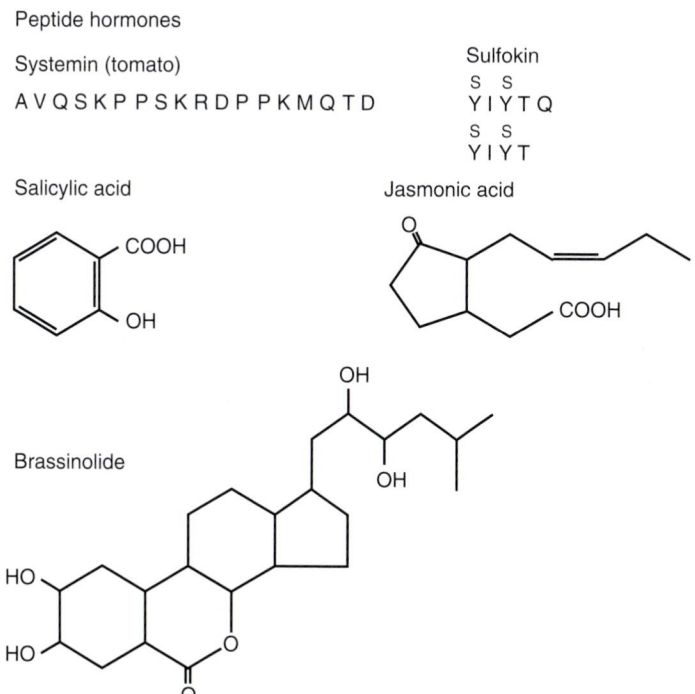

Peptide hormones

Systemin (tomato)

A V Q S K P P S K R D P P K M Q T D

Sulfokin

S S
Y I Y T Q
S S
Y I Y T

Salicylic acid

Jasmonic acid

Brassinolide

Fig. 7.6 Structures of some other plant hormones. The peptide hormones systemin and sulfokin (which is sulphated on the tyrosine residues) are synthesized from longer precursors. Both salicylic acid and jasmonic acid are involved in plant defence. Volatile derivatives can be produced in which the carboxyl group is methylated ($-COOCH_3$). Brassinolide, a steroid lactone, is a brassinosteroid hormone.

However, in many cases, these peptides may act at the cell surface, performing an important role in signalling positional information (Section 9.2.2). Whilst this is a fascinating area of plant biology, these substances cannot be considered plant growth hormones without evidence of transport within the plant. Nevertheless, there are undoubtedly many other peptide hormones in plants awaiting discovery.

RNA

Recently, small RNA molecules have been identified in the phloem which may act as transportable signals regulating gene expression. To date, these have been shown to be important in plant defence responses to viruses and in the silencing of synthetic genes introduced by genetic modification, both of which may be considered 'foreign' genes. This phenomenon is called post-transcriptional gene silencing (PTGS) and involves small double-stranded RNA molecules, known as short interfering RNAs (siRNAs), which are approximately 25 nucleotides long, and which mediate the destruction of specific target mRNAs. In addition, a number of groups have identified a separate class of microRNAs (miRNAs) which may regulate normal plant development. These are single-stranded RNAs of 21–22 nucleotides which are complementary to the genes encoding many regulatory proteins. Recently, the importance of miRNAs in leaf development has been demonstrated (Palatnik *et al.* 2003; Section 9.5.1). As many hundreds of these miRNAs have been identified it seems possible that they may represent a more general means of regulating gene expression, and hence these small RNAs may be considered to be plant growth regulators (Hunter & Poethig 2003).

Salicylic acid/jasmonic acid

These two compounds have been considered together as although they are structurally quite different they appear to interact closely to coordinate plant defence responses. Jasmonic acid (JA) is a 12-carbon fatty acid derivative of linolenic acid whereas salicylic acid (SA) is a simple aromatic ring with a hydroxyl and carboxylic acid sidegroup (Fig. 7.6).

SA is involved in the expression of many pathogenesis-related (PR) proteins which have antifungal and antibacterial activities in response to pathogen attack. JA has been implicated in wound responses such as those caused by herbivores. It has been proposed that many SA-induced genes are repressed by JA and vice versa, and in this way a more 'directed' response is achieved without the diversion of resources into unnecessary protein production. Ethylene also plays a role in this coordinated gene expression (reviewed in Kunkel & Brooks 2002). Some pathogens may exploit this interaction – the bacterium *Pseudomonas syringae* produces corotanine, an analogue of JA. This compound may repress the expression of the PR proteins, allowing infection to proceed. The complexity of signalling mechanisms in plant defence is beginning to be unravelled using a combination of mutants (reviewed by Glazebrook 2001) and gene expression profiling (Section 7.4.5), allowing interactions between these pathways to be elucidated.

Both SA and JA can be methylated *in vivo* and are volatile in this form. In the laboratory it has been demonstrated that gene expression in neighbouring plants can be induced by these volatile compounds, although this remains to be convincingly demonstrated in field conditions. Plants contain many volatile compounds, a number of which have the potential to act as signalling molecules. Interestingly, this signalling may extend to organisms other than plants. Compounds released from herbivore-damaged plants have the potential to attract predators and parasites of the herbivores (see Farmer 2001).

Brassinosteroids

More than 40 different brassinosteroid molecules have been found within plants (Fig. 7.6). These are all thought to be synthesized from the precursor campesterol. Brassinosteroids have been implicated in many roles including stem elongation, pollen tube growth, leaf bending and unrolling, stimulation of ethylene production, tracheary element differentiation and cell elongation. Plants with mutations in brassinosteroid metabolism are dwarf and exhibit unusual development in darkness (Section 10.5). The transport of these molecules has not been well documented, but clearly they have a profound effect on plant development.

Other plant growth hormones will undoubtedly be found and, in time, the boundaries between the traditional hormones and the plethora of signalling molecules which exist within, and pass between, the cells of plants will become more blurred.

7.3 | Detection and quantification of hormones in plants

As hormones are normally present in plants at very low concentrations, precise quantification is technically challenging. For example, vegetative tissue normally contains between a few μg kg^{-1} and 350 μg kg^{-1} auxin, although the concentrations of other hormones might be somewhat higher. Some tissues which are particularly rich sources of hormones may contain much more: maize (*Zea mays*) grains contain 80 mg kg^{-1} auxin whilst the GA content of liquid endosperm of the cucurbit *Echinocystis* has been estimated at 470 mg kg^{-1}. However, in general, concentrations are still quite low. In one series of experiments 1300 kg of pea seeds were used to obtain a final yield of 9 mg ABA!

The quantification of plant growth hormones is further complicated by the diverse forms in which they exist; as stated, there are over 100 different GAs which vary enormously in their potency.

With the appreciation that interactions between plant growth hormones are central to their actions, and that subcellular localization will also have a profound effect on the concentration perceived by the cell, it can be seen that measuring the hormone content of plant tissue in a physiologically meaningful manner is not a trivial undertaking.

Before the advent of modern analytical techniques, bioassays were used extensively to identify and quantify plant growth hormones. Plant extracts were prepared and partially purified before being applied to the test plant material such as that described for auxin or GA. Bioassays are extremely sensitive and have the advantage that they measure biological activity in relatively crude extracts. However, they suffer from a number of major problems. As well as being time-consuming to set up, they cannot readily distinguish between different plant hormones which have similar activities. Also, although they measure biological activity, they cannot distinguish between a small amount of a very potent hormone and a larger amount of a less potent one. Such assays are also subject to interference from other compounds which may enhance or attenuate the effect.

Bioassays have been largely superseded by analytical techniques. Developments in laboratory instrumentation allow the processing of tiny amounts of material whilst minimizing losses – essential for accurate measurements. Analytical procedures rely on the efficient separation of the different plant hormones coupled with extremely sensitive detection systems (which can routinely quantify pg amounts of a substance). High-performance liquid chromatography (HPLC) involves the passage of partially purified plant extracts in a liquid phase over an absorbing solid surface packed into a column. Different hormones will interact to different extents with the substrate, causing them to emerge at different times from the column. The retention time can then be compared with known standards of

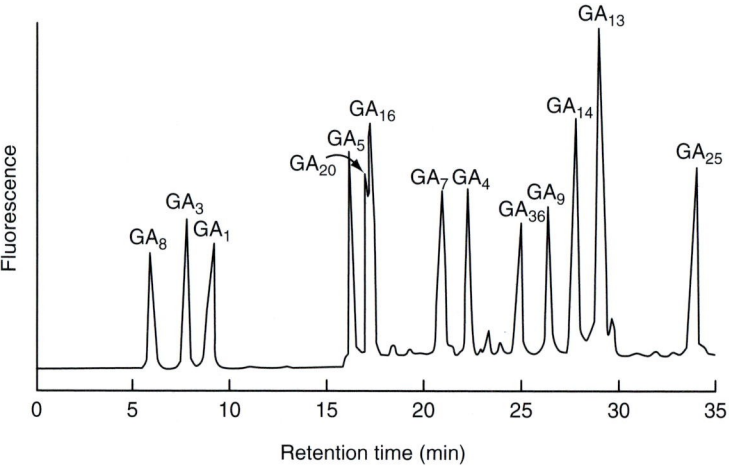

Fig. 7.7 Separation of gibberellins (as esters) by HPLC. The gibberellin esters were detected by their fluorescence. From Crozier (1981).

purified hormone. Figure 7.7 shows the separation of different GAs which have been converted to esters and their detection by fluorescence.

In a related technique, volatile derivatives of hormones are prepared and passed in a gaseous phase over a liquid surface. Gas–liquid chromatography (GLC) is extremely rapid and provides excellent separation of many hormones. This procedure also has the advantage that the emerging gas stream can be coupled directly to a detector system. By automating such an approach it is possible to analyse hundreds of samples within a day. The most widely used detector is the mass spectrometer, which can determine the molecular mass of a substance with great precision. The fragmentation pattern of compounds as they pass into the mass spectrometer can also aid in their identification. The combination of gas chromatography and mass spectrometry (GC–MS) allows the accurate quantification of a wide range of plant hormones from just a few milligrams of plant material.

These procedures require partially purified extracts to be prepared as otherwise the assay becomes swamped by the vast range of different compounds found within a typical plant cell. In some cases simple physicochemical properties can be used such as differing solubility in simple solvents. However, the use of antibodies provides one of the most powerful techniques which can be used either to detect hormones directly or as an initial purification step prior to analytical procedures.

To prepare antibodies, plant hormones are coupled to a protein (as immune responses are generally weak against small molecules) and injected into a rabbit or a mouse. The animal mounts an immune response against the hormone and produces antibodies which will recognize components (called epitopes) of the hormone with great specificity and bind them with high affinity. If the serum of an animal is used directly it will contain a mixture of antibodies which cross-react with different epitopes of the hormone – these are known as polyclonal antibodies as they derive from many different immune cells. However, it is also possible to isolate individual antibody-producing cells and grow them in culture. As each cell line produces only a single antibody, these monoclonal antibodies will recognize only a single epitope. If many such cell lines are analysed it is possible to identify antibodies which recognize an epitope shared by all the members of a hormone family, or they may be extremely specific and recognize only a single family member. As the cell lines can be grown in culture, essentially unlimited amounts of the antibody can be produced and characterized in great detail without the need to inject further animals.

Appropriate antibodies can then be immobilized on a solid support and used to separate a specific class of hormones (e.g. glycosylated cytokinins) from crude plant extracts. They can then be eluted and analysed by GC-MS. The antibodies can also be used to detect the amount of hormone directly. A number of related techniques exist. In

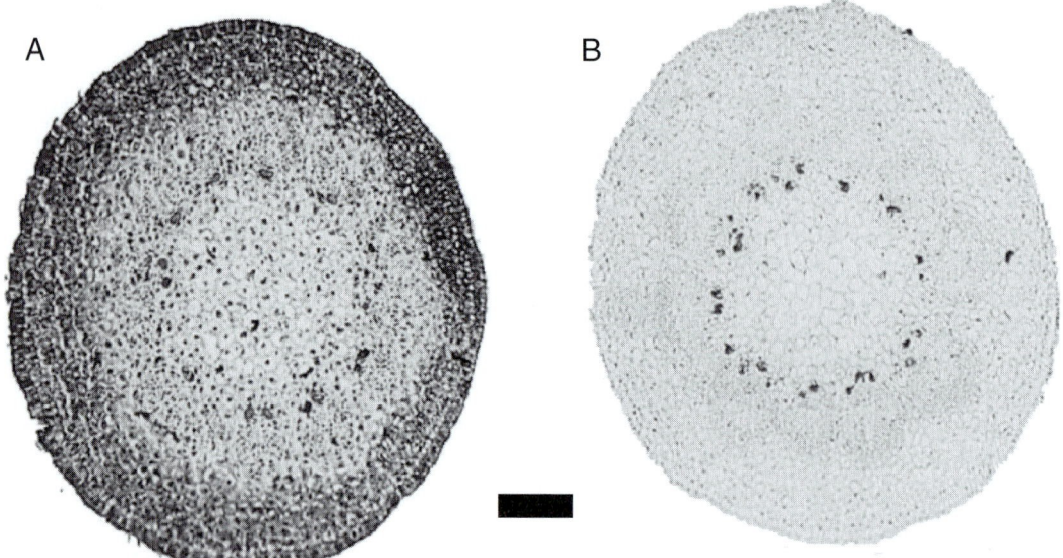

Fig. 7.8 Immunolocalization of indole-3-acetic acid in the gynophore of peanut. Peanut (*Arachis hypogaea*) produces its seeds on a specialized organ called the gynophore (Section 12.3). A cross-section through the gynophore has been incubated with an antibody raised against indole-3-acetic acid and then visualized using a second antibody linked to alkaline phosphatase. Areas containing IAA are dark in colour (A). In a control experiment the anti-IAA antibody was omitted (B). Scale bar = 0.25 mm. (Reproduced from Moctezuma & Feldman (1999). © Springer-Verlag, 1999.

radioimmunoassay (RIA) a plant extract is mixed with a purified, radioactively labelled hormone and a limited amount of a specific antibody. The efficiency with which the plant extract competes with the radioactive standard allows the amount of hormone to be determined. In enzyme-linked immunosorbent assays (ELISA), rather than use a radioactively labelled hormone as a standard, a purified hormone is linked to an enzyme (typically alkaline phosphatase) which provides an extremely sensitive and rapid assay. Antibodies linked to enzymes can be used to localize hormones in sections of plant material, allowing their cellular and subcellular distribution to be visualized (Fig. 7.8) (Moctezuma & Feldman, 1999). Transgenic plants containing a hormone-responsive promoter fused to a reporter gene provide a useful means of visualizing cells which are *responding* to a plant growth hormone. See Section 12.3.2 for an example of how this technique has been used to visualize auxin gradients involved in tropic curvatures.

7.4 | How do plant hormones cause responses?

Hormones act as internal signals within the plant. In exactly the same way as environmental signals, they must be perceived and then initiate a series of responses. The steps between the initial perception and the final response are known as **signal transduction** – another set of chemical changes, this time occurring inside the cell, which alter the biochemistry and/or patterns of gene expression of that cell. These changes may then in turn act as signals initiating yet further responses. Figure 7.9 shows this diagrammatically, although a simple linear pathway must be considered a gross oversimplification.

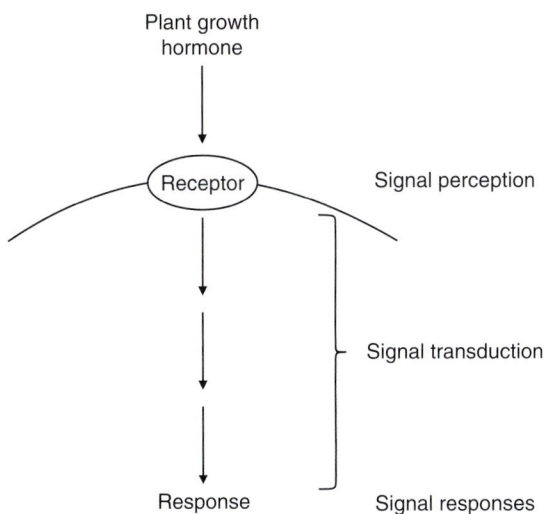

Plant growth hormone

Receptor — Signal perception

Signal transduction

Response — Signal responses

Fig. 7.9 A simple signal transduction pathway. Signal transduction pathways are rarely as simple as the linear model shown here. However, a plant hormone will interact with a specific receptor (although not necessarily located in a membrane) triggering a cascade of events which result in responses.

In the third edition of this book, published in 1984, the statement was made '*Not for one single hormone do we know one single biochemical reaction involving the hormone.*' Certainly this is no longer the case. In the past few years there has been stunning progress in many areas of plant physiology resulting from the marriage of traditional physiology/biochemistry with modern molecular genetic techniques. We can now, at least in outline, trace the signal transduction pathways for some plant hormones, although much remains to be understood. There is still the challenge of understanding how these interact with each other to generate appropriate responses to multiple signals.

It is beyond the scope of this text to consider the current knowledge regarding the action of all the plant hormones. The reader is referred to a series of excellent reviews (see Complementary reading) covering these topics in detail. However, some selected examples are presented to illustrate how these advances have been made and how some signal transduction pathways may operate.

7.4.1 Hormone receptors

It has long been assumed that the first step of signal perception is the binding of a hormone to a specific receptor. A receptor for a hormone is expected to have a number of characteristics:

- It should bind the hormone with an affinity similar to that required to initiate the response *in vivo*.
- It should be located in an appropriate region of the cell (often, but not always, the plasma membrane).
- Binding of the hormone to the receptor should alter the receptor's properties in some way to initiate a signal transduction pathway.

However, many other proteins within the cell which are not receptors may share some of these properties. For example, proteins which transport, degrade and conjugate auxin might all be expected to bind auxin. An auxin-binding protein (ABP1) has been isolated from the coleoptile membranes of maize, and a number of pieces of evidence point to this being an auxin receptor: antibodies against ABP1 block auxin-induced responses in cultured cells, and over-expression of this protein in tissues which do not normally respond to auxin makes them responsive. A small proportion of this protein is located in the plasma membrane, although the majority is found in the endoplasmic reticulum (ER); hence it has been suggested that the receptor shuttles between the ER and the cell surface. If plants are produced which lack ABP1 they die during embryogenesis, indicating an essential role for ABP1 in plant development.

Although proteins which bind ethylene, cytokinin, ABA and GA have all been identified in cell extracts, proving that these are receptors has been exceptionally difficult and has had only limited success. Given that the abundance of the receptors is expected to be low (as hormones generally act at very low concentrations) perhaps this is not surprising. However, in some cases, the use of mutant plants has proved to be spectacularly successful in identifying hormone receptors and other components of the signal transduction pathway.

7.4.2 Mutagenesis

The small cruciferous plant *Arabidopsis thaliana* (thale cress) has been the subject of intensive study and many hundreds of mutants relating to many aspects of plant physiology have been isolated. This plant has been selected as a model organism as it has a small genome (the sequence of which is now known in its entirety), it has a rapid life cycle, produces thousands of seeds and is amenable to molecular genetic techniques. If seeds are exposed to a mutagen (X-rays, chemicals or random insertions of small pieces of DNA) lesions are generated in the genome which will alter the function or expression of proteins. The seedlings are grown to maturity and the plants allowed to self-pollinate so that the seeds of the next generation will contain none, one or two copies of the defective gene. The resulting plants can then be screened to see if they exhibit alterations in response to plant hormones.

As already discussed (Section 7.2.8), seedlings exposed to ethylene in darkness exhibit a characteristic triple response. By examining mutagenized populations of seeds a number of different sorts of mutations have been identified. These include:

- seedlings which show the triple response in the absence of applied ethylene: *constitutive triple response (ctr)* and *ethylene overproducing (eto)*
- seedlings which do not exhibit a triple response in the presence of applied ethylene (Fig. 7.5): *ethylene resistant (etr)* and *ethylene insensitive (ein)*
- seedlings which show only part of the triple response: *hookless (hls)* – which lack an apical hook

These mutants result from lesions in different aspects of ethylene metabolism, not just signal transduction. The *eto* mutant, as its name implies, overproduces ethylene and is a mutation in the regulation of ethylene biosynthesis. Inhibitors (e.g. silver ions) of ACC synthase, a key enzyme in the ethylene biosynthetic pathway, restore the plants to a normal appearance. Most of the mutants identified, though, are in ethylene perception or the signal transduction pathway. Many of the genes which are mutated in these plants have now been cloned and sequenced and, in combination with physiological and biochemical measurements, we can now begin to understand some of their functions, allowing a model of the ethylene signal transduction pathway to be built (Fig. 7.10).

7.4.3 The ethylene signal transduction pathway

The details of the ethylene signal transduction pathway are beyond the scope of this text but an outline, and key concepts, are discussed below (see Wang *et al.* 2002 for a review). A number of ethylene receptors have been isolated from *A. thaliana*, and related proteins have been isolated from other species such as tomato. These are

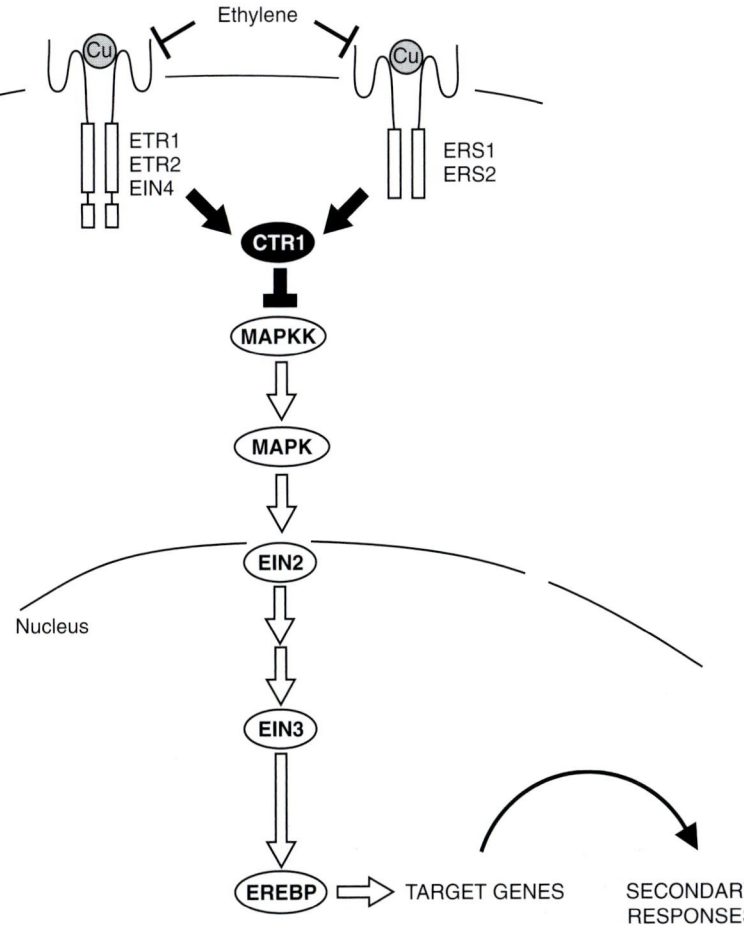

Fig. 7.10 A simplified model of the ethylene signal transduction pathway. Several components of the signal transduction have been omitted for clarity. These include gene products which have been identified by mutagenesis but which cannot yet be placed on this pathway and several branch points downstream of EIN2. Activation steps are shown as arrows (⇨) whilst inhibitory steps are shown as bars (▭). A family of copper-binding ethylene receptors exist (ETR1, ETR2, EIN4, ERS1, ERS2) which form dimeric complexes. In the absence of ethylene (solid symbols), the receptor complex phosphorylates, and activates CTR1. This blocks further activation of the signal transduction pathway. Mutations in CTR1 therefore lead to constitutive activation of the ethylene signalling pathway leading to a *constitutive triple response* phenotype. When ethylene binds to the receptor complex, the phosphorylation activity is repressed. CTR1 is inactivated and downstream kinases are released from inhibition, activating other components of the signal transduction pathway. These include up to 30 DNA-binding proteins (ethylene response element binding proteins – EREBP) that bind to specific motifs in the promoters of ethylene-regulated genes. These, in turn, trigger secondary responses. Redrawn and modified from Solano & Ecker (1998).

Box 7.2

Mutations on the ETR1 protein result in plants becoming *ethylene resistant* (*etr*) phenotype. The unmutated, functional protein is referred to as ETR1. See Appendix.

copper-binding proteins and have a number of hydrophobic regions typical of proteins located in a membrane. Recent studies have shown that the ETR1 receptor (Box 7.2) is located in the endoplasmic reticulum (Chen *et al.* 2002). When ethylene binds to an ethylene receptor, the receptor interacts with other proteins in the signal transduction pathway. Protein phosphorylation is a common component of signal transduction pathways in many organisms and a number of proteins involved in the ethylene and other hormone transduction pathways are known to be kinases or phosphatases. In the *absence* of ethylene, the receptor activates CTR1. This leads to the *repression* of the activity of the protein EIN2. When ethylene binds to the receptor this repression is released and gene expression is activated, leading eventually to the observed responses. Hence mutations in the early components of the chain, such as CTR1, lead to the pathway being permanently activated, causing the constitutive triple-response phenotype.

Although the term 'triple response' suggests that ethylene causes just three responses, in fact many more genes will be activated (Section 7.4.5). It is thought that part of the ethylene signal transduction chain consists of a series of DNA-binding proteins which will bind to specific sequences in the promoters of ethylene-responsive genes causing them to be expressed. In this way the activation of just a few ethylene-response-element binding proteins (EREBPs) can activate many different genes. Similar binding proteins and sequences have been found for other hormone response pathways such as auxin, ABA and GA, and this is likely to be a common feature of all signal transduction pathways leading to alterations in gene expression.

The mutagenic approach has been widely applied to studies of all the plant growth hormones and has uncovered components of many signal transduction pathways. However, relatively few receptors for hormones other than ethylene have been discovered in this way. Where putative receptors have been found (e.g. a cytokinin receptor which shares many features with the ethylene receptors) hormone binding has not yet been demonstrated. This failure to identify receptors might result from the presence of multiple receptor genes encoding proteins with complementary functions; hence lesions in one receptor will be compensated for by others, the mutations may be lethal, or the situation may require a major reassessment of how some growth hormones are perceived.

7.4.4 Rapid events following hormone application

Although relatively few hormone receptors have been identified, a wide range of events which are activated soon after the application of hormones to plant cells have been discovered. One of the best-studied systems has been guard cells and their role in the closure of stomata following exposure to ABA. Stomata provide an excellent experimental system as the guard cells are readily accessible at the leaf surface and the response is rapid and readily quantifiable. The closure of stomata by ABA is extremely rapid and is not thought to involve changes in gene expression, but many other plant responses to this hormone certainly will. It is thought that elevated ABA concentrations initiate a series of rapid ion movements, culminating in the movement of certain ions and solutes out of the vacuole of the guard cells and into the apoplast. This results in the loss of water from the cells via osmosis, a loss of turgor, and the closure of the stomata.

When ABA is applied to guard cells, ion channels are activated which result in a rapid influx of Ca^{2+}. ABA does not act directly on Ca^{2+} channels, rather binding of ABA to a receptor is considered to activate an enzyme which rapidly synthesizes a small signalling molecule, cyclic adenosine 5′-diphosphate ribose (cADPR) (Wu *et al.* 1997, Leckie *et al.* 1998). The Ca^{2+} influx occurs within seconds, and is transient, making measurements difficult. However, the Ca^{2+} concentration within the guard cell can be visualized by injecting the cells with a Ca^{2+}-sensitive fluorescent dye such as Fura-2. The colour of Fura-2 fluorescence depends upon the Ca^{2+} concentration. Alternatively, plants can be genetically modified to express proteins which emit light, or whose fluorescent properties change, upon binding Ca^{2+}, allowing the Ca^{2+} concentration in the cytoplasm to be determined. Application of ABA causes a wave of Ca^{2+} to appear in the cytoplasm within seconds, apparently released from the cell wall in some cases and the vacuole in others (McAinsh *et al.* 1992). The increase in cytosolic Ca^{2+} causes ion channels which allow K^+ to enter the cell to be inactivated whilst channels which allow Cl^- and malate to leave the cell are activated. The loss of Cl^- and malate from the cell causes a depolarization of the guard cell plasma membrane and this causes the activation of outward K^+ channels. The loss of ions from the cell results in the loss of turgor leading to stomatal closure.

Allen *et al.* (2001) demonstrated that guard cells respond to precise patterns of oscillations in intracellular Ca^{2+}. Guard cells, isolated from genetically modified *Arabidopsis* plants expressing the Cameleon protein, were alternately washed in buffers containing high or low concentrations of Ca^{2+}, allowing the intracellular Ca^{2+} content to be altered in a very precise manner. It was shown that a series of Ca^{2+} spikes, or transient increases in concentration, were required to close the stomata, and that these needed to be of a particular length and duration (Fig. 7.11). These oscillations can therefore be considered as a digital signal which controls stomatal aperture.

Changes in Ca^{2+} concentration are also thought to activate a protein kinase cascade leading to the activation of gene expression. These genes, although not involved in the control of stomatal aperture, are thought to be responsible for the other effects of ABA such as increased stress tolerance.

7.4.5 Hormone-induced changes in gene expression

Long-term changes in plant growth and development which occur in response to plant growth hormones will require changes in gene

Box 7.3

Both methods for visualizing the Ca^{2+} concentration use proteins isolated from the jellyfish *Aequorea victoria*. Aequorin is a luminescent protein which, in the presence of the compound coelenterazine, emits blue light when it binds Ca^{2+}. This can be detected using an ultra-sensitive camera. In the living jelly-fish this blue light is absorbed by a second protein, green fluorescent protein (GFP), which, as its name implies, fluoresces green. GFP itself has been genetically modified so that its fluorescence alters in a Ca^{2+}-dependent manner. This engineered version is evocatively called 'Cameleon' and allows intracellular Ca^{2+} concentrations to be determined without having to apply coelenterazine to the plant. The development of these, and other biosensors, is having a dramatic impact on all areas of biology.

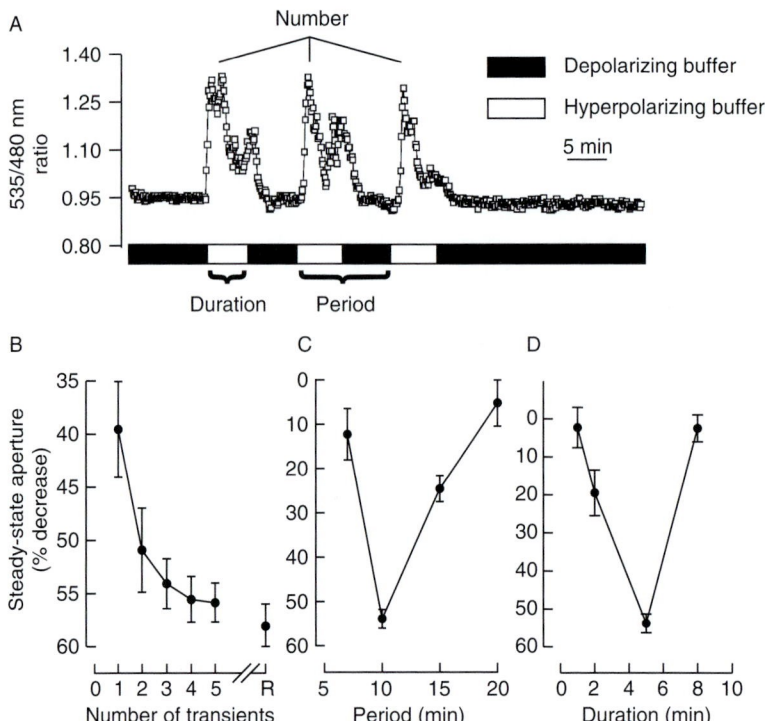

Fig. 7.11 Stomatal aperture is regulated by the number, duration and period of transient changes in cytosolic calcium concentration. The $[Ca^{2+}]_{cyt}$ was altered by exposing epidermal strips to depolarizing and hyperpolarizing buffers alternately. $[Ca^{2+}]_{cyt}$ was monitored using a calmodulin-linked Ca^{2+}-sensitive variant of green fluorescent protein called yellow cameleon 2.1. A typical experimental trace is shown in (A). Increasing numbers of $[Ca^{2+}]_{cyt}$ transients cause a progressive decrease in steady-state stomatal aperture (B). When exposed to three transients, the greatest response was observed when the transients were spaced at 10-minute intervals (C) and were of 5 minutes' duration (D). Redrawn and modified from Allen et al. (2001). © Nature Publishing Group.

expression, and many mutagenic studies have identified DNA-binding proteins and transcription factors as components of signal transduction pathways. However, even a plant with a small genome such as *Arabidopsis thaliana* contains approximately 30 000 genes, many of which may be regulated in complex ways by plant growth hormones. Traditional methods for measuring gene expression are restricted to the analysis of a few genes at a time. However, recent advances in methodology, borrowing techniques from the computer industry, allow the expression of *all* the genes in the genome to be measured simultaneously. Such an approach can be used to identify whole classes of genes which are regulated by a plant growth hormone (and it obviously has applications in many areas of plant biology).

This **genomic** approach requires many, if not all, of the genes in the genome to be identified and thus is restricted to a handful of plant species at present. However, the power of these methods, and the rapid increase in the rate at which sequence information can be obtained, means that this situation will change rapidly. A number of related methods are being used to profile changes in gene expression but the underlying principles are the same and are outlined in Fig. 7.12.

Firstly, genes must be identified from the species to be studied. This can be achieved by sequencing the entire genome and using computer-based methods to attempt to identify genes, or by isolating and sequencing mRNAs from a wide range of different tissues. In

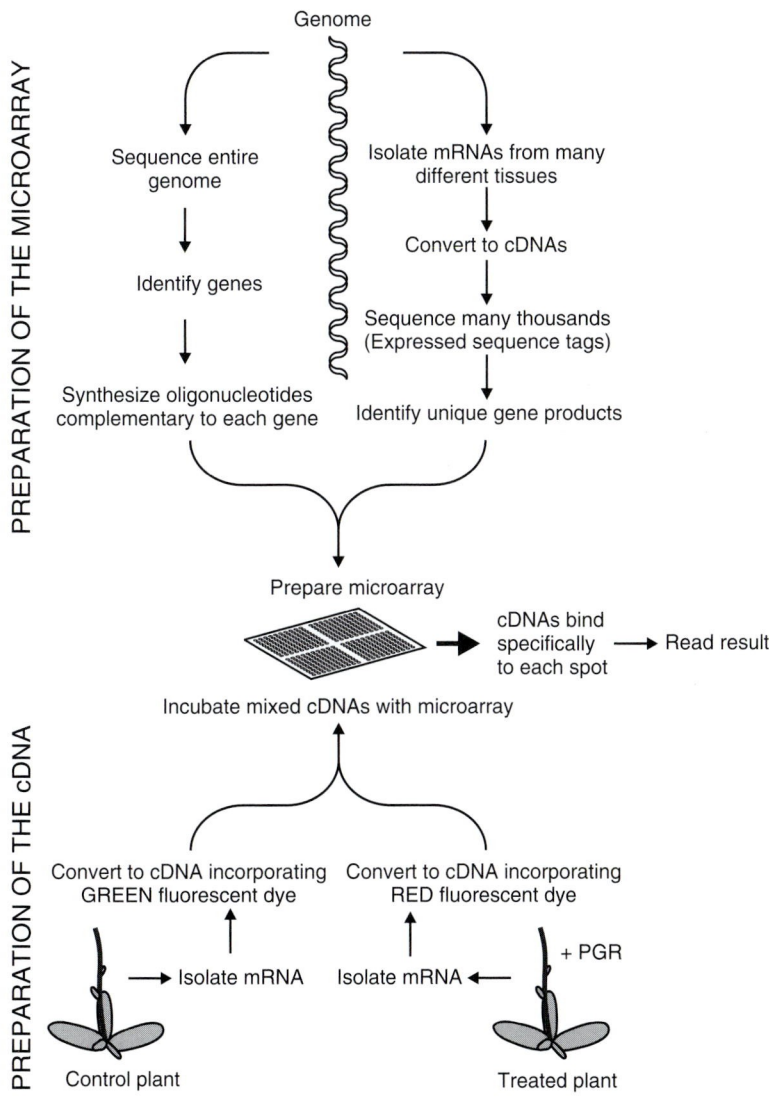

Genome

PREPARATION OF THE MICROARRAY

Sequence entire genome

Identify genes

Synthesize oligonucleotides complementary to each gene

Isolate mRNAs from many different tissues

Convert to cDNAs

Sequence many thousands (Expressed sequence tags)

Identify unique gene products

Prepare microarray

cDNAs bind specifically to each spot → Read result

Incubate mixed cDNAs with microarray

PREPARATION OF THE cDNA

Convert to cDNA incorporating GREEN fluorescent dye

Convert to cDNA incorporating RED fluorescent dye

→ Isolate mRNA   Isolate mRNA ←   + PGR

Control plant   Treated plant

Fig. 7.12 Microarray analysis of hormone-regulated gene expression. The expression of many thousands of genes can be determined simultaneously using microarray analysis.

fact, these methods of gene identification are complementary. Small fragments of DNA (known as expressed sequence tags – ESTs) corresponding to each gene are then placed as a grid (a microarray) on a glass slide or other substrate. RNA is prepared from plant tissue which has been treated with a growth hormone or from an untreated control. The two RNA samples are copied into complementary DNAs (cDNAs) using the enzyme reverse transcriptase. During this process a fluorescent dye is incorporated so that cDNAs from control plants fluoresce green whilst those from treated plants fluoresce red. Equal amounts of the differently labelled cDNAs are then mixed and incubated with the microarray. The labelled cDNAs will hybridize specifically to the corresponding gene fragments immobilized on the array. The more cDNA present in the sample, the more label will be bound. A scanner is used to measure the amount of red and green dye bound to each spot. Abundantly expressed genes will produce a strong signal whilst weakly expressed genes will produce a low signal. If the expression of a particular gene is unaffected by the hormone treatment, the intensities of the red and green signals will be the same. However, if there is more red dye bound than green, we can deduce that the expression of that gene is induced by the hormone, whilst the opposite situation indicates repression.

This approach allows genome-wide patterns of gene expression to be identified in a manner impossible with traditional techniques. For example, Schenk *et al.* (2000) examined the coordinated expression of 2375 *A. thaliana* genes in response to ethylene, salicylic acid, methyl jasmonate and the pathogen *Alternaria brassicola*. These studies showed that while some genes were differentially regulated by SA and JA (i.e. if expression was increased by JA, it was reduced by SA and vice versa), others were regulated independently or were co-induced.

The analysis of genome-wide patterns of gene expression using microarrays has provided a powerful tool with which to study plant responses to environmental and developmental signals, but it is important to remember that it is protein, not RNA, that provides the majority of activities. Hence genomic approaches are being extended to study protein expression (proteomics) and metabolites (metabolomics) and yet more 'omic' technologies are being developed. Whilst still in their infancy, these methods are increasingly being seen as the best way to study the sheer complexity of plant responses to the environment. The next challenge, undoubtedly, is to combine these results to form an integrated picture of plant function – an area known as bioinformatics.

Although the description of the signal transduction pathways above is complex, they are far from completely understood. However, the ways in which plant cells respond to plant hormones are gradually being revealed and we can, at last, begin to study the molecular basis of plant growth and development.

Complementary reading

The *Annual Review of Plant Biology* and the *Annual Review of Plant Physiology and Molecular Biology* have regular reviews of our current understanding of plant hormones.

Hedden, P. & Proebsting, W. M. Genetic analysis of gibberellin biosynthesis. *Plant Physiology*, **119** (1999), 365–70.

Howell, S. H., Lall, S. & Che, P. Cytokinins and shoot development. *Trends in Plant Science,* **8** (2003), 453–9.

Kepinski, S. & Leyser, O. Ubiquitination and auxin signaling: a degrading story. *The Plant Cell,* Supplement (2002), S81–95.

Leung, J. & Giraudat, J. Abscisic acid signal transduction. *Annual Review of Plant Physiology and Plant Molecular Biology*, **49** (1998), 199–222.

Leyser, O. Molecular genetics of auxin signaling. *Annual Review of Plant Biology*, **53** (2002), 377–98.

Lindsey, K. Plant peptide hormones: the long and the short of it. *Current Biology*, **11** (2001), R741–3.

Lindsey, K., Casson, S. & Chilley, P. Peptides: new signalling molecules in plants. *Trends in Plant Science*, **7** (2002), 78–83.

Milborrow, B. V. The pathway of biosynthesis of abscisic acid in vascular plants: a review of the present state of knowledge of ABA biosynthesis. *Journal of Experimental Botany*, **52** (2001), 1145–64.

Mok, D. W. S. & Mok, M. C. Cytokinin metabolism and action. *Annual Review of Plant Physiology and Plant Molecular Biology*, **52** (2001), 89–118.

Richards, D. E., King, K. E., Ait-ali, T. & Harberd, N. P. How gibberellin regulates plant growth and development: a molecular genetic analysis of gibberellin signaling. *Annual Review of Plant Physiology and Plant Molecular Biology*, **52** (2001), 67–88.

Wang, K. L. C., Li, H. & Ecker, J. R. Ethylene biosynthesis and signaling networks. *The Plant Cell*, **14** (2002), S131–51.

Wilkinson, S. & Davies, W. J. ABA-based chemical signalling: the co-ordination of responses to stress in plants. *Plant, Cell and Environment*, **25** (2002), 195–210.

References

Allen, G. J., Chu, S. P., Harrington, C. L. *et al.* (2001). A defined range of guard cell calcium oscillation parameters encodes stomatal movements. *Nature*, **411**, 1053–57.

Beveridge, C. A., Weller, J. L., Singer, S. R. & Hofer, J. M. I. (2003). Axillary meristem development: budding relationships between networks controlling flowering, branching, and photoperiod responsiveness. *Plant Physiology*, **131**, 927–34.

Chen, Y. F., Randlett, M. D., Findell, J. L. & Schaller, G. E. (2002). Localization of the ethylene receptor ETR1 to the endoplasmic reticulum of *Arabidopsis. Journal of Biological Chemistry*, **277**, 19861–6.

Chrispeels, M. J. & Sadava, D. E. (2003). *Plants, Genes and Crop Biotechnology*, 2nd edn. Sudbury, MA: Jones and Bartlett.

Crozier, A. (1981). Aspects of the metabolism and physiology of gibberellins. *Advances in Botanical Research*, **9**, 33–149.

de la Riva, G. A., Gozález-Cabrera, J., Vázquez-Padrón, R. & Ayra-Pardo, C. (1998). *Agrobacterium tumefaciens*: a natural tool for plant transformation. *Electronic Journal of Biotechnology*, **1** (3) www.ejbiotechnology.info/content/vol1/issue3/full/1

Farmer, E. E. (2001). Surface-to-air signals. *Nature*, **411**, 854–6.

Glazebrook, J. (2001). Genes controlling expression of defense responses in *Arabidopsis*: 2001 status. *Current Opinion in Plant Biology*, **4**, 301–8.

Hunter, C. & Poethig, R. S. (2003). miSSING LINKS: miRNAs and plant development. *Current Opinion in Genetics & Development*, **13**, 372–8.

James, W. O (1943). *An Introduction to Plant Physiology*, 4th edn. Oxford: Clarendon Press.

Kunkel, B. N. & Brooks, D. M. (2002). Cross talk between signaling pathways in pathogen defense. *Current Opinion in Plant Biology*, **5**, 325–31.

Leckie, C. P., McAinsh, M. R., Allen, G. J., Sanders, D. & Hetherington, A. M. (1998). Abscisic acid-induced stomatal closure mediated by cyclic ADP-ribose. *Proceedings of the National Academy of Sciences (USA)*, **95**, 15837–42.

Martin, D. N., Proebsting, W. M. & Hedden, P. (1997). Mendel's dwarfing gene: cDNAs from the Le alleles and function of the expressed proteins. *Proceedings of the National Academy of Sciences (USA)*, **94**, 8907–11.

McAinsh, M. R., Brownlee, C. & Hetherington, A. M. (1992). Visualizing changes in cytosolic-free Ca^{2+} during the response of stomatal guard cells to abscisic acid. *The Plant Cell*, **4**, 1113–22.

McGurl, B., Pearce, G., Orozcocardenas, M. & Ryan, C. A. (1992). Structure, expression, and antisense inhibition of the systemin precursor gene. *Science*, **255**, 1570–3.

Moctezuma, E. & Feldman, L. J. (1999). Auxin redistributes upwards in gravi-responding gynophores of the peanut plant. *Planta*, **209**, 180–6.

Palatnik, J. F., Allen, E., Wu, X. *et al.* (2003). Control of leaf morphogenesis by microRNAs. *Nature*, **425**, 257–63.

Palme, K. & Galweiler, L. (1999). PIN-pointing the molecular basis of auxin transport. *Current Opinion in Plant Biology*, **2**, 375–81.

Peng, J. R., Richards, D. E., Hartley, N. M. *et al.* (1999). 'Green revolution' genes encode mutant gibberellin response modulators. *Nature*, **400**, 256–61.

Sasaki, A., Ashikari, M., Ueguchi-Tanaka, M. *et al.*(2002). Green revolution: a mutant gibberellin-synthesis gene in rice. New insight into the rice variant that helped to avert famine over thirty years ago. *Nature*, **416**, 701–2.

Schenk, P. M., Kazan, K., Wilson I. *et al.* (2000). Coordinated plant defense responses in *Arabidopsis* revealed by microarray analysis. *Proceedings of the National Academy of Sciences (USA)*, **97**, 11655–60.

Scutt, C. P., Zubko, E. & Meyer, P. (2002). Techniques for the removal of marker genes from transgenic plants. *Biochimie*, **84**, 1119–26.

Skoog, F. & Miller, C. O. (1957). Chemical regulation of growth and organ formation in plant tissues cultured *in vitro*. *Symposium of the Society of Experimental Biology*, **11**, 118–31.

Solano, R. & Ecker, J. R. (1998). Ethylene gas: perception, signaling and response. *Current Opinion in Plant Biology*, **1**, 393–8.

Sorefan, K., Booker, J., Haurogne, K. *et al.* (2003). MAX4 and RMS1 are ortho-logous dioxygenase-like genes that regulate shoot branching in *Arabidopsis* and pea. *Genes & Development*, **17**, 1469–74.

Swain, S. M. & Olszewski, N. E. (1996). Genetic analysis of gibberellin signal transduction. *Plant Physiology*, **112**, 11–17.

Thimann, K. V. (1937). On the nature of inhibitors caused by auxin. *American Journal of Botany*, **24**, 407–12.

Wu, Y., Kuzma, J., Marechal, E. *et al.* (1997). Abscisic acid signaling through cyclic ADP-Ribose in plants. *Science*, **278**, 2126–30.

Chapter 8

Cell growth and differentiation

8.1 | Introduction

A mature plant is a complex organism made of many different organs, tissue types and cell types. The plant develops from a single cell, the zygote (fertilized egg) which first divides to form an embryo within a seed. By the time the embryo is mature, it already contains distinct meristems, from which the entire plant will develop upon germination. The meristems remain potentially capable of producing new cells throughout the life of the plant, and all of the complex organized structures of the plant develop from these apparently simple meristems by a combination of cell division, cell expansion and cell differentiation, as well as programmed cell death in some cases. Plant cells being immobile, migration of cells, as occurs in animal embryos, plays no part. This is the process of **morphogenesis** (morpho = form, genesis = origin) briefly touched on in Chapter 6. It is now necessary to consider in depth the manner in which meristems give rise to vegetative and reproductive structures.

The first stage in the formation of any plant structure is production of new cells by cell division. Determination of the position, direction, number and timing of the divisions is the first control stage in the morphogenetic process. Once the cells are formed, the morphogenetic process continues with expansion in a determined direction and to a controlled size. This is accompanied by cellular differentiation. This chapter covers cell division in the meristems, the control of the cell cycle, and control of organelle division, as well as the mechanism of plant cell expansion and some general facets of cell differentiation. The manner in which meristems give rise to specific vegetative and reproductive structures, and the differentiation of specific cell types, is discussed in the following chapters.

8.2 | Meristems and cell division

Meristems are found throughout the plant. The **apical meristems** give rise to primary growth – the shoot apical meristem (SAM) forming the above-ground structures (Section 9.2) and the root apical meristem (RAM) forming the below-ground structures (Section 9.6). The development of **secondary meristems** leads to branching whilst the **cambial meristems**, found outside of the apical regions, allow secondary growth leading to thickening, and hence strengthening, of structures. Not all of the meristems are active at any one time and, indeed, the entire plant may enter a dormant state, whether as an embryo (Section 11.14) or as a mature plant, with the re-initiation of cell division requiring specific environmental and internal signals. Similarly, in a growing plant, growth hormones such as auxin, produced in the apical meristem, may generate signals that inhibit the development of the secondary meristems in the axillary buds, leading to **apical dominance** (Section 7.2.2).

Although growth normally occurs from the division of cells within meristems, many (but not all) differentiated plant cells retain the ability to de-differentiate and regenerate an entire plant. This ability is termed **totipotency**. Although this may be possible in animals under exceptionally artificial circumstances (such as those used to clone Dolly the sheep from a differentiated udder epithelial cell), this ability is commonplace in plants. It is widely exploited in agriculture and horticulture to regenerate entire plants from cuttings or single cells (Section 7.2.5). This is a fundamental requirement for the genetic modification of many plants. Figure 8.1 shows an outline of the different processes essential to plant development. All of these may be occurring at any time during a plant's life cycle, even at an early stage such as a developing seedling.

8.2.1 Polyploidy

Not all cells remain totipotent. Cultured plant cells require auxins, cytokinins, sucrose and inorganic nutrients for growth. Alterations in the relative concentrations of auxin and cytokinin can cause the undifferentiated callus cells to generate shoots or roots but this ability is often lost after prolonged culture. The reasons underlying this are uncertain. In some cases a correlation between a loss of capacity to generate new organs and the development of polyploidy has been observed. This may result from the abnormal growth conditions of cultured cells but it is also true that separate callus cultures initiated from the same plant may differ markedly in their ability to

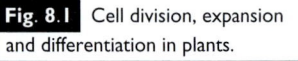

Fig. 8.1 Cell division, expansion and differentiation in plants.

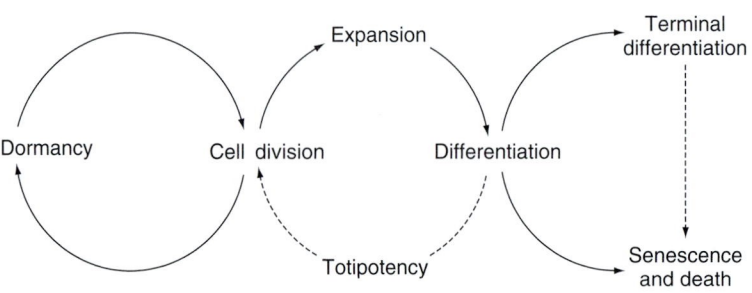

generate new organs and that different levels of ploidy are commonly present in different tissues. The generation of polyploidy is thus a normal, although not essential, part of cell development. Cortical cells and endodermal cells are frequently tetraploid whilst vessel unit cells in the metaxylem are tetraploid, octoploid or of higher ploidy. Not all cells within the same tissue will exhibit the same ploidy.

The physiological consequences of polyploidy are unclear but clearly it must be a regulated process. A link has been made between ploidy and the ability to increase cell expansion further. In developing metaxylem cells of *Arisaema triphyllum* there is a clear correlation between nuclear volume (a measure of DNA content and hence ploidy levels) and cell volume (Fig. 8.2). The requirement for a certain amount of genetic material to support a given cell size has also been observed in phloem cells. Long, primary phloem fibres are often multinucleate whereas shorter, secondary fibres remain uninucleate. Commercially many crops are polyploid – e.g. strawberry (*Fragaria ananassa*), banana (*Musa* spp.) – as these tend to exhibit increased fruit size.

8.2.2 The cell cycle

If the majority of plant cells have the potential to divide, what prevents most of them from doing so? At the heart of the answer to this question is the regulation of the **plant cell cycle**. Plant growth hormones such as auxins, cytokinins and gibberellins (see Chapter 7) are potent stimulators of cell division in many tissues, and appropriate nutritional conditions, such as the availability of sucrose, must also exist. Other plant hormones, e.g. ABA, can act to inhibit cell division. The precise ways in which plant growth hormones regulate the cell cycle are not fully understood. However, the control of the cell cycle

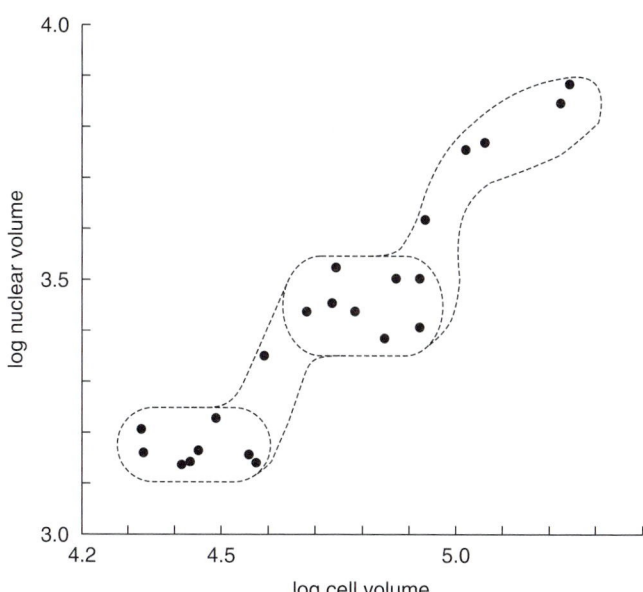

Fig. 8.2 Nuclear volume and ploidy in metaxylem cells of *Arisaema triphyllum*. Plot of log nuclear volume against log cell volume for developing metaxylem unit cells. The clustering of the groups (indicated by enclosure in the dotted lines) indicates a periodicity in the growth of cell and nucleus. Redrawn from List (1963).

has been extensively studied in many organisms and the fundamental processes appear to be similar in all eukaryotes. Cellular components involved in the plant cell cycle will function in yeast although, obviously, the factors regulating cell division are quite different.

The cell cycle is summarized in Fig. 8.3. It involves a complex and highly coordinated set of biochemical interactions, with checkpoints to limit progression unless certain conditions are met. Cell division occurs in the mitotic stage, where the nucleus and then the entire cell divides (cytokinesis). Cell growth and expansion can continue throughout the other phases. The G_1 phase is a preparation for the replication of the chromosomes (which occurs in the S phase). The cell enlarges, organelles enlarge and divide, and other cell components are synthesized. Cells which are fully differentiated are often arrested at this stage.

During the S phase the chromosomes are replicated (which can be measured by the uptake of radioactive ^{32}P or ^{3}H-thymidine). The chromosomes consist of a double helix of deoxyribonucleic acid (DNA) closely associated with basic proteins called **histones**. The DNA is wound around the histones, allowing extremely efficient packing – essential to fit the long DNA molecule into a relatively tiny nucleus. For example the DNA encoding the maize genome is approximately 2.5 *metres* long and fits into a nucleus only 10 *micrometres* in diameter. The DNA is replicated by the enzyme DNA polymerase, which uses the genetic information in one strand of the

Fig. 8.3 The plant cell cycle. Non-dividing cells are normally arrested in phase G_1, in which cell expansion and differentiation occur. Cell division occurs in the M phase.

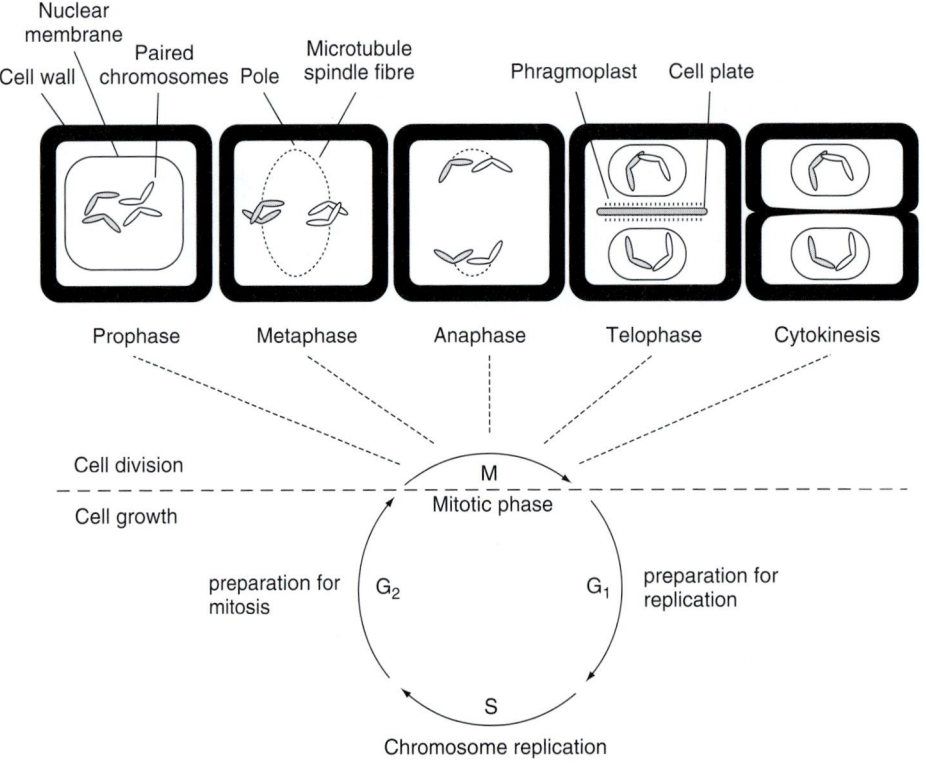

double helix to direct the synthesis of a second, complementary strand. Although DNA polymerase synthesizes the new strands rapidly (50 bases per second) and replication occurs from multiple sites, complete replication takes several hours. The length of the genome varies significantly between plants. In *Arabidopsis thaliana* it is rather small (180 million base pairs) whilst *Lilium longiflorum* has a genome almost 30 times larger. DNA polymerase is a remarkable enzyme, as it copies the genetic information very accurately (error rates have been estimated at approximately one error per 10^5-10^6 base pairs), but even this low rate would introduce many errors in each division of the cell. The fidelity of the enzyme is improved by a 'proof-reading' reaction. If a mistake is made, DNA polymerase destroys part of the recently synthesized strand and resynthesizes, improving the error rate to approximately one error per 10^9 base pairs. Once chromosome replication is complete the cell enters the G_2 phase – preparation for mitosis. It now contains twice the normal complement of genetic information.

Mitosis is often divided into different phases (Fig. 8.3), although this is a continuous process. During prophase the paired chromosomes (**chromatids**) condense and become visible by light microscopy. The nuclear envelope disappears, and in metaphase the chromosomes align along the central axis of the cell. Individual chromosomes are attached to the poles of the cell by spindle fibres composed of microtubules. In anaphase these microtubules act to pull the chromosomes apart, which travel to the poles of the cell. In telophase the separate chromosomes become surrounded by the nuclear envelope and cell wall material is laid down in the centre of the cell. A series of microtubules (the **phragmoplast**) directs the synthesis of the new cell wall resulting in the separation of the two cells (**cytokinesis**). Cytokinesis does not always occur (e.g. in phloem fibres), leading to the formation of multinucleate cells.

The entire cycle can occur in 16–24 hours but often the G_1 phase is extremely extended. Although the progression from G_1 to S represents an important checkpoint governing the cell cycle, replication of the DNA in the S phase does not guarantee that the cell will divide. Although the majority of cells which are not dividing are arrested in the G_1 phase, the cell cycle can also be arrested in G_2. For example, the pericycle cells of the root are arrested in G_2 and hence contain double the usual number of chromosomes (i.e. they are *polyploid*).

8.2.3 The control of the cell cycle

The progression of the cell cycle is controlled by a series of interacting proteins, called cyclins (Cycs) and cyclin-dependent kinases (CDKs), the activity of which is controlled in many cases by phosphorylation and dephosphorylation (see Stals & Inzé 2001 and Rossi & Varotto 2002 for reviews). Two key control points have been identified in plants and these are illustrated in a simplified form in Fig. 8.4.

Fig. 8.4 The regulation of the plant cell cycle. There are two checkpoints in the plant cell cycle at the G_1/S and G_2/M transitions. The transition from G_1 to S requires the protein E2F to be in the free, active state – if bound to Rb (retinoblastoma protein) it is inactive. Binding of E2F to Rb is controlled by the phosphorylation status of Rb which, in turn, is controlled by the activity of kinases and phosphatases. In the presence of auxin, cytokinin, gibberellic acid, brassinosteroids and sucrose, a kinase activity is stimulated. This hyperphosphorylates Rb and causes it to dissociate from E2F. The kinase can be inhibited by ABA. Transition from G_2 to M is also governed by the phosphorylation status of specific amino acids in proteins forming a regulatory complex which, in turn, is regulated by cytokinin. Modified from Stals & Inzé (2001) and Rossi & Varotto (2002).

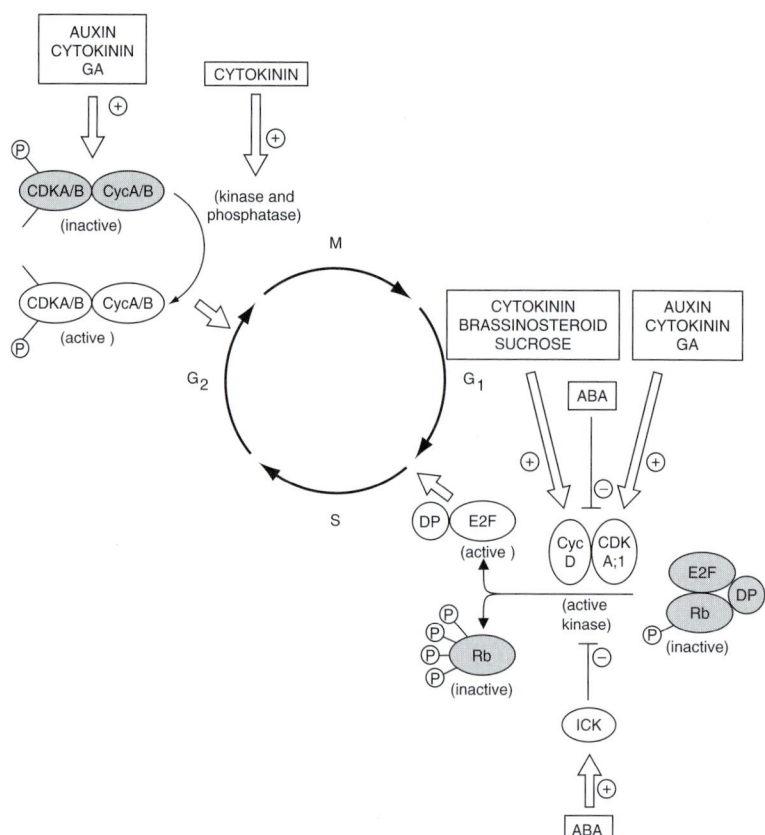

Box 8.1

Plants contain many Cycs and CDKs. These have been placed into families on the basis of similarity of amino acid sequences and recently renamed (Joubès et al. 2000).

The transition from G_1 to S depends upon a series of genes, the expression of which requires the transcription factor E2F to be active. The activity of E2F is regulated by a second protein, Rb. This is a retinoblastoma protein which binds to, and inactivates, E2F. Mutations of this protein in mammals lead to cancers of the eye as it is a *tumour suppressor*, inhibiting cell division. The binding of Rb to E2F depends upon the extent to which Rb is phosphorylated and this, in turn, is controlled by the proteins CDKA;1 and CycD. Together these form a kinase which will hyperphosphorylate Rb, causing it to disassociate from E2F. The expression of CDKA;1 and CycD is stimulated by growth-promoting hormones (auxin, cytokinin and GA) and sucrose, and inhibited by ABA; ABA also inhibits cell division by stimulating the expression of an inhibitor (ICK) of the kinase. Hence plant growth hormones interact to regulate cell division. The precise effect of the plant hormones depends upon the tissue type – auxin promotes CDKA;1 expression in roots whereas cytokinin represses it.

However, as discussed above, the transition from G_1 to S is not sufficient to ensure that cell division will occur. A second checkpoint has been identified controlling the G_2 to M transition. This step requires the activities of kinases and phosphatases which are regulated by auxin, cytokinin and GA.

At present the signal transduction pathways by which plant hormones regulate these enzyme activities are not known but this is a very active area of research. In addition to the *induction* of enzyme activities (whether by increasing transcription or post-transcriptional processes), the *destruction* of proteins is equally important in controlling the cell cycle. Some plants which have mutations in their response to auxin (e.g. *axr1* – *auxin resistant 1; tir1* – a mutant which shows altered responses to inhibitors of auxin transport) have lesions in the biochemistry of ubiquitin, a small protein which is important in targeting other proteins for degradation in a large, multimeric complex called the proteasome (Leyser *et al.* 1993, Kepinski & Leyser 2002). Recently, plant responses to other growth hormones including cytokinin, brassinosteroids and gibberellic acid have been shown to involve components of this protein degradation pathway.

8.3 | Mitochondrial and plastid division

The prokaryotic origin of plastids and mitochondria is now widely accepted, although relatively little is known about the processes controlling their division. Proplastids are present in meristematic cells and these divide and differentiate, in a tissue-specific manner, to form the photosynthetic chloroplasts or specialized chromoplasts, etioplasts and leucoplasts. Plastids divide by constriction in a manner similar to prokaryotes and there is evidence that some components of the prokaryotic division machinery exist in plants. However, there must be fundamental differences in plastid division as their environment and structure differs markedly from that of prokaryotic cells. Firstly, plastids are located within the plasma membrane of the plant cell and their division, eventual size and numbers are likely to be coordinated with nuclear and cell division. Also, plastids contain more membranes; the chloroplast, for example, is bounded by an outer and inner envelope membrane and has internal thylakoid membranes.

Mutants which exhibit **a**ltered **r**eplication of **c**hloroplasts (*arc* mutants) have been identified in which chloroplast division has been disrupted. For example, the mesophyll cells of the *arc*6 mutant of *Arabidopsis* contain only two, enormously enlarged, chloroplasts, rather than the normal complement of 200. Remarkably, the relationship between the total cell volume and plastid volume is maintained in these plants (Pyke *et al.* 1994). Nuclear-encoded homologues of genes involved in prokaryotic cell division have been found in plants. In *Escherichia coli* mutations in the *FtsZ* gene impair cell division and result in the formation of filaments (*fts* = filamentous temperature sensitive). Plant homologues of *FtsZ* are found at the constriction points of dividing chloroplasts (Fig. 8.5G) and mutagenesis of these genes disrupts chloroplast division (Fig. 8.5A–F). It is proposed that a multiple ring structure forms which constricts the envelope and thylakoid membranes until the chloroplast is divided in half (Fig. 8.5H) (Mori *et al.* 2001). A molecular link between cell division in prokaryotic cyanobacteria and chloroplast division in plants has recently been demonstrated by Fulgosi *et al.* (2002). They describe a nuclear-encoded protein (ARTEMIS) located at the inner chloroplast membrane necessary for chloroplast division. Inactivation of the homologous gene in *Synechocystis* impairs cell division but this can be restored with the equivalent gene from *Arabidopsis*. Some components of chloroplast division machinery are therefore known to be homologous with that of prokaryotes whereas others are likely to be unique to plants.

Even less is known about the control of mitochondrial division in plants although, again, this must be a highly regulated process. A *FtsZ* homologue has recently been found in plant mitochondria and a dynamin-like protein is found at the tips and constriction sites of dividing mitochondria. Mutagenesis of this protein perturbs mitochondrial division, with long filaments and tubular structures forming (Arimura & Tsutsumi 2002). Similar dynamin rings have now been found at the constriction sites of dividing chloroplasts, and mutations in dynamin-like proteins are found in some *arc* mutants (Gao *et al.* 2003).

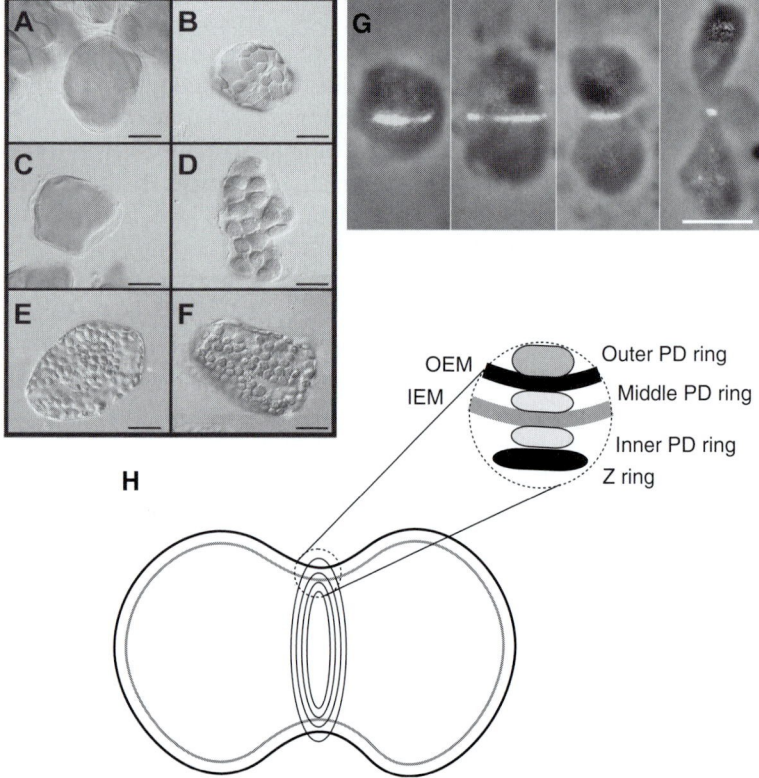

Fig. 8.5 The regulation of chloroplast division. Phenotypes of transgenic *Arabidopsis* plants expressing antisense constructs of AtFtsZ1 (A, B) or AtFtsZ2 (C, D). Cells from wild type plants are shown in (E) and (F) (Osteryoung *et al.* 1998). In the dividing chloroplasts of *Lilium longiflorum* (G) the FtsZ2 protein is shown as a bright ring at the constriction point (Mori *et al.* 2001). © Japanese Society of Plant Physiologists. Scale bar = 25 μm (A–F) and 4 μm (G). A model of the proteins found at the constriction point of dividing chloroplasts is shown in (H) (redrawn from Miyagishima *et al.* 2001). The plastid is constricted by the plastid division (PD) and Z rings which are concentrically arranged on each side of the inner (IEM) and outer (OEM) envelope membranes.

8.4 | Cell expansion: mechanism and control

As cells in the meristematic regions divide, they are pushed away and undergo massive expansion. Root-tip meristematic cells which develop into root parenchyma may expand 30- to 100-fold, whilst the water-storing cells of water melon (*Citrullus vulgaris*) increase in volume 350 000-fold and sometimes several million-fold. The direction in which this expansion occurs has a profound effect on the shaping of organs and the direction of growth. Expansion which occurs in all directions equally may fill spaces (e.g. in fruits) with loosely packed spherical cells or result in the production of a mass of unstructured cells called **callus** (Section 7.2.5). The formation of callus is an important component of much plant tissue culture but can also occur naturally such as at wound sites or when plants become infected with plant hormone-producing pathogens – e.g. *Agrobacterium tumefaciens*, which causes crown gall disease. However, in most cases cell expansion is not uniform but proceeds in a regulated and directional manner. Asymmetrical cell expansion can cause rapid directional growth such as that observed in tropic curvatures (Chapter 12) and is thus important in rapid plant responses to environmental signals.

Typically (but not always) the majority of cell expansion occurs after cell division has ceased. For example, in the spadix of *Arum maculatum* there is an initial phase of cell division followed by a phase of cell expansion (Fig. 8.6). However, in potato (*Solanum tuberosum*) tubers cell division and cell expansion occur concurrently as the tuber grows from 40 mg to 200 g. The cell number may increase 300-fold in this period whilst cells expand 10-fold. Growth beyond 200 g largely results from cell expansion.

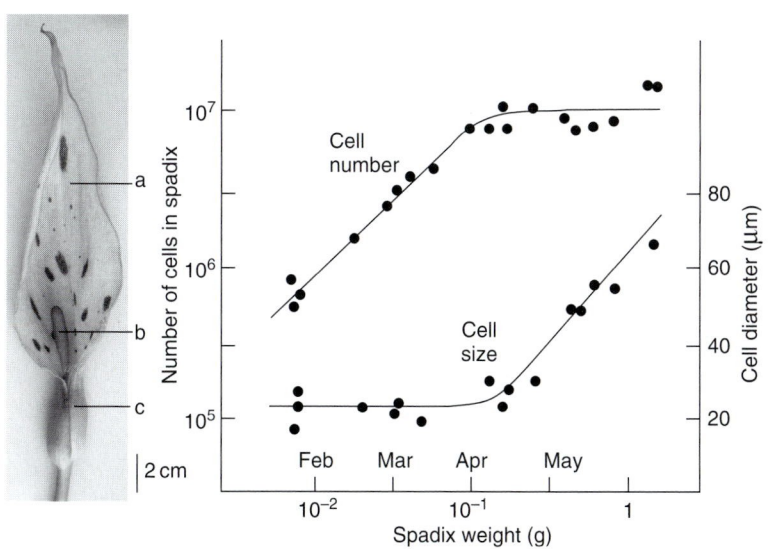

Fig. 8.6 Cell number and cell division in the spadix of *Arum maculatum*. The spadix exhibits an extraordinarily rapid rate of respiration, which is thermogenic, volatilizing amines and indoles which attract insect pollinators. a = spathe; b = spadix; c = enclosure containing the flowers. Within the spadix the stages of cell division and cell expansion are distinctly separate. From Simon & Chapman (1961). Photograph from www.bioimages.org.uk, courtesy of Malcolm Storey.

Cell expansion results from the uptake of water and solutes. Vacuoles appear and coalesce to form a central vacuole which can comprise 98% of the total volume of a mature cell, and new cell wall material is continuously synthesized. The direction in which the cell expands is governed by the orientation of the cellulose microfibrils present within the cell wall. A random orientation of microfibrils results in a spherical cell but normally the distribution is not random, directing cells to take up characteristic shapes. Figure 8.7 shows intact epidermal cells of onion (*Allium cepa*), which maintain their typically elongated shape owing to the presence of the cell wall. If the cell wall is removed by digestion with cellulases and pectinases, the resulting **protoplasts** are spherical as they are surrounded only by the flexible plasma membrane. The orientation of the cellulose microfibrils deposited in the cell wall is controlled by the orientation of microtubules within the cell. Disruption of microtubule structure using drugs such as colchicine leads to disordered cell expansion.

8.4.1 Auxin and the control of cell expansion

For many years auxin has been regarded as *the* hormone that promotes and controls cell expansion. However, it is now recognized that other plant hormones such as gibberellins and cytokinins may also stimulate cell enlargement. In leaf and cotyledon tissue, for example, cytokinins can stimulate expansion, and gibberellins lead to cell elongation in germinating embryo radicles. Nevertheless, the effects of auxin have been studied most intensively.

Cell wall stretching

Quite early observations revealed that auxin has an effect on the wall properties of elongating cells. During cell expansion, particularly

Fig. 8.7 The cells of onion (*Allium cepa*) epidermis are rectangular as their shape is maintained by the rigid cellulose cell wall (A). Removal of the cell wall by digestion with cellulase and pectinase results in the formation of spherical protoplasts (B).

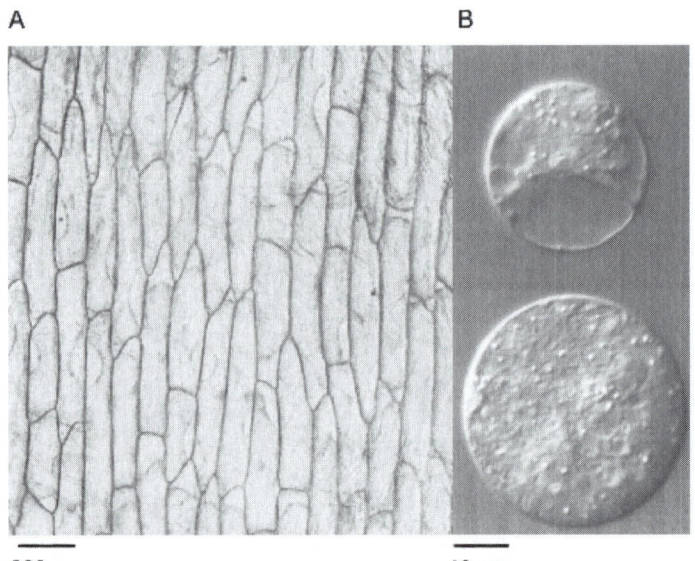

A   B

200 µm   10 µm

early in the process, changes occur in the physical properties of the wall and middle lamella which permit an increase in cell area without increase in turgor pressure and irrespective of simultaneous synthesis of new cell wall material. However, soon after cell expansion is initiated the synthesis of new cell-wall polysaccharides, including cellulose, is stimulated. Heyn in 1931 first reported values for the **elasticity** (reversible stretching) and **plasticity** (irreversible stretching) of sections of plasmolysed, etiolated oat coleoptiles. The coleoptile section was held horizontally and weights were applied. The bending of the coleoptile, and its recovery upon removal of the weights, were measured (Fig. 8.8). Auxin pretreatment increases the plasticity.

The growing cell wall is a complex system (see Cosgrove 2000 for a review). The structural framework is provided by cellulose microfibrils which are coated with heteroglycans (hemicellulose – xyloglucans) embedded in a matrix of neutral and acidic polysaccharides (pectins). The matrix also contains a number of proteins (both structural and enzymes), hydrophobic compounds and inorganic ions. Both cellulose and xyloglucan are polymers of β-1,4-glucosyl residues – xyloglucan has additional mono-, di- and tri-saccharide sidechains of other sugars (xylose, galactose and fucose). The wall constituents are held together by extensive hydrogen bonding with xyloglucan forming crosslinks between cellulose microfibrils (Fig. 8.9).

During cell extension this structure must be disrupted, allowing the cellulose microfibrils to slide past each other, and then the bonds must reform in new positions. A key component of this process is the **acidification** of the cell wall.

The acid growth theory

Auxin promotes the acidification of the cell wall by stimulating the activity of H^+ ATPases located in the plasma membrane which pump protons from the cytoplasm to the apoplast – this is the **acid growth theory**. The mechanism by which this stimulation occurs is not fully understood but is not thought to be a direct interaction of auxin with the ATPase. The fungal toxin **fusicoccin**, which is also a potent stimulator of H^+ ATPase activity, binds to a 30 kDa protein which is

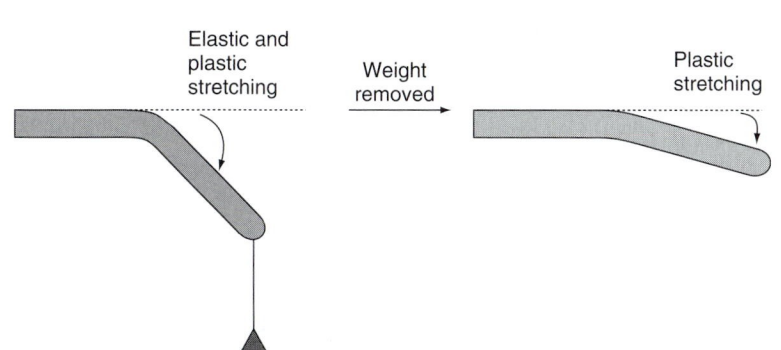

Elastic and plastic stretching

Weight removed

Plastic stretching

Fig. 8.8 Elastic and plastic stretching. A weight attached to a coleoptile causes bending as a result of reversible elastic stretching and irreversible plastic stretching.

Fig. 8.9 A possible model of the structure of the plant cell wall. A number of models exist. In this one cellulose microfibrils (grey) are linked by xyloglucan chains (black), some of which are buried within the microfibrils (dotted lines). Pectin is found interspersed with the xyloglucan chains. (A) face view; (B) section view. Adapted from Cosgrove (2001).

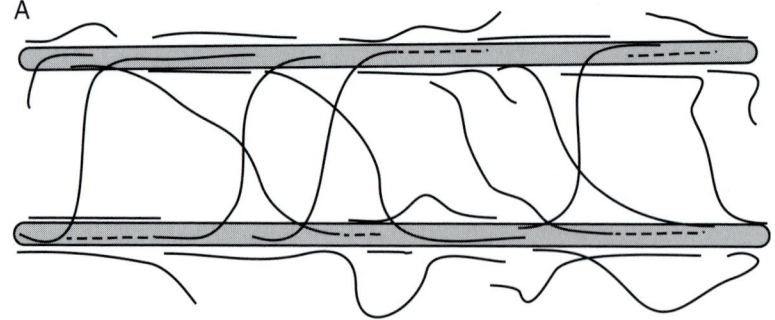

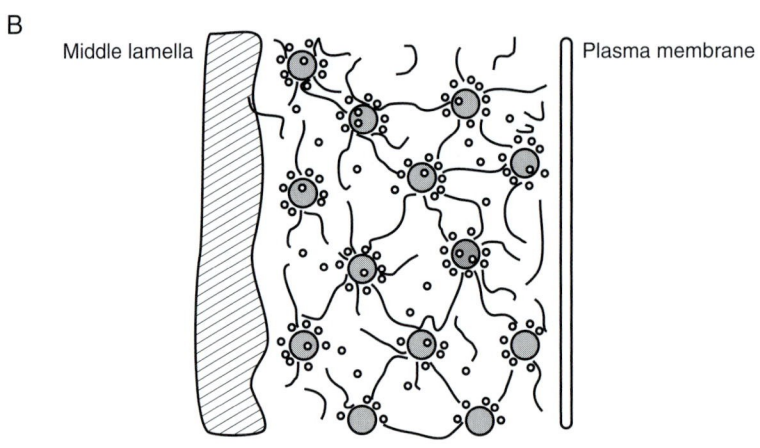

Fig. 8.10 Cell wall extension in oat (*Avena sativa*) coleoptiles in response to low pH, fusicoccin and auxin. Sections of *Avena* coleoptiles were incubated for 30 minutes with 10 μM auxin, 10 μM fusicoccin in pH 7 buffer, or 10 μM potassium phosphate–citrate buffer pH 3 without hormones. Redrawn from Cleland (1977).

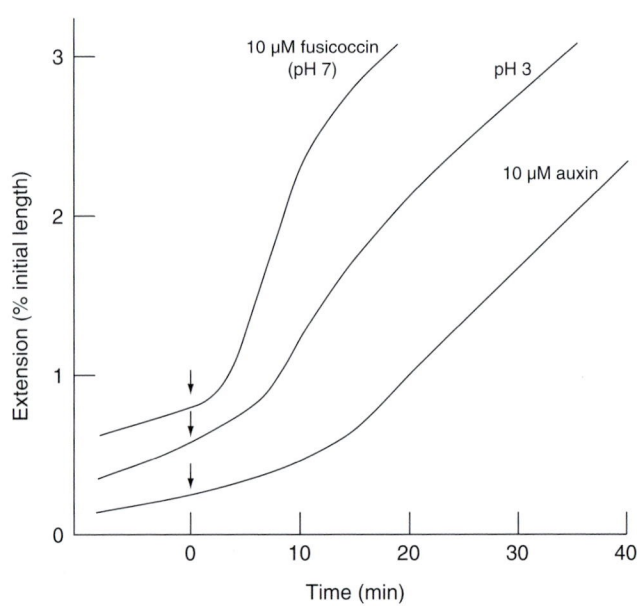

a member of a family of regulatory proteins called **14-3-3** proteins. This complexes with the H⁺ ATPase to stimulate activity, although the effect on growth is not as long-lived as that observed with auxin. Excised plant organs elongate more rapidly, and exhibit increased cell wall plasticity, when placed into buffers of low pH (Fig. 8.10).

The acid growth is not a direct consequence of low pH on the cell wall polysaccharides but rather is mediated by a series of proteins called **expansins**. If a section of cucumber hypocotyl is excised, frozen, and then thawed, the cells within it are killed but the cell wall still retains active proteins. An abraded section (to remove the impermeable cuticle) of the hypocotyl is clamped into an extensometer, in which extension in response to a small weight is measured using an electronic position transducer (Fig. 8.11). Changing the pH of the buffer surrounding the section from pH 6.8 to 4.5 results in an increase in extension although, again, this is not as long-lived as that caused by auxin, presumably as auxin-induced alterations in gene expression are not induced by low pH. If the enzymes in the hypocotyl are denatured by a brief heat treatment this acid-induced growth is abolished. However, it can be restored if proteins extracted from the walls of non-heat-treated hypocotyls are added. This **bioassay** was used to identify two proteins necessary for this response, which are called expansins (McQueen-Mason *et al.* 1992).

Plants contain many different expansins whose roles are unclear. However, they are likely to be involved in numerous aspects of plant

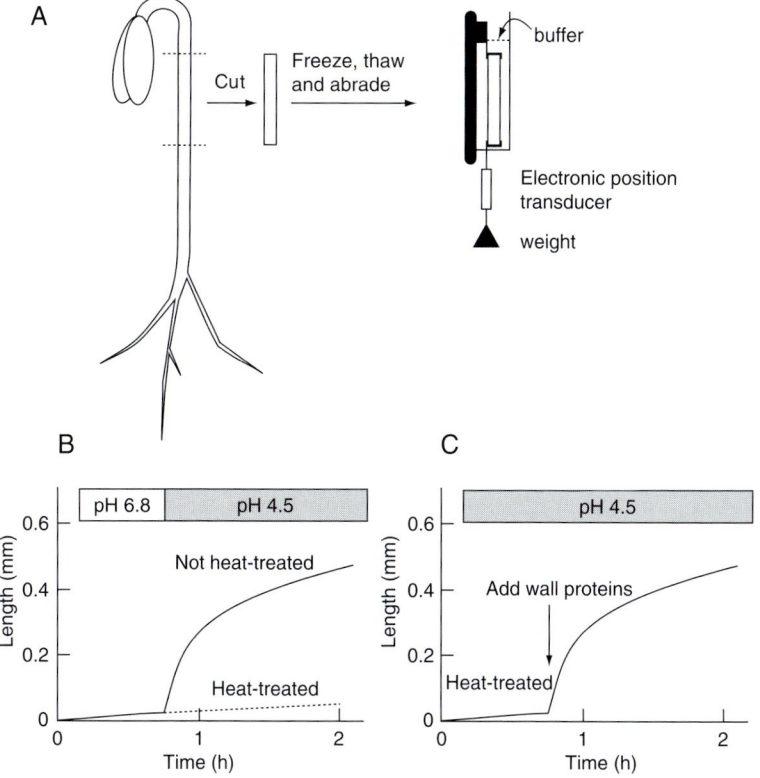

Fig. 8.11 The extensometer and its use for the bioassay of expansins. (A) Sections of cucumber hypocotyl are excised and killed by freezing. They are then thawed and the surface abraded to allow solutions to enter the section. The section is placed in a buffer and a small weight attached. Increases in the length of the section are measured using an electronic position transducer. (B) Transferring sections from a buffer of near-neutral pH to one of acidic pH results in cell extension. This response is abolished in heat-treated sections as key proteins have been inactivated by denaturation. (C) acid growth can be restored to heat-treated hypocotyl sections by the addition of isolated cell wall proteins. Redrawn from Cosgrove (1998).

growth involving cell wall softening, whether in elongating hypo-cotyls, growing pollen tubes or ripening fruit. This, as will be seen in Section 9.3, may play an important role in the control of plant development. The molecular mechanisms by which these enzymes act are not understood and undoubtedly other enzymes are necessary to remo-del the cell wall architecture. These include **endo-1, 4-β-D-glucanases** (EGases), which can cut the backbone of cellulose and xyloglucan, **xyloglucan endotransglycosylases** (XET), which cut xyloglucans and transfer the newly formed ends to other xyloglucans, hence alter-ing branching patterns, and also **glycosidases**, which alter the length of the polysaccharide chains. This must be a highly regulated process (and will involve many other synthetic and degradative enzymes) other-wise the cell wall would thin and fail. In more mature cells the degree of cross-linking present may act to prevent cell expansion.

Components of the cell wall released during these processes may act as signals which are perceived by the cell, initiating further responses. Certainly degraded cell wall components are important in activating plant responses to many pathogens, although the pre-cise signal transduction chain activated may depend upon the exact chemical nature of the groups released as well as other signals perceived by the plant.

8.5 | Cell differentiation

Cellular differentiation can be defined as *the development of specialized structures and functions.* Undoubtedly cell division, expansion and differentiation are closely linked and coordinated processes. Differentiated cells arise from the cells present in the meristems (both primary and secondary) in a position-dependent manner. All cells, including those present in the meristems, are exposed to different internal environments and there are gradients present in these environmental components whether plant growth hormones, nutrients, dissolved gases or signals derived from surrounding cells. Hence cells have a **polarity**.

Developmental polarities were observed by Vöchting in 1878, who described how willow shoots form roots at their physiologi-cally basal ends and buds at their apical ends regardless of orien-tation. Similar phenomena can be observed in root segments of dandelion (*Taraxacum officinale*) induced to sprout roots and shoots. Gradient-dependent differentiation can be demonstrated in blocks of undifferentiated callus cells induced to differentiate by the application of sucrose and auxin in a wedge to one side of the block. This method of application generates both vertical and horizontal gradients within the block and rings of vascular cells develop.

The polarity of a cell is determined very early during develop-ment. In the fertilized ovum of flowering plants, gymnosperms and pteridophytes a polarity exists defining the position of the root and shoot poles *prior to the first cell division*, i.e. a gradient exists within the cytoplasm of a single cell that is established by its microenvironment. These gradients establish polarity within the meristem and define the initial plane of cell division by controlling the location and orientation of the new cell wall formed within the dividing cell. Within a meristem different cells will divide in different planes, generating the pattern which will lead to the development of a fully differentiated organ. The integration of these processes is discussed in later chapters, together with differentiation of specific cell types in both vegetative and floral tissues.

References

Arimura, S. & Tsutsumi, N. (2002). A dynamin-like protein (ADL2b), rather than FtsZ, is involved in *Arabidopsis* mitochondrial division. *Proceedings of the National Academy of Sciences (USA)*, **99**, 5727–31.

Cleland, R. E. (1977). The control of cell enlargement. *Symposium of the Society for Experimental Biology*, **31**, 101–15.

Cosgrove, D. J. (1998). Cell wall loosening by expansins. *Plant Physiology*, **118**, 333–9.

Cosgrove, D. J. (2000). Loosening of plant cell walls by expansins. *Nature*, **407**, 321–6.

Cosgrove, D. J. (2001). Wall structure and wall loosening: a look backwards and forwards. *Plant Physiology*, **125**, 131–4.

Fulgosi, H., Gerdes, L., Westphal, S., Glockmann, C. & Soll, J. (2002). Cell and chloroplast division requires ARTEMIS. *Proceedings of the National Academy of Sciences (USA)*, **99**, 11501–6.

Gao, H. B., Kadirjan-Kalbach, D., Froehlich, J. E. & Osteryoung, K. W. (2003). ARC5, a cytosolic dynamin-like protein from plants, is part of the chloroplast division machinery. *Proceedings of the National Academy of Sciences (USA)*, **100**, 4328–33.

Joubès, J. Chevalier, C., Dudits, D. *et al.* (2000). CDK-related protein kinases in plants. *Plant Molecular Biology*, **43**, 607–20.

Kepinski, S. & Leyser, O. (2002). Ubiquitination and auxin signaling: a degrading story. *The Plant Cell*, Supplement 2002, S81–95.

Leyser, H. M. O., Lincoln, C. A., Timpte, C., Lammer, D., Turner, J. & Estelle, M. (1993). *Arabidopsis* auxin-resistance gene Axr1 encodes a protein related to ubiquitin-activating enzyme E1. *Nature*, **364**, 161–4.

List, A. jr. (1963). Some observations on DNA content and cell and nuclear volume growth in the developing xylem cells of certain higher plants. *American Journal of Botany*, **50**, 320–9.

McQueen-Mason, S., Durachko, D.M. & Cosgrove, D. J. (1992). Two endogenous proteins that induce cell wall expansion in plants. *The Plant Cell*, **4**, 1425–33.

Miyagishima, S. Y., Takahara, M., Mori, T., Kuroiwa, H., Higashiyama, T. & Kuroiwa, T. (2001). Plastid division is driven by a complex mechanism that involves differential transition of the bacterial and eukaryotic division rings. *The Plant Cell*, **13**, 2257–68.

Mori, T., Kuroiwa, H., Takahara, M., Miyagishima, S. & Kuroiwa, T. (2001). Visualization of an FtsZ ring in chloroplasts of *Lilium longiflorum* leaves. *Plant and Cell Physiology*, **42**, 555–9.

Osteryoung, K. W., Stokes, K. D., Rutherford, S. M., Percival, A. L. & Lee, W. Y. (1998). Chloroplast division in higher plants requires members of two functionally divergent gene families with homology to bacterial ftsZ. *The Plant Cell*, **10**, 1991–2004.

Pyke, K. A., Rutherford, S. M., Robertson, E. J. & Leech, R. M. (1994). *arc6*, a fertile *Arabidopsis* mutant with only 2 mesophyll cell chloroplasts. *Plant Physiology*, **106**, 1169–77.

Rossi, V. & Varotto, S. (2002). Insights into the G1/S transition in plants. *Planta*, **215**, 345–56.

Simon, E. W. & Chapman, J. A. (1961). The development of mitochondria in *Arum* spadix. *Journal of Experimental Botany*, **12**, 414–20.

Stals, H. & Inzé, D. (2001). When plant cells decide to divide. *Trends in Plant Science*, **6**, 359–64.

Chapter 9

Vegetative development

9.1 | Introduction

As described in Chapter 8, plant growth results from a combination of cell division, elongation and differentiation, initiated from groups of cells known as meristems. During embryogenesis two meristems are formed – the shoot apical meristem which gives rise to the shoot system, and the root apical meristem which forms the root system. As these primary meristems develop, they give rise to more apical meristems which will form side branches or lateral roots, and lateral meristems which result in an increase in girth (Fig. 9.1).

Together, environmental and internal signals control the rate of growth, the activation of new meristems, and the differentiation of cells and tissues, producing the plant body within the framework of the basic 'body plan' of the plant. Meristem activity must be under precise control to generate the specific structures of the plant, but at the same time must be flexible enough to respond to environmental signals. Some of the controlling signals, and the genetic systems on which they act, are discussed below, with reference to the formation of the vegetative organs of the plant. The formation of reproductive structures is covered in Chapter 11.

9.2 | The structure and activity of the shoot apical meristem

9.2.1 Organ initiation

During the vegetative phase of growth, the shoot apical meristem (SAM) produces stem, leaves and axillary buds in units known as **phytomers**. Each phytomer consists of a leaf (or leaves) joined to a stem node, the associated axillary bud (or buds) in the leaf axil(s), and the length of internode down to the next node (Fig. 9.1). The time interval between successive leaf initiations is termed the **plastochron**, of very variable length according to species, plant age and environmental conditions, but taking at least a few days. The organ primordia

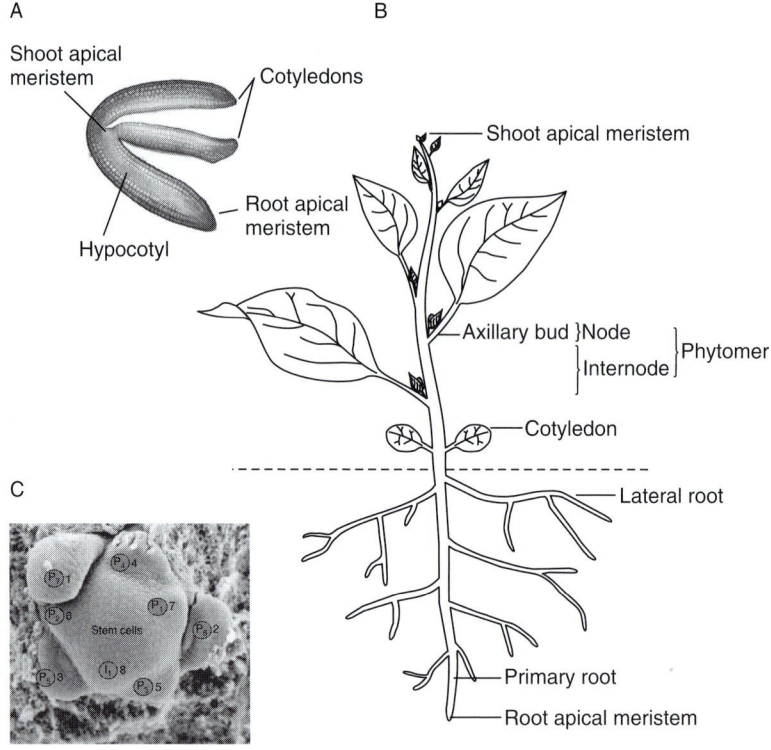

Fig. 9.1 Location of meristems within the plant. (A) An embryo of *Arabidopsis thaliana* showing the position of the shoot and root apical meristems (from J. Haseloff, Cambridge). (B) A diagram of a typical dicotyledon showing the locations of meristems within the mature plant. (C) The shoot apical meristem of *Arabidopsis* showing leaf primordia developing in a phyllotactic spiral. The primordia have been numbered in the order in which they have developed (from Clark 2001). © Nature Publishing Group.

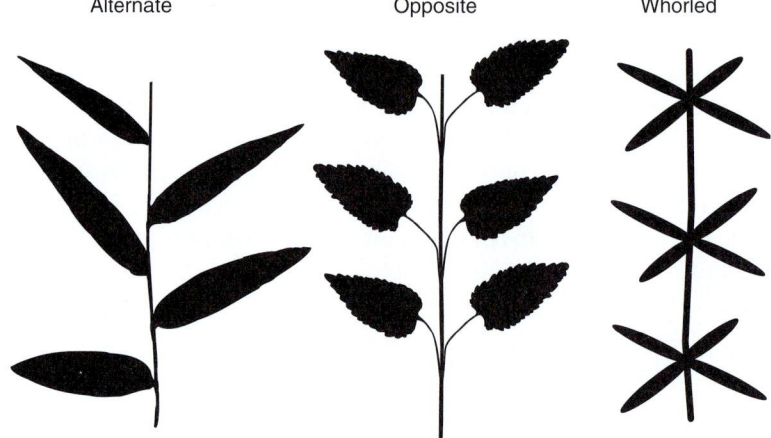

Fig. 9.2 Phyllotaxis. An alternate phyllotaxis has one leaf per node. If leaves are arranged such that each is inserted at a 135° angle to the preceding leaf, a spiral arrangement is produced which minimizes self-shading. If leaves are inserted at 180° to each other, they form vertical rows and are said to be *distichous*. An opposite phyllotaxis has two leaves per node. If pairs of leaves are inserted at 90° to the preceding pairs the arrangement is *decussate*. A whorled phyllotaxis has three or more leaves per node.

can be numbered sequentially; the convention is that the primordium which is about to develop is called I_1, the previous one (just developed) is P_1 (although some authors start at zero); as new primordia develop I_1 becomes P_1 whilst P_1 becomes P_2 and so on. The spacing of the leaf primordia is very precisely controlled and is characteristic for species; the arrangement is called **phyllotaxis**. In the majority (85%) of flowering plants the primordia are regularly spaced around the SAM in a spiral, but the phyllotaxis can be alternate, opposite or whorled (Fig. 9.2).

The phyllotactic pattern can change according to the developmental stage of the plant; for example, juvenile ivy produces leaves with an alternate phyllotaxis with leaves at a 180° angle, whereas mature plants exhibit a spiral phyllotaxis. All the patterns result from the precise positioning, and direction, of cell divisions in the SAM.

Microscopic analyses of the structure and activity of the SAM from many different plants show that it can be divided into several distinct regions (Fig. 9.3A). At the very centre and apex of the SAM is the **central zone** (CZ), containing **stem cells**, which are the progenitors of all cells developing from the SAM. If plants are fed radioactively labelled thymidine, the incorporation of radiolabel into newly synthesized DNA can be used to determine the rate of cell division. Such experiments have shown that the cells in the CZ divide very slowly, maintaining the size of the CZ, as well as forming the **peripheral zone** (PZ). Cells in the PZ, although derived from the CZ, behave quite differently. They divide very rapidly and give rise to the **organ primordia** which develop to form the leaves. Underlying the central and peripheral zones is the **rib meristem** (RM) which divides to form the pith of the stem. As this grows it pushes the cells above it upwards and outwards.

Within the SAM distinct cell layers can also be recognized. Typically there are three layers in dicotyledonous plants – L_1, L_2 and L_3 (Fig. 9.3B) – whilst monocotyledonous plants have two – L_1 and L_2. Cells of dicotyledonous plants within L_1 and L_2 characteristically divide **anticlinally** – that is, the newly formed cell walls are perpendicular to the surface so that the cells are produced parallel to the surface. Together these cells are called the **tunica**. In contrast, the cells in layer L_3 (the **corpus**) divide in a more random manner. It

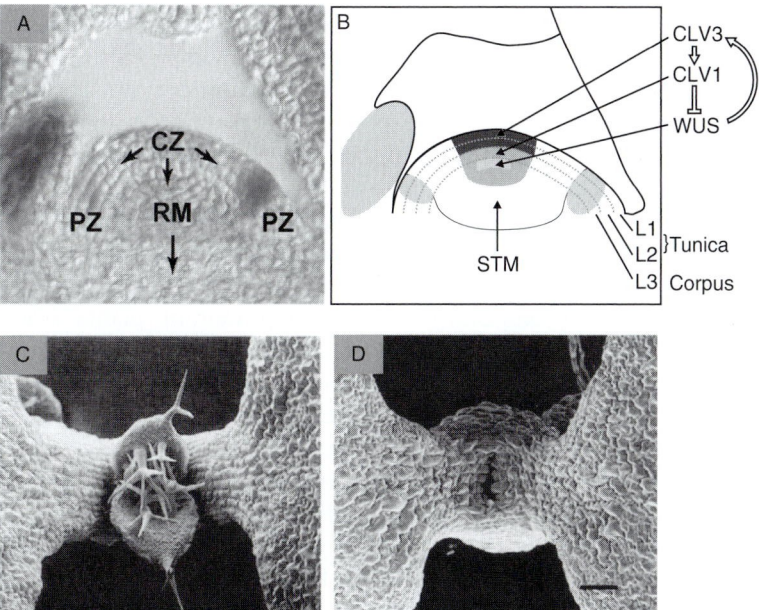

Fig. 9.3 The structure of the *Arabidopsis* shoot apical meristem. (A) The SAM consists of a central zone (CZ), a peripheral zone (PZ) and underlying rib meristem (RM). Dividing cells are pushed away from the meristem in the direction of the arrows. (B) A diagram of the SAM showing the three cell layers L1, L2 and L3 typical of dicotyledons. The expression and interaction of proteins (WUS, CLV1, CLV3, STM) involved in the control of meristem development are indicated. The light grey shaded areas indicate regions in which organ primordia are developing (adapted from Bowman & Eshed 2000). (C) A developing meristem of a wild type *Arabidopsis* plant. The cotyledons are visible at either side and leaves, with surface trichomes, are developing in the centre. (D) The meristem has been lost in the *stm* mutant (from Long *et al.* 1996). (C) and (D) © Nature Publishing Group.

is possible to track which layers give rise to which cell types in the mature organ by marking them. Exposure of the SAM to colchicine will make some cells polyploid whilst other mutations can prevent the development of chlorophyll (this occurs naturally in variegated plants). One can then recognize the progeny of particular cells by their ploidy or pigmentation. Such experiments show that the cells in L_1 give rise to the epidermis whilst the rest of the leaf is derived from L_2 and L_3 (L_2 only in monocotyledons). These cell layers are largely clonally isolated from each other, because of the orientation of cell divisions, and hence the term **cell lineage** is sometimes used to define which cell types develop from which layer. However, this lineage is not fixed; rather cells develop according to their *positions*. This **position-dependent development** is an important concept in many aspects of plant development.

Cell marking experiments show that very occasionally a cell will be pushed from one layer into another. In such circumstances the cell will develop according to its new position rather than according to the type characteristic of the layer from which it originated, which is why, perhaps, the term lineage is misleading. Such cell displacements are quite rare but demonstrate the position-dependent nature of cell development. This flexibility of development extends to later stages of organ formation. If cells within layer L_2 are marked by polyploidy their growth rate is essentially the same as that of cells with normal ploidy. Under these conditions it is possible to demonstrate that a considerable number of cells within the leaf are derived from L_2. However, if cells are marked by making them achlorophyllous, then they grow more slowly. Although proportionately fewer cells within the mature leaf will be derived from layer L_2, increased growth from L_3 will have compensated (reviewed by Meyerowitz 1997).

9.2.2 What regulates the size of the shoot apical meristem?

Clearly a balance must exist between cells dividing to form new organs and maintaining the size of the original meristem. The SAM will continue to produce cells throughout the plant's lifetime (except during periods of dormancy or if damaged) but it remains the same size regardless of the eventual size and age of the mature plant. This is quite remarkable given that trees may reach over a hundred metres in height and grow for hundreds of years. Our understanding of this balance between cell recruitment into organ primordia, the PZ and CZ, has been revolutionized by studying mutant plants – e.g. *Arabidopsis*, maize (*Zea mays*) and petunia (*Petunia hybrida*) – where this process has been perturbed. At the heart of the balance appears to be communication between the different zones of the meristem. In the *clavata1* (*clv1*) mutation of *Arabidopsis*, the meristem enlarges as a result of the accumulation of undifferentiated cells whilst in the *shoot meristemless 1* (*stm1*) mutant, cells are not maintained in the meristem which is lost during embryogenesis. In the *wuschel* (*wus*) mutant (German – tousled hair) the meristem is lost after a few organs have been formed. In the wild-type plant the gene which has been inactivated in the *clv1* mutant encodes a protein (CLV1) which acts to reduce the number of cells in the meristem, whilst STM and WUS act to increase the number of meristematic cells. These genes interact antagonistically and, together with many other genes, act to maintain the size and structure of the SAM (Fig. 9.3C–D) (Clark *et al.* 1996).

The expression of these genes within the SAM has been investigated, as have the proteins which they encode (Fig. 9.3B). CLV1 is primarily expressed in the L_3 layer of the CZ and encodes a leucine-rich repeat (LRR) receptor kinase. Another gene, *CLAVATA3*, whose loss also leads to meristem enlargement, is primarily expressed in layers L_1 and L_2 of the CZ. The protein encoded by *CLV3* is small, and most probably secreted from the cell. It is currently hypothesized that CLV3 may act as a short-range signal detected by CLV1. The interaction between CLV3 and CLV1 represses the expression of WUS, reducing the size of the meristem. However, as WUS is required for the expression of CLV3, a feedback loop is set up which maintains the meristem at the correct size. STM is expressed throughout the meristem and is thought to maintain the meristematic cells in an indeterminate state, as well as playing a role in the regulation of CLV3 expression. Although many of the details remain to be resolved, it is clear that the SAM is controlled by a network of interacting genes and signals passed over short distances. Hence when a cell is displaced, from one position within the SAM to another, it will receive a new set of signals and develop in a different manner.

Box 9.1

A detailed description of gene, mutant and protein nomenclature is given in the Appendix.

9.3 | Organ formation

As the cells in the PZ divide they are pushed away from the apex of the meristem and form organ primordia. The signals which regulate this process are not well understood but, again, networks of interacting genes are likely to be important. It is possible that one of the very earliest steps in primordium formation occurs in response to alterations in cell wall properties which facilitate cell expansion. A number of lines of evidence support this:

(1) Irradiation of the SAM with gamma rays inhibits cell divisions but a bulge, indicating cell expansion, still appears on the flanks of the SAM in the position where the next primordium would have developed.

(2) Plants with mutations in cell division still initiate organ primordia normally.

(3) Application of the cell wall loosening expansins (Section 8.4, Fleming *et al.* 1997) can induce the formation of leaf primordia in tomato (*Lycopersicon esculentum*). Localized expression of expansins can reverse the phyllotaxy in tobacco (*Nicotiana tabacum*).

The outermost layer L_1 of the SAM is thought to be under tension, restraining the layers below which are under compression. Within the SAM, organs are generated in a phyllotactic spiral (Fig. 9.1C). Organ primordia P_1 and P_2 have developed most recently and the next organ to develop would normally be placed at position I_1. However, Fleming *et al.* (1997) showed that if a small bead coated with expansin is placed at position I_2, an organ primordium is generated there and development at position I_1 is inhibited. This suggests that the release of tension at specific sites within the SAM is an important factor regulating the spatial development of organ primordia. In these early experiments, the organ primordia were green and possessed trichomes, but did not generate a vascular system. However, only the outermost layer of cells within the meristem was exposed to the expansin protein. The technique was developed further by creating transgenic tobacco plants which contained an expansin gene linked to a tetracycline-inducible promoter (Pien *et al.* 2001). When small beads of lanolin containing tetracycline were placed on the tobacco SAM at position I_2 (Fig. 9.4A), expansin gene expression was induced in a localized manner but within several cell layers of the meristem. An organ primordium developed within 72 hours of this treatment and developed into a morphologically normal leaf. Control plants, treated in the same way but with the tetracycline replaced by buffer (Fig. 9.4B) showed organ development at I_1 as normal. Remarkably, generating a new organ in this way reversed the phyllotactic spiral of organs generated subsequently, (Fig. 9.4C) showing that organ development is regulated in a position-dependent manner.

After initiation the primordia must grow rapidly, expanding several thousand-fold to become a complete organ. Beneath the leaf primordium the procambium forms. This divides and differentiates to form the primary phloem and xylem which will connect the

Fig. 9.4 Expansins cause leaf primordia to develop in the tobacco shoot apical meristem and lead to a reverse phyllotaxy. Transgenic tobacco (*Nicotiana tabacum*), plants containing a tetracycline-inducible expansin gene were produced allowing localized induction of expansin gene expression. Scanning electron micrograph (SEM) of an apex from transgenic plants in which a small pellet of lanolin containing tetracycline (A) or buffer (B) was placed at position I2 on the meristem (m) between the leaf primordia P1 and P2. Local induction of expansin led to the development of a bulge after 72 h (A) which eventually developed into a normal leaf, but reversed the phyllotaxy (C). Initially leaves (P3, P2, P1) were formed in an anticlockwise direction. Leaves formed subsequent to I'2 (I'3 – I'7) have been formed in a clockwise direction. Control plants treated with buffer developed a bulge at position I1 as normal (B) and phyllotaxy was unaffected (not shown). Bar = 150 μm. Modified from Pien *et al.* (2001). © 2001 National Academy of Sciences, USA.

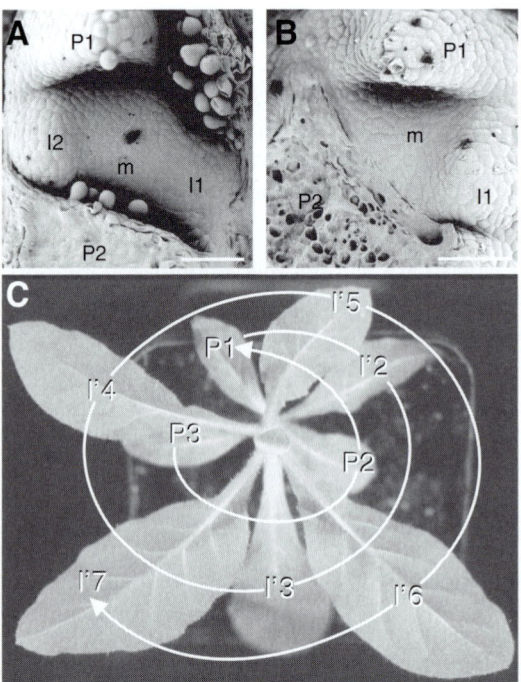

developing vascular network of the leaf (Section 9.5.3) with the vascular system of the rest of the plant.

9.4 | Secondary growth

Although many studies of plant development have focused on relatively small plant species amenable to laboratory use, it is important to remember that larger plants, in particular trees, dominate our landscape. Their stature allows them to compete efficiently for sunlight. A small plant is formed mainly or entirely of *primary* tissue, defined as tissue formed by cell divisions in the apical meristems and their immediate vicinity. The primary stem of a typical dicotyledonous plant is shown in cross-section in Fig. 9.5A. The primary phloem and xylem are clearly visible as vascular bundles, arranged in a ring towards the outside of the stem. In contrast, the vascular bundles of a monocotyledonous plant are scattered throughout the stem (Fig. 9.5B). Once the primary tissues have matured, they cannot contribute to an increase in size. In those dicots which form woody stems, secondary growth leads to an increase in girth – and strength.

Secondary growth results from cell divisions in a lateral meristem. A layer of procambial cells, left undifferentiated between the phloem and xylem, begins to divide and becomes the **fascicular cambium** (from the Latin *fasces*, bundle), whilst resumption of divisions in a layer of cortical cells between the bundles forms the **interfascicular cambium**; when these join together the resulting cylinder of dividing, meristematic cells is known as the **vascular cambium** (VC). The majority of cell divisions within the VC occur parallel to the axis of the organ. As a cell within the VC divides, one daughter cell remains

Box 9.2

Monocots do not produce a vascular cambium. Palms are monocots; they may grow to an impressive size, but their mode of growth is quite unlike that of dicot trees.

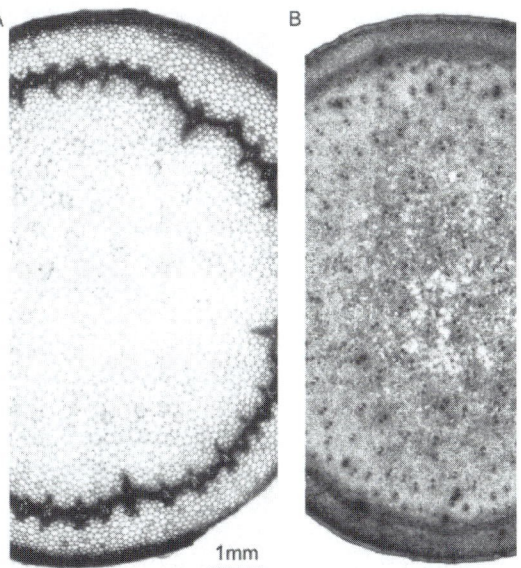

Fig. 9.5 Cross-sections through stems of dicotyledonous and monocotyledonous plants. The vascular bundles in the stem of the dicotyledonous plant *Begonia* (A) are arranged in a ring, whilst in the monocotyledonous bamboo (*Bambusa* sp.) (B) they are distributed throughout the stem.

1mm

meristematic, the other differentiates. Cells to the outside of the VC differentiate to form secondary phloem, whilst those to the inside form secondary xylem, i.e. wood. To keep pace with the increasing girth, the cambial cells occasionally divide radially so that the cambial cylinder itself expands. Some cell divisions also occur perpendicular to the organ's axis, forming the transverse rays of parenchymatous cells which run through the phloem and xylem and deliver nutrients and water to the VC. The VC is active in temperate climates for only a limited period in spring and early summer, and each year's growth can be distinguished in transverse section as an annual (growth) ring.

The primary cortex and epidermis cannot accommodate the expanding secondary tissues. Another secondary meristem, the **cork cambium**, develops near the stem surface and divides to form the bark (periderm) which consists of tightly packed cork cells in which suberin has been deposited. This substance is both gas- and watertight and effectively seals the stem; the original epidermis and cortex are isolated, stretched, torn and die. Oxygen, however, still enters the stem through lenticels ('little eyes') developed under the original stomata in the epidermis. The cork underlying lenticels has large intercellular spaces and allows gas exchange.

Cambial activity is under both environmental and developmental control. During periods of active cell division the auxin and GA content of the cambial regions is high and a concentration gradient is formed around the VC (Moyle *et al.* 2002). Auxin, transported from developing leaves, is required to maintain the VC in a meristematic state and is also a signal triggering xylem development. Other plant growth hormones, and positional cues, also regulate cell differentiation. VC activity is stimulated by mechanical stress leading to the formation of **tension wood**, which will support branches and leaning trunks. The duration of the growing season is controlled by factors such as water availability, photosynthate supply and photoperiod, which will, in turn, affect growth hormone supply to the VC. As growth ceases the VC enters a quiescent state in which it is no longer responsive to auxin. A period of chilling is required to overcome this.

9.5 | Development of the leaf

9.5.1 Shaping by cell division and expansion

A leaf is a complicated organ containing several different cell types arranged according to an underlying pattern. In leaves which are orientated close to the horizontal there are at least three different developmental gradients which influence cell development (Fig. 9.6), whilst in leaves which are held closer to the vertical (e.g. many monocot leaves) the dorsiventral gradient is much less apparent, presumably as a result of both sides of the leaf experiencing a

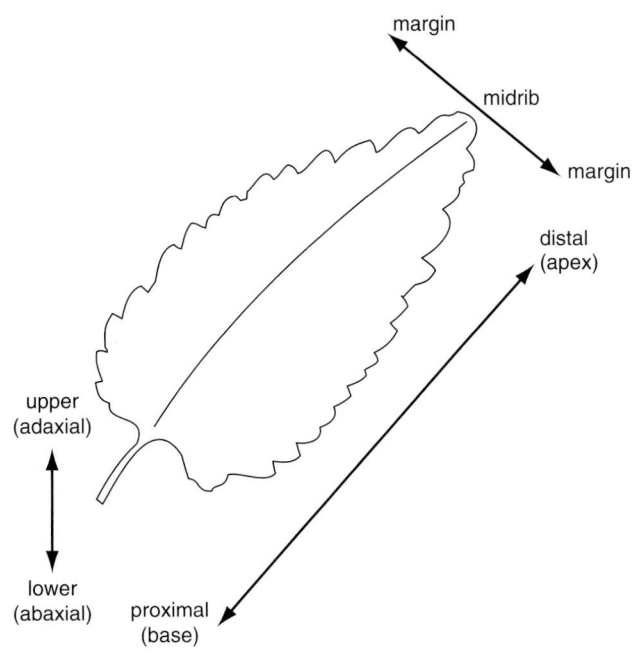

Fig. 9.6 The major developmental gradients in a dicotyledonous leaf.

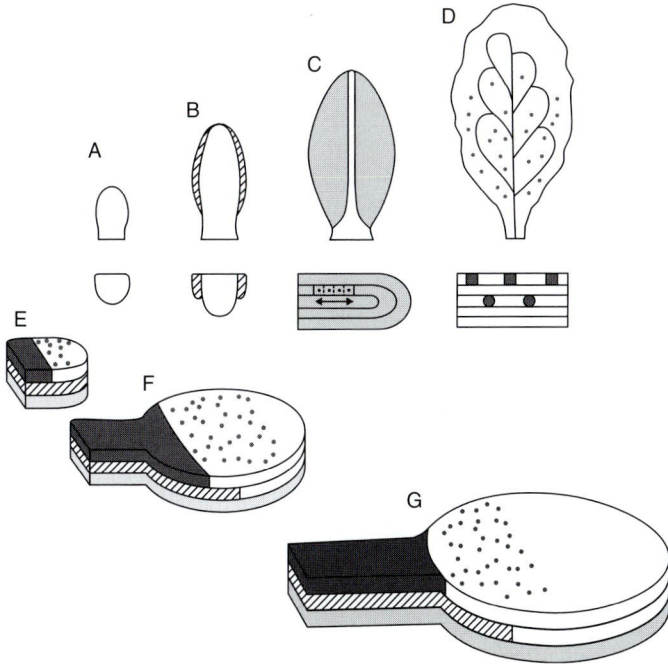

Fig. 9.7 Development in a dicotyledonous leaf. Surface and cross-sectional views: (A) leaf primordium; (B) blade morphogenesis associated with marginal meristem activity (hatched); (C) blade expansion associated with plate meristem activity (light grey); (D) cell pattern formation and cell differentiation within tissue layers, including stomatal and vein precursors (dark grey). Diagrams summarizing tissue layer-specific patterns of cell cycling in developing leaves of *Arabidopsis*: (E) day 4; (F) day 8; (G) day 12. Cycling related to stomatal guard cell formation (dark grey dots) and differentiation of vascular tissue (light grey circles and lines) is superimposed on the more general spatial patterns of cell proliferation within the adaxial epidermal (dark grey) and palisade mesophyll (hatched) layers. Adapted from Donnelly *et al.* (1999). With permission from Elsevier.

relatively similar environment. In monocot leaves, which develop from a basal meristem, the majority of cell division occurs first, followed by cell expansion and differentiation. In dicot leaves these processes overlap to a much greater extent. However, in both cases, many of the underlying principles governing leaf development are likely to be shared.

Different stages in the development of a typical dicot leaf are shown in Fig. 9.7 (Donnelly *et al.* 1999). The flattened structure of the leaf blade results from the majority of cell divisions being anticlinal, with new cells being formed parallel to the leaf surface. The number of periclinal divisions (which will increase leaf thickness) is much lower. Very early in primordium development cell divisions form a **plate meristem** which divides to form the leaf blade. In maize leaves the L_1 layer of the SAM divides to form the epidermis. As all L_1 cell divisions are anticlinal, the epidermis is only one cell thick. In contrast, the L_2 layer divides once periclinally to form two layers, the innermost of which divides again, eventually forming the vascular bundles, bundle sheath cells and the innermost mesophyll layer.

The cell divisions leading to increased leaf thickness are regulated by both developmental and environmental factors. In the *fat* mutant of tobacco, eight mesophyll cell layers form instead of the more usual four. The number of mesophyll layers in the leaves of many plants also responds to irradiance. In leaves of *Chenopodium album* grown under low irradiance (shade leaves) only a single palisade mesophyll layer forms whilst those grown under high irradiance (sun leaves) contain more (Yano & Terashima 2001) (Fig. 9.8). Not all species respond in this way. Maple (*Acer* spp.), for example, develops a thicker leaf in response to elevated irradiance but this results from an increase in the depth of cells of the palisade mesophyll layer rather than an increase in cell number. Leaf development is, therefore,

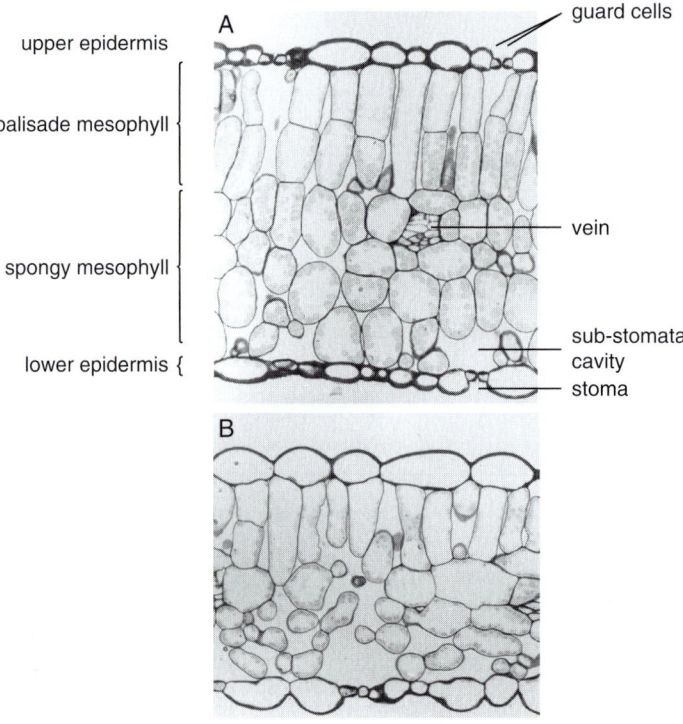

Fig. 9.8 Sun and shade leaves of *Chenopodium album*. Leaves were grown at 360 μmol m^{-2} s^{-1} (A) and 60 μmol m^{-2} s^{-1} (B). At high irradiance two palisade mesophyll cell layers have developed, in contrast to plants grown at low irradiance, in which only a single layer exists. Scale bar = 50 μm. From Yano & Terashima (2001), photographs courtesy of Professor Yano. © Japanese Society of Plant Physiologists.

upper epidermis

palisade mesophyll

spongy mesophyll

lower epidermis {

guard cells

A

vein

sub-stomatal cavity

stoma

B

plastic in that it can respond to environmental conditions but is also determinate, growth ceasing when a final size and structure has been attained.

In the majority of species leaf development follows a basipetal gradient, i.e. cell differentiation and maturation start at the apex and advance towards the base. This gradient is reflected in many aspects of leaf development, including cell division. Figure 9.9 shows the patterns of cell division in developing leaves of *Arabidopsis*. To identify those cells which are undergoing cell division, the promoter of a cyclin gene, whose expression increases during mitosis (Section 8.2.3), has been fused to a reporter gene, β-glucuronidase (GUS). Dividing cells therefore express GUS and can be stained blue (Donnelly *et al.* 1999). Initially cell divisions occur throughout the leaf but then become limited to the more basal regions of the leaf blade until finally they are restricted to the petiole. Although the majority of cell divisions forming the leaf tissue follow this pattern, cell divisions in specialized cells, such as those giving rise to the vasculature or the guard cells, continue for much longer. In the *cin* mutant of *Antirrhinum*, the timing of the cessation of cell division is subtly altered, with division continuing somewhat longer in the middle of the leaf than in the wild-type. This small alteration in timing has a dramatic effect on leaf morphology – rather than developing as a flat lamina, the leaf becomes highly crinkled (Nath *et al.* 2003).

Box 9.3

β-glucuronidase is a bacterial enzyme not normally found in plants. It cleaves 5-bromo-4-chloro-3-indolyl-β-D-glucuronic acid (abbreviated to X-gluc for obvious reasons!) to form a blue precipitate.

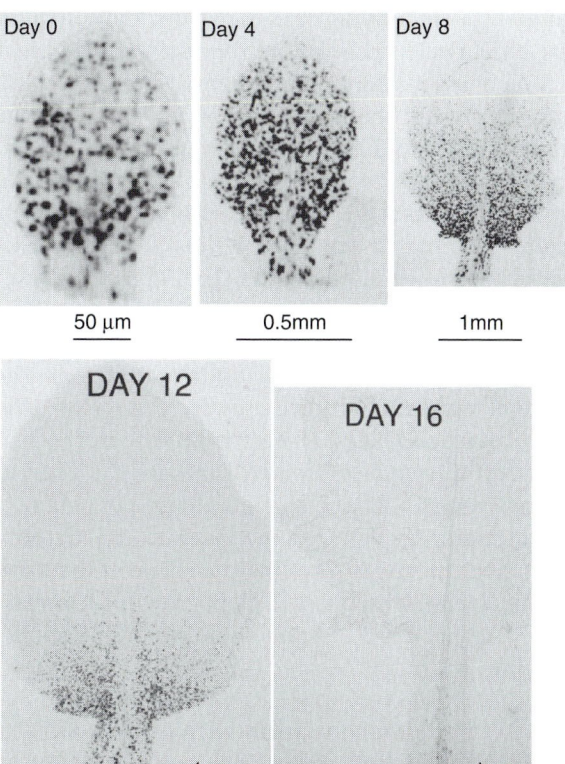

Fig. 9.9 Cell division in developing leaves of *Arabidopsis*. Leaves of transgenic plants containing a cyclin promoter fused to β-glucuronidase (GUS) were stained for activity at different points during development. Dividing cells stain a dark colour. As the leaves develop cell division becomes increasingly restricted to the base of the leaf although divisions leading to guard cell formation continue. From Donnelly *et al.* (1999). With permission from Elsevier.

Interestingly, it has been demonstrated recently that the expression of related genes in *Arabidopsis* is controlled, at least in part, by microRNAs (Palatnik *et al.* 2003, Section 7.2.9).

9.5.2 What drives leaf blade expansion?

Cell expansion is controlled by a combination of cell wall extensibility, which is a highly regulated process (Section 8.4), and cell turgor. The extensibility of the cell wall can be defined in terms of a **yield threshold**. Increases in turgor pressure above this threshold will result in expansion; hence both are important. Osmotic effects will therefore play an important role in driving leaf expansion. Many ion channels exist in the plasma membrane, which regulate ion movement into, and out of, the apoplastic space. As the volume of the apoplastic space is small compared to the rest of the cell volume, the movement of a relatively small number of ions will have a profound effect on the apoplastic osmotic potential, and thus provides a sensitive means of controlling the rate of cell expansion.

Clearly, for a flat leaf blade to develop (although not all leaves are flat!) the processes of cell division and expansion must be closely coordinated. It is thought that the epidermis plays an important role in restricting the expansion of the underlying tissues whilst the driving force may come from the mesophyll layers and/or the vasculature. Evidence for the restrictive nature of the epidermis comes from observations that removal of the epidermis from leaf discs allows the underlying cells to expand and cell divisions in epidermal pavement cells cease before those of the palisade mesophyll cells.

A compact mesophyll tissue will result when the epidermis stops growing whilst that of the underlying tissue continues. On the other hand more rapid epidermal tissue growth will pull the mesophyll cells apart and form air spaces. In the extreme case of the *argentum* mutant of pea the epidermis stops growing first but the mesophyll cells continue to expand, eventually causing them to buckle and separate from the epidermis. The large air space formed gives the leaves a silvery appearance (*argent* = silver).

The size and shape of the leaves are controlled by a combination of environmental and developmental factors. The SAM undergoes several **phase changes**, initially producing **juvenile** leaves and then **mature** leaves. Once the SAM is in the mature phase it can respond to signals which will initiate the **reproductive** phase. In some species both the shape and the phyllotaxis of the leaves may differ markedly between the juvenile and mature phases (Fig. 9.10). The mechanisms controlling the shape of the leaves in the different phases include responses to plant growth hormones (especially gibberellic acid, Section 7.2.3) and photoreceptors such as phytochrome (see Chapter 10).

9.5.3 Cell differentiation within the leaf

In a developing leaf certain cell layers are readily identifiable at a very early stage – the epidermis is present in the meristem and trichomes

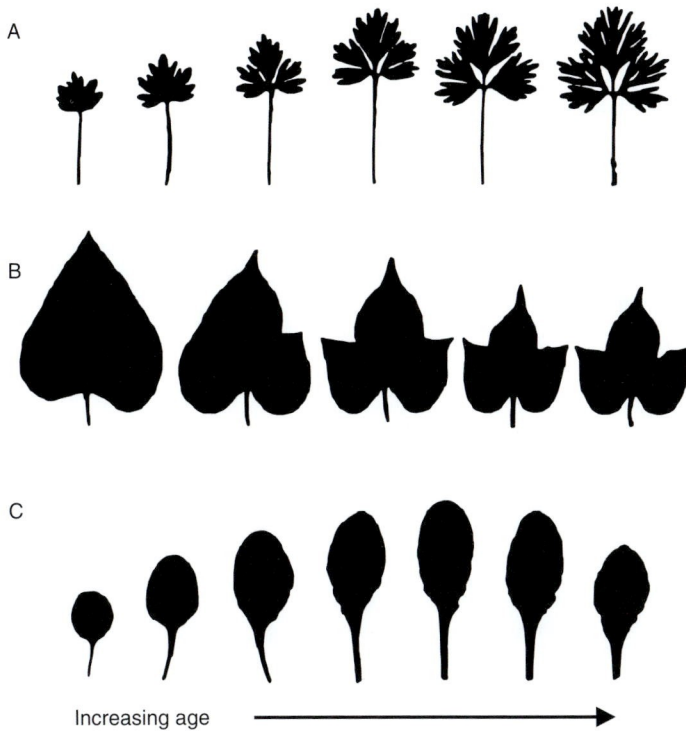

Fig. 9.10 Leaves produced by juvenile and adult plants are often of different shapes. (A) The first six leaves of *Delphinium ajacis*. (B) The first five leaves of *Ipomoea caerulea*. (C) Leaves 2–8 of *Arabidopsis thaliana* (ecotype Columbia). From Kerstetter & Poethig (1998). Drawings are not to scale.

Increasing age

develop very early. However, many cell types differentiate only much later in leaf development. Again, the theme is one of position-dependent development onto which a degree of environmental regulation is superimposed.

Guard cells

Consider the case of the guard cells which bound the stomata and regulate gas exchange into, and out of, the leaf. These develop as a result of cell division within the epidermis. The majority of epidermal cell divisions cease at the same time (approximately day 8 in *Arabidopsis*) but a few, which will go on to form the guard cells, continue to divide until day 20 (Donnelly *et al.* 1999).

Both environmental and developmental signals control which epidermal cells divide to form guard cells. For example, in the leaves of dicotyledonous plants, there are generally more stomata on the abaxial (lower) surface of the leaf than the adaxial (upper) surface and there is always at least one, usually more, epidermal cells between stomata. Also, the spacing of stomata is regulated – too few and there will be insufficient CO_2 entering the leaf to support photosynthesis, too many and water losses would become excessive. Environmental factors can strongly influence the number of stomata which develop including irradiance and atmospheric CO_2 concentration.

The development of a guard cell in a dicotyledonous leaf is shown in Fig. 9.11. An asymmetric division of a meristemoid cell gives rise to a large and a small daughter cell. The small daughter cell retains

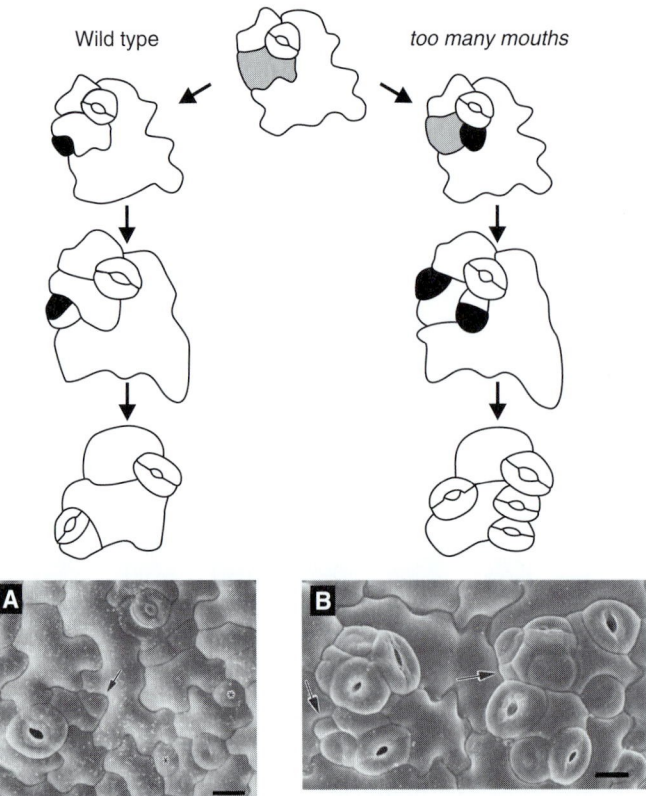

Fig. 9.11 Cell divisions leading to guard cell formation. Stomatal pattern development in wild type *Arabidopsis* (A) and the *too many mouths (tmm)* mutant (B). An asymmetric division of a meristemoid mother cell (grey) produces a meristemoid (black). In the wild type plant the orientation of this division is such that the meristemoid is placed away from the pre-existing stoma, whereas in *tmm* the plane of division is random (Nadeau & Sack, 2002). The lower panel shows cryoscanning electron microscopy of leaf surfaces. Meristemoids are indicated by arrowheads, correctly located in the wild type and incorrectly in *tmm*. Scale bar = 10 μm. From Geisler *et al.* (2000).

meristemoid activity and usually undergoes two further cell divisions to form first a guard mother cell and then the guard cells themselves. The first cell division is always orientated so that the meristemoid is placed away from an existing stoma, i.e. there are position-dependent signals controlling the orientation of the plane of cell division. Other signals may also prevent asymmetric divisions in cells adjacent to two existing stomata or precursors. A number of mutants have been described in which stomatal patterning is disrupted. In the *Arabidopsis* mutant *too many mouths (tmm)* the plane of asymmetric division is random and stomata form in clusters. The gene mutated in *tmm* has been cloned and found to encode a LRR receptor kinase similar to CLV1 (Section 9.2.2). Another mutant, *stomatal density and patterning 1 (sdd1)* also shows aberrant patterning. This mutation arises from a lesion in a gene encoding a protease, and it is tempting to speculate that this may cleave other proteins to release signalling peptides (analogous to CLV3).

The environmental regulation of stomatal density is also the subject of extensive research. In many species, stomatal density (and index) decreases with increasing atmospheric CO_2 or decreased irradiance. Lake *et al.* (2001) showed that these environmental signals could be detected in mature leaves and generate a long-distance signal which influenced stomatal number in developing leaves. They did this by placing the mature leaves of an *Arabidopsis* plant in an airtight cuvette so that the CO_2 environment of these leaves could be controlled independently from developing leaves (Fig. 9.12). If the mature leaves were placed at elevated CO_2 (720 µmol mol^{-1}), whilst the rest of the plant was maintained at ambient CO_2 (360 µmol mol^{-1}), the stomatal density and index of the developing leaves was reduced. The reverse happened when the mature leaves were placed at ambient CO_2 and the developing leaves were placed at elevated CO_2, i.e. the stomatal index of the developing leaves is influenced by the atmospheric environment of the mature leaves. Similar observations were made in response to reduced irradiance. The authors speculate that because the developing leaves are sheathed by other leaves, a more accurate picture of the light and atmospheric environment will be perceived by mature leaves growing out in the open.

Recent studies have supported the long-standing hypothesis that diffusible signals may play a role in regulating stomatal development. Several mutants of *Arabidopsis* with lesions in cuticle biosynthesis genes have altered stomatal and trichome spacing. It has been hypothesized that these alterations either directly affect the synthesis of a signalling molecule or the diffusion of a signalling molecule within the cuticle (Bird & Gray 2003).

Box 9.4

Two measures of stomatal number are commonly used: *stomatal index*, which is the proportion of epidermal cells which are stomata, and *stomatal density*, which is the number of stomata per unit area of leaf surface.

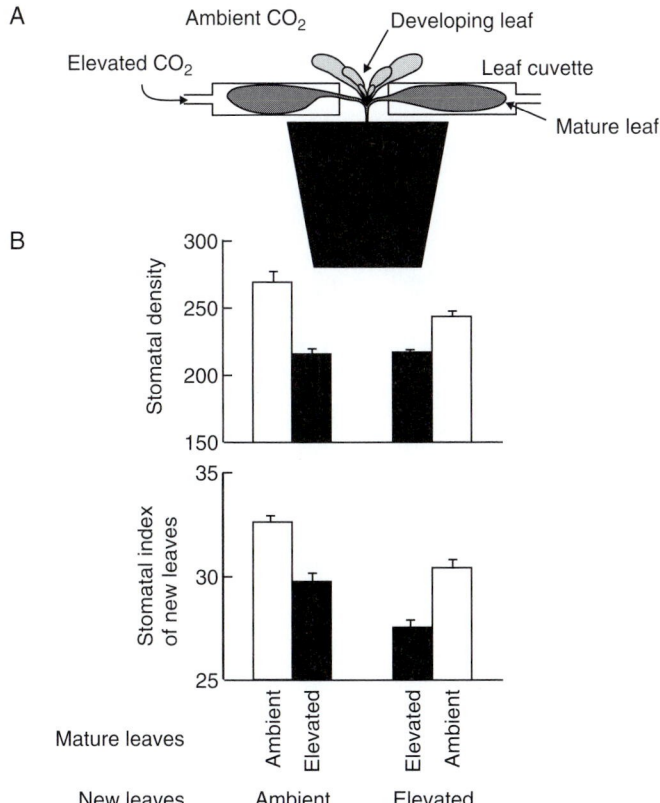

Fig. 9.12 Stomatal number in developing leaves is regulated by signals from older leaves. (A) *Arabidopsis* (Columbia, Col-0) plants were grown for 4 weeks under ambient CO_2 (360 µmol mol^{-1}). The mature leaves were then enclosed in transparent airtight cuvettes under either 720 (elevated) or 360 (ambient) µmol mol^{-1} CO_2. Subsequent leaves developed outside the cuvette under ambient CO_2. Plants were maintained in cuvettes for 7 to 9 days and the stomatal density (number of stomata per mm^2) and index ((no. of stomata/no. of stomata + no. of epidermal cells) × 100) of the last three leaves to develop were measured. (B) Stomatal index and density for new leaves under ambient CO_2 when mature leaves inside cuvettes are supplied with elevated CO_2. Both stomatal density and index are reduced in new leaves if the supply of CO_2 is increased to the mature leaves. The reverse experiment is also shown: mature leaves inside cuvettes are supplied with ambient CO_2 whilst external CO_2 is elevated. Results are shown for the abaxial leaf surface, although the same response is also observed on the adaxial surface. From Lake *et al.* (2001). © Nature Publishing Group.

The leaf vascular system

The vasculature of the leaf also develops sequentially. In the leaves of monocotyledonous plants there is a central midvein and a series of parallel longitudinal veins connected by short, fine transverse veins. In dicotyledons, a reticulate pattern develops; a central vein forms first, followed by secondary, tertiary and quaternary veins (Fig. 9.13). It is thought that the positions in which the vascular system develops within the leaf are controlled, at least in part, by auxin (Section 7.2.2). A model, the 'canalization of auxin flow hypothesis', has been proposed that attempts to explain how the sequential pattern of vein development is controlled (Sachs 1981, 1991). The apices of young leaves are known to be rich sources of auxin. Cells exposed to high concentrations of auxin respond by elongating and expressing specialized membrane-bound auxin transport proteins at their basal ends. This results in auxin being transported to a neighbouring cell, which responds in a similar manner as a result of the localized increase in auxin content. In this way a row, or file, of auxin-transporting cells is produced in which auxin concentrations are high, whilst auxin is removed from adjacent cells. Past a certain threshold, the localized high concentration of auxin acts as a signal which triggers vascular development, causing cell division and differentiation, the cell files generating sieve tubes and tracheary elements. This process is repeated as the leaf expands, leading to the sequential formation of a network of veins. Addition of auxin transport inhibitors disrupts this process (Fig. 9.13) and many mutants in vascular patterning, including some with lesions in auxin transport or perception, have been

Fig. 9.13 Leaf vein reticulation in developing leaves of *Arabidopsis*. (A) Vascular patterning in leaves of *Arabidopsis* grown in the presence of no, low or high concentrations of the auxin transport inhibitor, NPA. (B) A model of auxin transport within these leaves showing how canalization is blocked by NPA. When auxin transport is not inhibited, auxin produced in the developing leaf margin is transported away by neighbouring cells. These cells respond to the elevated concentration of auxin by increasing their conductive capacity. Past a certain threshold concentration, auxin causes vascular differentiation. Weak inhibition impairs this process and multiple vascular strands develop within the leaf. In plants grown at the highest concentrations of NPA, vascular development is greatly restricted and largely confined to the cells in the immediate vicinity of the leaf margin. Scale bar = 0.5 mm. From Berleth *et al.* (2000), modified from Mattsson *et al.* (1999).

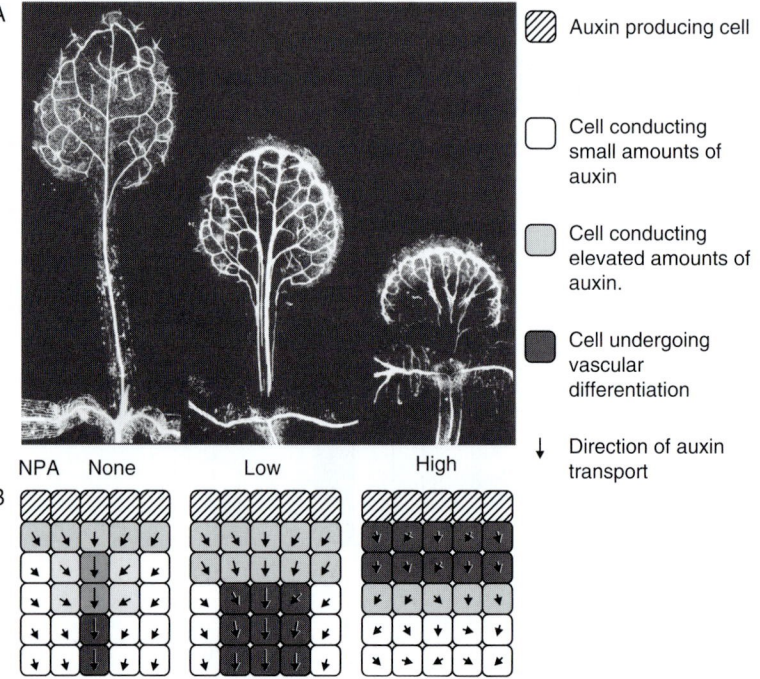

identified. This same procedure can operate in the mature leaf, allowing the vascular system to regenerate in response to wounding. As the development of the vascular system is directed by auxin flow, continuity with the rest of the plant is ensured.

Kranz anatomy

In maize, which exhibits Kranz anatomy, the expression of Rubisco in mature leaves is restricted to the bundle sheath cells in the vicinity of the veins whilst the mesophyll cells further away accumulate C_4 enzymes. Both developmental (spatial) and environmental (light) signals are thought to elicit this pattern. In darkness Rubisco is transcribed in all of the mesophyll cells (although they are not photosynthetic). In high light a substance is thought to be released from the vasculature which represses Rubisco expression in nearby mesophyll cells (but does not affect Rubisco expression in the bundle sheath cells). In the husk leaves, where the veins are much more widely spaced, mesophyll cells far away from the veins do not receive this signal, express Rubisco, and remain C_3 (Fig. 9.14).

9.5.4 Compound leaves

Many dicot leaves are compound (e.g. tomato, pea) although the leaves of some monocots are also compound. How is a compound leaf formed? Division into the different leaflets occurs very early –

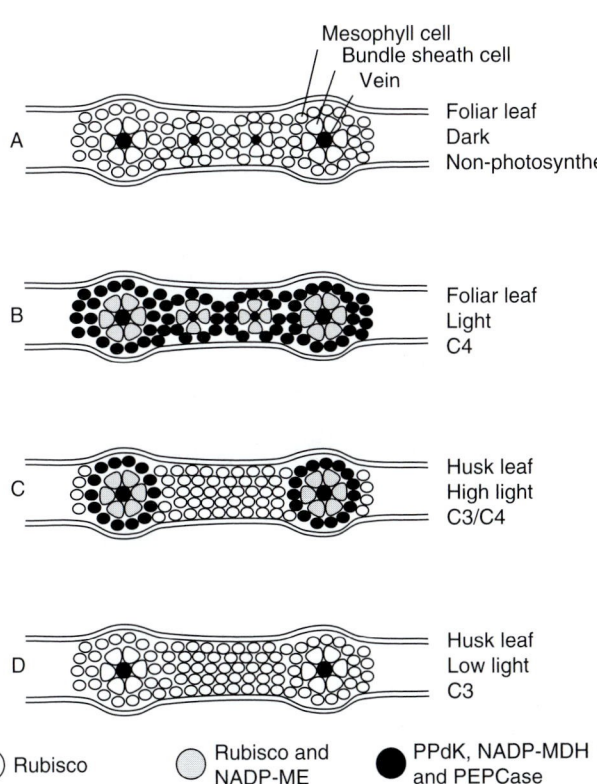

Fig. 9.14 The spatial regulation of photosynthetic metabolism within maize (*Zea mays*) leaves is developmentally and environmentally controlled. (A) Foliar leaves grown in darkness express Rubisco but are not photosynthetic. (B) If illuminated, Kranz anatomy and C4 photosynthesis develop. Rubisco expression is confined to the bundle sheath cells surrounding the veins whilst enzymes required for CO_2 concentration accumulate in mesophyll cells. In husk leaves the veins are more widely spaced. (C) Under high light C4 metabolism develops in cells proximal to veins, but cells distal develop as C3. (D) Under low light all photosynthetic cells in husk leaves exhibit C3 metabolism. NADP-ME = NADP-malic enzyme; PPdk = pyruvate, inorganic phospate dikinase; NADP-MDH = NADP-malate dehydrogenase; PEPCase = Phosphoenolpyruvate carboxylase. Redrawn from Langdale & Nelson (1991).

plastochron (Section 9.2.1) 3–4 in tomato – and subsequent development of each leaflet is not dissimilar to that of simple leaves. It is thought that differences in the timing of gene expression lead to these different morphologies. In simple leaves the expression of a class of genes known as *KNOX1* (which includes the shoot meristemless gene, Section 9.2.2) declines in primordium P_0; in fact this decline is an accurate predictor of where the next primordium will form. However, in the compound leaves of tomato, this decline does not occur immediately upon primordial development, rather it persists and multiple subprimordia develop. If tomato plants carry a mutation which makes them produce less of the protein encoded by *KNOX1*, then the leaves become much more simple in character, whilst transgenic plants over-expressing the protein produce thousands of tiny leaflets (Fig. 9.15) (Hareven *et al.* 1996). It should be noted, however, that over-expressing *KNOX1* genes in species with simple leaves does not lead to the development of compound leaves – hence other signals must be involved. In addition, Bharathan *et al.* (2002) found in a survey of *KNOXI* expression of various vascular plants that although continued *KNOXI* gene expression was correlated with the formation of complex leaf primordia, this was not always reflected in the final morphology of the mature leaf. In spite of this it is still truly remarkable that altering the expression patterns of a single gene can have such a profound effect on leaf development.

Recent studies have shown that the *KNOX1* mRNA can be transported long distances throughout the plant via the phloem (Kim *et al.* 2002). In addition, the protein itself can move between cells within the

Box 9.5

The maize mutant *Knotted1* exhibits aberrant leaf development. As more genes which controlled different aspects of development were identified and cloned, their similarities led to them being grouped as Knotted1-like homeobox genes (*KNOX*).

Fig. 9.15 Compound leaves of tomato (*Lycopersicon esculentum*) and the impact of alterations in *Knotted-1* expression. (A) The leaf of tomato is compound, and is composed of a number of petiolated leaflets attached to the rachis. Folioles appear between the leaflets. (B) Over-expression of the maize *Knotted-1* gene results in a super-compound leaf consisting of hundreds of leaflets branched many times. However, the overall dimension of the leaf is similar to that of the wild type plant. Mutations can produce pseudo-simple leaves as seen in the *entire* mutant (C) and the simple leaf of the *lanceolate* mutant (D) of tomato. From Hareven *et al.* (1996).

leaf. Hence our models of regulatory networks of genes controlling plant development will have to consider that identification of the cells in which a gene is expressed may not necessarily indicate the cell types in which it acts.

9.6 | The structure and activity of the root apical meristem

9.6.1 Zones of cell division

The root meristem, which forms during embryogenesis, gives rise to the **primary** root from which the lateral roots develop. The development of the root system is coordinated with the development of the shoot as the root : shoot ratio is highly regulated (Chapter 6).

Unlike the shoot meristem, the root apical meristem (RAM) is two sided – it produces cells above which will form the main body of the root and also cells below to form the root cap. While the main body of the root elongates, the root cap maintains a constant size, the outermost cells senescing and being sloughed off as the root pushes through the soil. Also, the RAM does not produce lateral organs. Microscopic analyses and measurements of cell division using radioactive nucleotides show that the RAM contains different zones. At its heart is a **quiescent centre**, which, like the central zone of the shoot apical meristem, contains cells which divide very slowly. In maize roots the cells in the quiescent centre divide every 170 hours compared to 10–25 hours in neighbouring cells. The size of this centre varies between species. In maize it consists of between 1000–1500 cells whilst in *Arabidopsis* it is composed of just 4 cells (see Kerk & Feldman 1995). A quiescent centre is thought to be essential for meristem activity: if it is removed it will redevelop before the rest of the meristem. It is considered to play an important role in coordinating cell division and differentiation.

Surrounding the quiescent centre are actively dividing cells forming meristematic zones. These divide to form files of cells which elongate, and then differentiate, to form the root. Development is **stereotyped**, i.e. a specific cell in the root meristem undergoes a defined series of cell divisions, forming **initials** which then divide further to form specific cell types. The development of these cells depends upon their position. If cells are destroyed, new cell divisions will occur to replace them. Although the RAM differs from the SAM in a number of ways, it is likely that the size and development of cells within the RAM are regulated by interactions between networks of genes presumably in a manner similar to that of the SAM. Homologues of CLV3 and WUS have been found in the apices of *Arabidopsis* and rice roots (see Casamitjana-Martínez *et al.* 2003 and Kamiya *et al.* 2003). The reader is referred to Casson & Lindsey (2003) for a recent review of root development.

Figure 9.16 shows the organization of an *Arabidopsis* primary root (Dolan *et al.* 1993). The root can be divided into different, overlapping

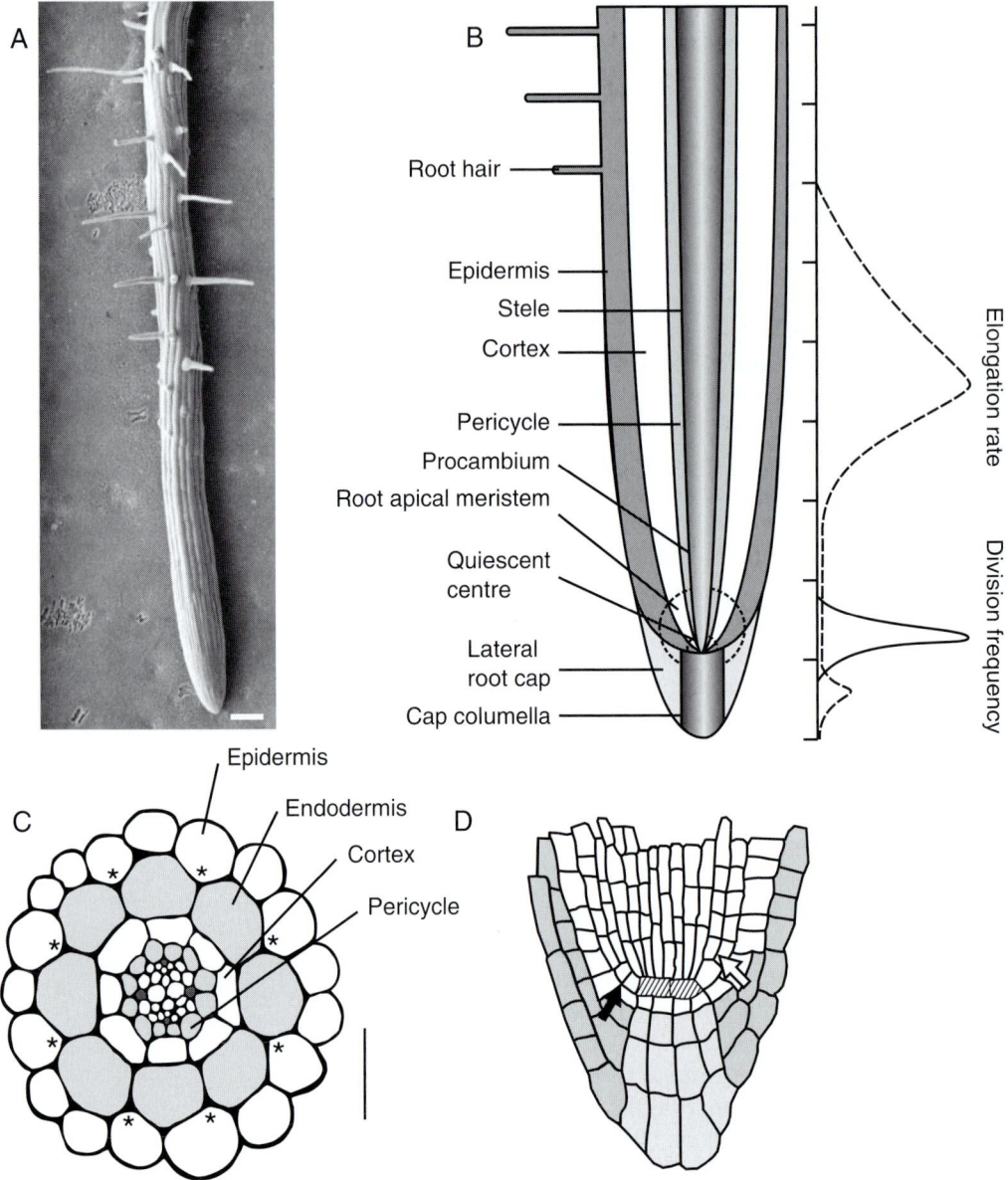

Fig. 9.16 Root structure. (A) Exterior view of a developing root of *Arabidopsis thaliana*. Scale bar = 100 μm. (B) Generalized diagram showing the typical structure of a developing root, indicating zones of cell division, expansion and differentiation. Ticks on the axis are at 1 mm intervals for maize (*Zea mays*) and 50 μm intervals for *Arabidopsis*. (C) Transverse section through an *Arabidopsis* root taken 1 mm behind the tip. Cells have been shaded for clarity. The protoxylem and protophloem are shaded dark grey and lie at 90° to each other. Epidermal cells overlying two adjacent cortical cells (indicated by *) have the potential to develop into a root hair. Scale bar = 25 μm. (D) Longitudinal section through an *Arabidopsis* root tip. The columella and lateral root cap are shown in grey. The cells at the quiescent centre are hatched. The black arrow shows a cortex initial and its daughter cell. The white arrow shows the next developmental step, in which a lateral division has given rise to the cortex and endodermis. (A) from Dr Liam Dolan and (C) from Dolan *et al.* (1993). © Company of Biologists Ltd. (B) redrawn from Jacobs (1997). (D) redrawn from van den Berg *et al.* (1997). © Nature Publishing Group.

zones. At the base of the root is the meristematic zone, approximately 250 µm long, overlaid by the root cap. The majority of cell division occurs in this region. Distal to this is the zone of cell elongation, again approximately 250 µm long, and then a zone of differentiation. At this early stage of root development the structure of an *Arabidopsis* root is very defined and consists of layers of cells containing consistent numbers of cells. Figure 9.16C shows a cross-section through the root 1 mm behind the root tip. Four layers of cells, the epidermis, cortex, endodermis and pericycle, surround the stele which contains the developing vascular system. Two protoxylem and two protophloem elements are visible arranged at 90° to each other. At this stage of development there are usually 8 cortical, 8 endodermal and 12 pericycle cells per cross-section.

All of these files of cells develop from a few, defined cell divisions in the meristems which give rise to precisely positioned initials (Fig. 9.16D). Four sets of initials have been described giving rise to:

(1) the columellar root cap
(2) the lateral root cap and epidermis
(3) the endodermis and pericycle
(4) the stele (procambium) or central cylinder of the root

For example, a meristematic cell divides to form a cortex-endodermal initial (Fig. 9.16D). This divides to form two daughter cells, one of which acts as a replacement for the initial, the second of which divides asymmetrically to form a cortical and an endodermal cell. In common with many other plant developmental processes, cell divisions leading to root development are precisely controlled in orientation, symmetry and position. Again, our understanding of these processes has been greatly aided by the identification, and characterization, of mutants in which aspects of these processes are disrupted. Further information about these mutants can be found in Benfey *et al.* (1993).

9.6.2 Cell differentiation within the root

It is thought that cell differentiation in the root is also position-dependent. Some epidermal cells (trichoblasts) will differentiate to form root hair cells, which are important in increasing surface area for water and nutrient uptake. Only those cells overlying anticlinal cell walls between adjacent cortical cells become trichoblasts (indicated by * in Fig. 9.16C). Similarly, the development of the protoxylem and protophloem is related to the orientation of adjacent pericycle cells with the overlying endodermal cells. The precise way in which this position- and orientation-dependent cell development is controlled remains to be resolved but localized signals are likely to be involved.

As the root grows it expands in diameter. The outer three layers are lost and secondary thickening results from division of cells in the stele. Thin-walled (parenchymal) cells located between the primary xylem and phloem divide to form a vascular cambium which, as in stems, forms phloem on the outside and xylem on the inside. Outside the phloem, in many plants, is a layer of thickened fibre cells derived from the pericycle, and outside this a cork cambium forms the periderm by anticlinal cell divisions.

9.6.3 Lateral root formation

The elaborate and highly branched root structures of mature plants result from the development of lateral roots. These develop in response to a range of developmental and environmental signals. In contrast to the shoot, where branch primordia (buds) are formed at the very apex, lateral root primordia are initiated a little way back from the tip. The primordia are, moreover, formed deep inside the root, not superficially. Cells in the pericycle are triggered to start dividing again and form a meristem similar to that of the primary root which generates the lateral root (Fig. 9.17). Although lateral root structure is similar to that of the primary root, the number of cells in each layer is not as strictly controlled.

The rate at which lateral roots are formed is strongly influenced by nutrients (particularly nitrogen) (Fig. 9.17) and water availability, and in response to damage of the primary root. Plant growth hormones are also important. Auxins promote lateral root formation and the mutants *rooty* and *superroot* which, as their names imply, have greater numbers of lateral roots, have high indole acetic acid contents (Boerjan *et al.* 1995, King *et al.* 1995). Auxin-resistant mutants exhibit reduced lateral root formation. A nitrate-inducible *Arabidopsis* gene (*ANR1*) has been described by Zhang & Forde (1998) which regulates lateral root expansion (but not initiation). When the roots

Fig. 9.17 Lateral root formation in *Arabidopsis* is triggered by nitrate and requires ANR1. As *Arabidopsis* roots encounter a region enriched in NO_3^-, lateral roots are stimulated to develop. If the expression of the gene *ANR1* is repressed, roots develop normally but do not proliferate in the nutrient-enriched zone. From Zhang & Forde (1998).

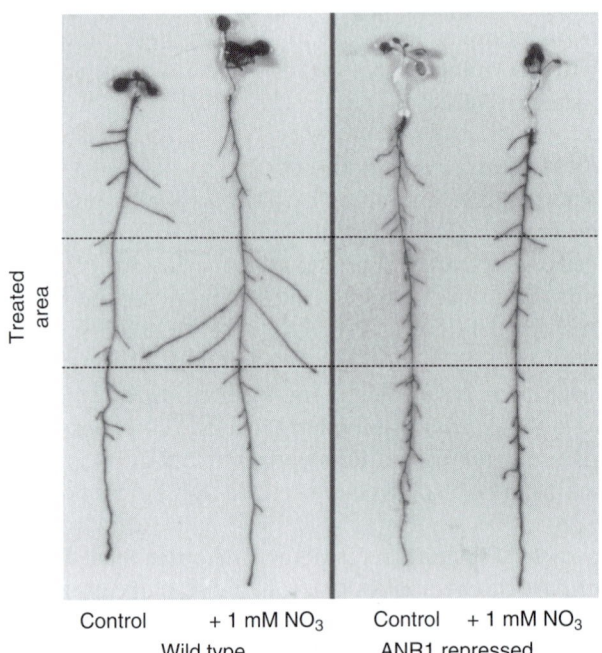

Treated area

Control   + 1 mM NO_3 Control   + 1 mM NO_3

Wild type   ANR1 repressed

of nitrate-starved *Arabidopsis* plants reach a nitrate-rich area, the expression of *ANR1* is induced and lateral root expansion is stimulated. If transgenic plants are produced in which the expression of *ANR1* is suppressed, lateral root initiation still occurs but nutrient-induced expansion is absent. The *ANR1* gene encodes a DNA-binding protein which shares homologies with proteins which regulate many diverse aspects of plant development. In this example an environmental signal, NO_3^-, regulates the expression of a transcription factor which presumably, in turn, regulates the expression of the numerous genes required for lateral root expansion.

References

Benfey, P. N., Linstead, P. J., Roberts, K., Schiefelbein, J. W., Hauser, M.-T. & Aeschbacher, R. A. (1993). Root development in *Arabidopsis*: four mutants with dramatically altered root morphogenesis. *Development*, **119**, 57–70.

Berleth, T., Mattsson, J. & Hardtke, C. S. (2000). Vascular continuity, cell axialisation and auxin. *Plant Growth Regulation*, **32**, 173–85.

Bharathan, G., Goliber, T. E., Moore, C., Kessler, S., Pham, T. & Sinha, N. R. (2002). Homologies in leaf form inferred from KNOXI gene expression during development. *Science*, **296**, 1858–60.

Bird, S. M. & Gray, J. E. (2003). Signals from the cuticle affect epidermal cell differentiation. *New Phytologist*, **157**, 9–23.

Boerjan, W., Cervera, M. T., Delarue, M. *et al.* (1995). Superroot, a recessive mutation in *Arabidopsis*, confers auxin overproduction. *The Plant Cell*, **7**, 1405–19.

Bowman, J. L. & Eshed, Y. (2000). Formation and maintenance of the shoot apical meristem. *Trends in Plant Science*, **5**, 110–15.

Casamítjana-Martinez, E., Hofhuis, H. F., Xu, J., Liu, C. M., Heidstra, R. & Scheres, B. (2003). Root-specific *CLE19* overexpression and the *sol1/2* suppressors implicate a CLV-like pathway in the control of *Arabidopsis* root meristem maintenance. *Current Biology*, **13**, 1435–41.

Casson, S. A. & Lindsey, K. (2003). Genes and signalling in root development. *New Phytologist*, **158**, 11–38.

Clark, S. E. (2001). Cell signalling at the shoot meristem. *Nature Reviews, Molecular Cell Biology*, **2**, 276–84.

Clark, S. E., Jacobsen, S. E., Levin, J. Z. & Meyerowitz, E. M. (1996). The CLAVATA and SHOOT MERISTEMLESS loci competitively regulate meristem activity in *Arabidopsis*. *Development*, **122**, 1567–75.

Dolan, L., Janmaat, K., Willemsen, V. *et al.* (1993). Cellular organisation of the *Arabidopsis thaliana* root. *Development*, **119**, 71–84.

Donnelly, P. M., Bonetta, D., Tsukaya, H., Dengler, R. E. & Dengler, N. G. (1999). Cell cycling and cell enlargement in developing leaves of *Arabidopsis*. *Developmental Biology*, **215**, 407–19.

Fleming, A. J., McQueen-Mason, S., Mandel, T. & Kuhlemeier, C. (1997). Induction of leaf primordia by the cell wall protein expansin. *Science*, **276**, 1415–18.

Geisler, M., Nadeau, J. & Sack, F. D. (2000). Oriented asymmetric divisions that generate the stomatal spacing pattern in *Arabidopsis* are disrupted by the *too many mouths* mutation. *The Plant Cell*, **12**, 2075–86.

Hareven, D., Gutfinger, T., Parnis, A., Eshed, Y. & Lifshitz, E. (1996). The making of a compound leaf: genetic manipulation of leaf architecture in tomato. *Cell*, **84**, 735–44.

Jacobs, T. (1997). Why do plant cells divide? *The Plant Cell*, **9**, 1021–9.

Kamiya, N., Nagasaki, H., Morikami, A., Sato, Y. & Matsuoka, M. (2003). Isolation and characterization of a rice WUSCHEL-type homeobox gene that is specifically expressed in the central cells of a quiescent center in the root apical meristem. *The Plant Journal*, **35**, 429–41.

Kerk, N. M. & Feldman, L. J. (1995). A biochemical model for the initiation and maintenance of the quiescent center: implications for organization of root meristems. *Development*, **121**, 2825–33.

Kerstetter, R. A. & Poethig, R. S. (1998). The specification of leaf identity during shoot development. *Annual Review of Cell Developmental Biology*, **14**, 373–98.

Kim, J. Y., Yuan, Z., Cilia, M., Khalfan-Jagani, Z. & Jackson, D. (2002). Intercellular trafficking of a *KNOTTED1* green fluorescent protein fusion in the leaf and shoot meristem of *Arabidopsis*. *Proceedings of the National Academy of Sciences (USA)*, **99**, 4103–8.

King, J. J., Stimart, D. P., Fisher, R. H. & Bleecker, A. B. (1995). A mutation altering auxin homeostasis and plant morphology in *Arabidopsis*. *The Plant Cell*, **7**, 2023–37.

Lake, J. A., Quick, W. P., Beerling, D. J. & Woodward, F. I. (2001). Signals from mature to new leaves. *Nature*, **411**, 154.

Langdale, J. A. & Nelson, T. (1991). Spatial regulation of photosynthetic development in C4 plants. *Trends in Genetics*, **7**, 191–6.

Long, J. A., Moan, E. I., Medford, J. I. & Barton M. K. (1996). A member of the KNOTTED class of homeodomain proteins encoded by the *STM* gene of *Arabidopsis*. *Nature*, **379**, 66–9.

Mattsson, J., Sung, R. Z. & Berleth, T. (1999). Responses of plant vascular systems to auxin transport inhibition. *Development*, **126**, 2979–91.

Meyerowitz, E. M. (1997). Genetic control of cell division patterns in developing plants. *Cell*, **88**, 299–308.

Moyle, R., Schrader, J., Stenberg, A. *et al.* (2002). Environmental and auxin regulation of wood formation involves members of the *Aux/IAA* gene family in hybrid aspen. *The Plant Journal*, **31**, 675–85.

Nadeau, J. & Sack, F. D. (2002). Control of stomatal distribution on the *Arabidopsis* leaf surface. *Science*, **296**, 1697–1700.

Nath, U., Crawford, B. C. W., Carpenter, R. & Coen, E. (2003). Genetic control of surface curvature. *Science*, **299**, 1404–7.

Palatnik, J. F., Allen, E., Wu, X. *et al.* (2003). Control of leaf morphogenesis by microRNAs. *Nature*, **425**, 257–63.

Pien, S., Wyrzykowska, J., McQueen-Mason, S., Smart, C. & Fleming, A. (2001). Local expression of expansin induces the entire process of leaf development and modifies leaf shape. *Proceedings of the National Academy of Sciences (USA)*, **98**, 11812–27.

Poethig, R. S., (1997). Leaf morphogenesis in flowering plants. *The Plant Cell*, **9**, 1077–87.

Sachs, T. (1981). The control of the patterned differentiation of vascular tissues. *Advances in Botanical Research*, **9**, 152–262.

Sachs, T. (1991). *Pattern Formation in Plant Tissues*. Cambridge: Cambridge University Press.

Sieburth, L. E. (1999). Auxin is required for leaf vein pattern in *Arabidopsis*. *Plant Physiology*, **121**, 1179–90.

Sinha, N. (1999). Leaf development in angiosperms. *Annual Review of Plant Physiology and Plant Molecular Biology*, **50**, 419–46.

van den Berg, C., Willemsen, V., Hendriks, G., Weisbeek, P. & Scheres, B. (1997). Short-range control of cell differentiation in the *Arabidopsis* root meristem. *Nature*, **390**, 287–9.

Yano, S. & Terashima, I. (2001). Separate localization of light signal perception for sun and shade type chloroplast and palisade tissue differentiation in *Chenopodium album*. *Plant Cell Physiology*, **42**, 1303–10.

Zhang, H. & Forde, B. G. (1998). An *Arabidopsis* MADS box gene that controls nutrient-induced changes in root architecture. *Science*, **279**, 407–9.

Chapter 10

Photomorphogenesis

10.1 | Introduction

Light is critically important to plants. The majority of them are photosynthetic and light provides the energy source required for growth. However, light is equally important for the normal development of plants as an **information medium**. In the environment light is a very complex and dynamic signal. It varies in quantity, quality (colour) and direction over timescales ranging from seconds to months (Fig. 10.1). These different variables can indicate the passing of the seasons, the availability of new habitats for growth or the presence of neighbouring vegetation which may compete for resources. Therefore it is not surprising that many aspects of plant growth and development are strongly influenced by light. The plant, too, is a complicated and ever-changing system, and the response of a plant to a given set of environmental conditions will depend upon its developmental state. As discussed in Chapter 9, plants pass through a juvenile state where their response to environmental signals differs from that of mature plants. Likewise, signals which stimulate a mature plant to flower may cause the seed of the same species to germinate – radically different developmental pathways. Similarly, plant responses are species-specific. Whilst a fast-growing weed such as *Chenopodium album* will respond to shaded conditions (i.e. low light) by elongating rapidly, rainforest tree seedlings can persist under a vegetation canopy for many years and commence rapid growth only when a gap opens in the forest canopy.

The complex nature of light as an environmental signal and the complex ways in which plants respond to it are reflected in the multitude of **photoreceptors** which have been identified. We are now beginning to understand how some of these photoreceptors work and interact with each other to coordinate plant development. Much of this research has, necessarily, been performed under highly artificial conditions allowing the roles of single (or at most, a few) photoreceptors to be studied. However, the information derived from these experiments is increasingly being applied to plants growing in a natural or agricultural environment.

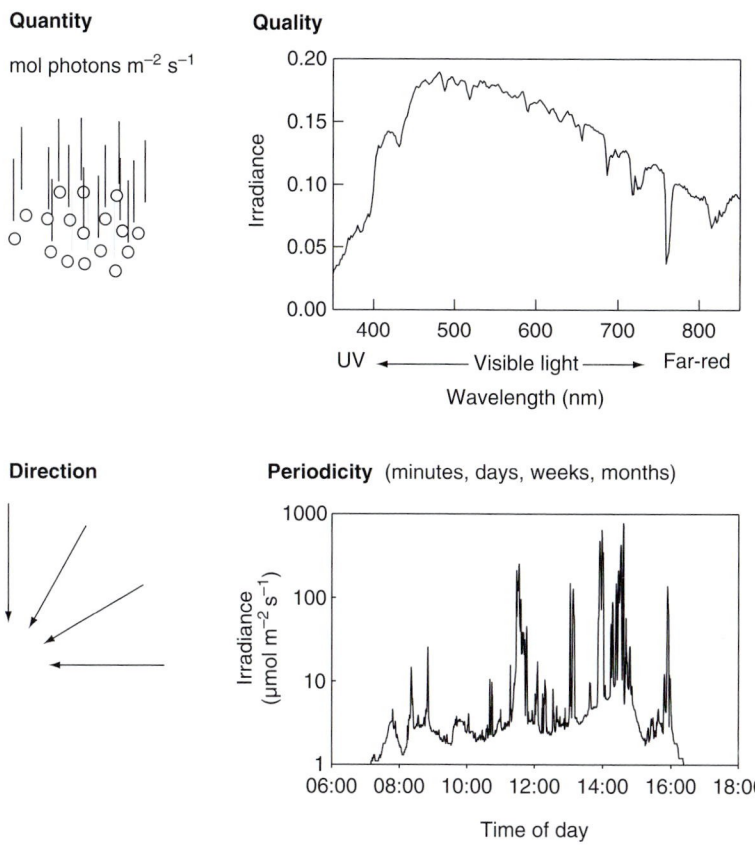

Quantity

mol photons m^{-2} s^{-1}

Quality

Irradiance

UV ← Visible light → Far-red

Wavelength (nm)

Direction

Periodicity (minutes, days, weeks, months)

Irradiance (μmol m^{-2} s^{-1})

Time of day

Fig. 10.1 Light is a complex and dynamic signal. Plants respond to many different aspects of the light environment. The quantity of light falling on a plant can be expressed in many different ways, but plant biologists generally refer to the number of photons falling per unit area per unit time. Light quality is also an important determinant of plant responses. Photosynthetically active radiation (PAR) is roughly similar to visible light and extends from 400 nm to 700 nm. Plants also respond to UV and far-red light. The light quality spectrum is that of full sunlight taken in Sheffield, UK. The variation in irradiance over the course of a day was measured in a tropical rainforest. Irradiance is very low most of the day (note the logarithmic scale) but increases as light intermittently penetrates gaps in the forest canopy. The direction at which light falls upon a plant and the way in which it varies over time is also important. Rainforest data courtesy of Dr A. Leakey.

10.2 | The switch from etiolated to de-etiolated growth

A very simple, yet informative, experiment is to grow two sets of the same species of a plant in light and in darkness (Fig. 10.2 and Fig. 10.8). The two sets of plants may then be scarcely recognizable as belonging to the same species. Whilst the light-grown individuals will be green, sturdy, with expanded leaves and of moderate height, the specimens grown in the dark will be **etiolated** – unpigmented, very tall, thin and weak with rudimentary or folded leaves. A plant which has been grown in darkness and then exposed to light will switch from etiolated to de-etiolated growth – the control of development by light being termed **photomorphogenesis** (*photo* = light, *morpho* = form, *genesis* = origin). Light regulates many aspects of plant development, inhibiting internode elongation, promoting leaf expansion (dicotyledons) or leaf unrolling (monocotyledons), promoting chlorophyll synthesis and chloroplast development and stimulating the synthesis of secondary products such as anthocyanin pigments. These processes can be initiated at light energies far

Fig. 10.2 Etiolated and de-etiolated growth maize plants (*Zea mays*). The etiolated maize plant (A) is pale and achlorophyllous. The leaves have emerged from the coleoptile but development is limited. Root development is also poor. In contrast the light-grown maize plant (B) is de-etiolated. The leaves are green and have expanded as photosynthesis fuels growth. Root development is extensive.

below that necessary for photosynthesis, in many cases by brief exposure only, and can occur in plants where the photosynthetic apparatus has not yet developed.

Light is perceived by a series of photoreceptors, the best studied of which are the **phytochromes**. However, plants contain several distinct families of photoreceptors in addition to phytochrome – these include the blue/UV-A photoreceptors (**cryptochromes**), poorly characterized UV-B photoreceptors and the **phototropins** (involved in phototropic responses – see Chapter 12).

10.3 | Phytochrome and photomorphogenesis

10.3.1 The discovery of phytochrome

Our understanding of plant photoreceptors has been revolutionized in recent years by the use of molecular genetic techniques coupled with careful biochemical and physiological measurements. By isolating and characterizing the genes encoding the different photoreceptors, and by producing mutant and transgenic plants with lesions in photo-perception, it has been possible to build a framework describing some of the mechanisms by which plants respond to light. However, this current revolution is based upon a long, and distinguished, history of studying plant responses to light.

The existence of phytochrome was first deduced from studies of certain varieties of lettuce (*Lactuca sativa*) seeds which require light for germination. These seeds are induced to germinate by brief (a few minutes) exposure to red light of low irradiance, whilst a similar exposure to far-red light is inhibitory to germination. Careful measurements made at many different wavelengths allowed the **action spectra** of these two responses to be established (Fig. 10.3). The stimulation of germination was found to have a maximum in the red at 660 nm whilst the inhibition of germination by far-red light had a maximum at 730 nm. A key breakthrough came with the discovery that alternating exposure to red and far-red light can be given for many cycles and whichever wavelength is given last determines the response. These observations led Hendricks and Borthwick, working in Beltsville, Maryland in the 1950s, to postulate the existence of a photoreceptor named **phytochrome**, existing in two photointercon-vertible forms. One form, termed P_R, absorbs red light and is converted to the other form P_{FR}. The spectral properties of P_{FR} are such that it absorbs light with a peak in the far-red and under far-red illumination it is converted to P_R again (Fig. 10.4). P_{FR} also reverts slowly to P_R when plants are placed in complete darkness. Many models of phytochrome action have proposed that P_{FR} is the active form, whilst P_R is inactive. However, as will become apparent later in this chapter, such a simple interpretation may not always be entirely accurate.

Subsequently phytochrome was isolated from plant tissues. It is a low-abundance, blue-green chromoprotein and, in solution, shows photoreversibility between the P_R and P_{FR} forms with absorption

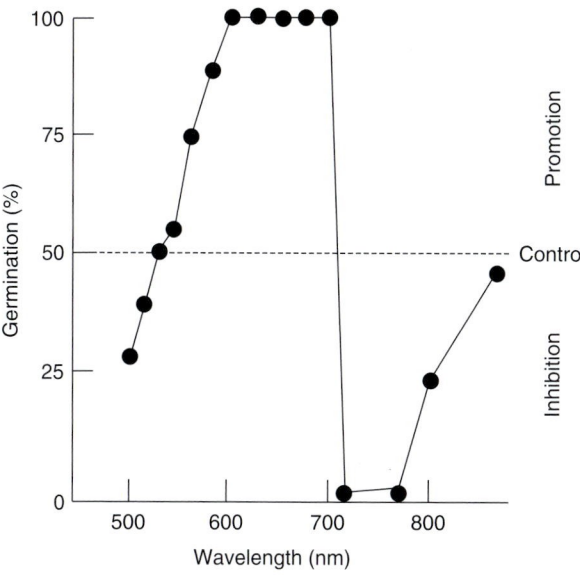

Fig. 10.3 The action spectrum of germination of Grand Rapids lettuce (*Lactuca sativa*). Approximately 50% of untreated, control seeds placed in darkness germinate. This increases to 100% when seeds are illuminated with red light, and falls close to zero if they are illuminated with far-red light.

maxima at the same peaks as required for the physiological response (Fig. 10.5). At its heart is a chromophore composed of a linear tetra-pyrrole molecule (Section 10.7.1). With specially designed equipment it is possible to detect the change in the phytochrome absorption spectrum on illumination in an intact etiolated seedling, although similar changes are masked in mature tissues which contain chloro-phyll with its strong absorption peak in the red.

10.3.2 Low-fluence responses

The control of the germination of lettuce seeds by phytochrome was termed a **low-fluence response** (LFR) because it is governed by exposures to low fluences of red, or far-red, light. Many aspects of plant development exhibit low-fluence phytochrome responses including seed germination, seedling development, internode elonga-tion and flowering in short day plants. Other key features of the LFR were determined using such experimental systems. As already described, low-fluence phytochrome responses exhibit **photorever-sibility**. However, far-red light can reverse the effect of red light only if given within a certain period of time known as the **escape time** (i.e. escape from phytochrome control). Likewise there is a delay – the

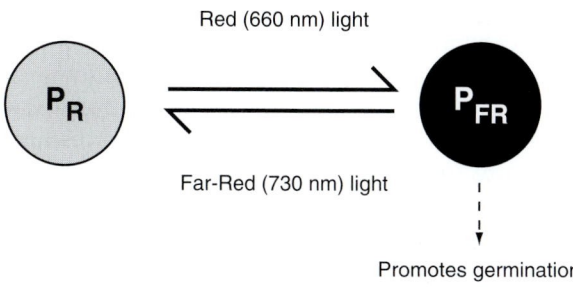

Fig. 10.4 Phytochrome exists in two photointerconvertible states. Phytochrome is synthesized in the dark in the P_R form which predominantly absorbs red light. Exposure to red light results in conversion to the P_{FR} form which triggers many phytochrome-regulated responses. The P_{FR} form can be converted back into the P_R form by exposure to far-red light.

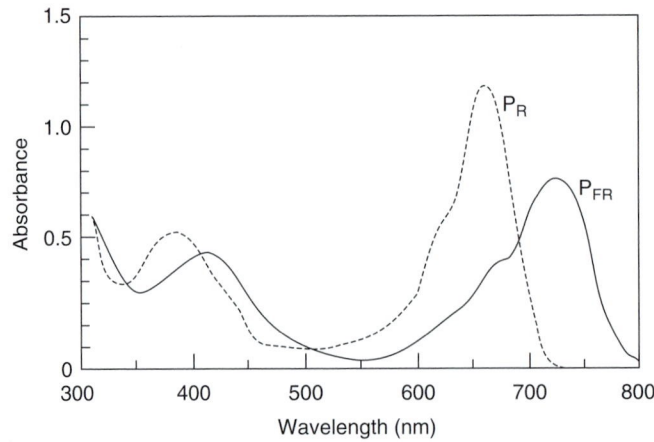

Fig. 10.5 The absorption spectra of phytochrome in the P_R and P_{FR} forms. P_R absorbs light predominantly in the red with an absorption peak near 660 nm. P_{FR} absorbs light at longer wavelengths, with an absorption peak near 730 nm, although significant amounts of red light are also absorbed. From Hartmann (1966).

lag time – following exposure to light before the response is observed. This might be a matter of a few minutes or hours, as in the case of the etiolation/de-etiolation responses, or may be days or even weeks in the case of floral induction or seed germination. Also it is the total number of photons that the sample receives, rather than the rate at which they are received, which determines whether a LFR occurs. A brief exposure to relatively intense red light has the same effect as exposure to dimmer red light for a longer period of time. This characteristic is called **reciprocity**, and low-fluence phytochrome responses are often reported in μmol photons m^{-2}, i.e. without a time component.

However, as the number of studies examining plant responses to phytochrome expanded it became apparent that not all phytochrome responses had the characteristics of a low-fluence response. Two other types of response, the **very-low-fluence response** and the **high-irradiance response**, were also described.

10.3.3 Very-low-fluence responses

The very-low-fluence responses (VLFR) are observed only in dark-grown seedlings or imbibed seeds. They are triggered by very low fluences (hence the name!), brief exposures to light, and exhibit reciprocity. However, they differ from the LFR in that either red or far-red light triggers the same response. For example, dormant *Arabidopsis* seeds can be triggered to germinate by red light but this is prevented by far-red light – a classic LFR. However, after 48 hours imbibition these same seeds exhibit a VLFR. They will germinate when exposed to either red or far-red light at fluences 100–1000 times lower than that required for the LFR. So how can we understand very-low-fluence responses in terms of phytochrome acting as a photointerconvertible switch?

Figure 10.5 shows the absorption spectra of P_R and P_{FR}. Although the different forms of phytochrome have absorption maxima in the red and far-red, the peaks are quite broad and overlap. So although P_R is preferentially converted to P_{FR} by red light, a small proportion will

Box 10.1

Irradiance is the amount of light falling on an object directly from above, whilst **fluence** takes into account that this light may come from many different angles.
Intensity refers to the amount of light emitted from a light source. It is therefore appropriate to use the terms fluence or irradiance when discussing the light falling on a biological sample.

also be converted by far-red light. In dark-grown plants phytochrome is synthesized as P_R but even brief exposure to dim far-red light will convert a small amount of P_R to P_{FR}. Even though some of this will quickly be re-converted to the P_R form, enough P_{FR} is present to trigger a very-low-fluence phytochrome response. In this case phytochrome is acting as a detector of the amount of light falling on the sample without discriminating between different colours of light. Such responses are important in the very earliest responses of plants to light.

10.3.4 The photostationary state

The concept of phytochrome switching between different forms applies to more than just the low- and very-low-fluence responses. Under any illumination conditions phytochrome will exist in both forms, although the precise proportion depends upon the precise spectral properties of the light. An equilibrium is established called the **photostationary state**. With monochromatic illumination of 660 nm and 730 nm, the proportions of P_R and P_{FR} at photoequilibrium are approximately as indicated in Table 10.1.

The proportion of the total phytochrome which is present in the P_{FR} form is termed ϕ – the Greek letter ϕ 'phi' is often used in biology as a shorthand for 'proportion'. Although these proportions may be constant, individual phytochrome molecules will be continually switching between the two forms. This switch is not instantaneous, and a significant proportion of the phytochrome molecules may exist in an intermediate state. There is increasing evidence that such intermediates may play an important role in phytochrome responses.

10.3.5 The high-irradiance responses

Another class of phytochrome-mediated responses, the high-irradiance responses (HIR), are maximized by exposure to continuous illumination, although repeated pulses of light can trigger such responses if given frequently enough. H. Mohr, working with mustard seedlings (*Sinapis alba*), found that cotyledon expansion was elicited by brief red illumination and that this was prevented by brief exposure to far-red light (a classic low-fluence response). However, if exposure to far-red light was continued for 2–3 hours the cotyledons proceeded to expand even more than under red light. This is a far-red high-irradiance response (FR-HIR).

Table 10.1 The proportion of phytochrome (ϕ) in the P_R and P_{FR} states under different illumination conditions.

	P_R	P_{FR}	ϕ
660 nm	20%	80%	0.8
730 nm	97%	3%	0.03

If dark-grown seedlings of most dicotyledonous plants are exposed to continuous far-red light there is a strong inhibition of seedling elongation. Typically the maximal inhibition results at far-red wavelengths (FR-HIR, Fig. 10.6) but this is species-dependent, with some plants exhibiting maxima in the red (R-HIR). As seedlings green these maxima often shift, indicating the operation of more than one inhibitory mechanism. Current theories favour cycling between P_R and P_{FR} as central to the operation of the HIR, but the mechanisms underlying this remain unclear. The peaks in the UV/blue of the action spectrum shown in Fig. 10.6 are thought to result from the action of specific UV-blue light photoreceptors (Section 10.4) which interact with phytochrome to coordinate plant development.

10.3.6 What is the significance of far-red light?

Low ratios of red : far-red light are a signal of vegetation shade
Full sunlight contains approximately equal amounts of all visible wavelengths of light, including red and far-red components (Fig. 10.1). However, plant leaves are rich in chlorophyll, which strongly absorbs blue and red light, green light to a lesser extent, but is relatively inefficient at absorbing far-red light. Fig. 10.7 shows the spectral composition of sunlight after it has passed through one or two leaves. Absorption of light by the canopy reduces the total irradiance markedly and has greatly reduced the amount of red light relative to far-red light. This shift in red : far-red ratio from values close to 1 in full sunlight to values nearer 0.1 is a strong signal indicating vegetation shade. This signal is not restricted to light falling directly from above. Light reflected from neighbouring vegetation will also be enriched in the far-red and has the potential to act as a signal of future competition for resources.

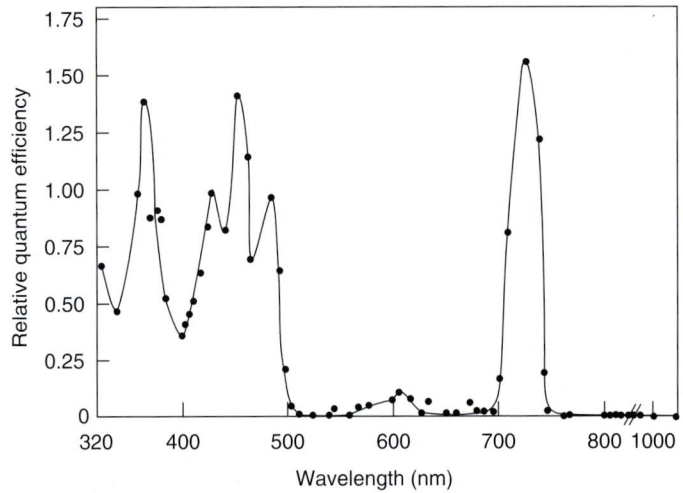

Fig. 10.6 Action spectrum for the inhibition of hypocotyl growth in *Lactuca sativa*. Hypocotyl elongation is inhibited by far-red light in a high irradiance response (HIR). In addition, blue light is also effective: the action spectrum in the 400–500 nm region shows the typical three peaks of many blue-light responses. From Hartmann (1967).

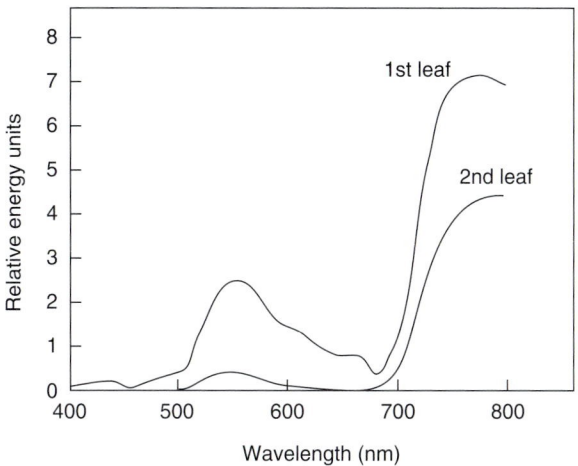

Fig. 10.7 The calculated spectral distribution of midday daylight transmitted through one or two leaves of sugar beet (*Beta vulgaris*). Data from Smith (1973).

Shade tolerance vs. shade avoidance

The response of plants to the red : far-red ratio is both species-specific and highly dependent upon developmental state. Consider the example of a small-seeded plant such as *Arabidopsis*. Seeds of *Arabidopsis* have limited nutrient reserves, which will support growth for only a few days. Once these reserves are exhausted the seedling must photo-synthesize if it is to continue to grow. The low-fluence response of these seeds stimulates germination if at, or near, the soil surface unless the sunlight has been filtered through a vegetation canopy and is thus enriched in far-red light. If the seed does germinate, etiolation must be considered a 'last chance' response – the hypo-cotyl elongates rapidly until the cotyledons are above the soil. The HIR is then triggered even if the seedling is shaded. De-etiolation, resulting from exposure to far-red light, is thus a **shade-tolerance** response.

Some plants, e.g. dog's mercury (*Mercurialis perennis*), can tolerate shade throughout their lives and do not respond greatly to alterations in the ratio of red : far-red light. However, mature plants of many species exhibit a **shade-avoidance** response if exposed to elevated far-red light. Typically, the shade-avoidance response results in an increase in apical dominance, an increase in stem growth as a result of internode elongation, increased petiole elongation and accel-erated flowering. The morphological changes may enable the plant to grow out of the shade of other plants. These responses can be triggered by a decrease in the ratio of red : far-red light falling from above, but can also be triggered by far-red reflected from surrounding vegetation. A single, brief exposure to far-red light can also trigger the shade-avoidance response if given at the end of the photoperiod (the 'end-of-day far-red effect', EOD-FR). Mature *Arabidopsis* plants exhibit a clear shade-avoidance response with increased petiole extension and accelerated flowering, although, as this plant grows as a rosette, the internodes do not elongate. Clearly, the shade-tolerance response of seedlings and shade-avoidance response of mature plants are antagonistic; hence a switch between the two must occur as the plant develops. The role of phytochrome in these responses is becoming increasingly clear, and, as seen in Section 10.6.3, these responses can be manipulated to control plant development in natural environments.

10.3.7 Phytochrome is encoded by a multigene family

This plethora of phytochrome responses is difficult to interpret if phytochrome is a single entity. However, it has become apparent that plants contain multiple phytochromes. Work with etiolated oat seedlings (*Avena sativa*) indicated the presence of light-labile ('Type I') and light-stable ('Type II') forms. Then Sharrock and Quail (1989) showed that *Arabidopsis thaliana* contains five phytochrome genes, which they named *PHYA*, *PHYB*, *PHYC*, *PHYD* and *PHYE*, and showed that whilst etiolated seedlings of *Arabidopsis* contained phyA, B and C, upon illumination there was a rapid decline in phyA. This indicated that phyA was light-labile, corresponding to 'Type I' phytochrome described in oat, whilst phyB and C were light stable ('Type II'). PhyD and E were also demonstrated to fall into the latter class in subsequent studies. Although the phyA content of seedlings declines rapidly upon illumination, mature plants exhibit responses attributable to phyA action; therefore a small amount must still remain in light-grown plants.

Box 10.2

See Appendix for a discussion of gene and protein nomenclature.

The expression of phyA is regulated by light by both transcriptional and post-transcriptional processes. The transcription of the *PHYA* gene decreases markedly (but is not abolished) upon illumination. At the post-transcriptional level phyA protein is readily degraded in the light after its photoconversion to the phyA$_{FR}$ form.

Most plants contain multiple phytochromes, which appear to have arisen by gene duplication. For example tomato (*Lycopersicon esculentum*) contains five (or perhaps six) phytochrome genes. The nomenclature has become somewhat confused, but comparisons of phytochrome amino acid sequences have indicated that four major subfamilies, each with one or more members, can be identified in herbaceous dicots. These subfamilies have been named *PHYA*, *PHYB*, *PHYC* and *PHYE*, based on comparisons with the genes from *Arabidopsis*. *PHYD* is closely related to *PHYB* and is placed in the same subfamily. In some species, e.g. black cottonwood (*Populus trichocarpa*), not all of the subfamilies have been detected (Howe *et al.* 1998). Monocots also contain multiple phytochromes, although an extensive survey of grasses by Mathews and Sharrock (1996) found evidence of only the *PHYA*, *B* and *C* subfamilies.

Unravelling the roles of these different phytochromes is a challenging task but has been greatly facilitated by the use of mutant plants which have lesions in their phytochrome genes. Likewise the use of transgenic plants where the content of an individual phytochrome has been increased has been highly informative. Screening for *Arabidopsis* plants which exhibited elongated hypocotyls when grown under white light revealed a number of different mutants, the *hy* mutants. The mutants *hy1*, *hy2* contain phytochrome apoprotein but do not contain the chromophore – these are mutations affecting tetrapyrrole synthesis. Therefore these mutants are often considered to lack all functional phytochrome, although some residual activity may remain. The *hy3* mutant lacks *PHYB* and the *hy5* mutant has a lesion in the phytochrome signal transduction chain rather than in a phytochrome gene itself (Section 10.7). Although many of these initial mutations exhibited lesions in aspects of phytochrome responses, because the screen was performed under white light other photomorphogenic mutants were also identified. The power of this approach was amply demonstrated when it was found that the *hy4* mutant contained a lesion in the blue-light receptor which had previously proved impossible to identify (Ahmad & Cashmore 1993). More directed screens using continuous far-red illumination identified plants with lesions in the *PHYA* gene, and other molecular genetic approaches have been used to identify *Arabidopsis* plants with lesions in *PHYC, D* and *E*. Such mutants allow the functions of individual phytochromes to be assessed, and many plants with lesions in the phytochrome signal transduction chain have also been isolated (Møller *et al.* 2002).

Our understanding of the phytochrome gene family is currently most advanced in *Arabidopsis* (see the paper by Sullivan & Deng, under *Complementary reading*, for a review of photoreceptors and *Arabidopsis* development), but mutations have also been found in other species which are often more useful in ecophysiological studies including pea (*Pisum* spp.), tomato, cucumber (*Cucumis sativus*), tobacco (*Nicotiana tabacum*), *Brassica napus* and *Sorghum bicolor*.

10.4 | UV-A/blue light photoreceptors (cryptochrome)

Traditional biochemical approaches were used firstly to isolate the phytochrome protein. This strategy succeeded because the unique spectral properties of phytochrome enabled it to be identified in subcellular fractions even though it is a very low abundance protein. This approach failed with the UV-A/blue light photoreceptor as many other proteins within the cell, which have no role in photoperception, absorb blue light owing to bound cofactors. However, the identification of a mutant (*hy4*) which had a lesion in blue-light responses allowed the gene encoding a UV-A/blue photoreceptor to be identified. This mutant was renamed *cry1* as cryptochrome (*crypto* = hidden) is an alternative name for this class of photoreceptors.

The *Arabidopsis CRY1* gene has been sequenced and encodes a protein with similarity to a class of microbial DNA repair enzymes called DNA photolyases. Whilst biochemical measurements of the plant protein showed no evidence of DNA repair activity, these enzymes are known to bind a flavin cofactor (flavin adenine dinucleotide) which absorbs blue light, consistent with its proposed role as a blue-light receptor. It has also been suggested that cry1 binds a second chromophore – a pterin (5,10 methenyltetrahydrofolate). The identification of *CRY1* in *Arabidopsis* quickly led to the identification of a second, related gene called *CRY2*, indicating that the cryptochromes are encoded by a small, multigene family. Likewise, two related cryptochrome genes have recently been identified in tomato.

Plants contain other, unrelated, blue-light receptors in addition to the cryptochromes. As described in Chapter 12, blue-light phototropic curvatures, chloroplast and stomatal movements are stimulated, in part, by phototropin(s) and a separate, as yet unidentified, photoreceptor most probably mediates responses to UV-B.

10.5 | Genes controlling etiolated growth

Clearly, the switch from etiolated to de-etiolated growth requires many aspects of plant development to change. The emphasis in this chapter so far has been on the development of plants in the light (photomorphogenesis) but it is equally important to remember that seedlings are following a developmental pathway when growing in darkness. This is sometimes referred to as **skotomorphogenesis** (*skoto* = dark). Just as plants with mutations in light perception have proved invaluable in understanding photomorphogenesis, valuable insights into skotomorphogenesis have been obtained by isolating plants which exhibit some, or all, of the characteristics of de-etiolated plants even though they have not been exposed to light (Fig. 10.8). A number of laboratories have isolated such mutants, calling them *det* (*de-etiolated*) and *cop* (*constitutively photomorphogenic*). A third class of mutants, *fus* (*fusca*), also develop in a de-etiolated manner when

Fig. 10.8 Mutants defective in brassinolide biosynthesis develop abormally in the dark. (A, from left to right) Wild-type *Arabidopsis* seedlings grown in darkness for 10 days are etiolated with a long, thin hypocotyl, a small root system and a pronounced apical hook. The *det2* mutant grown under the same conditions exhibits many of the characteristics of a light-grown, de-etiolated plant: the hypocotyl is short and thick, the apical hook has opened and the cotyledons have expanded; the root system is also more developed. Applying brassinolide to the *det2* mutant restores the wild-type phenotype. (B) The same plants after 12 days' growth in the light. The *det2* mutant (middle) is dwarf but can be rescued by brassinolide. From Li *et al.* (1996).

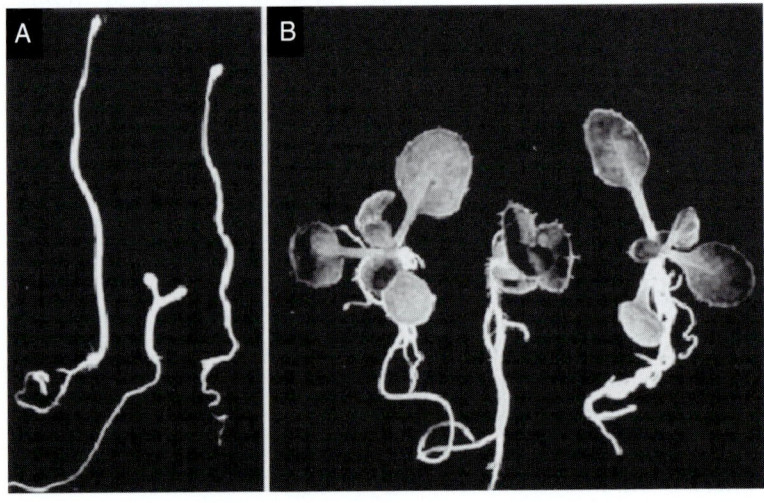

grown in darkness but were originally isolated on the basis of increased anthocyanin accumulation when grown in the light.

Many of these mutants (involving lesions in at least 11 different genes) show an almost complete photomorphogenic development in total darkness. They have short hypocotyls, the cotyledons open and true leaves form, chloroplasts develop and many light-regulated genes are expressed. It is quite remarkable that these plants can complete their entire life cycle in darkness, from germination to flowering and seed production, if provided with a suitable nutrient source (e.g. glucose). They do not green under these conditions, however, as the biosynthesis of chlorophyll includes a light-dependent step. The *det*, *cop* and *fus* genes are obviously fundamental in maintaining the seedling in an etiolated state. Other mutations result in only partial photomorphogenesis in darkness.

Some of the genes which have been mutated in these plants have been identified and sequenced, providing valuable insights into this aspect of plant development. The *det2* mutant contains a lesion in the gene encoding a cytochrome P450. This class of enzymes plays a central role in many aspects of secondary metabolism and plants contain many different cytochrome P450s (at least 30 in *Arabidopsis*). This particular cytochrome P450 is required for brassinosteroid biosynthesis (section 7.2.9) and the phenotype of this mutant reverts to that of the wild type if supplied with appropriate brassinosteroid precursors. *Det2* mutants grown in the light are severely dwarfed, indicating that these compounds are important in both dark- and light-regulated developmental pathways. A number of other dwarf plants have since been found to have lesions in other parts of the brassinosteroid biosynthetic pathway. Interestingly, wild-type plants treated with cytokinins also develop in a de-etiolated manner when grown in darkness, although the significance of this is not yet clear. Other *det/cop/fus* mutations are thought to be components of a signal transduction chain involved in both skoto- and photomorphogenesis. The role of some of these is described in more detail in section 10.7.2, and the reader is referred to recent reviews (by Schwechheimer & Deng, and Hardtke & Deng; see *Complementary reading* for a more detailed explanation.)

10.6 | Unravelling photomorphogenesis

Given that plants contain so many light-responsive pathways, the ability to inactivate individual components of specific pathways selectively provides a powerful tool with which to study these aspects of plant biology. This approach can be readily extended to examine multiple pathways by crossing different mutants together and selecting offspring which have lesions in two, three or even more photoreceptors. This is amply demonstrated in the examples described in the following section.

10.6.1 Phytochrome A and B regulate seed germination in *Arabidopsis*

As described in section 10.3.3, the germination of recently imbibed seeds of *Arabidopsis* is controlled by a low-fluence phytochrome response (LFR) whilst imbibition for 48 hours results in a shift to control by a very-low-fluence response (VLFR). In phyB-deficient mutants the initial photoreversible LFR is abolished. This tells us not only that phyB is responsible for the LFR but also that the other phytochromes (phyA, phyC–E) cannot substitute for it. With phyA-deficient mutants the situation is reversed. The VLFR is lost, showing that this response requires phyA. Since we now have seeds which show just one or other response we can measure the properties of the low- and very-low-fluence responses independently (Fig. 10.9). The VLFR is triggered by 1–1000 nmol m^{-2} (at 660 nm) and is not photoreversible. The action spectrum closely resembles the absorption spectrum of phyA$_R$, which is consistent with the view that absorption of light by phyA$_R$ leads to the formation of phyA$_{FR}$, which triggers the response. In contrast, the LFR requires much more light (10–1000 µmol m^{-2}), is stimulated by red light (550–690 nm), and reversed by far-red light (700–800 nm).

In this example the different responses can be attributed quite readily to the action of either phyA or phyB, and there appears to be little interaction with either each other or cryptochrome. However, in many cases, interactions within and between the families of

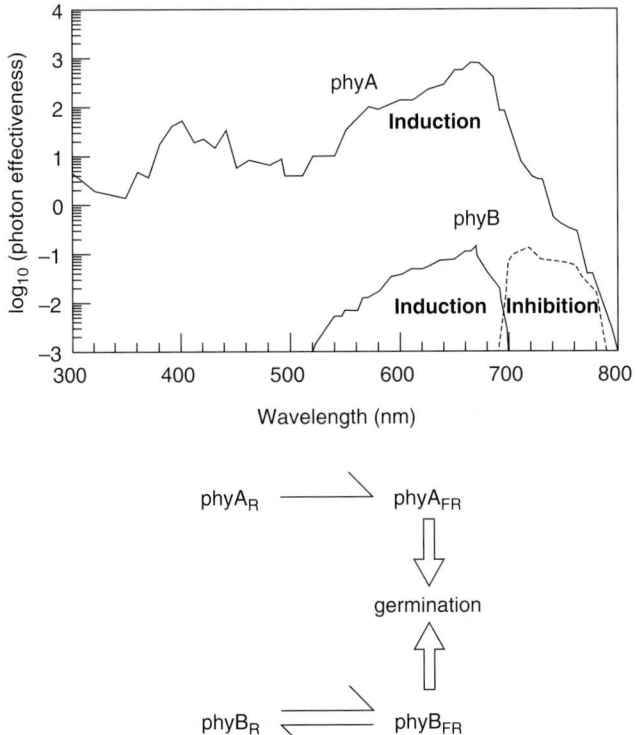

Fig. 10.9 Phytochrome A and B regulate seed germination in *Arabidopsis*. In the absence of phyB, the action spectrum for phyA-dependent germination is revealed. All wavelengths of light induce germination, although the response is optimal in the red (note the log scale). This is a characteristic very-low-fluence response. In the absence of phyA, the action spectrum for phyB-dependent germination is apparent. This is a low-fluence response: red light stimulates germination whilst far-red light is inhibitory. The different responses to phyA and phyB can be separated only in the mutant plants. From Shinomura *et al.* (1996). © 1996 National Academy of Sciences, USA.

photoreceptors is the norm. The responses of seedlings to light are complex, and our understanding of the regulation of de-etiolation is far from complete.

10.6.2 Cryptochrome and phytochrome interact to regulate de-etiolation

When *Arabidopsis* seedlings are grown under continuous far-red light, a high-irradiance response causes them to de-etiolate. However, phyA-deficient plants are etiolated in all respects under these conditions, clearly identifying phyA as an essential component of the HIR signal transduction pathway.

For seedlings grown under continuous red light the situation is more complex. Some aspects of red-light-induced de-etiolation, such as the inhibition of hypocotyl elongation, can be attributed to phyB. However, other responses, such as the opening of the apical hook, can be mediated by phyA or phyB. Experiments under a variety of illumination conditions have shown that apical hook opening (and other de-etiolation responses) are controlled by **multiple redundant pathways** (Table 10.2).

In an attempt to unravel the roles of different photoreceptors in the control of de-etiolation responses Neff & Chory (1998) examined the development of *Arabidopsis* seedlings with lesions in phyA, phyB and cry1. The single mutants were also crossed together so that additional plants with double and triple mutations could be studied. Whereas the studies of seed germination measured an 'all-or-nothing' response (i.e. a seed germinated or it did not), de-etiolation is much more complex, consisting of multiple responses, each of which may be quantified precisely.

Wild-type plants which have been grown in white light for five days have short hypocotyls, an open apical hook and open, expanded cotyledons which have accumulated chlorophyll and anthocyanin. In contrast, dark-grown plants have a much longer hypocotyl, a well-developed apical hook and folded, unexpanded cotyledons without chlorophyll and anthocyanin (Table 10.3).

The triple mutant lacks cry1, phyA and phyB. When grown in white light some de-etiolation responses are entirely abolished. For example, the apical hook, cotyledon expansion and anthocyanin

Table 10.2 | Redundant interactions between phyA, phyB and cry1. Apical hook opening can be achieved under various light regimes, mediated by different photoreceptors and pathways.

Illumination	Photoreceptors
Far-red	phyA, (phyB[a])
Red	phyA, phyB
Blue	phyA, phyB, cry1

[a]partial response only

Table 10.3 | Phenotypes of wild-type and photoreceptor mutants of *Arabidopsis* when grown in the light or dark. In many ways (but not all), plants lacking phytochromes A, B and cryptochrome 1 grown in the light resemble dark-grown plants. Data from Neff & Chory (1998)

Genotype	Angle (degrees)		Cotyledon area (mm^2)	Hypocotyl length (mm)	Chlorophyll content (μg seedling^{-1})	Anthocyanin content (Units seedling^{-1})
	Hook	Cotyledon				
Wild type (light)	172	206	2.2	3.1	0.76	0.75
Wild type (dark)	132	15	0.2	9.7	not detected	not detected
phyAphyBcry1 (light)	132	38	0.2	13.9	0.19	not detected

content are the same as those of a wild-type plant grown in darkness, which suggests that these three photoreceptors control the development of these features. However, the cotyledons have opened slightly and a small amount of chlorophyll is still produced, indicating that other photoreceptors have a minor role in de-etiolation (other studies have implicated phyD). The hypocotyl is even longer than that of an etiolated wild-type plant, probably because limited photosynthesis can occur, providing additional energy for growth.

This approach allows the potential role of individual photoreceptors to be investigated, although care must be taken in interpreting the results. When one photoreceptor has been inactivated by mutagenesis, another may take over its function. However, just because a photoreceptor *can* substitute for another is not necessarily proof that it *does* fulfil the same role in a normal plant. Other studies have shown that the interaction between photoreceptors may be complex. For example, the action of one photoreceptor may be greatly enhanced by the activity of another, beyond that expected from simple additive effects.

10.6.3 Phytochrome responses throughout the life cyle

Wild-type seedlings of *Arabidopsis* exhibit shade tolerance (i.e. they initially de-etiolate when exposed to far-red light) whilst mature plants exhibit shade avoidance (the leaf petioles elongate and flowering is accelerated under far-red light). Therefore the response of the plant clearly depends upon its developmental state. The photoreceptor mutants have helped us to understand how these different responses are controlled at different stages of the plants' life cycle.

When a seedling first develops it contains relatively high amounts of phyA and phyB. Therefore the seedling will de-etiolate under red light (a LFR mediated by phyB) and far-red light (a HIR mediated by phyA) and is thus shade tolerant. However, continued illumination leads to a decline in phyA content. The seedling will remain de-etiolated under red light (due to the continued action of phyB) but will now etiolate under far-red light; i.e. it now exhibits shade avoidance.

A phyB-deficient plant will exhibit a constitutive shade-avoidance response both as a seedling and a mature plant. Under white light the *phyB* mutant has elongated petioles (Fig. 10.10) but if exposed to a 10-minute pulse of far-red light at the end of the day, a more extreme shade-avoidance response is triggered, indicating that other photoreceptors are still responsive. If phyD is also mutated, the petioles become further elongated under white light, whilst mutations in phyE cause the rosette habit to be lost. The role of these minor phytochromes is normally masked by phyB.

Although mature plants of many species do not exhibit shade tolerance, it can be re-established using genetic engineering techniques. Although the phyA content of a mature plant is normally low, an artificial gene can be made where a strong, unregulated promoter is fused to the coding region of the *PHYA* gene. When this *over-expression* construct is introduced into plants they will now contain much more phyA than normal. Vierstra and coworkers (Ballaré *et al.* 1994) used this approach to produce shade-tolerant tobacco plants. Wild-type tobacco plants normally show a strong shade-avoidance response when they are illuminated with far-red light from the side,

phyAphyB

phyAphyBphyD

phyAphyBphyE

White light   +EOD FR

Fig. 10.10 Shade-avoidance responses in phytochrome-deficient mutants of *Arabidopsis*. The *phyAphyB* double mutant shows a partial shade-avoidance response when grown under white light for 60 days. The leaves are smaller than in wild-type plants, have elongated petioles and are held more upright. This is due primarily to the absence of phyB. If these plants are exposed to 10 minutes far-red light at the end of each day (EOD-FR), a strong response is still observed, indicating that other phytochromes can also detect this signal; the leaves are even smaller and the internodes etiolate. This response is masked in plants containing *PHYB* and is apparent only in the mutant. Plants which also lack *PHYD* show etiolated petioles when grown under white light, whilst plants lacking *PHYE* show internode extension under these conditions. Scale bar = 5 cm. From Whitelam *et al.* (1998).

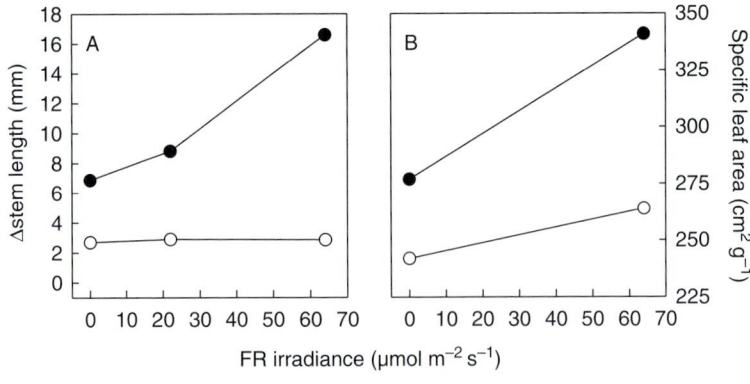

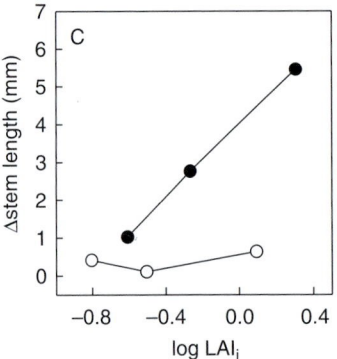

or are grown in dense canopies. The internodes elongate and the specific leaf area (the area of leaf per unit dry weight) increases; i.e. the leaves become thinner. These responses were abolished in transgenic plants which contained five times as much phyA as the wild type (Fig. 10.11). However, increasing the phyA content by this much resulted in dwarf plants, presumably because they became hypersensitive to far-red light. Smith and coworkers (Robson *et al.* 1996) found that transgenic tobacco plants which contained lower amounts of oat phyA developed normally under white light, but showed a reduced shade avoidance response (Fig. 10.12). This ability to manipulate plant

A White light   B White light + FR

WT   2.5x   3x   5x   WT   2.5x   3x   5x

responses subtly to shade has obvious agricultural applications as it allows plants to be grown closer together.

10.7 | Phytochrome signal transduction

So far our discussion of photomorphogenesis has focused on signal perception (i.e. the photoreceptors) and the responses observed. Clearly, a complex set of signal transduction pathways must link the two. Although our understanding of these pathways is far from complete, we can trace certain components of them. Our understanding of phytochrome signal transduction is considerably more advanced than that of cryptochrome, so this will be discussed further.

10.7.1 How does phytochrome detect light?

The first step of photoperception by phytochrome is the absorption of a photon by the chromophore, but how does this initiate subsequent responses? The chromophore is a linear tetrapyrrole molecule, covalently attached to the phytochrome apoprotein, and it undergoes a reversible conformational change when it absorbs light. In P_R, each bond linking the pyrrole groups is in the *cis* configuration (Fig. 10.13). Upon conversion to P_{FR}, one of these bonds shifts to the *trans* configuration, i.e. part of the chromophore has moved.

We do not know precisely how this movement initiates subsequent events, but it alters the absorption spectrum, and presumably other properties, of the protein. We do know that both phyA and phyB interact with other proteins, and that both possess a kinase activity – they can phosphorylate themselves (**autophosphorylation**) and other proteins. A number of proteins have been identified which interact with phytochrome, including PIF3 (phytochrome interacting factor 3). The gene encoding PIF3 was also found, in an independent set of experiments, to be mutated in plants which exhibited impaired responses to red and far-red light. PIF3 is, therefore, likely to be an important component of the phytochrome signal transduction chain. Mutagenic studies have identified many plants with lesions in phytochrome signal transduction chains, indicating that many proteins play a role in signal transduction. In some cases mutations affect responses to just phytochrome A or B, whereas other mutants affect responses to both phytochromes. In addition, microinjection and pharmacological studies have implicated a number of small molecules and other proteins in early signalling processes including Ca^{2+}, cyclic AMP and calmodulin (a protein which binds Ca^{2+}). Again, the role of these compounds is not yet understood but, by analogy with other systems, they are likely to have a role in very rapid responses occurring immediately following phytochrome activation.

Fig. 10.13 The phytochrome chromophore undergoes a *cis–trans* isomerization as phytochrome is converted from the P_R to P_{FR} form. The phytochrome chromophore is a linear tetrapyrrole molecule covalently linked to the phytochrome apoprotein via a cysteine residue (Cys). One of the bonds linking the pyrrole rings moves as phytochrome is illuminated (arrow).

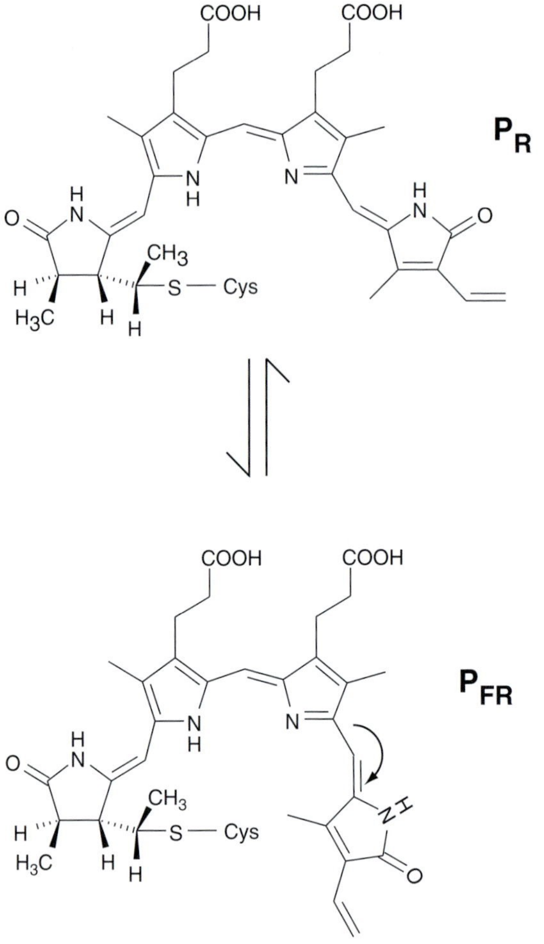

10.7.2 Phytochrome A and B move into the nucleus upon illumination

Many of the changes in plant development which occur during photomorphogenesis result from changes in gene expression. Many hundreds, if not thousands, of genes are light-regulated, which is not surprising given that light triggers the development of the photosynthetic apparatus and the expression of many enzymes of metabolism. The majority of these light-regulated genes are located in the nucleus, although a proportion of the photosynthetic apparatus is encoded in the chloroplast genome. The regulation of chloroplast gene expression is quite different from that of nuclear-encoded genes, and the reader is referred to a paper by Stern *et al.* (see *Complementary reading*) for a review.

Early studies proposed that a proportion of phytochrome was associated with membranes, but this is now thought either to be an artefact resulting from the preparation of cell extracts or to represent a pool of physiologically irrelevant phytochrome in the process of being degraded. Current theories favour a model where phytochrome A and B are initially located in the cytoplasm and then move into the nucleus upon activation. Studies of this movement are possible because phytochrome is still physiologically active even if fused to another protein. Genetically modified plants have been produced where the coding region of *PHYA* or *PHYB* has been fused to a **reporter gene**. One such reporter gene encodes green fluorescent protein (GFP) which fluoresces green when illuminated with blue light, allowing its position to be determined by microscopy. When plants are grown in darkness, phyA::GFP fusions (or phyB::GFP fusions) are found in the cytoplasm. When treated with light which will trigger a phytochrome response, the fusion proteins move into the nucleus. The movement of phyB::GFP is rapid (it occurs within 15 minutes), red : far-red reversible and requires the presence of the chromophore. The phyA::GFP fusion protein enters the nucleus under illumination conditions which would trigger VLFR and HIR. Having entered the nucleus the phytochromes accumulate in small patches or 'speckles'. In contrast, phyC, D and E are constitutively localized in the nucleus, although speckle formation results from illumination with red light and can be reversed by far-red light.

At least two pathways have been identified by which phytochrome can stimulate nuclear gene expression, and others undoubtedly exist. In both cases, phytochrome enters the nucleus and interacts with other proteins, and these then stimulate gene expression by binding to short stretches of DNA (light-responsive-elements – LREs) found in the promoters of light-responsive genes. In this way the expression of many different genes can be stimulated by the action of a few signal transduction chains.

The first pathway involves a number of proteins whose role was first identified in mutagenic studies (Fig. 10.14). These include COP1, other members of the *cop/det/fus* family, and HY5. In the dark, a large multiprotein complex (referred to as a **signalosome**) is found in the nucleus which includes the COP1 protein. This interacts with HY5 and prevents HY5 from binding to the LREs in the promoters of light-regulated genes. In the absence of HY5, transcription does not occur. When phytochrome is activated, it enters the nucleus and breaks down the association between HY5, COP1 and the signalosome. HY5 binds to the LREs and the transcription of light-regulated genes is stimulated, triggering de-etiolation. COP1 leaves the nucleus and enters the cytoplasm, although the rest of the signalosome remains.

This model elegantly explains the phenotypes of the various mutants. Plants which lack phytochrome or HY5 will be etiolated in the light as the transcription of the light-regulated genes is not stimulated. On the other hand, the absence of COP1, or other components of the signalosome, causes the plants to be constitutively de-etiolated as HY5 is always able to stimulate transcription.

The second pathway involves PIF3. As well as interacting with phytochrome, PIF3 will also bind to another class of LRE. Although PIF3 will bind to the LRE in both the light and the dark, it forms

Box 10.3

Green fluorescent protein (GFP) is a naturally fluorescent protein which is found in the Pacific northwestern jellyfish (*Aequorea victoria*). The advantage of such a reporter gene is that the protein it encodes can be visualized without destroying the plant material, allowing the same cells to be studied over time. Other approaches have used reporter genes such as *β-glucuronidase* (GUS), which allows cells to be stained blue. However, the staining process is destructive so it is not possible to follow the movement of the protein directly over time.

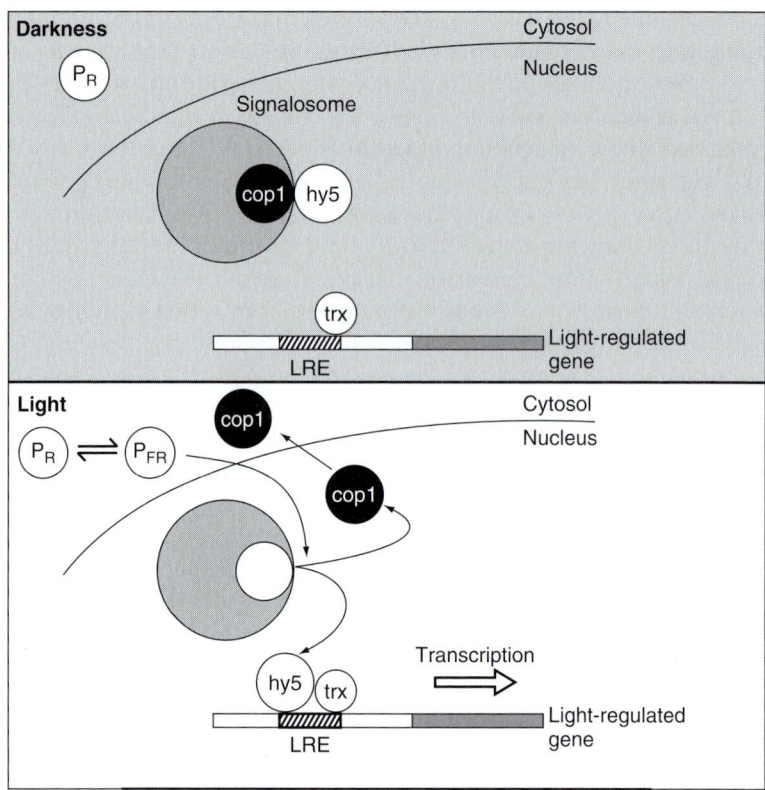

Fig. 10.14 Signalling and photomorphogenesis. A number of signalling pathways have been identified which result in light-activated gene expression, one of which is shown here. In the dark, phytochrome is in the P_R form and located in the cytosol. In the nucleus a large multiprotein complex, known as a signalosome, binds HY5. Light-regulated genes are not active. Upon illumination, some phytochrome is converted to the P_{FR} form. This enters the nucleus and disrupts the interaction between a component of the signalosome, COP1, and HY5. HY5 is now free to bind to DNA sequences known as light-responsive elements (LREs) found in the promoters of light-regulated genes, triggering transcription.

a complex with phytochrome only in the P_{FR} form. This complex of PIF3 and P_{FR} is thought to regulate the recruitment of other protein factors required for transcription.

These models go some way to illustrate how phytochrome can trigger gene expression, but undoubtedly much remains to be discovered. Many other proteins can interact with these signal transduction chains and some light-regulated genes are regulated by other pathways. However, the principles of these models are likely to be repeated in other signal transduction chains.

10.7.3 Photoreceptors' interactions with plant growth hormones

Although the emphasis of this chapter has been on the perception of light by photoreceptors and the ways in which these signals affect gene expression, and subsequently plant development, photomorphogenesis represents a profound change in plant development. It is not surprising therefore that plant growth hormones play a central role in mediating many of these responses. We have already seen how cytokinin and brassinosteroids influence skotomorphogenesis. Recently, a protein induced by cytokinin treatment (ARR4) has been shown to bind to PhyB and stabilize the FR form (Sweere *et al.* 2001) and over-expression of ARR4 results in plants hypersensitive to red light. This demonstrates a direct link between a photoreceptor and a plant growth hormone.

Another example is provided by the interaction between the shade-avoidance response and ethylene. Ethylene production in sorghum (*Sorghum bicolor*) is under circadian control, the peak amplitude of which is strongly increased by shade conditions. A phyB mutant of sorghum that exhibits a constitutive shade-avoidance response also produces much more ethylene than normal, even when grown under full sunlight (Finlayson *et al.* 1999). In tobacco, Pierik *et al.* (2003) have demonstrated that many features of the shade-avoidance response can be stimulated by exposing plants to ethylene, and that ethylene-insensitive tobacco plants show a delayed shade-avoidance response when grown at high densities or when exposed to EOD-FR light. The molecular basis of this interaction is yet to be understood.

As auxin and gibberellic acid are important in stimulating plant growth, it is to be expected that there will be many interactions between these plant growth hormones and light. The interaction of auxin and light in stimulating phototropic curvatures is considered further in Chapter 12. The reader is referred to papers by Halliday & Fankhauser and Morelli & Ruberti (see *Complementary reading*) for more extensive reviews of photoreceptor–hormone interactions.

As our understanding of the signal transduction chains of individual photoreceptors improves, the interactions between them will become clearer. The challenge will still remain of integrating this detailed molecular information into a broader ecophysiological context.

Complementary reading

Halliday, K. J. & Fankhauser, C. Phytochrome-hormonal signalling networks. *New Phytologist*, **157** (2003), 449–63.

Hardtke, C. S. & Deng, X. W. The cell biology of the COP/DET/FUS proteins. Regulating proteolysis in photomorphogenesis and beyond? *Plant Physiology*, **124** (2000), 1548–57.

Morelli, G. & Ruberti, I. Light and shade in the photocontrol of *Arabidopsis* growth. *Trends in Plant Science*, **7** (2002), 399–404.

Quail, P. H. Phytochrome photosensory signalling networks. *Nature Reviews, Molecular Cell Biology*, **3** (2002), 85–93.

Schwechheimer, C. & Deng, X. W. The COP/DET/FUS proteins: regulators of eukaryotic growth and development. *Seminars in Cell & Developmental Biology*, **11** (2000), 495–503.

Smith, H. Physiological and ecological function within the phytochrome family. *Annual Review of Plant Physiology and Plant Molecular Biology*, **46** (1995), 289–315.

Stern, D. B., Higgs, D. C. & Yang, J. J. Transcription and translation in chloroplasts. *Trends in Plant Science*, **2** (1997), 308–15.

Sullivan, J. A. & Deng, X. W. From seed to seed: the role of photoreceptors in *Arabidopsis* development. *Developmental Biology*, **260** (2003), 289–97.

References

Ahmad, M. & Cashmore, A. R. (1993). HY4 gene of *A. thaliana* encodes a protein with characteristics of a blue-light photoreceptor. *Nature*, **366**, 162–6.

Ballaré, C. L., Scopel, A. L., Jordan, E. T. & Vierstra, R. D. (1994). Signalling among neighbouring plants and the development of size inequalities in plant populations. *Proceedings of the National Academy of Sciences (USA)*, **91**, 10094–8.

Finlayson, S. A., Lee, I. J., Mullet, J. E. & Morgan, P. W. (1999). The mechanism of rhythmic ethylene production in sorghum: the role of phytochrome B and simulated shading. *Plant Physiology*, **119**, 1083–9.

Hartmann, K. M. (1966). A general hypothesis to interpret 'high energy phenomena' of photomorphogenesis on the basis of phytochrome. *Photochemistry and Photobiology*, **5**, 349–66.

Hartmann, K. M. (1967). Ein Wirkungsspektrum der Photomorphogenese unter Hochenergiebedingungen und seine Interpretation auf der Basis des Phytochroms (Hypokotylwachstumshemmung bei *Lactuca sativa L.*). *Zeitschrift für Naturforschung*, **B 22**, 1172–5.

Howe, G. T., Bucciaglia, P. A., Hackett, W. P., Furnier, G. R., Cordonnier-Pratt, M. M. & Gardner, G. (1998). Evidence that the phytochrome family in black cottonwood has one PHYA locus and two PHYB loci but lacks members of the PHYC/F and PHYE subfamilies. *Molecular Biology and Evolution*, **15**, 160–75.

Li, J. M., Nagpal, P., Vitart, V., McNorris, T. C. & Chory, J. (1996). A role for brassinosteroids in light-dependent development of *Arabidopsis. Science*, **272**, 398–401.

Mathews, S. & Sharrock, R. A. (1996). The phytochrome gene family in grasses (Poaceae): a phylogeny and evidence that grasses have a subset of the loci found in dicot angiosperms. *Molecular Biology and Evolution*, **13**, 1141–50.

Møller, S. G., Ingles, P. J. & Whitelam, G. C. (2002). The cell biology of phytochrome signalling. *New Phytologist*, **154**, 553–90.

Neff, M. M. & Chory, J. (1998). Genetic interactions between phytochrome A, phytochrome B, and cryptochrome 1 during *Arabidopsis* development. *Plant Physiology*, **118**, 27–36.

Pierik, R., Visser, E. J. W., De Kroon, H. & Voesenek, L. (2003). Ethylene is required in tobacco to successfully compete with proximate neighbours. *Plant, Cell and Environment*, **26**, 1229–34.

Robson, P. R. H., McCormac, A. C., Irvine, A. S. & Smith, H. (1996). Genetic engineering of harvest index in tobacco through overexpression of a phytochrome gene. *Nature Biotechnology*, **14**, 995–8.

Sharrock, R. A. & Quail, P. H. (1989). Novel phytochrome sequences in *Arabidopsis thaliana*: structure, evolution, and differential expression of a plant regulatory photoreceptor family. *Genes and Development*, **3**, 1745–57.

Shinomura, T., Nagatani, A., Hanzawa, H., Kubota, M., Watanabe, M. & Furuya, M. (1996). Action spectra for phytochrome A- and B-specific photoinduction of seed germination in *Arabidopsis thaliana*. *Proceedings of the National Academy of Sciences (USA)*, **93**, 8129–33.

Smith, H. (1973). Light quality and germination: ecological implications. In *Seed Ecology*, ed. W. Heydecker. London: Butterworths. pp. 219–31.

Sweere, U., Eichenberg, K., Lohrmann, J. *et al.* (2001). Interaction of the response regulator ARR4 with phytochrome B in modulating red light signaling. *Science*, **294**, 1108–11.

Whitelam, G. C., Patel, S. & Devlin, P. F. (1998). Phytochromes and photomorphogenesis in *Arabidopsis*. *Philosophical Transactions of the Royal Society of London B*, **353**, 1445–53.

Chapter 11

Reproductive development

11.1 | Introduction

Reproductive development of flowering plants has been studied for many hundreds, if not thousands, of years. This is not surprising, given the importance of flowering, fruiting and seed setting in agriculture. Society also has a fascination with producing ever more diverse flowers for horticultural purposes. The rose is the oldest known domesticated flower and its popularity endures today; over 103 million roses are sent for Valentine's day in the USA alone, with the global trade in all cut flowers exceeding \$4 billion annually. Moreover, since cut flowers are desired at all seasons, control of the time of flowering has great commercial value. Hence a study of the reproductive processes of flowering plants is of great economic importance as well as enabling us to understand the functioning of plants in their natural ecosystems.

11.2 | Juvenility and 'ripeness to flower'

Vegetative growth eventually leads to a transition to reproductive development. However, plants will not flower, nor respond to environmental stimuli which ensure subsequent flowering, until they have completed a certain period of vegetative growth and reached 'ripeness to flower'. A plant can therefore be considered to pass through three growth phases:

(1) **juvenile** – in which it will not flower
(2) **mature** – in which appropriate environmental stimuli will evoke flowering
(3) **reproductive** – in which flowering actually takes place

Flowering occurs as a result of a reprogramming of the development of the shoot apical meristem (SAM : Section 9.2). Rather than initiating stems, leaves and axillary buds, a reproductive (or floral) meristem gives rise to an inflorescence, i.e. flowers and their subtending bracts. The sharpness of the transitions between each phase, their duration, and the extent to which these phases can coexist within regions of a single plant vary very widely between species. Given the diversity of flowering plants, it is not possible to provide an all-encompassing description of these processes, so representative examples are discussed to provide an overview of the underlying mechanisms.

A number of environmental factors have been identified which are essential to, or greatly hasten, the transition to reproductive growth. These have been extensively studied in a selected number of species, which are often those which show the most clear-cut responses. However, many species do not require a precise set of environmental stimuli and will flower under almost any conditions compatible with continuing growth. In this chapter we will see how temperature, daylength, endogenous plant rhythms, size and age interact to regulate floral evocation, and how this leads to subsequent flower development. Although molecular approaches have revolutionized our understanding of reproductive development, much is still unknown, especially in the signals which lead to the reprogramming of the SAM. Reproductive development will here be interpreted to encompass not only flower evocation but also flower development, the formation of haploid sex cells, fertilization and the growth and ripening of fruit, including the development of the embryo and seed structures from the fertilized ovule.

11.3 | The control of flowering by daylength and temperature

11.3.1 The discovery of the photoperiodic induction of flowering

The experimental study of the role of light in controlling flowering dates from the work of W.W. Garner and H.A. Allard, first published in 1920. These workers noticed that the Maryland Mammoth variety of tobacco (*Nicotiana tabacum*) failed to flower and set seed during the summer despite its vigorous vegetative growth. However, rootstocks which were transferred to the glasshouse readily gave rise to small flowering plants during the winter months. They also found that successive spring sowings of soybean (*Glycine max (soja)*) all came into flower at the same time, and that these plants also flowered, whilst still quite small, if grown in a glasshouse during the winter. No significant promotive effects on the flowering of these plants could be traced to irradiance, moisture or temperature within the glasshouse. Garner and Allard therefore examined the effect of daylength by extending the natural day with artificial light or shortening the day by placing plants in light-proof cabinets for the required time. The outcome of these studies was a demonstration that these varieties of tobacco and soybean required a period of exposure to *short days* in order to flower and that, after an appropriate period of short days, they would flower irrespective of the subsequent daylength. This discovery has led to the concept of **photoperiodic induction** of flowering.

Garner and Allard extended these studies to other species and varieties of plants and found that some plants had no special daylength requirements (**day-neutral plants**), others were **short-day plants** (SDP) whilst others were **long-day plants** (LDP). A SDP is defined as a species which is induced to flower by daylengths *shorter* than a critical value, whereas a LDP is induced by daylengths *longer* than a critical value. The critical value is not absolute, but varies according to species, so that 'short' and 'long' days for individuals may overlap. Since these pioneering studies it has become clear that the situation is somewhat more complex. Some plants have an absolute requirement for short-day (or long-day) induction of flowering, whilst in others flowering is only hastened by inductive photoperiods. Some plants require exposure to a combination of daylengths, e.g. short days followed by long days. As seen in Section 11.5, there are also complex interactions between temperature and daylength. In natural ecosystems, the photoperiodic response determines the time of year at which inducible species flower. LDPs sense the lengthening days in spring and early summer; SDPs respond to the shortening daylengths of later summer and autumn. Photoperiodic requirement for flowering is characteristic of plants native to temperate and arctic habitats, where the daylength changes considerably with seasons. Near the equator there is little change in daylength and tropical plants are generally not controlled by photoperiod. Species growing over a wide latitude range may possess photoperiodic ecotypes varying in critical daylength, and plant species may be unable to flower when grown far from their native latitude. Growers of ornamental plants are, of course, able to use photoperiodic induction to stimulate flowering in glasshouses at will.

The majority of studies have focused on plants which exhibit a clear-cut response to daylength. Two classic examples are cocklebur (*Xanthium pennsylvanicum*, syn. *X. strumarium*) which is an obligate short-day plant and the annual strain of henbane (*Hyoscyamus niger*), an obligate long-day plant.

11.3.2 Night length, rather than day length, regulates flowering

When grown under normal conditions, all plants have a light requirement for flowering since they need photosynthetic products to grow and develop. However, plants with no daylength requirement can be brought to flowering in darkness if supplied with sugar. The need for 'high irradiance' light for flowering is therefore a photosynthetic requirement rather than specifically related to flower induction. *Xanthium strumarium* will flower when exposed to daylengths shorter than 15.5 h (the **critical daylength** for this species). As, in a natural environment, the combined day and night periods are always of 24 h duration, this observation does not discriminate between a response to daylength (i.e. < 15.5 h) or a response to the length of the night (i.e. > 8.5 h). If plants are grown under artificial conditions it is possible to alter the lengths of the light and dark periods independently. Such an approach demonstrated that it is actually the length of the *night* which is important in inducing flowering in *X. strumarium*. Therefore a short-day plant could more appropriately be referred to as a long-night plant, although the reference to daylength has become fixed in the literature. Conversely, long-day plants are really short-night plants. The long-day plant *Hyoscyamus niger* will flower only when the dark period does not exceed 13 h, i.e. when the dark period is shorter than a critical value.

Recognition of the importance of the dark period followed from the demonstration that a SDP such as *Xanthium*, placed under inductive conditions, will still flower if the light period is interrupted by brief periods of darkness. However, brief periods of illumination during the dark period will prevent flowering if the uninterrupted dark period no longer exceeds 8.5 h. In this species a short 'night-break' of just 5 min duration is sufficient to interrupt the dark period. Measurements of the action spectrum of this response identified phytochrome as the photoreceptor. Red light (640–680 nm) was most effective, whilst the effect of red light could be nullified by brief exposure to far-red light (710–740 nm), indicating a low-fluence response (Section 10.3). The same response was observed in the SDP Biloxi soybean. In LDP, such as barley (*Hordeum vulgare*), night-breaks are also perceived by phytochrome although, in this case, they induce flowering. As seen in Section 10.6, mutants in specific photoreceptors have proved to be a valuable tool with which to unravel the perception of light and have demonstrated that other photoreceptors, such as cryptochrome, also have an important role in the regulation of flowering.

11.3.3 Circadian rhythms

The time at which a night-break is given is important in determining its effectiveness. Figure 11.1 shows the response of the SDP red goosefoot (*Chenopodium rubrum*) (inducible by a single photoperiod) to 2 min night-breaks of red light given at different times during a single 72 h dark period. The plants were then placed in continuous light and flowering scored. The night-breaks most effectively inhibited flowering when given at around 6, 33 or 60 h from the beginning of the dark period, i.e. the times when the plant on a normal 24 h inductive cycle would have been in darkness. In contrast, night-breaks given at, or near, 18 and 46 h into the photoperiod were ineffective, these being times when the plant would normally have been in the light. The responsive periods occurred at approximately 24 h intervals (actually slightly longer). In this case photoinduction is interacting with an endogenous plant rhythm which persists even when the plant is placed into constant conditions. These rhythms are called **circadian** rhythms (*circa* = about, *diem* = day) and are found in organisms as

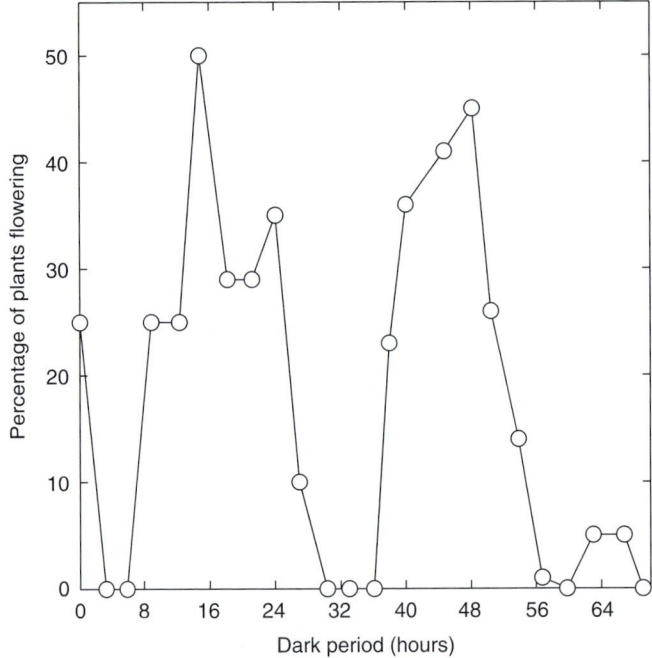

Fig. 11.1 Flowering in *Chenopodium rubrum* is sensitive to night length and circadian rhythms. *C. rubrum* plants were subjected to 2 min of red light at various times during a 72 h dark treatment. Plants were then returned to continuous light and scored for flowering. Modified from Cumming *et al.* (1965).

diverse as slime moulds, plants, insects and humans. Many aspects of plant development are controlled by such rhythms, including leaf and flower movement, growth and gene expression as well as flowering (McClung 2001). In this case the 'biological clock' determines the plant's response to light over the daily cycle.

11.3.4 Daylength is sensed in the leaves

Photoperiodic induction occurs in the leaves, and the flowering stimulus moves in the phloem from the leaves to the meristems where flowers are to be initiated. If *Xanthium* plants are defoliated so that only a single leaf remains, exposure to appropriate photoinductive conditions will still lead to flowering. Likewise, if just one leaf is induced, whilst the others are held under non-inductive conditions, flowering results. The flowering stimulus can be transmitted from an induced to a non-induced plant of the same species by grafting and, in some cases, transmission is possible between species, even if these have different photoperiodic requirements; e.g. the induction of flowering may be stimulated in a SDP (non-induced) by transmission from a LDP (induced), and transmission may occur between day-neutral and day-sensitive plants. Such experiments have led to the hypothesis that induced leaves produce a supposed flowering hormone, **florigen**, which induces flowering. This concept has now lasted over 60 years but the compound itself has steadfastly resisted identification, leading to the development of more sophisticated, or alternative, hypotheses (Section 11.5).

The precise characteristics of the signal produced by induced leaves, and the response of the plant to this signal, vary between species. *Xanthium* is quite unusual in that flowering can be induced by a single inductive cycle but, in a population of *Xanthium* plants, three inductive cycles are required to ensure that 100% of the plants produce flowers. Further inductive cycles increase the number of flowers which are formed and the speed at which flowering takes place. Once a *Xanthium* plant has been induced to flower, leafy shoots formed subsequently can still transmit the flowering stimulus to non-induced plants by grafting; hence the flowering signal pervades the entire plant, rather than being localized in the induced leaves.

A rather different situation is seen in the SDP *Perilla ocymoides*, a member of the mint family. Here, the induced state persists once the plant is returned to non-inductive conditions, but it is localized in the leaves which were exposed to the inductive conditions. As these leaves age and die, the plant becomes vegetative once more. Biloxi soybeans require continuous inductive conditions for flowering and the formation of flowers ceases very rapidly when the plants are transferred to non-inductive conditions.

These results are difficult to interpret in terms of a single stimulus which promotes flowering. It is clear, however, from both physiological and genetic studies, that the induction of flowering is a multifactorial process, with both stimulatory and inhibitory processes interacting in a complex manner to regulate flowering.

11.3.5 Vernalization

Some plants require a prolonged cold treatment (**vernalization**) before they will flower. This requirement is shown by species native to climates with a cold winter and is thought to prevent premature flowering in the autumn, when the flowers and developing fruits would be damaged by frost. Species needing vernalization do not flower when grown in a warm climate. The biennial form of *Hyoscyamus niger* has an absolute requirement for cold treatment; if overwintered at too high a temperature, it will remain vegetative indefinitely. In other species, vernalization is not an absolute requirement, but hastens flowering. Winter rye (*Secale cereale*) has no absolute cold requirement but when vernalized the winter varieties flower as rapidly as spring varieties. Both *H. niger* and winter rye are long-day plants, but cold requirements for flowering are also found in short-day and day-neutral plants. The temperature most effective for vernalization is at or near 6 °C, and the duration of treatment required for the maximal acceleration of flowering can range, according to species, from four days to three months. Vernalization can often be reversed if followed immediately by a period of high temperature, which suggests that there is a neutral temperature. For *H. niger* this neutral temperature is approximately 20 °C.

In certain of these cases, it has been shown that it is the stem apex which is the sensitive region, and that this has to reach the required maturity before the cold treatment is effective (e.g. biennial *H. niger*) whilst in other species (e.g. winter cereals) the cold requirement can be achieved by the embryo of an immature seed or in the moistened mature seed. Current theories of vernalization favour active mitotically dividing cells as the site of cold-temperature perception. Leaf cuttings of honesty (*Lunaria annua*) and field pennycress (*Thlaspi arvense*) which have not been exposed to vernalizing temperatures give rise to vegetative rosettes whilst cuttings which have been exposed to low temperatures give rise to flowering shoots. The cuttings contain dividing cells, but no stem apex, during the treatment. Burn *et al.* (1993) propose that vernalization depends on demethylation of DNA of mitotically dividing cells. This results in the activation of a gene, or genes, necessary for flowering. A number of lines of evidence support this model:

(1) Exposure of dividing cells to vernalizing temperatures causes demethylation of DNA. As methylation patterns are faithfully transmitted to the daughter cells during mitosis, this acts as a 'memory' of previous cold treatment. This is an **epigenetic** process as the DNA sequence of the genes is not altered.

(2) Treatment of unvernalized plants with methylation inhibitors such as 5-azacytidine results in early flowering.

(3) Transgenic, or mutant, plants with reduced activities of endogenous methylation enzymes flower early without a requirement for vernalization (Finnegan *et al.* 1998).

Some early studies had proposed the existence of a transmissible vernalizing chemical (called **vernalin**) but this concept has fallen from favour as, in the majority of cases, the results can be better explained by the model described above.

11.3.6 Plant hormones and flowering

A number of lines of evidence point to a role for plant hormones, especially gibberellic acid, in the induction of flowering, both in photoperiodic responses and in vernalization. It has been known for many years that the application of exogenous plant growth hormones to vegetative plants can trigger reproductive development in some species. For example, auxin applications are used commercially to induce flowering in the pineapple (*Ananas comosus*) and in the litchi tree (*Litchi chinensis*), although in pineapple the effect of auxin on flowering results from the stimulation of ethylene production. With other species auxin inhibits flowering, so the role of this plant hormone is not clear. In many species the requirement for vernalization can be overcome by treatment with GA. In *Thlaspi* the activity of a key enzyme of GA biosynthesis is induced in the stem apex (but not the leaves) by low temperatures. Further evidence for the role of GA in the induction of flowering comes from measurements of GA content and enzymes involved in GA biosynthesis in extracts prepared from spinach (*Spinacia oleracea*). Spinach plants maintained under short days remain vegetative and have a rosette growth habit. Extracts prepared from these plants suggest that a critical step of GA biosynthesis is inhibited. When exposed to long days, the content of active GA increases, the stems elongate and flowers are produced. It appears that the synthesis of a number of enzymes of GA biosynthesis is under photoperiodic control.

Again, the response of a plant to GA is species specific. In grapevines (*Vitis* spp.) it is thought that the tendrils which support the vine's climbing habit are modified inflorescences which are prevented from completing their development by GA. In a GA-insensitive dwarf grapevine inflorescences are produced instead of tendrils along an actively growing stem (Boss & Thomas 2002).

11.4 | Plant size and flowering

Our discussion of signals which control flowering has focused, so far, on plants which respond to temperature and daylength. In these plants the number of vegetative nodes which will develop from the shoot apical meristem (SAM) before it switches to reproductive development is strongly influenced by environmental signals. The great variation in the ages at which different species first flower can be attributed to differences in the length of the juvenile period.

However, many plants which grow at lower latitudes, where the seasonal variation in daylength and temperature is much less marked, flower only once they have attained a certain size. In many varieties of tobacco the SAM becomes committed to flowering only once a certain number of phytomers (internodes with their leaves: Section 9.2) have been produced (Fig. 11.2). The precise number depends upon the variety examined but is typically near 35. If the top of the plant is removed and re-rooted prior to this point, it will continue to grow in a vegetative manner until the appropriate number of nodes have been produced. The SAM is said to be **indeterminate**, as the fate of the cells within it is not fixed. It is capable of producing many more vegetative nodes and can, in fact, be maintained in the vegetative state indefinitely if continually removed and

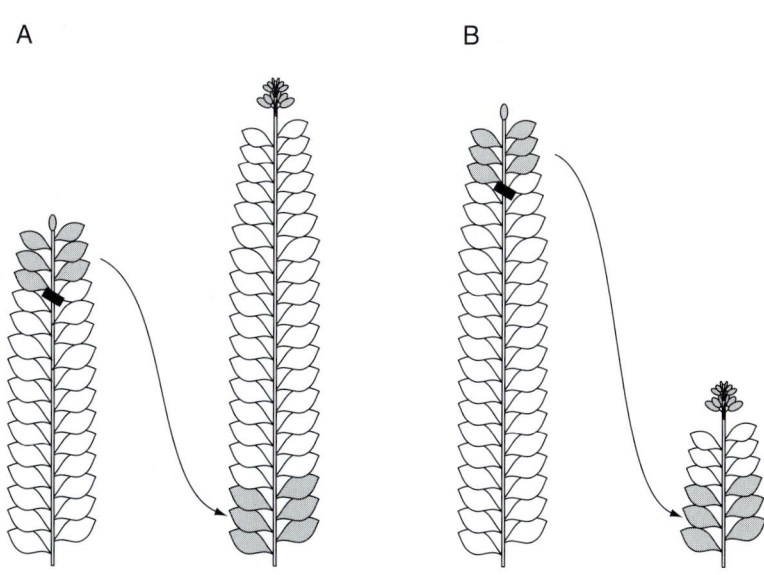

A   B

Fig. 11.2 Commitment to flowering in tobacco Wisconsin 38 is controlled by the number of phytomers. The tobacco (*Nicotiana tabacum*) becomes committed to flowering only once a specific number of phytomers have been produced. If the top of the plant is removed and re-rooted before this number is reached (A), vegetative growth continues. If the procedure occurs after this threshold has been exceeded (B), a few further vegetative phytomers develop and then flowering occurs.

re-rooted. However, once the plant has attained a certain size, the SAM becomes committed to forming reproductive tissues and its fate is now **determined.** If the top of the plant is removed and re-rooted now, only a few vegetative nodes will be produced prior to flowering. The signals which cause this reprogramming of SAM development are not known but include transmissible signals produced in the leaves.

In tobacco the SAM switches to reproductive development during the vegetative growth period of the plant. In maize (*Zea mays*) the situation is somewhat different in that the fate of the cells within the SAM is thought to be determined already during embryogenesis. During maize embryogenesis the SAM is formed together with 4–5 embryonic vegetative leaves. As the plant develops it will produce between 16 and 22 nodes (depending upon the variety) before a tassel is produced. The tassel consists of hundreds of closely packed nodes bearing clusters of flowers. It is possible to map the fate of different regions of the SAM, at a broad level, whilst still within the embryo, although the fate of individual cells is not fixed (Fig. 11.3). It is possible to make a maize plant produce twice as many vegetative nodes by removing the shoot apices and growing them in culture for a time. This extra vegetative growth results from cells in the upper region of the SAM, which would normally develop as the base of the tassel, now developing as vegetative nodes. There has been no extra production of nodes, rather the fate of a few existing nodes has been altered.

Fig. 11.3 A fate map of the maize (*Zea mays*) shoot meristem at the mature embryo stage. (A) A diagram of the maize shoot meristem. At this stage leaf 6 is just about to be initiated and leaf 5 is a small primordium. There are no clearly defined boundaries within the shoot, hence the domains indicated are only approximate. Nodes to which cells in each domain typically contribute are indicated. From McDaniel & Poethig (1988). (B) A mature maize plant showing leaves and tassels. Image from *How a Corn Plant Develops* (Special Report 48, Iowa State University of Science and Technology), reproduced with permission.

A

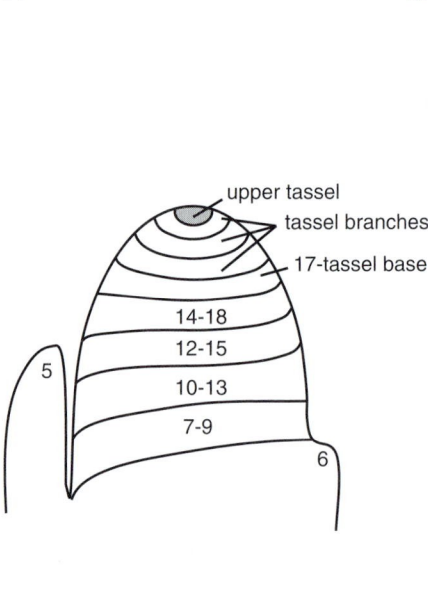

B

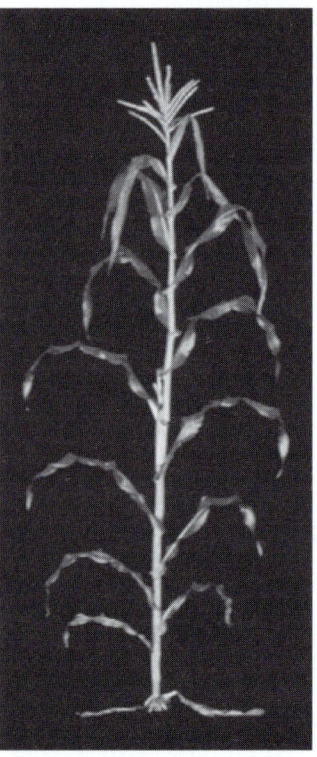

The cells at the top of the SAM have therefore undergone relatively few cell divisions when the shift to reproductive growth occurs. This minimizes the accumulation of errors in the DNA sequence which occur during cell division, and non-dividing cells are known to be less sensitive to damage. Therefore the cells which will give rise to the gametes have been protected.

11.5 | The regulation of floral induction is a multifactorial process

It is clear that floral induction is regulated by different endogenous and environmental signals which, together, cause flowering at an appropriate time. Unravelling this complex area of biology has been greatly aided by the study of mutant plants in which different aspects of floral induction have been affected. Although many such mutants have been isolated from many different species, relatively few species have been studied in depth. In this next section our current understanding of floral induction in *Arabidopsis thaliana* is discussed.

Thale cress (*A. thaliana*) is a facultative long-day plant and ecotypes from high latitudes also exhibit a facultative requirement for vernalization. Plants maintained under short days and in the absence of vernalizing temperatures will eventually flower, but the process is accelerated by long days and vernalization. A number of mutants have been isolated in which flowering time is altered. The late-flowering mutants have lesions in genes which *promote* the transition from vegetative to reproductive growth whilst the early-flowering mutants have lesions in genes which *repress* this transition. Analysis of these plants supports a model of floral induction where multiple, parallel pathways regulate flowering (Fig. 11.4) (reviewed in Mouradov *et al.* 2002). The model presented below is a considerable oversimplification of the regulatory networks operating as many aspects of, and interactions between, pathways have been omitted. However, the major regulatory features are illustrated.

As *Arabidopsis* will eventually flower irrespective of environmental stimuli a pathway for the **autonomous promotion of flowering** must exist. A number of genes involved in this process have been identified. Plants with mutations in these genes flower later than wild-type plants but flowering is still accelerated by long days and vernalization, indicating that these factors act on different systems. Genes involved in the repression of flowering have also been identified. The most extreme of these are the *embryonic flowering (emf)* mutants. When seeds of these mutant plants germinate they immediately produce an inflorescence rather than the normal vegetative rosette of leaves (Fig. 11.5).

The *FLC* (*FLOWERING LOCUS C*) gene also encodes an important component of the repressive pathway. Transgenic plants which over-express *FLC* flower late, whilst mutations in this gene lead to

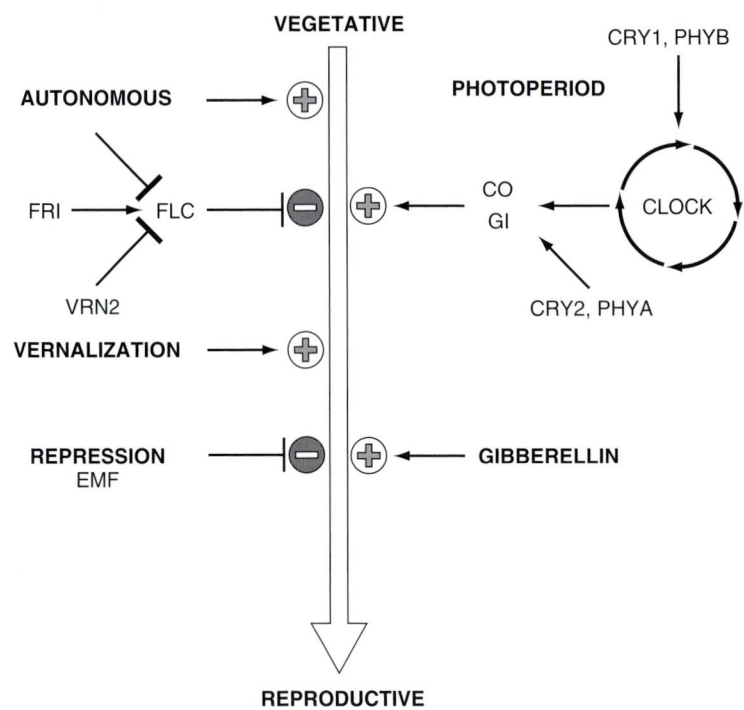

Fig. 11.4 Multiple pathways regulate the transition from vegetative to reproductive growth. In *Arabidopsis thaliana* both environmental and internal signals interact. In this simplified diagram, many interactions between signalling pathways have been omitted for clarity.

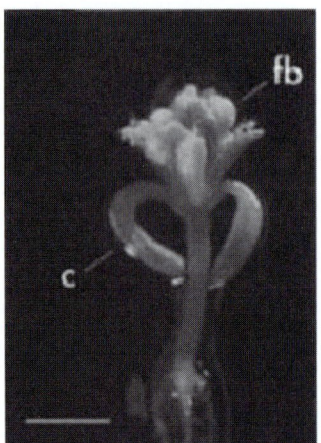

Fig. 11.5 The *emf 2–3* mutant of *Arabidopsis*, 17 days after sowing. The seedling possesses two cotyledons, c, but a flower bud, fb, has been produced instead of the first true leaves. Scale bar = 1.3 mm. From Yang et al. (1995). With permission from Elsevier.

early flowering as the repressive signal is lost. Vernalization is thought to repress the expression of *FLC*, allowing flowering to proceed. A number of genes regulate this vernalization response forming a regulatory network. One such gene, *FRIGIDA* (*FRI*), is found in vernalization-sensitive ecotypes and promotes the expression of *FLC*, thus repressing flowering. Other genes, including *VRN1* and *VRN2*, are required for the perception of the vernalizing temperatures and lead to a decrease in *FLC* expression. Most ecotypes which are not vernalization-sensitive, and hence flower early, have lesions in the *FRI* gene which disrupt its function, although a few have lesions in the *FLC* gene. The role of DNA methylation in this signal transduction pathway, as discussed in Section 11.3.5, remains to be resolved, and vernalization can trigger flowering in a *FLC*-independent manner.

Models which seek to explain the photoperiodic induction of flowering by long days must also incorporate the role of the circadian clock. It is thought that this circadian clock consists of a series of proteins which regulate each others' expression forming an oscillating system. This central **circadian oscillator** is synchronized with the photoperiod by the action of two photoreceptors, phy B and cry 1, which perceive high-irradiance red and blue light respectively (Chapter 10). The output of the circadian clock controls the expression of many other genes including, but not limited to, those which regulate the photoperiodic induction of flowering. Mutations in these output genes cause late flowering under long days (LD) but have little effect on flowering time under short days (SD). Such mutations include the *co* (*constans*) and *gi* (*gigantea*), which, as their names imply, cannot perceive photoperiodic stimuli and grow very large under long days. Two other photoreceptors, cry 2 and phyA, also regulate flowering under long days but their role is more apparent under low-fluence conditions and they are thought to act largely independently of the circadian clock.

Other factors are also important in the regulation of floral induction. Many mutants with lesions in gibberellic acid synthesis or sensitivity have altered flowering time. The supply of carbohydrate has also been implicated in the transition from vegetative to reproductive growth, which is an attractive idea given the strong demand of reproductive tissues for carbon to fuel growth. Plants with mutations in carbohydrate metabolism often flower late, especially those with alterations in starch metabolism.

Although the mutational approach has proven successful in defining many aspects of floral induction, much still remains to be discovered. For example, the nature of the transmissible signals (i.e. florigen) is still elusive and the interactions between genes are still being investigated. However, although florigen has not been identified, analysis of plants with alterations in flowering should eventually provide an insight into the nature of this signal. For example, the *indeterminate 1* mutant of maize is particularly intriguing as it has a prolonged vegetative growth phase. The gene responsible for this phenotype has been isolated, and it encodes a transcription factor which is expressed in young leaves, rather than in the meristem. It has been proposed that it may have a role in synthesizing mobile signals which lead to flowering, or could potentially migrate itself, as has been demonstrated for a number of other transcription factors (see Colasanti & Sundaresan 2000).

Homologues of many of the genes described in *Arabidopsis* have been, or are likely to be, found in other plants. For example, many varieties of green pea (*Pisum sativum*) are also facultative long-day plants which respond to vernalization. Both early- and late-flowering mutants have been described with lesions in the autonomous induction, repression and photoperiodic sensing pathways, and the models of floral induction devised for pea are broadly similar to those of *Arabidopsis*. Pea has a significant advantage over *Arabidopsis* in that grafting studies are more simple and strong evidence for mobile stimulants and inhibitors of flowering has been obtained, although the genes have yet to be identified.

11.6 | Floral development

The output of the regulatory pathways which sense environmental and developmental signals results in the SAM entering the reproductive phase of growth. An inflorescence meristem is produced on which secondary floral meristems are borne; it is these which will develop to form the flowers. The inflorescence meristem may be determinate, eventually producing a terminal flower, or indeterminate, depending upon the species examined.

Early studies focused on the structural changes which occur in the meristem during this transition. It was recognized very early that reproductive development and reproductive organs appear to be modified forms of vegetative growth. Goethe in 1790 suggested that flowers were modified shoots and that the floral organs were modified leaves. Indeed the cell layers which can be distinguished in the vegetative SAM persist in the floral meristems, and genes such as *CLV* and *STM* (Section 9.2.2), which regulate the size of the vegetative meristem, fulfil a similar role in the floral meristem.

Studies on a number of 'model systems' including snapdragon (*Antirrhinum majus*) and, inevitably, *Arabidopsis* have begun to show how meristem fate and floral development are controlled. Although these plants produce flowers which are superficially quite different, many common features can be identified which are reflected in a similarity in the underlying molecular mechanisms. These studies have been extended to encompass the vast diversity of flowering plants and, again, many common features have emerged.

11.6.1 The development of the inflorescence

Figure 11.6 shows a flowering *Arabidopsis* plant. In the vegetative period of growth, the internodes of *Arabidopsis* are very short, leading to the rosette growth habit. During early reproductive development, the

Fig. 11.6 *Arabidopsis thaliana* in flower. (A) The plant produces leaves in a rosette during vegetative growth. When reproductive growth is initiated, the internodes elongate to form the stem of the primary inflorescence on which secondary inflorescences develop. Cauline leaves and flowers are borne on the inflorescence. (B) A close-up of the apex of an inflorescence. Flowers at different stages of development can be seen at the apex. Siliques containing developing seeds are found lower down the inflorescence.

SAM becomes an inflorescence meristem which produces a few cauline leaves and phytomers with much longer internodes. Floral meristems are borne on the flanks of the inflorescence meristem and develop to form the flowers. Later, secondary inflorescences develop which also bear flowers. The regulation of inflorescence and floral development is controlled by a complex network of interacting genes, many of which have been identified in mutagenic studies. Although a detailed description of the processes so far identified is beyond the scope of this text, a number of regulatory pathways will be described which regulate different aspects of reproductive development. The reader is referred to reviews for further information (e.g. Coen 2001, Ng & Yanofsky 2001).

For the inflorescence to develop properly, the terminal inflorescence meristem must be maintained in an indeterminate state whilst the floral meristems, which are determinate, form flowers. Mutagenic studies have identified a number of key genes which regulate this process. The *TERMINAL FLOWER 1* (*TFL1*) gene of *Arabidopsis* maintains the inflorescence meristem in an indeterminate state whilst the *LEAFY* (*LFY*) and *APETALA1* (*AP1*) genes are important in the early stages of floral development. Mutations in *TFL* cause the inflorescence meristem to differentiate into a flower whilst mutations in *LFY* or *AP1* cause inflorescences to develop in the place of flowers (Fig. 11.7). These genes are thought to regulate each other so that expression of *TFL* occurs early, and in the centre of the apical meristem, preventing the expression of *LFY/AP1*, whilst the opposite is true on the flanks (Fig. 11.8).

Genes with homology to the *Arabidopsis* genes *TFL/LFY/AP1* have been found in many other species, and mutations in these genes give rise to similar phenotypes. The *TFL* homologue of *Antirrhinum* is called *CENTRORADIALIS* (*CEN*) and mutations in *CEN* also cause the inflorescence meristem to differentiate into a flower. Homologues of

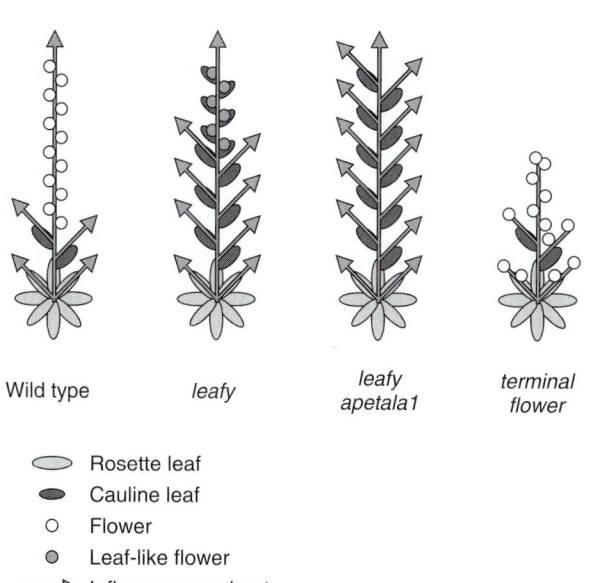

Wild type   *leafy*   *leafy apetala1*   *terminal flower*

- Rosette leaf
- Cauline leaf
- ○ Flower
- ◉ Leaf-like flower
- ⟶ Inflorescence shoot

Fig. 11.7 Mutations affecting reproductive development in *Arabidopsis*. In the wild-type plant a primary inflorescence, bearing cauline leaves and flowers, emerges from the rosette. Secondary inflorescences also develop. In the *leafy* mutant more secondary inflorescences develop and flowers are leaf-like. The *leafy apetala1* double mutant produces few or no flowers and all the axillary buds on the primary inflorescence develop as secondary inflorescences. In the *terminal flower* mutant, inflorescence development is limited as the meristems differentiate to produce flowers. Redrawn from Weigel *et al.* (1995).

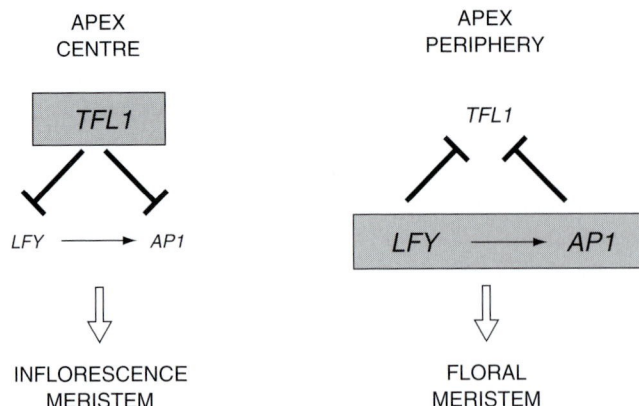

APEX
CENTRE

APEX
PERIPHERY

Fig. 11.8 Shoot and flower meristem development are separated in the shoot apex. In the centre of the shoot apex, *TFL1* is up-regulated prior to *LFY* and *AP1*. The activity of *TFL1* maintains the development of the shoot and represses the activity of *LFY* and *AP1*. At the apex periphery, the expression of *LFY* and *AP1* result in the development of flowers and repress the expression of *TFL1*. Therefore floral meristems develop from the apex periphery whilst an indeterminate shoot meristem develops from the apex centre. Adapted from Ratcliffe *et al.* (1999).

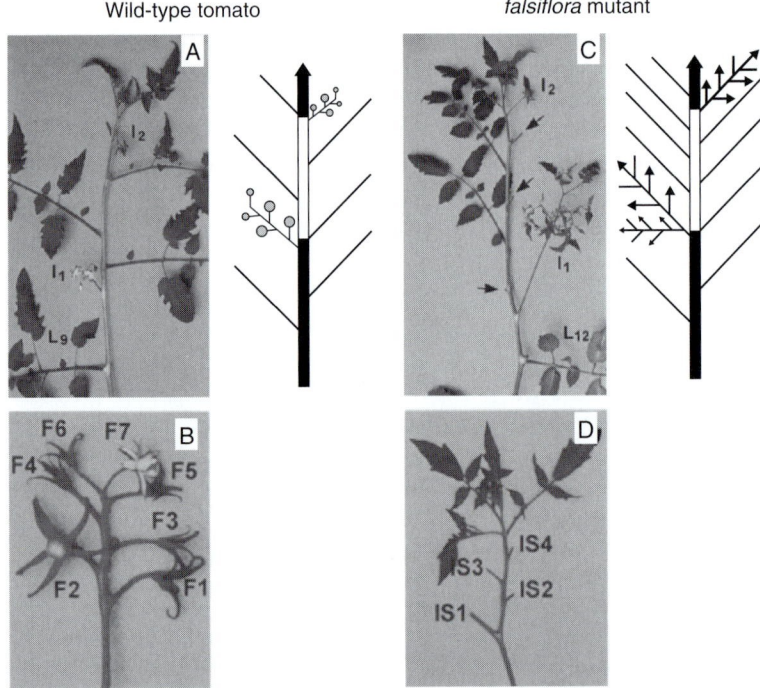

Fig. 11.9 The *falsiflora* mutant of tomato (*Lycopersicon esculentum*). (A) In wild-type tomato, flowers develop on inflorescences, I, borne on the main stem and leaves, L, continue to be produced. (B) A close-up view of an inflorescence of wild-type tomato bearing flowers, F. (C) In the *falsiflora* mutant, flowers are replaced by secondary inflorescence shoots and leaves; arrows indicate where some shoots have been removed for clarity. (D) A close-up of an inflorescence of the *falsiflora* mutant, showing the conversion of flowers into shoots and leaves; secondary inflorescence shoots, IS, have been removed for clarity. In the accompanying diagrams, shoots which will continue to produce leaves are shown as lines with arrowheads; flowers are shown as circles. From Molinero-Rosales *et al.* (1999).

LFY have been found in *Antirrhinum*, pea, tobacco, petunia and tomato, and in the conifer Monterey pine (*Pinus radiata*). Figure 11.9 shows the *falsiflora* mutant of tomato, in which inflorescences develop in the place of flowers. The *LFY* homologue of *P. radiata* is appropriately named *NEEDLY*, and can function in *Arabidopsis* even though *P. radiata* is a gymnosperm.

The expression of the regulatory genes *TFL1*, *LFY* and *AP1* is controlled by environmental conditions forming a link between floral induction and floral development. Both long days and treatment with GA increase the expression of *LFY* which, in turn, stimulates the expression of *AP1*. The autonomous pathway is thought to act on both *LFY* and *AP1* directly. *AP1* and *LFY* encode transcription factors which have been shown to regulate the expression of many other genes which control flower development, as discussed below. If transgenic plants are produced in which *LFY* (or *AP1*) is expressed throughout the life cycle of the plant, rather than just in response to inductive conditions, the vegetative growth phase is shortened and the plants flower early. This is being exploited to accelerate flowering in a wide range of plants as diverse as the cereal rice (*Oryza sativa*) and trees such as aspen (*Populus tremula*). Transformation of aspen trees, so that they express *LFY* continuously, causes them to flower after just 18 months rather than the usual 8–20 years (Coupland 1995). This raises the possibility of breeding trees on a much reduced timescale, which has great implications for forestry.

11.6.2   The development of floral structures

The diversity of shape and colours of flowers provides an attractive system in which to study pattern development in plants. Over the last two decades our understanding of the processes which govern the expression of these patterns has increased enormously, and it has been one of the most rapidly moving and exciting areas of plant biology. Initial studies on model systems have now been expanded to examine a variety of flowering plants, providing an insight not only into how the flowers are formed but also into how these structures may have evolved.

Figures 11.10 and 11.11 show the flowers of *Antirrhinum* and *Arabidopsis*. Although these flowers appear superficially quite different, the underlying patterns are similar in many ways. In each case the different organs have developed acropetally from the floral meristem (i.e. the youngest organs are closest to the apex of the meristem). They are arranged in concentric whorls upon a pedicel, although the organs of flowers of some species have a spiral arrangement. The order of the organs is the same in each case: outermost are the sepals, then petals, stamens and carpels. Together the sepals are known as the calyx whilst the petals form the corolla. In combination these non-reproductive organs are known as the perianth. The male part of the flower, the androecium, consists of stamens, which are composed of a filament topped by the anther. The anthers contain the pollen sacs in which the pollen grains develop. Finally the female part of the flower, the gynoecium, is formed by the carpels. These consist of the ovaries, containing the ovules, and an extension of the ovary wall known as the style, on top of which is the stigma.

Fig. 11.10 Inflorescence and flowers of *Antirrhinum majus*. (A) Flowers are borne on an inflorescence. (B) The wild-type flower exhibits bilateral symmetry. (C) Mutations in the *cycloidea* variety result in the development of radially symmetrical flowers. Images (A) and (B) taken by R. Carpenter (courtesy of John Innes Centre, Norwich), (C) by D. Bock (supplied by W. E. Lönnig, Max-Planck Institute, Köln).

Arabidopsis flowers have four green sepals, four white petals, six stamens, two of which are shorter than the others, and a gynoecium which consists of a two-chambered ovary capped by stigmatic papillae and containing 30–50 ovules. The *Arabidopsis* flower is radially symmetrical (except for the stamens) and is said to be **actinomorphic** as it possesses more than one plane of symmetry. The *Antirrhinum* flower also consists of four concentric whorls of organs arranged in the same order. There are five green sepals, five petals fused to form a tube with five lobes, five stamens, the uppermost of which aborts early in development, and an ovary composed of two carpels. The arrangement is bilaterally symmetrical and is said to be **zygomorphic,** possessing a single plane of symmetry.

Zygomorphy is thought to have arisen from actinomorphy several times during evolution, and most probably helps to maintain pollinator specificity, often in flowers pollinated by bees. It is found in many families of angiosperms including the Fabaceae (includes pea and soybean), Lamiaceae (mints), Scrophulariaceae (*Antirrhinum* and the common toadflax, *Linaria vulgaris*) and, perhaps most elaborately, in the Orchidaceae. Zygomorphic development is controlled by just a few genes, mutations in which lead to a reversion to an actinomorphic pattern of development (Fig. 11.10C). One such gene,

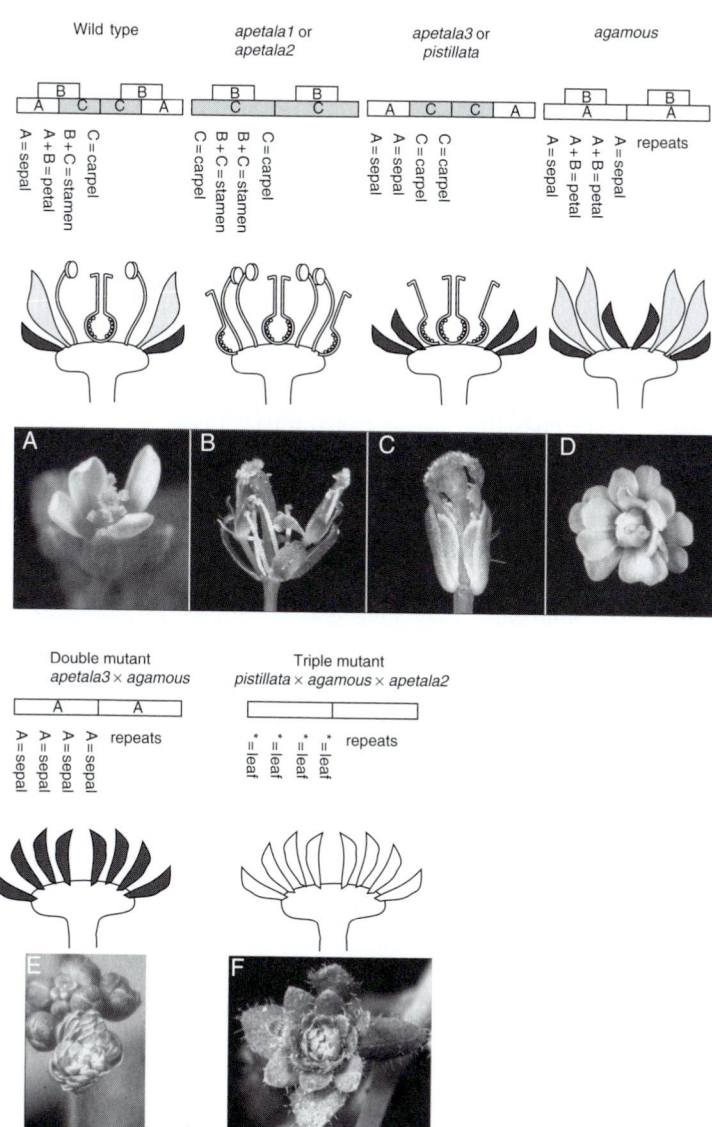

Fig. 11.11 The ABC model of floral development and homeotic mutations of *Arabidopsis* flowers. The upper panel shows wild-type and single-gene homeotic mutations of *Arabidopsis thaliana* flowers. Mutations in single floral identity genes affect organs in adjacent whorls and can be described by the 'ABC model' of flowering. In the wild-type flower the A, B and C activities act, in combination, to direct the wild-type pattern of organ development. The ABC model accurately predicts the pattern of organs observed in double and triple mutants (lower panel). In the absence of A, B and C activities, whorls of leaf-like organs are produced. (A) wild type; (B) *apetala1*; (C) *pistillata*; (D) *agamous*; (E) *apetala3* × *agamous*; (F) *pistillata* × *agamous* × *apetala2*. Images of flowers courtesy of E. Meyerowitz. Photographs A–D by J. L. Reichmann, (E) and (F) by J. Bowman.

CYCLOIDEA (CYC), has been isolated from *Antirrhinum* and encodes a transcription factor.

11.7 | Pattern development in flowers

Although flowers are complex structures composed of different organs and many different cell types, mutagenic studies have been extremely valuable in identifying the way in which the basic underlying patterns are expressed. Some of the most enlightening mutations with respect to the development of patterns within flowers are those in which **organ identity** has been altered. These are known as

Box 11.1

One of the very first descriptions of a plant with a mutation in a gene controlling zygomorphy, in 1744, was by the famous Swedish naturalist Carl Linnaeus, responsible for the introduction of the binomial system for naming species. Common toadflax (*Linaria vulgaris*) normally produces zygomorphic flowers but one of his students found an example which was normal in every way except that it produced actinomorphic flowers. As the classification system of Linnaeus relied heavily on floral morphology, this should have placed the plant in an entirely different taxonomic group. Linnaeus recognized that this plant may have arisen from the common toadflax but it disturbed him sufficiently to name it *Peloria* (Greek – monster) and to write 'This is certainly no less remarkable than if a cow were to give birth to a calf with a wolf's head'. Fortunately for our peace of mind, we now know that this form has a mutation in a gene homologous to *CYC* (Cubas *et al.* 1999).

homeotic mutants (*homeo* = like) as one part of the flower has been transformed into another part. Homeotic mutants have been found in many organisms including insects, fish and mammals.

Fig. 11.11 shows some of the homeotic mutations which have been described in *Arabidopsis*, together with a diagram to show how the identity of the organs within individual whorls has been modified. It can be seen that these mutations affect organ development in *adjacent* whorls of organs. Analysis of these mutants led to the development of the 'ABC' model of flowering (Coen & Meyerowitz 1991). There are a number of features of this model which, together, account for many aspects of floral patterning.

11.7.1 The ABC model of floral patterning

The model proposes that three distinct activities are expressed in the flower but that these activities are restricted to adjacent pairs of whorls. Activity A is expressed in whorls 1+2, activity B in whorls 2+3 and activity C in whorls 3+4. Several different genes may be necessary for each 'activity' to be expressed; hence mutations in more than one gene may give rise to the same phenotype. The combination of activities present in any particular whorl determines which organ will develop there (Table 11.1, Fig. 11.11). The model also proposes that activity A and activity C are mutually antagonistic, i.e. activity A represses activity C and vice-versa. The *apetala2* mutant lacks activity A so activity C is now expressed in all four whorls. Likewise, the *agamous* mutant lacks activity C, so activity A is now expressed in all four whorls. The model also proposes that if a carpel is present in whorl 4, further whorls of organs do not develop.

This model can account for the homeotic mutants which are observed in many plant species, and it is supported by a number of lines of evidence. If mutants are crossed together so that two, or even three, of the activities are absent, the phenotypes of the double and triple mutants can largely be predicted by the model. It is interesting that in the absence of activities A and C an intermediate sepal/petal structure is produced in whorls 2 and 3. In the absence of all three

Table 11.1 | The ABC model of flower development predicts the *activities* of homeotic genes to be restricted to specific whorls of organs. In most cases the expression of these genes is also restricted to these same whorls. APETALA 2 is an exception – expression occurs in all whorls, but activity is restricted, in the wild-type plant, to whorls 1 and 2.

			B PISTILLATA, APETALA3		
		A APETALA1, APETALA2		C AGAMOUS	
	A		A + B	B + C	C
Whorl	1		2	3	4
Organ	Sepal		Petal	Stamen	Carpel

activities, leaf-like structures are produced, supporting the assertion of Goethe made 200 years ago.

Many of the genes have been cloned and are found to encode DNA-binding proteins, which is consistent with a role in activating cascades of other genes necessary for organ formation. Analysis of the expression of these genes has confirmed that they are expressed in pairs of whorls within the developing flower and has confirmed the antagonistic nature of activity A and C expression. The majority of the genes which lead to homeotic mutations in flowers show homology to each other and are called MADS-box genes; this unusual name is derived from the names of the first four genes which were isolated which share this homology (Ng & Yanofsky 2001).

These genes all share common motifs thought to be involved in DNA-binding, transcriptional activation and interactions with other proteins. Since the first MADS-box genes were isolated from *Arabidopsis* and *Antirrhinum* many others have been discovered. *Arabidopsis* contains over 100 related MADS-box genes. The function of all of these genes is not known but some have roles in pattern formation in other vegetative parts of the plant, e.g. controlling root architecture.

The similarities of these genes raise intriguing questions about their origins. They most probably arose from a series of gene duplications and subsequent DNA sequence divergence. Homologous genes have been found in conifers and ferns, showing that they are not restricted to angiosperms and evolved prior to the development of flowering plants. This is not surprising if pattern development in flowers is just one component of pattern development within the whole plant.

Mutagenic studies, although immensely useful, have their limitations. If a plant contains several different genes which can fulfil the same role, the function of a mutated gene will be 'hidden' by the activities of the others. One approach to overcome this problem is referred to as 'reverse genetics'. Rather than isolating mutants and then identifying the inactivated genes, genes are identified, inactivated and the impact on the plant determined. This complements the mutagenic approach and will form a major emphasis of future studies.

The power of this approach has been demonstrated in studies of other MADS-box genes and flower patterning. For example, three genes which share homology with the *AGAMOUS* gene, and which are expressed in flowers, have been isolated from *Arabidopsis*. These genes have been named *SEPALLATA* (*SEP*) *1/2/3* (formerly called *AGAMOUS-LIKE*, *AGL*) and the inactivation of each individual gene produces subtle changes, at most, in flower morphology. However, if these plants are crossed to provide plants in which all three genes are inactivated, their function is revealed. Simultaneous inactivation of *SEP1*, *SEP2* and *SEP3* results in flowers with sepals in all four whorls (i.e. like the *apetala3* × *agamous* double mutant: see Fig. 11.11), indicating that their function is required for the expression of the B and C activities. The importance of these genes, especially *SEP3*, was recently

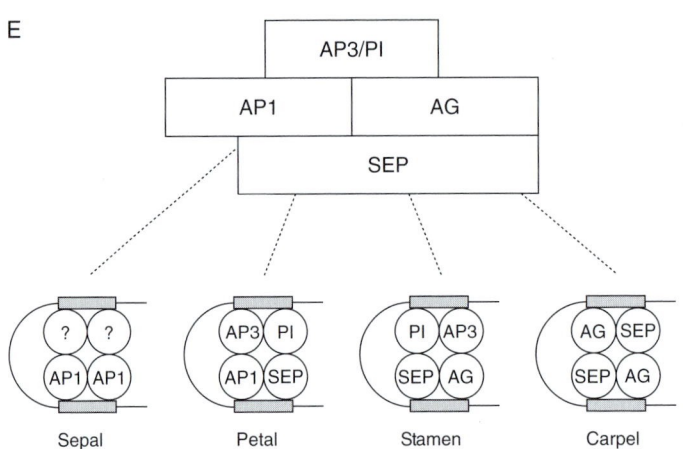

Fig. 11.12 The role of *SEPALLATA* genes in floral development. Inactivation of the *SEPALLATA1, 2* and *3* genes in *Arabidopsis* converts a wild-type flower (A) into a flower in which all organs develop as sepals (B). If transgenic plants are produced (C) which constitutively express B-class floral-development genes (*PISTILLATA* and *APETALA3*) and *SEPALLATA3*, which is necessary for B gene function, the cotyledons, C, are unaffected by the transformation but the true leaves develop as petaloid organs. A few stamens, S, and a terminal flower, TF, are produced. Scale bar = 1 mm. (D) Constitutive expression of B-class genes, *SEP3* and the C-class gene *AGAMOUS* results in the development of staminoid organs. Both flowers, F, and cauline leaves, CL, are affected. Scale bar = 0.5 mm. (E) Current models of flower development propose that multimeric complexes of homeotic proteins include SEPALLATA as a component. (B) courtesy of S. Pelaz and M. Yanofsky, (C) & (D) from Honma & Goto (2001). © Nature Publishing Group. (E) adapted from Theißen (2001).

demonstrated in a landmark paper by Honma and Goto (2001). The first true leaves of transgenic *Arabidopsis* plants constitutively expressing *PI*, *AP3* and *SEP3* were transformed into petaloid organs, whilst the cauline leaves of transgenic plants expressing *PI*, *AP3*, *SEP3* and *AG* were transformed into staminoid organs (Fig. 11.12). SEP3 was shown to be an important component required for the interaction between the PI, AP3 and AG proteins, and activity of the multimeric complex. This paper demonstrates that it is possible to convert vegetative tissue into reproductive tissue by constitutively expressing homeotic genes.

It is important to remember that the 'ABC' model is just that – a model. Not all species will necessarily behave in the same way, and other factors must control the formation of the whorls and the expression of the individual activities. However, the success of this model, and its longevity in what is a rapidly developing field, imply that it is essentially correct. As our understanding of flower development grows, these models become increasingly sophisticated. For example, Theißen (2001) proposed a 'quartet' model of interactions between homeotic proteins which seeks to integrate the observed mutations, expression patterns and interactions between these proteins (Fig 11.12).

11.7.2 Helical flowers

So far our discussion of flower patterning has been restricted to flowers in which relatively few organs are formed in a circular arrangement within whorls. However, in many species, believed primitive,

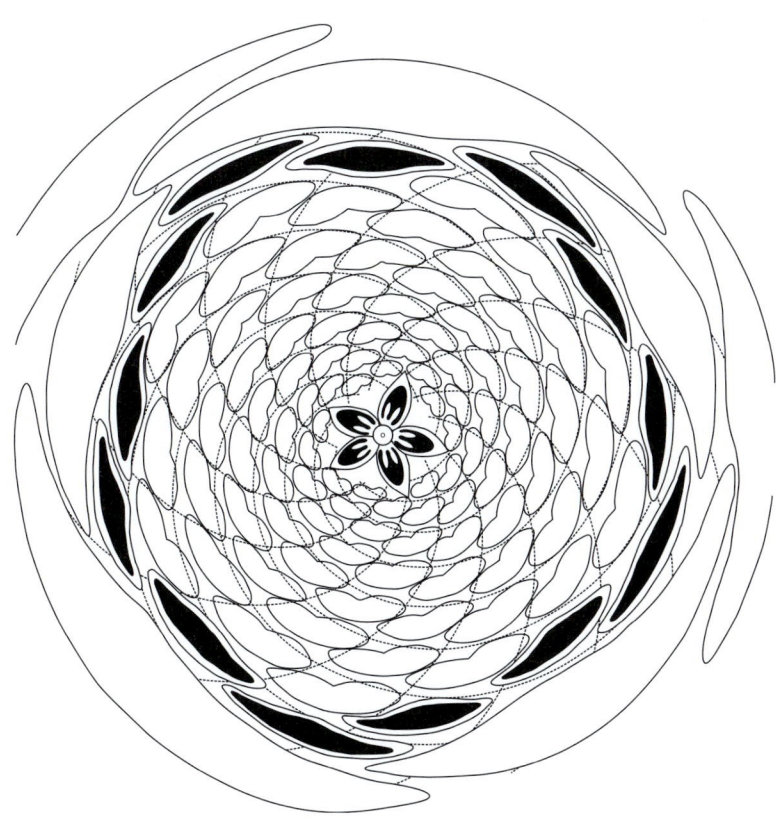

Fig. 11.13 A diagram of a *Helleborus niger* flower. The flowers contain organs in a helical arrangement. The organs are produced sequentially, their position regulated by either temporal or spatial patterns. This leads to spiral patterns (dotted lines) apparent within the flower.

the organs are produced in a helical arrangement and can be very numerous. Figure 11.13 shows a diagram of a flower of *Helleborus niger* which consists of 5 sepals, 13 petals (tubular nectaries), 91 stamens and 5 carpels. These organs are produced from primordia which develop sequentially on the flanks of the floral meristem. The fate of individual primordia might be controlled temporally (as they develop one after another) or spatially. The latter view is favoured, as the transition from one type of organ to the next coincides with a complete circuit of the meristem. One consequence of this is that organs are produced in regular, numerical sequence often corresponding to existing mathematical patterns, e.g. the Fibonacci sequence, a feature which was not lost on early observers. It will be interesting to see whether the expression of flower-patterning genes in these species follows a spatial or temporal pattern.

Box 11.2

The Fibonacci sequence is 0, 1, 1, 2, 3, 5, 8; i.e. add the previous two numbers to determine the next number in the sequence.

The development of the flower results in the formation of the gametes. The male gametes are contained within the pollen whilst the female gametes are contained within the ovules.

11.8 | The formation of pollen

Pollen is formed within the anthers, which are borne on top of the filaments, which act as a conduit for water and nutrients. Within the developing anther, a diploid sporophytic cell divides to form the tapetal initials and a sporogenous mother cell (Fig. 11.14) which forms many pollen mother cells (microspore mother cells). The tapetal cells line the inside of the pollen sac and are essential for pollen development whilst the pollen mother cells undergo two meiotic divisions to form a tetrad of haploid cells – the **microspores**. These cells are contained within a callose-rich wall and are released by the action of callase produced by the tapetal cells. The microspores then undergo an asymmetric mitotic division to form a **pollen grain** consisting of two cells: a larger **vegetative cell** which completely encloses a smaller **generative cell**.

The asymmetry of this division is important in determining the fate of the daughter cells. Treatment with colchicine, a drug which disrupts microtubule formation, results in this asymmetry being lost. In some species the microspore will then develop sporophytically, rather than gametophytically, to form a haploid callus or microspore embryo, from which it may be possible to generate an entire haploid plant. This procedure, although technically difficult, is of great use in plant breeding.

The generative cell divides mitotically to form two 'sperm' nuclei. In about 70% of plants (including the Solanaceae and Liliaceae) this division occurs after pollination, whilst in 30% of plants (including the Brassicaceae and Poaceae) it occurs within the pollen sac. The

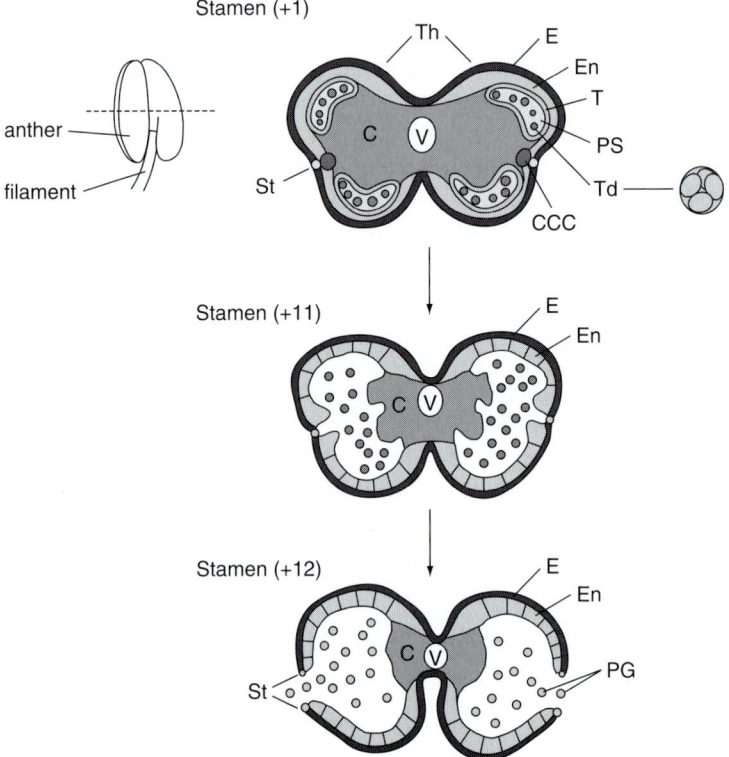

Fig. 11.14 An overview of pollen formation in anthers. The anthers are part of the stamens, borne on top of the filaments. Diagrams of cross-sections through the anthers at different stages of pollen development are shown. C = connective, E = epidermis, En = endothecium, PS = pollen sac, T = tapetum, CCC = circular cell cluster, St = stomium, V = vascular bundle, Td = tetrad, PG = pollen grain. The lines drawn through the endothecium at stages 11 and 12 represent fibrous cell wall bands. Redrawn from Goldberg et al. (1993).

Fig. 11.15 Scanning electron micrographs of pollen grains. The walls of pollen grains are highly resistant to degradation. Differences in surface patterns, which often allow individual species to be identified, are of great use in palynology. (A) Morning glory (*Ipomoea purpurea*); (B) sunflower (*Helianthus annuus*); (C) prairie hollyhock (*Sildalcea malviflora*); (D) castor bean (*Ricinus communis*). Scale bars = 10 μm. Micrographs courtesy of L. Howard and C. Daghlian, Dartmouth College, USA.

wall of the pollen grain consists of a pectin- and cellulose-rich intine and an exine which contains sporopollenin. This material is extremely resistant to degradation within the environment and is often elaborately sculpted, allowing the parent species to be readily identified (Fig. 11.15). Preserved, or fossilized, pollen grains are scientifically extremely useful in fields as diverse as vegetative history, forensics, archaeology and climatology.

11.9 | The formation of the embryo sac

The scientific term angiosperm, for flowering plants, is derived from the Greek *angeion* (vessel) and *sperma* (seed), as the seed develops within a fruit. The gynoecium contains the ovules in each of which a **female gametophyte** or **embryo sac** containing the egg cell (and eventually the embryo) develops. There is considerable diversity in the manner in which the egg cells are produced, but the most common, occurring in 70% of species, is illustrated here (Fig. 11.16); this is known as the *Polygonum* type after the species in which it was first described.

Early in ovule development a **megasporocyte** is formed within the nucellus; the megasporocyte forms the embryo sac whilst

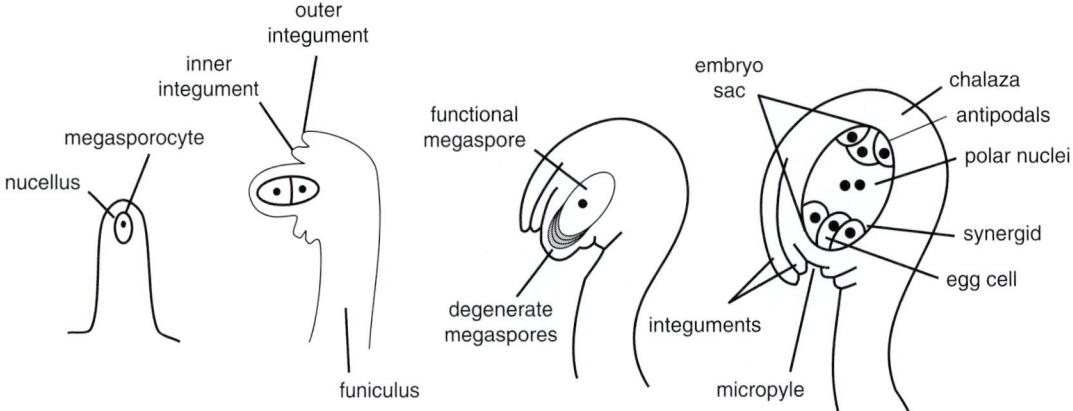

MEGASPOROGENESIS				MEGAGAMETOGENESIS			
mega-sporocyte	meiosis I	meiosis II	functional megaspore	mitosis	mitosis	mitosis	mature gametophyte

Fig. 11.16 *Polygonum*-type embryogenesis. Redrawn from Reiser & Fischer (1993).

protective integuments grow to surround the nucellus, leaving a small gap, the **micropyle**; the funiculus provides a vascular supply for the ovule. The megasporocyte undergoes two successive meiotic divisions to form four haploid cells. The one closest to the chalazal end enlarges to become the functional megaspore, the other three degenerate and are crushed. A series of three successive mitotic divisions of the megaspore then occurs, forming the female gametophyte composed of seven cells containing a total of eight nuclei (Fig. 11.16) with the egg cell at the basal end. Already a polarity has been established, which is important in the future development of the embryo.

Over 15 other types of gametophytic development have been described. In *Arabidopsis* the polar nuclei fuse to form a diploid secondary endosperm nucleus whilst the antipodal cells degenerate. In contrast, in maize the polar nuclei only partially fuse and the antipodal cells proliferate to about 100. The *Polygonum* type is termed monosporic since only one megaspore nucleus produces the embryo sac; in the *Allium* type (bisporic) two nuclei contribute, and in the *Adoxa* type (tetrasporic) all four nuclei from the meiotic division survive to form the embryo sac.

Although the ovules are normally developed in unpollinated flowers, further development may occur after pollination. In some orchids, the ovule primordia are suspended at a premeiotic stage or are even entirely absent in unpollinated flowers, developing only once pollination has occurred. In the orchid *Phalaenopsis*, ovule formation is initiated only after pollination, and fertilization occurs 80 days later. This extreme delay is thought to be an adaptation to the low probability of pollination. The large investment in megagametogenesis occurs only once pollen is available. Although the orchids are unusual in the extent of this delay, some post-pollination ovule development has been reported in maize, barley and *Nicotiana*.

11.10 | Pollination

11.10.1 Pollen tube growth

Pollination is the transfer of pollen from the anther to the stigma. This transfer may occur by physical means (e.g. wind) or animal vectors, and floral structures are often elaborate, enhancing these processes. In a **compatible** system the pollen hydrates and germinates to form a **pollen tube**. This process can be quite rapid. For example, maize pollen will germinate 5 min after deposition and the pollen tube grows at up to 1 cm h^{-1}. This rapid growth is necessary as the pollen tube must grow down a style which may be 50 cm long – an adaptation to wind pollination which requires the stigma to stand out from the plant.

Figure 11.17 shows pollen tubes and a diagram of a typical pollen tube. Growth occurs from the tip, behind which is a dense cytoplasm rich in vesicles. These vesicles fuse with the plasma membrane of the pollen tube to provide cell wall material for rapid growth. The pollen tube wall is double-layered; the outer layer consists of pectin, hemicellulose and cellulose whilst the inner layer (absent at the tip) is rich in callose. Many of the proteins and RNAs necessary for germination and pollen tube growth are already present in the mature pollen grain as treatment with transcription inhibitors (and, in some species, protein synthesis inhibitors) does not block these processes. The cytoplasm moves down the pollen tube as it grows, being sealed off behind it with callose plugs, and large vacuoles develop. In this way the two sperm nuclei and the vegetative nucleus are transferred in the pollen tube down the style until the tube tip enters the embryo sac via the micropyle.

The growing pollen tube is in intimate contact with the transmitting tissue of the style. The pollen tube is chemotropic and many compounds present in the style influence its growth, including minerals, simple sugars, polysaccharides, proteins and lipids. Calcium ions are essential for pollen tube growth, and can be visualized using calcium-sensitive fluorescent dyes (Section 7.4.4). This approach has shown that there is a steep gradient of Ca^{2+} at the growing pollen tube tip. Ca^{2+} is thought to enter the pollen tube via Ca^{2+} channels at the tip and is then pumped into the endoplasmic

Fig. 11.17 Pollen tube growth. Fluorescence micrographs showing pollen tubes within the style of *Nicotiana alata* after compatible pollination, stained for callose. (A) The upper segment of the pistil which includes the stigma; the pollen grains and tube walls stain for callose. (B) Section within the style; the callose plugs stain brightly. (C) The major components of the pollen tube involved in growth. (A) and (B) from Professor Elizabeth Williams, CSIRO Division of Horticulture; (C) from Franklin-Tong (1999).

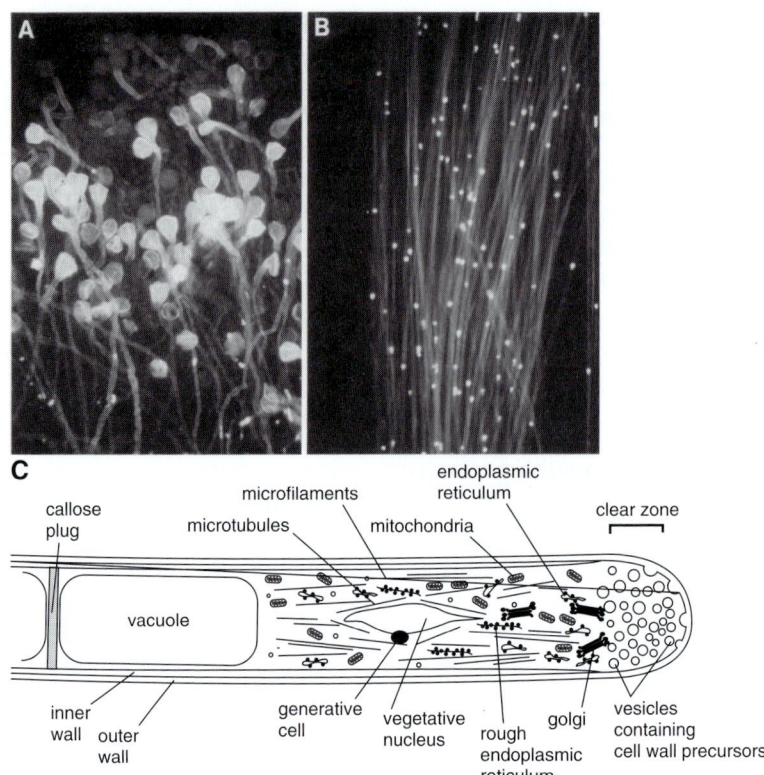

reticulum by Ca^{2+}-ATPases, thus maintaining the gradient. Oscillations in this gradient occur which coincide with oscillations in the growth rate, and alterations in the direction of tube growth are associated with redistribution of Ca^{2+}. It is thought that Ca^{2+} is an important component of a signal transduction pathway that determines when, and where, secretory vesicles fuse to the membrane surface resulting in growth at that time and location.

11.10.2 Pollination and self-incompatibility

Most flowering plants favour out-breeding, cross-fertilization between different individuals, which maintains genetic diversity and aids survival in a changing environment. This can be achieved in various ways. In some species **cross-pollination** is favoured, or made obligatory, by the production of separate male and female flowers, borne on the same plant (**monoecious** species, e.g. oak, *Quercus* spp., and mulberry, *Morus* spp.) or on separate individuals (**dioecious** species, e.g. yew, *Taxus* spp., and willow, *Salix* spp.). In many species the male and female organs are present in the same flower (**hermaphrodite** species), but the physical structure of the flower may prevent self-pollination, or the pollen and the gynoecium may mature at different times.

A very important mechanism for ensuring cross-fertilization is **self-incompatibility**, defined by de Nettancourt in 1977 as '*the inability of a fertile hermaphrodite seed plant to produce zygotes after self-pollination*'. This mechanism is exhibited by about 50% of angiosperms, and is thought to have evolved several times. Many cultivated varieties have, however, had the trait bred out of them as self-compatibility allows the production of homozygous plants for the selection of desirable traits. Self-incompatibility is under genetic control and, in most species, is regulated by a cluster of tightly linked genes located at a single locus – the S locus. Variants at this locus are known as **haplotypes** whilst variants in the individual genes are referred to as **alleles.** Although a single locus is most usual, self-incompatibility is controlled by two separate loci in some grasses whilst in sugar beet (*Beta vulgaris*) and meadow buttercup (*Ranunculus acris*) there are four.

Two forms of self-incompatibility have been described and are illustrated in Fig. 11.18. In **gametophytic self-incompatibility** pollen landing on an incompatible stigma will hydrate and germinate but the pollen tubes do not develop properly if the S allele of the haploid pollen matches an S allele of the style. In **sporophytic self-incompatibility** the incompatible pollen will not hydrate or germinate if the S allele of the diploid parent plant which produced the pollen matches that of the recipient. These self-incompatibility systems require the stigma or style to distinguish 'self' from 'non-self' pollen and therefore there must be an interaction between both male and female determinants of the response.

Gametophytic self-incompatibility

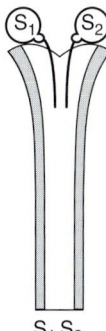 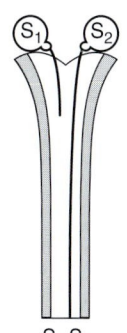 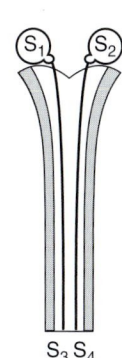

$S_1 S_2$ $S_1 S_3$ $S_3 S_4$

Sporophytic self-incompatibility

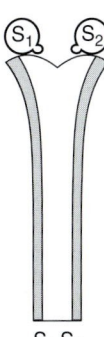 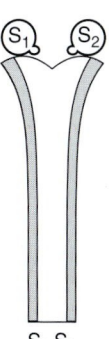 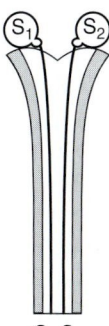

$S_1 S_2$ $S_1 S_3$ $S_3 S_4$

Fig. 11.18 Gametophytic and sporophytic self-incompatibility. In this example, pollen has been produced by an S_1S_2 plant, hence individual haploid pollen grains contain either the S_1 or S_2 allele. In gametophytic self-incompatibility pollen tube growth is arrested, generally in the style, if the allele of the individual pollen matches either allele of the diploid style. In sporophytic self-incompatibility, pollen tube growth is determined by the genotype of the parent, rather than the individual pollen grain. Hence a pollen grain containing the S_2 allele will not develop on an S_1S_3 stigma as the pollen parent shared an allele with the recipient. Growth is arrested at the papillar surface.

Sporophytic self-incompatibility

In *Brassica* species a number of genes at the S locus have been cloned which are important in regulating self-incompatibility. The best studied of these are *SLG* (*S LOCUS GLYCOPROTEIN*) and *SRK* (*S RECEPTOR KINASE*). Both of these genes are expressed in the stigma and hence are part of the female component of the process. These genes are highly polymorphic and over 80 alleles have been described. Plants which have mutations in *SRK* become self-compatible, although the phenotype of the pollen is unaffected, confirming its role in the stigma. Another gene, *SCR* (*S locus cysteine rich*), has been shown to be a male determinant of self-incompatibility. This gene encodes a small, cysteine-rich protein and is expressed in the anthers. Again this protein is highly polymorphic and, to date, 21 variants have been described. Interestingly, SCR is closely related to a class of proteins involved in plant defence reactions – another system where recognition is an important process.

It is thought that SCR in the pollen binds to SRK in the stigma, which stimulates the kinase activity of SRK, initiating a signal transduction pathway which leads to the rejection of the incompatible pollen. This process involves other genes, some of which are also located in the S locus. This is a highly localized process as nearby compatible pollen is unaffected. One current theory is that the incompatible reaction stimulates the expression of an aquaporin in the stigmatic papillar cells. This acts as a water channel and causes the papillar cells to take up water, preventing the hydration of the pollen. A highly localized response is essential, as many different sorts of pollen may land on the stigma simultaneously.

Gametophytic self-incompatibility

Two different systems have been described which result in gametophytic self-incompatibility (Newbigin *et al.* 1993). In the Solanaceae (including tobacco, petunia, tomato, potato) the female determinant encodes a RNA-degrading enzyme or RNase known as the S-RNase. This is produced in the style and, if taken up by the pollen, would degrade pollen RNAs essential for pollen tube growth. The male determinant of incompatibility has not yet been identified but two alternatives have been proposed. Either a protein is present in the pollen which would specifically take up the RNase (the 'gatekeeper' hypothesis) or an inhibitor is produced which inactivates the RNase in compatible pollen (the 'inhibitor' hypothesis). Current theories favour the inhibitor hypothesis as it has been demonstrated that both compatible and incompatible pollen take up the RNase.

In the Papaveraceae (poppies) the female determinant is quite different from that of the Solanaceae. The S locus encodes a small protein (the S protein) which triggers Ca^{2+} release in the pollen, inhibiting pollen tube growth. The male determinant has not yet been identified.

Originally it was proposed that the difference between sporophytic and gametophytic self-incompatibility systems might arise from differences in the timing of the expression of the male determinant. If the male S genes were expressed before meiosis then all the pollen would contain both parental S gene products (and hence be sporophytic). However, if the male S genes were expressed after meiosis then each pollen grain would contain only one parental S gene product (and hence be gametophytic). Unfortunately this attractive hypothesis was not supported when the expression of the male *SCR* gene was examined in the sporophytic *Brassica* system as the gene is expressed after meiosis. It is thought, therefore, that SCR proteins are exchanged between the different haploid pollen, although this awaits experimental confirmation.

11.10.3 Male sterility and the formation of F1 hybrids

In agriculture, the ability to control self-pollination is of great economic benefit. When two homozygous (true-breeding) parental plants are crossed the heterozygous offspring (the F1 hybrids) often exhibit **hybrid vigour** or **heterosis**. A large proportion of crops such as maize, sunflower, sorghum, cotton and many vegetables are grown as hybrids as they typically produce a 15–20% increase in yield. From the point of view of a seed supplier there is also the advantage that the offspring of the hybrid plants will not breed true, compelling the farmer to purchase new seed each year.

The production of F1 hybrid seed requires that the maternal plants do not self-pollinate, and a number of methods have been used to prevent pollen production. One of the most widespread is the use of **cytoplasmic male sterility**, in which a mutation in the mitochondrial genome prevents the production of viable pollen. This approach has been used to produce hybrid maize, rice, sorghum, sunflower and millet. Unfortunately, in maize the mutation also causes plants to become susceptible to the fungal pathogen *Cochliobolus heterostrophus* (southern corn leaf blight). The widespread use of maize cytoplasmic male sterility caused this disease to become epidemic in the USA in 1970, resulting in an estimated $1 billion in losses.

Maize produces separate male and female flowers, and physical emasculation is now widely used for the production of hybrid seed. Alternate rows of the male and female parental lines are sown in a remote field. The tassels (which produce pollen) of the prospective female parents are mechanically removed so that cross-pollination from the neighbouring male plants is ensured. The males are removed after pollination so that they do not contaminate seed harvested from the field. Obviously this is a labour-intensive process and the low yields of the parental lines require a large area to be devoted to seed production. This approach is possible in maize because of the structure of the flowers but is impractical, on the large scale needed for seed production, for many commercially important species.

Recently molecular genetic techniques have been employed to create male sterile plants. Anther development requires the expression of many different genes, some of which are expressed only at this time during the life cycle of the plant. If the promoter of such a gene is fused to a gene encoding a cell toxin, those cells will self-destruct as the anthers develop, thus preventing the production of viable pollen.

Figure 11.19 shows transgenic tobacco and oilseed rape plants which contain a tapetal-cell specific promoter fused to *barnase* (an RNase). The tapetal cells are selectively destroyed and, as these are essential for pollen development, the plant is made male-sterile. This approach has the advantage that it has the potential to be applied to many different species. The F1 hybrid will also be male-sterile as it will contain one copy of the toxin. This has been seen as a potential advantage as it would prevent the spread of pollen from transgenic plants to neighbouring fields or wild relatives. However, in many cases pollination is necessary for the production of the crop (e.g. seeds) so the RNase must be inactivated in the hybrid. This can be achieved by engineering the male plant to contain *barstar* – an inhibitor of the RNase.

The use of F1 hybrids has proven benefit to agriculture but, in many areas, their use is controversial. Farmers cannot save seed and become wholly reliant on the seed companies who may be unable, or unwilling, to supply seed at an affordable price. Likewise, the

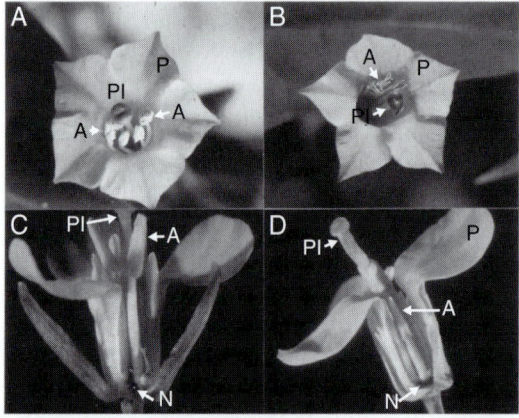
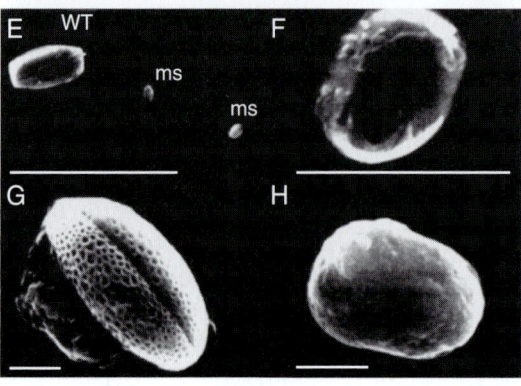

Fig. 11.19 Male sterile flowers produced by expression of RNase in the tapetal cells. Transgenic tobacco (*Nicotiana tabacum*) (A, B) and oilseed rape (*Brassica napus*) plants (C, D) were produced which expressed an RNase (barnase) in the tapetal cells. In the wild-type plants (A, C) pollen is visible on the anthers, but it is essentially absent from the transgenic plants (B, D). A comparison of pollen grains from wild-type and transgenic tobacco plants (E) shows that pollen development is severely impaired in the male-sterile, ms, transgenic plants. A higher magnification of the male sterile pollen grain is shown in (F). Whilst the wild-type oilseed rape plants produce large pollen grains with a sculpted surface (G), the transgenic plants produce smaller, non-viable 'grains' (H). Scale bars E = 100 μm, F, G, H = 10 μm. From Mariani *et al.* (1990). © Nature Publishing Group.

development of open-pollinated varieties may suffer as the hybrids are commercially more attractive. When the added controversy of genetically modified crops is included, this area of agriculture is likely to remain contentious for years to come.

11.11 | Embryo formation

11.11.1 Fertilization

Once the pollen tube enters the embryo sac via the micropyle, it pushes into one of the synergids, which degenerates. The synergids are thought to be very metabolically active, having many cell wall ingrowths (the filiform apparatus) and containing many mitochondria, and are thought to be the origin of the signals which guide pollen tube growth. The male gametes ('sperm cells') are released from the pollen tube and move first to the chalazal end of a remaining synergid. Then one of the sperm cells fuses with the egg to form the diploid **zygote**, which will develop to form the embryo. The other sperm cell fuses with one or more of the nuclei in the central cell which will then develop to form a triploid **endosperm**. Thus a **double fertilization** occurs. Of the maternal tissues, the integuments develop to form a hard, protective seed coat or **testa** whilst the ovary will form the fruit.

11.11.2 Embryogenesis

The zygote undergoes a series of cell divisions which form the embryo. The embryo is aligned to the chalazal–micropyle axis, which is thought to provide positional cues which determine the polarity of embryo development. The embryo develops as a result of a series of stepwise cell divisions in defined planes passing through a series of recognizable stages known as **globular**, **heart** and **torpedo** because of their shapes. The next section describes the development of the *Arabidopsis* embryo, although other patterns of embryo development occur in other species (Fig 11.20).

An initial, asymmetrical cell division of the zygote produces the **two-cell** stage, composed of a smaller apical cell and a larger basal cell. The apical cell will divide many times to form the greater part of the embryo whilst the basal cell undergoes just a few divisions to form the **suspensor,** which connects the developing embryo to the maternal tissues, and the **hypophysis,** which will form the root meristem. Further well-defined and orientated cell divisions then result in the formation of the **octant** and then **dermatogen** stages of globular embryo development. Even though the embryo consists of just a few cells at this point, it is possible to predict which cells will form specific tissues in the mature embryo based upon their position. For example the outermost, the **protoderm,** will develop into the epidermis whilst the innermost cells will form the vasculature.

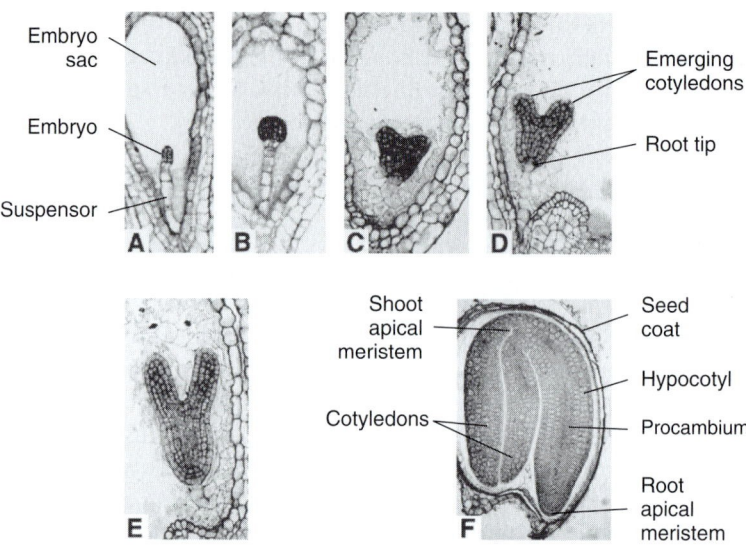

Fig. 11.20 Embryogenesis in *Arabidopsis thaliana*. (A) Two-cell embryo stage; (B) globular embryo; (C) early heart-stage embryo with cotyledons beginning to emerge; (D) late heart-stage embryo; (E) torpedo stage; (F) mature embryo with folded cotyledons. From Mordhorst et al. (1998).

Further cell divisions result in the embryo developing bilateral symmetry and entering the heart stage of development. The uppermost parts develop to form the cotyledons (first leaves) and the position of the shoot apical meristem is already defined. Genes such as *SHOOT MERISTEMLESS (STM)* are important in this process. Many other genes have been identified which are necessary to guide the development of the embryo. These include several homeotic genes, mutations in which lead to alterations in the body plan of the embryo, with specific cell layers developing in an inappropriate manner. A full description of these genes is beyond the scope of this text, but many of the principles governing embryo development are shared with other aspects of plant development. As the embryo develops further, the cotyledons expand and many physiological changes occur which will allow the embryo to enter a dormant state and become tolerant to desiccation.

The development of the cotyledons within the embryo varies greatly between species. One of the most obvious differences is that the embryos of the Monocotyledonae contain a single cotyledon (the scutellum in Poaceae) whilst those of the Dicotyledonae typically contain two. The developing embryo is nourished by the surrounding endosperm. The endosperm may persist within the mature seed, surrounded by the aleurone layer, whilst the cotyledons remain relatively small (in the Poaceae including maize and other cereals) or the endosperm may be entirely absorbed (e.g. in the bean *Phaseolus vulgaris* and pea *Pisum sativum*) so that most of the seed is composed of the cotyledons, the commonest situation in dicots. Usually the cotyledons or endosperm act as a store of carbohydrates, lipid and protein which will fuel the development of the seedling upon germination. In some species, other parts of the embryo serve as storage organs.

11.11.3 Apomixis

Although our discussion of reproduction has naturally focused on the development of the embryo following pollination, in some species the embryo may develop without pollination – a process known as **apomixis**. During gamete development the usual reduction in chromosome number does not occur, allowing the diploid egg cell to develop into an embryo without pollination in a process known as **parthenogenesis.** Apomixis is found in certain *Citrus* species, orchids and grasses as well as mango (*Mangifera indica*) and some members of the Asteraceae (including dandelion *Taraxacum officinale*). From an agricultural viewpoint, apomixis has great potential as it would allow heterozygous plants, exhibiting hybrid vigour, to breed true without the laborious techniques required to produce F1 hybrid seed. Unfortunately, apomixis is rare in most crop species and transgenic procedures to form apomictic crops have yet to be commercially developed. Even then there is the problem of endosperm development to be overcome, as this requires pollination in most apomictic species (an exception is the Asteraceae).

11.12 | Seeds and nutrition

The seed is an embryo plant provided with a nutrient store and protective coats by the parent plant. The value of the seed in plant survival is attested by the fact that the possession of seeds has evolved independently in five living divisions of the plant kingdom, as well as in some groups now extinct. Containing highly concentrated nutrient stores, the seed has also been of indispensable value in human nutrition and the development of civilization. As seeds form an important staple food source for much of the world's population, considerable research has been devoted to the synthesis, and subsequent mobilization, of nutrient reserves within seeds, especially the endosperm of monocotyledonous plants (mainly cereals). There is a complex genetic interaction between the embryo (1 : 1 male : female), endosperm (1 : 2 male : female) and maternal tissues (0 : 2 male : female) as a balance between the genotypes of these different tissues must be established for normal seed development to occur. If these relationships between the ratios of male and female parental genes are disturbed, for example by crossing together plants of different ploidy, seed development is disrupted. It is thought that the differences in ploidy and genetic makeup of the tissues in the developing seed play an important role in ensuring that nutrients flow from the maternal tissue, via the endosperm, to the embryo. In addition, maternal genes may be 'imprinted', most probably as a result of methylation of DNA within their promoters, providing an additional level of regulation of expression.

The developing seed (and fruit) are a strong sink for carbohydrate, which is delivered as sucrose via the phloem. Sucrolytic enzymes (invertase and/or sucrose synthase) metabolize the sucrose, thus forming, and maintaining, a concentration gradient into the developing tissue. Maize provides an attractive experimental system in which to study seed development, as lesions affecting kernel development are simple to observe as well as maize being an agriculturally important crop. Many such mutations have been identified and include *miniature1 (mn1)*, *shrunken1 (sh1)* and *shrunken2 (sh2)*. As their names imply, the kernels of maize plants bearing these lesions do not develop properly. The *mn1* mutation results in a loss of invertase activity. This enzyme is located at the base of the endosperm where it hydrolyses sucrose, supplied from the maternal tissue, to glucose and fructose which are taken up by the developing endosperm. These are then used to synthesize starch and other compounds. In the absence of invertase, nutrient flow to the endosperm is disrupted. The *sh1* and *sh2* mutations result in a loss of activity in sucrose synthase and ADP glucose pyrophosphorylase, which are important enzymes of starch biosynthesis.

There is great diversity in the protein, lipid and carbohydrate content of seeds (Table 11.2). Starch-storing seeds are of great economic importance and include cereals (e.g. rice, maize, wheat and barley) and dicotyledonous plants (e.g. pea and bean). In contrast the seeds of pumpkin, sesame and oilseed rape (known as canola in the USA) are very rich in lipids. In most seeds a significant proportion of the total dry matter is protein – up to 40% in soybean – hence its importance in vegetarian diets. Seeds may also be important sources of dietary vitamins and minerals.

Starch is stored as granules within amyloplasts, either in the endosperm (e.g. in the Poaceae or castor bean) or in the cotyledons

---

Table 11.2 Composition of different seeds. Data derived from the US Department of Agriculture. Values have been rounded to the nearest whole number.

Species		% dry weight		
		Carbohydrate	Lipid	Protein
Rice	*Oryza sativa*	86	3	9
Wheat	*Triticum aestivum*	84	2	12
Maize	*Zea mays*	83	5	11
Barley	*Hordeum vulgare*	81	3	14
Pea	*Pisum sativum*	69	2	26
Flax	*Linum usitatissimum*	38	37	21
Soybean	*Glycine max (soja)*	34	21	40
Sesame	*Sesamum indicum*	25	52	19
Almond	*Prunus dulcis*	21	53	22
Pumpkin	*Cucurbita* sp.	19	49	26
Oilseed rape	*Brassica napus*	11	49	32

(e.g. pea, bean). Starch is a polymer of glucose and exists both as an essentially linear form known as amylose (linked by α1–4 linkages) and a highly branched form, amylopectin (branches are introduced as additional α1–6 linkages). Synthesis occurs within the plastid where the enzyme ADP glucose pyrophosphorylase synthesizes ADP-glucose. This is the substrate for starch synthase, which catalyses the synthesis of amylose, and starch branching enzyme, which, as its name suggests, introduces branches into the growing starch polymer. Together these enzymes lead to the formation of a multi-layered starch grain up to 50 μm in diameter.

Fatty acids, which are the precursors of oil, are also synthesized in plastids. Two carbon units are added sequentially to an acyl carrier protein, building up an saturated fatty acid chain. Desaturase enzymes may introduce double bonds forming mono- or polyunsaturated fatty acids, and other modifications may occur; for example, the fatty acids of castor oil are hydroxylated. Storage fatty acids are very varied, and over 300 different forms have been found in seed oils. Much interest has been focused on the degree of saturation of the fatty acids, since there is a general perception that unsaturated fatty acids are 'good' for human health and saturated fats are 'bad'. Table 11.3 shows the proportions of saturated, monounsaturated and poly-unsaturated fatty acids in a number of economically important oily seeds. In species such as coconut, the majority of lipids are short-chain and saturated (principally 12:0 and 14:0) whilst sunflower oil is a rich source of mono- and polyunsaturated fatty acids. Fatty acids form the substrate for lipid (oil) synthesis which occurs on the endo-plasmic reticulum (ER) in the cytoplasm. As storage lipids are formed they accumulate in oil bodies which bud off the ER. These are bound by a single monolayer envelope and the lipids within are stabilized by proteins known as oleosins. Jojoba (*Simmondsia chinensis*) is unusual in that it stores liquid waxes (esters of C20, C22 and C24 long-chain fatty acids and alcohols) rather than oil in the seed cotyledons.

Both starch and oils have many commercial applications beyond culinary uses or as feedstocks for animals. Starch is used in the production of materials as diverse as nappy liners, adhesives and explosives whilst oils are used as fuels, lubricants and in the manufacture of detergents, plastics and nylon. The precise chemical properties of plant-derived starch or lipid is of great importance in determining their commercial uses. For example, very-long-chain fatty acids (VLCFA) such as erucic acid (22:1) can potentially be used in the production of plastics. Varieties of oilseed rape have been developed by traditional plant breeding programmes in which the erucic acid content approaches 60%. On the other hand, lauric acid (12:0) is widely used in detergent manufacture and there is considerable research effort in developing transgenic varieties of temperate crops which can act as a source of these short-chain fatty acids. Of course, this has implications for those countries whose economies depend on the export of coconut and palm kernel oil which are the current sources of these fatty acids.

Box 11.3

Many fatty acids have 'trivial' names based on plants from which they were first isolated or in which they are particularly abundant. However, it is often convenient to refer to fatty acids by the number of carbon atoms in the acyl chain and the number of double bonds present. Palmitic acid contains 16 carbon atoms and is saturated (i.e. contains no double bonds) and can be written as 16:0. Linoleic acid has 18 carbon atoms and 2 double bonds and can be written as 18:2. A Δ sign can be used to indicate the position of the double bond. Most plants produce oleic acid Δ^6 18:1 in which the double bond is at carbon-6 whilst members of the Apiaceae produce petroselinic acid Δ^9 18:1.

Table 11.3 Lipid content of plant oils. Values are the proportion of total lipid (%).

Fatty acid	Coconut (Cocos nucifera)	Peanut (Arachis hypogaea)	Soybean (Glycine max (soja))	Olive (Olea europaea)	Sunflower (Helianthus annuus)	Oilseed rape (Brassica napus)
4:0	0	0	0	0	0	0
6:0	0.6	0	0	0	0	0
8:0	7.5	0	0	0	0	0
10:0	6	0	0	0	0	0
12:0	44.6	0	0	0	0	0
14:0	16.8	0.1	0.1	0	0	0
16:0	8.2	9.5	10.3	11	5.9	4
18:0	2.8	2.2	3.8	2.2	4.5	1.8
20:0	0	1.4	0	0	0	0.7
22:0	0	2.8	0	0	0	0.4
24:0	0	0.9	0	0	0	0.2
Total saturated	86.5	16.9	14.4	13.5	10.3	7.1
16:1	0	0.1	0.2	0.8	0	0.2
18:1	5.8	44.8	22.8	72.5	19.5	56.1
20:1	0	1.3	0.2	0.3	0	1.7
22:1	0	0	0	0	0	0.6
Total monounsaturated	5.8	46.2	23.3	73.7	19.5	58.9
18:2	1.8	32	51	7.9	65.7	20.3
18:3	0	0	6.8	0.6	0	9.3
Total polyunsaturated	1.8	32	57.9	8.4	65.7	29.6

Seeds also contain protein in variable amounts (Table 11.2). All seeds contain structural and metabolic proteins vital for cellular function; but in addition, seed cells, especially in the storage organs, accumulate special proteins which serve as nutrient stores (particularly for N and S) and sometimes also as protectants. Seed storage proteins, which may represent 50% of a seed's total protein content, are found as single-membrane bound protein bodies, 1–20 μm in diameter, derived from the deposition of protein in (small) vacuoles, or, more rarely, in the rough ER. Three major classes of seed storage proteins have been identified, initially on the basis of their solubility in different solvents and more recently by comparison of the DNA sequences of the genes which encode them. **Albumins** are water soluble, **globulins** are soluble in dilute saline solutions, and **prolamins** (found in cereals) are soluble in alcohol. Any one seed contains a diverse range of seed storage proteins which are often named in a species-specific manner e.g. zeins in maize (*Zea mays*), phaseolins in French bean (*Phaseolus vulgaris*). Inclusions in the protein bodies (globoids) contain mineral stores as Ca, K and Mg salts of inositol hexaphosphate (phytin).

Because seeds form a major part of many diets, the amino acid composition of the seed storage proteins has a profound effect on the nutritional value of the seed. Proteins from legume seeds are poor in sulphur-containing amino acids (i.e. cysteine and methionine) whilst those from cereals tend to be poor in lysine, threonine and tryptophan. Therefore many animal feeds are often based on proteins from mixed sources, e.g. legume and maize. However, there is still often a need for additional supplements, such as synthetic methionine, which significantly increases costs.

There is a considerable research effort, using both traditional plant breeding and GM techniques, to modify the amino acid composition of seeds to improve their nutritional value. The strategies employed include altering the relative proportions of the different seed storage proteins, manipulating the amino acid sequence of individual proteins, and transferring seed storage proteins from one species to another. Whilst these approaches have met with some success, great care must be taken to assess other potential impacts on health. In 1993 Pioneer Hi-Bred International Inc. discontinued a research programme which sought to improve the nutritional quality of soybean as poultry feed by transferring a methionine-rich albumin from Brazil nut (*Bertholletia excelsa*). It was found that the Brazil nut protein caused allergic responses in some people, and although the soybeans were not intended for human consumption, the potential health risks resulting from any contamination were deemed too great. Whilst opponents of GM crops use this as an example of the dangers associated with transgenic plants, proponents point out that the rigorous safety testing mandatory for all transgenic plants identified this problem at an early stage before any commercial use occurred.

As seeds represent a compact source of nutrients, many contain anti-microbial, anti-fungal and anti-herbivore compounds for

Box 11.4

The relative proportions of individual seed storage proteins and their precise amino acid compositions affect other properties of seed-derived products, e.g. the gluten content of wheat grains has important effects on the elasticity and viscosity of doughs.

Box 11.5

Ricin achieved considerable notoriety when used during the Cold War to assassinate the Bulgarian dissident Georgi Markov. A pellet containing approximately 450 µg of ricin was shot into his leg using a gas-powered umbrella. Ricin and abrin are ribosome-inactivating proteins (somewhat morbidly abbreviated to RIPs) and nowadays are recognized as having therapeutic uses in targeted anti-cancer treatments.

protection. A wide range of such compounds has been identified (see Harborne 1994 for details) with many different modes of action. The glucosinolates, found in many brassicas, form toxic breakdown products and also impair iodine metabolism, leading to goitre. Legume seeds are often rich in non-protein amino acids (e.g. azetidine 2-carboxylic acid from *Convallaria majalis*) which impair function if incorporated into proteins. Other seeds accumulate protease inhibitors which will inhibit the digestive enzymes of herbivores. Some of the most potent toxins are seed proteins – abrin from *Abrus precatorius* and ricin from *Ricinus communis* are lethal at milligram doses.

11.13 | Fruit development

The fruit provides a suitable environment for the maturation of the seed, as well as having an important role in seed protection and seed dispersal. There is great diversity in fruit structures, which reflects these different roles. The fleshy fruits of tomato and grape develop from a single ovary, whilst blackberry and raspberry develop from multiple ovaries. Apples, pears and strawberries are largely receptacle tissue. Other species bear fruits which are dry. These include the pods of legumes (e.g. pea, bean), which split open to release the seed, capsules which disperse seeds (e.g. poppies, campions), and nuts, which do not split open until germination.

In spite of their great agricultural importance, our understanding of fruit development is far from complete and based largely on those species most amenable to study. The tomato fruit has been particularly well studied as it passes through a number of readily identifiable stages, many different varieties exist and, of course, it is economically important. Figure 11.21 shows the different stages of tomato fruit development. A limited amount of development occurs prior to fertilization but fruit set, and continued fruit development, will occur only after successful fertilization.

Early stages of tomato fruit development are characterized by a high rate of cell division, is accompanied by elevated auxin, cytokinin and gibberellic acid content. Auxin and gibberellic acid are thought to play an important role in fruit set, as application of these plant hormones to unfertilized fruit can often lead to **parthenocarpy** – the formation of seedless fruit. After this period of cell division, the fruit increases rapidly in size as a result of cell expansion. Many genes are expressed for proteins which remodel the cell walls during this period, including expansins, xyloglucan endotransglycosylases, endo-1,4-β glucanases and glycosidases (Section 8.4). Again, auxin and GA content are high and abscisic acid content is also elevated. As described below, ABA is important in seed maturation, and the highest ABA concentrations are found within, and immediately surrounding, the seed. At this point the fruit is full size, but is still green and unripe. The ovary wall has developed to form the fruit wall

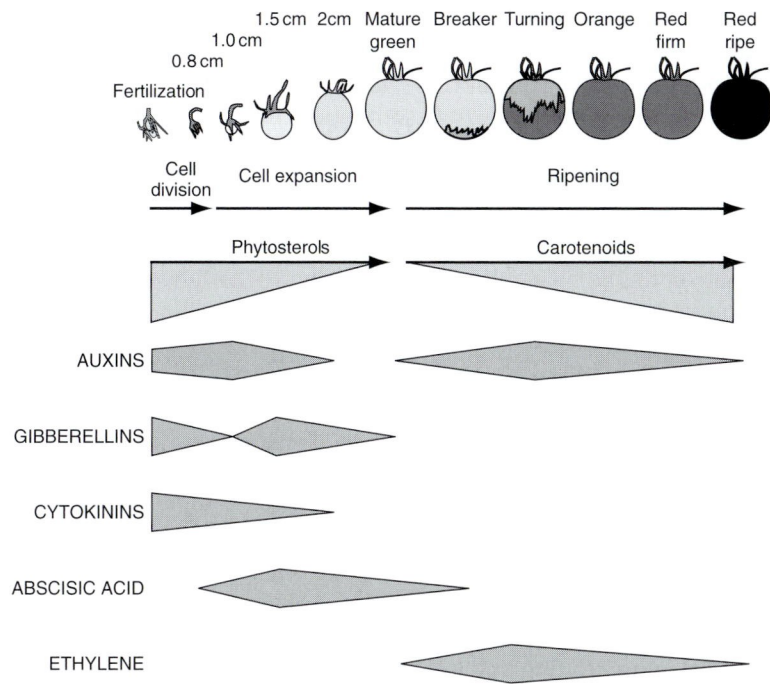

Fig. 11.21 Tomato (*Lycopersicon esculentum*) fruit development. As the fruits develop they undergo periods of cell division and cell expansion, followed by ripening. Fruits attain their full size at the 'mature green' stage but are unripe. The onset of ripening coincides with the first sign of colour development ('breaker' stage). The auxin, cytokinin and gibberellic acid content of tomato fruit is greatest during the early stages of development as the fruit increases in size. Abscisic acid increases later during fruit growth whilst ethylene is associated with ripening. Redrawn from Gillaspy et al. (1993).

or pericarp (composed of the endocarp, mesocarp and exocarp) covered by a thin cuticle. The septa of the carpels divide the fruit into locules and the seeds are attached to an axial placenta. The majority of the cells which make up the carpel are large, vacuolated and contain chloroplasts. These cells are photosynthetically active and express many photosynthetic genes. Unlike a leaf, the fruit is also a strong sink for carbohydrate, provided as sucrose by the rest of the plant. A concentration gradient is maintained into the fruit as a result of sucrose degradation, principally by the enzyme sucrose synthase (which, in spite of its name, generally acts to break down sucrose under physiological conditions, rather than to catalyse its synthesis).

The final stage of fruit development is ripening. In species such as grape and cherry this is a gradual process following earlier stages of fruit development. However, many other species, including tomato, apple and avocado, exhibit an abrupt increase in respiration at the onset of ripening – the **climacteric**. Many physiological changes occur in the fruit as it ripens. In tomato the chlorophyll content of the fruit declines whilst the carotenoid content increases, giving the fruit a red colouration. The chlorophyll-containing chloroplasts are converted into carotenoid-containing chromoplasts and photosynthetic capacity is lost. Sucrose is converted into glucose and fructose by the enzyme invertase, leading to an increase in sweetness and palatability. The fruit softens as enzymes are synthesized which degrade the pectin of the middle lamella, loosening the connections between cells. Again, many cell wall remodelling enzymes are expressed which alter wall plasticity.

One of the first genetically modified plants to be brought to market was the Flavr-Savr tomato. These plants had been modified so that they expressed less polygalacturonase, an important enzyme involved in cell wall degradation and fruit softening. This modification did not prevent the fruit from ripening but it did limit the degree of softening. Soft tomatoes are difficult to transport and so tomatoes are normally picked under-ripe. They are ripened artificially (see below) but do not develop the same flavour as vine-ripened tomatoes. The Flavr-Savr tomato could be picked later, thus improving the flavour (hence the name, although this is no excuse for the spelling!). Perhaps more important, from a commercial point of view, was the increase in the solids content of the tomatoes. This increased the yield of tomato pastes and purees – a market worth $2 billion annually in the USA. Pastes made from genetically modified tomatoes captured up to 60% of the market in the UK even though they were clearly identified as GM, but they were eventually withdrawn from sale in 1999 as a result of public and commercial attitudes towards GM foods.

Many aspects of fruit ripening are regulated by the plant hormone ethylene. Ethylene synthesis is controlled by two key enzymes, ACC synthase and ACC oxidase, the activities of which increase early in fruit ripening, leading to a massive increase in ethylene production. The fruits of tomato plants which have been genetically modified to prevent the expression of ACC synthase do not ripen and the *never-ripe* mutant of tomato has been identified as having a lesion in an ethylene receptor. Ethylene is thought to trigger many of the subsequent stages of ripening, including the climacteric, and thus placing ripe fruit in close proximity to unripe fruit stimulates the latter to ripen. Commercially many fruit are picked unripe and stored in low oxygen, CO_2-enriched atmospheres which prevent ethylene-induced ripening. The fruit can then be transported long distances and ripened on demand by the addition of ethylene-generating chemicals. It is this control of the ripening process which allows many fruits to be economically shipped around the world, providing a year-round supply.

11.14 | Seed dormancy

11.14.1 Late embryogenesis and the onset of dormancy

One of the principal reasons for the success of the angiosperms is their production of desiccation-tolerant seed. This allows survival in a metabolically inactive state during periods of unfavourable environmental conditions and facilitates widespread dispersal. The majority of angiosperm species produce desiccation-tolerant seed but, as discussed below, some do not, including many commercially important species.

A developing seed grows rapidly following fertilization as the embryo and surrounding tissues develop. This growth phase is associated with high concentrations of plant growth regulators such as auxin, cytokinin and especially gibberellic acid. Cell division within the embryo largely ceases at the heart stage once all of the essential structures have been formed, and further growth is largely as a result of cell expansion and the accumulation of storage reserves. The embryo has the *potential* to develop into a new plant relatively early during embryogenesis and indeed this can be achieved if the embryo is removed from the seed and cultured under appropriate conditions. However, further development is normally suppressed at least until the seed has been shed from the parent plant, and often for much longer.

As the seed matures the GA content falls and that of ABA rises. The water content also falls and metabolic activity declines, till in the dry seed it is practically at a standstill. A number of genes are expressed at this stage which are thought to be important in the development of desiccation tolerance including the late embryogenesis abundant (*LEA*) and dehydrin genes. The development of desiccation tolerance and dehydration is associated with seed dormancy. Dormancy is difficult to define precisely but can be considered as a state in which a viable seed fails to germinate under otherwise favourable conditions. Seeds which are shed from the parent plant in this state are said to exhibit **primary dormancy**, whereas seeds which become dormant after being shed from the parent, in response to adverse environmental conditions, are said to exhibit **secondary dormancy**.

In principle, dormancy can result from the absence of signals required for growth or the presence of inhibitory signals which prevent growth. In practice, a combination of the two is likely to operate to regulate the onset of dormancy, the breaking of dormancy and subsequent germination. Both the embryo itself and the surrounding seed coat have been implicated in the control of these processes. Seed-coat-imposed dormancy is common and has been proposed to operate by a number of mechanisms. These include the mechanical restraint of radicle elongation (e.g. water plantain, *Alisma plantago-aquatica*, shepherd's purse, *Capsella bursa-pastoris*), the prevention of water uptake (e.g. many Fabaceae) or oxygen uptake (e.g. cocklebur, *Xanthium*), the production of germination inhibitors, and the retention of inhibitors produced by the embryo. Embryo-imposed dormancy is thought to result from an interaction between GA and ABA; ABA prevents germination whilst GA stimulates germination. The concentrations of these plant growth hormones are, in turn, regulated by developmental state and environmental conditions.

The concentration of ABA contrasts with that of GA throughout seed development: initially ABA concentrations are low, they increase during seed filling and decline again during maturation drying. ABA-insensitive plants, or plants with lesions in ABA biosynthesis, often exhibit precocious germination or vivipary (germination whilst still attached to the parent plant). At least 10 *viviparous* mutants of maize have been described, most of which have lesions in ABA synthesis or response. The seeds of the *sitiens* mutant of tomato

Box 11.6

It has been proposed that fusing a *LEA* gene promoter to a toxin such as *barnase* could be used to prevent seed germination in a manner similar to that used to produce male-sterile plants. This has been termed 'Terminator Technology' and is a highly contentious area of transgenic plant research. The widespread application of this technology might be viewed as a means for seed companies to protect (and enforce) their intellectual property rights, a cynical way of forcing farmers to buy seed each year, or a useful means of preventing the escape of GM crops into the environment.

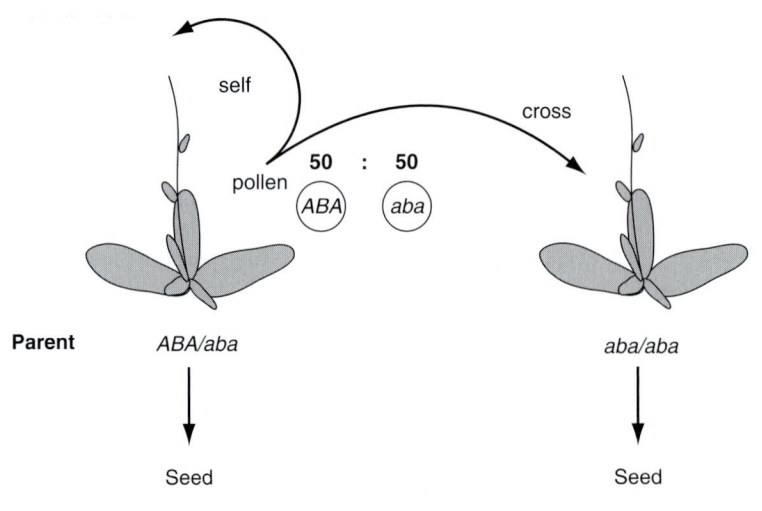

Fig. 11.22 ABA and dormancy. Homozygous *aba* mutants of *Arabidopsis thaliana* (*aba/aba*) are unable to produce abscisic acid unlike the heterozygote (*ABA/aba*). If a heterozygous plant is allowed to self-fertilize, all the maternal tissue will contain ABA but only 50% of the embryos will be able to produce ABA. As only 50% of the embryos exhibit dormancy, dormancy must require embryo-derived ABA rather than maternal ABA. This is confirmed if the pollen from the heterozygous plant is used to fertilize the homozygote. Again, 50% of the embryos can produce ABA and exhibit dormancy although no ABA is supplied from maternal tissue.

Parent	*ABA/aba*			*aba/aba*

Seed   Seed

Maternal tissue   **Maternal tissue**

ABA/aba   ABA produced   *aba/aba*   no ABA produced

Embryo   **Embryo**

50% *ABA/aba*   exhibit dormancy   50% *ABA/aba*   exhibit dormancy

50% *aba/aba*   no dormancy   50% *aba/aba*   no dormancy

germinate in over-ripe fruit and have an ABA content just 10% of that of wild-type tomato. Seeds of *Arabidopsis* plants with lesions in ABA production (ABA-deficient, *aba*) or which are insensitive to ABA (ABA-insensitive, *abi*) exhibit reduced seed dormancy whilst the double mutant (*aba*, *abi*) germinates precociously.

It is possible to manipulate the ABA content of the maternal and embryonic tissue independently by crossing mutants together (Figure 11.22). If a homozygous abscisic acid-deficient plant (*aba/aba*) is fertilized with pollen from a heterozygous *ABA/aba* plant, the maternal tissue will lack abscisic acid, 50% of the embryos will be *ABA/aba* and hence contain abscisic acid, whilst 50% will be *aba/aba* and lack abscisic acid. If maternal abscisic acid regulated dormancy, all of these seed would be expected to lack dormancy but, in fact, 50% exhibit dormancy, indicating that it is the embryonic ABA which is important. Likewise, if the cross is performed in the other direction, all of the maternal tissue will contain abscisic acid but still only 50% of the embryos exhibit dormancy.

11.14.2 Breaking dormancy

The role of GA in seed dormancy has been elucidated by the use of inhibitors and the study of plants with mutations in hormone synthesis. GA-deficient mutants of *Arabidopsis* require the application of exogenous GA before they will germinate. However, if these GA-deficient plants are crossed with ABA-deficient plants, their

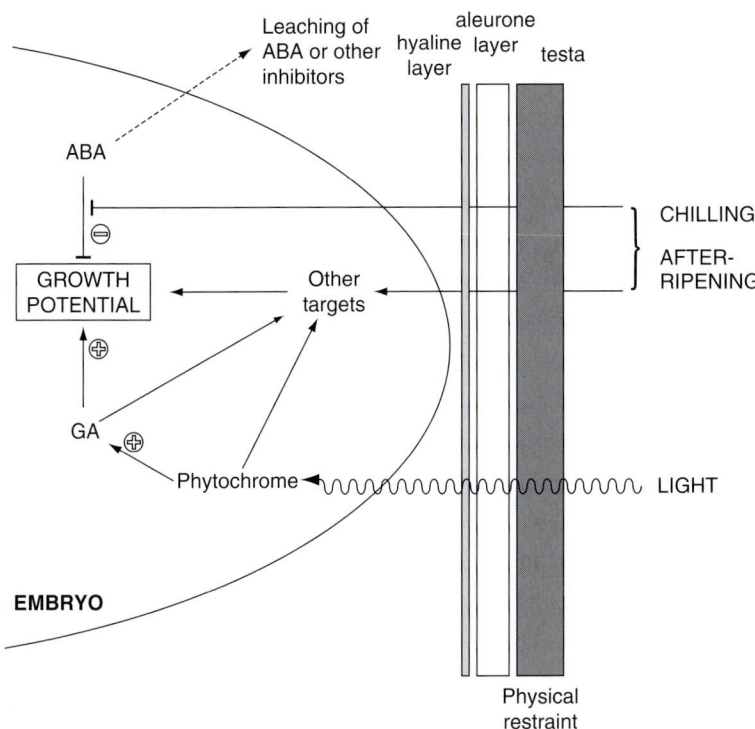

Leaching of ABA or other inhibitors

aleurone
hyaline layer
layer   testa

ABA

CHILLING

AFTER-
RIPENING

GROWTH
POTENTIAL

Other
targets

GA

Phytochrome

LIGHT

EMBRYO

Physical
restraint

Fig. 11.23 Breaking dormancy. Environmental factors and plant growth regulators interact to maintain dormancy or trigger germination. Redrawn from Debeaujon & Koornneef (2000).

germination is stimulated. This result demonstrates that it is the relative concentrations of ABA and GA which are important in maintaining, or breaking, dormancy. If a seed is treated with inhibitors of GA synthesis as it imbibes, germination is prevented, demonstrating that *de novo* GA synthesis is required, rather than the utilization of pre-formed GA. GA is an important signal which stimulates both cell division and the mobilization of seed reserves.

Figure 11.23 shows how environmental and endogenous signals may interact to regulate seed dormancy. Not all of these will operate in one species and they may have different relative importances depending upon how long the seed has been dormant. The growth potential of the embryo is stimulated by GA, and inhibited by ABA (and potentially other substances), with the seed coat acting as a physical restraint to germination as well as preventing the loss of inhibitory substances. Removal, or damage to, the seed coat (a process called **scarification**) will stimulate germination in many species, as will a period of **after-ripening** or cold treatment (**stratification**). These requirements allow seeds to germinate at an appropriate time. At temperate latitudes the cold requirement for dormancy breaking is fulfilled by overwintering; hence seeds will germinate in the spring rather than the autumn. In hot arid regions subject to periodic fires, seeds may need to be cracked open by fire (e.g. in the sumach, *Rhus ovata*) ensuring that seed germination occurs at an optimal time. Smoke alone has been identified as an important factor which triggers germination in many species found

Box 11.7

The terms stratification and vernalization are often misused. Stratification refers to the cold treatment of *seeds* stimulating germination whilst vernalization refers to the cold treatment of *plants* stimulating flowering. The term stratification comes from the commercial practice of spreading seeds between layers (strata) of moist substrate and exposure to low temperatures.

in fire-prone habitats. Following fire, the soil will be enriched with nutrients released in ash and competition from established plants will be reduced. The hard seeds of some desert shrubs are abraded when a sporadic rainfall floods the normally dried-up water courses, and torrents of water roll the seeds along the stream beds (e.g. ironwood, *Olneya*, and paloverde, *Cercidium*).

Some seeds germinate in response to the light environment. Seeds which germinate in response to light are termed **positively photoblastic** (e.g. willowherbs, *Epilobium* spp.; mistletoe, *Viscum album*) whilst those whose germination is inhibited by light are termed **negatively photoblastic** (e.g. many members of the Liliaceae). Others (e.g. birch, *Betula pendula*) are photoperiodic. Environmental factors may regulate the concentrations of ABA and GA, and the sensitivity to these hormones, or potentially act via other targets. In seeds which are light-responsive, phytochrome has been demonstrated to increase the expression of two key enzymes of GA biosynthesis.

In an agricultural environment it is highly advantageous for all the seed sown to germinate at the same time, and crops have been selected for prompt germination. However, many wild plants, including important agricultural weeds (e.g. wild oat, *Avena fatua*), produce seed with multiple dormancies. Only a small proportion of seed will germinate in any given year, thus establishing a persistent seed bank within the soil. This ensures that even if an entire year group is destroyed by unfavourable conditions, seed is available for germination in subsequent years.

11.14.3 Recalcitrant seed and vivipary

Although the production of desiccation-tolerant seed is widespread, members of at least 60 angiosperm families produce seeds which do not exhibit dormancy. These include tropical species such as cocoa (*Theobroma cacao*), wetland species (e.g. many mangroves) and large-seeded trees such as pedunculate (English) oak (*Quercus robur*). Seeds which do not enter a dormant state are often called 'recalcitrant' as viability is quickly lost upon storage.

Recalcitrant seeds germinate immediately after they are shed from the parent plant and seedlings are quickly established. For example, tropical dipterocarps such as *Shorea* exhibit a phenomenon known as 'mast fruiting'. In most years the trees in a given locality do not produce seeds, but once every 5–10 years vast numbers are produced. These quickly germinate and produce a large seedling bank which can survive in deep vegetation shade for many years, developing only when a gap in the canopy arises. It is thought that the production of so many seeds at one time is an adaptation to overwhelm the local population of predators, improving seedling survival. However, the infrequency of seed production and the difficulty in storing such seed severely complicate conservation efforts.

In some species, the embryo grows continuously throughout seed development and germinates whilst still attached to the parent plant.

Box 11.8

The presence of long-lived seed can lead to the reappearance of species assumed to be extinct. The great fen ragwort (*Senecio paludosus*) was thought to be extinct for many years, principally as a result of habitat loss, but a few plants reappeared on a ditch bank in Cambridgeshire in 1972, allowing a re-establishment programme to be started.

This is known as vivipary (live birth) or cryptovivipary (hidden live birth) if the seedling does not emerge from the fruit. Vivipary has been extensively investigated in mangrove. An interesting study by Farnsworth and Farrant (1998) showed that ABA levels were consistently lower in the embryos of viviparous mangrove species compared with that in related species which produce desiccation-tolerant seed, suggesting that reductions in ABA allowed vivipary, similar to that observed in mutants of maize.

11.15 | Germination and the resumption of growth

Germination commences with the uptake of water by the dry seed (**imbibition**) and is complete once the growing embryo emerges from the seed coat. The initial uptake of water is rapid. Macromolecules and structures are rehydrated and assume their functional forms; during this period, low-molecular-weight solutes can be lost from the seed, presumably because the lipids in the dehydrated membranes take a little time to pass from a leaky gel phase to the more impermeable hydrated semi-crystalline phase. Respiratory activity rises dramatically. As the seed coat may impair O_2 diffusion, germinating seeds often undergo anaerobic respiration and accumulate significant quantities of ethanol (Chapter 1) before the testa splits, but then the mitochondria are activated and mitochondrial replication is also stimulated. Protein synthesis commences, initially directed by pre-stored mRNAs, followed by transcription and translation of new mRNA as genes necessary for seedling development become active.

To fuel the growth of the embryo the seed reserves are mobilized, i.e. converted from their insoluble forms to soluble, transportable and/or metabolizable derivatives. The best-studied system is undoubtedly the cereal endosperm. During germination, enzymes such as **amylase** and **maltase** are produced which will break down the starchy endosperm to glucose. These are produced in the aleurone layer which surrounds the endosperm. Experiments with cereals demonstrated that the embryo produces GA, triggering this process. If the embryo is removed from the seed, no starch breakdown occurs even after prolonged periods of incubation. However, if the embryo is placed on the surface of an agar plate in close proximity to the rest of the seed, GA diffuses across the gap and starch degradation results. Application of GA also stimulates starch hydrolysis.

In the cotyledons of oil storing seeds, fatty acids are released from the oil bodies by lipoxygenases. The fatty acids enter the glyoxysomes (small organelles) where successive rounds of β-oxidation generate acetyl CoA, which are in turn used to form succinate in the glyoxylate cycle. This organic acid enters the mitochondria and hence the Krebs cycle. Oxaloacetate from the TCA cycle then acts as a substrate for sucrose synthesis (gluconeogenesis), which fuels further growth, and in due course photosynthetic activity is developed.

At this point we have come full circle. The seedling will now grow vegetatively and eventually enter the reproductive phase once more.

References

Boss, P. K. & Thomas, M. R. (2002). Association of dwarfism and floral induction with a grape 'green revolution' mutation. *Nature*, **416**, 847–50.

Burn, J. E., Bagnall, D. J., Metzger, J. D., Dennis, E. S. & Peacock, W. J. (1993). DNA methylation, vernalization, and the initiation of flowering. *Proceedings of the National Academy of Sciences (USA)*, **90**, 287–91.

Coen, E. (2001). Goethe and the ABC model of flower development. *Comptes Rendus de l'Académie des Sciences, Series III, Sciences de la Vie – Life Sciences*, **324**, 523–30.

Coen, E. S. & Meyerowitz, E. M. (1991) War of the whorls: genetic interactions controlling flower development. *Nature*, **353**, 31–7.

Colasanti, J. & Sundaresan, V. (2000). 'Florigen' enters the molecular age: long-distance signals that cause plants to flower. *Trends in Biochemical Sciences*, **25**, 236–40.

Coupland, G. (1995). Flower development: leafy blooms in aspen. *Nature*, **377**, 482–3.

Cubas, P., Vincent, C. & Coen, E. (1999). An epigenetic mutation responsible for natural variation in floral symmetry. *Nature*, **401**, 157–61.

Cumming, B. G., Hendricks, S. B. & Borthwick, H. A. (1965). Rhythmic flowering responses and phytochrome changes in a selection of *Chenopodium rubrum*. *Canadian Journal of Botany*, **43**, 825–53.

Debeaujon, I. & Koornneef, M. (2000). Gibberellin requirement for *Arabidopsis* seed germination is determined both by testa characteristics and embryonic abscisic acid. *Plant Physiology*, **122**, 415–24.

Farnsworth, E. J. & Farrant, J. M. (1998). Reductions in abscisic acid are linked with viviparous reproduction in mangroves. *American Journal of Botany*, **85**, 760–769.

Finnegan, E. J., Genger, R. K., Kovac, K., Peacock, W. J. & Dennis, E. S. (1998). DNA methylation and the promotion of flowering by vernalization. *Proceedings of the National Academy of Sciences (USA)*, **95**, 5824–9.

Franklin-Tong, V. E. (1999). Signaling and the modulation of pollen tube growth. *The Plant Cell*, **11**, 727–38.

Gillaspy, G., Ben-David, H. & Gruissem, W. (1993). Fruits: a developmental perspective. *The Plant Cell*, **5**, 1439–51.

Goldberg, R. G., Belas, T. P. & Sanders, P. M. (1993). Anther development: basic principles and practical applications. *The Plant Cell*, **5**, 1217–29.

Harborne, J. B. (1994). *Introduction to Ecological Biochemistry*. London: Academic Press.

Honma, T. & Goto, K. (2001). Complexes of MADS-box proteins are sufficient to convert leaves into floral organs. *Nature*, **409**, 525–9.

Mariani, C., De Beuckeleer, M., Truettner, J., Leemans, J. & Goldberg, R. B. (1990). Induction of male sterility in plants by a chimaeric ribonuclease gene. *Nature*, **347**, 737–41.

McClung, C. R. (2001). Circadian rhythms in plants. *Annual Review of Plant Physiology and Plant Molecular Biology*, **52**, 139–62.

McDaniel, C. N. & Poethig, R. S. (1988). Cell-lineage patterns in the shoot apical meristem of the germinating maize embryo. *Planta*, **175**, 13–22.

Molinero-Rosales, N., Jamilena, M., Zurita, S., Gómez, P., Capel, J. & Lozano, R. (1999). *FALSIFLORA*, the tomato orthologue of *FLORICULA* and *LEAFY*, controls flowering time and floral meristem identity. *The Plant Journal*, **20**, 685–93.

Mordhorst, A. P., Voerman, K. J., Hartog, M. V. *et al.* (1998). Somatic embryogenesis in *Arabidopsis thaliana* is facilitated by mutations in genes repressing meristematic cell divisions. *Genetics*, **149**, 549–63.

Mouradov, A., Cremer, F. & Coupland, G. (2002). Control of flowering time: interacting pathways as a basis for diversity. *The Plant Cell*, **14**, S111–30.

Newbigin, E., Anderson, M. A. & Clarke, A. E. (1993). Gametophytic self-incompatibility systems. *The Plant Cell*, **5**, 1315–24.

Ng, M. & Yanofsky, M. F. (2001). Function and evolution of the plant MADS-box gene family. *Nature Reviews Genetics*, **2**, 186–95.

Pelaz, S., Ditta, G. S., Baumann, E., Wisman, E. & Yanofsky, M. F. (2000). B and C floral organ identity functions require *SEPALLATA* MADS-box genes. *Nature*, **405**, 200–3.

Ratcliffe, O. J., Bradley, D. J. & Coen, E. S. (1999). Separation of shoot and floral identity in *Arabidopsis*. *Development*, **126**, 1109–20.

Reiser, L. & Fischer, R. L. (1993). The ovule and the embryo sac. *The Plant Cell*, **5**, 1291–1301.

Theißen, G. (2001). Development of floral organ identity: stories from the MADS house. *Current Opinion in Plant Biology*, **4**, 75–85.

Weigel, D. (1995). The genetics of flower development: from floral induction to ovule morphogenesis. *Annual Review of Genetics*, **29**, 19–39.

Yang, C.-H., Chen, L.-J. & Sung, Z. R. (1995). Genetic regulation of shoot development in *Arabidopsis*: role of the *EMF* genes. *Developmental Biology*, **169**, 421–35.

Chapter 12

Growth movements

12.1 | Introduction

Although it is a general perception that plants do not move very much, or very quickly, this is true only when seen from a human perspective. If we view the world using time-lapse photography we quickly become aware that all plants are, more or less, in continuous motion. This should not come as a surprise when one considers that plants cannot uproot themselves and relocate to a new environment to maintain suitable conditions; they must orientate their organs, largely by growth, to optimize their interactions with the non-uniform environment which surrounds them. We tend to take it for granted that shoots (usually) grow upwards into the air and roots grow down into the ground; leaves spread out and turn to the light; flowers take up specific orientations. All this positioning is the result of differential growth, **growth movements**, in precise and complex responses to environmental stimuli, especially light and gravity. Mutants which lack some of these responses are unable to grow normally; e.g. mutant shoots unable to respond to gravity lie on the ground and in the field would be overgrown and perish. Growth movements, imperceptible as they are to instantaneous observation, are vital to the plant. In addition to the relatively slow growth movements, more rapid, visible movements are exhibited by specialized plant organs.

12.2 | Nastic responses

Nastic responses are movements unrelated to the orientation of the triggering stimulus and include some really rapid movements in response to touch (thigmonasty or seismonasty). When touched, the sensitive plant (*Mimosa pudica*) responds by folding its leaflets together and lowering the whole leaf, all within just a few seconds. This response probably affords some protection against herbivores: when disturbance by an animal causes the leaves to fold up, the plant no longer looks very appetizing. The leaves of the insectivorous

Venus flytrap (*Dionaea muscipula*) respond to touch by the two leaf halves closing together within half a second – a response which is sufficiently rapid to trap a fly (Fig. 12.1). More commonplace, and much slower, are the nastic 'sleep movements' of leaves, shown by many species, which occur throughout the diurnal cycle. Leaves are raised and leaflets unfold by day, whilst by night the leaves are lowered and the leaflets fold up. Numerous flowers open by day and close at night – or vice versa (see van Doorn & van Meeteren 2003 for a review). These movements are under the control of a circadian rhythm (Section 11.3.3) and will persist for several days even when the plant is placed under constant illumination or in darkness.

A common feature of each of these responses is that they result from changes in cell turgor. Leaf movements usually result from the expansion and contraction of cells within the pulvinus – a thickening at the base of the petiole where it joins the stem. Rapid movements of ions and other solutes cause these changes in cell turgor in a manner similar to that in guard cells regulating stomatal aperture. In the Venus flytrap the leaf closes as cells on the outside of the trap expand whilst those on the inside contract, forcing the two halves together rapidly. These responses do not involve growth; there is no cell division, permanent (plastic) cell expansion or differentiation (Chapter 8), which is why these responses can occur so quickly and are fully reversible. Nastic **growth** movements, however, also do occur, e.g. the presence of light (irrespective of direction) can induce a higher growth rate on one side of a leaf, and hence curving.

Although the nastic movements of plants have long fascinated biologists, and demonstrations of thigmonasty are a useful means of convincing sceptical students that plants are indeed capable of rapid

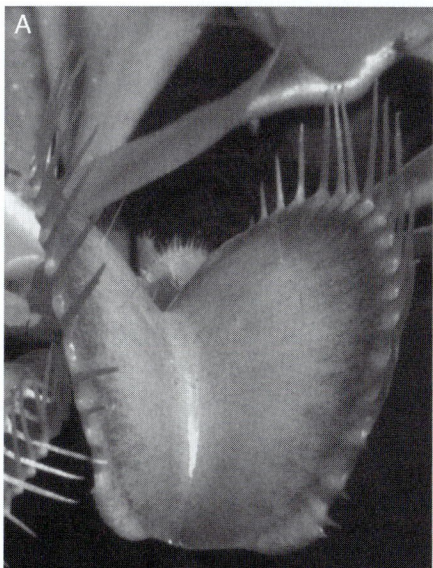
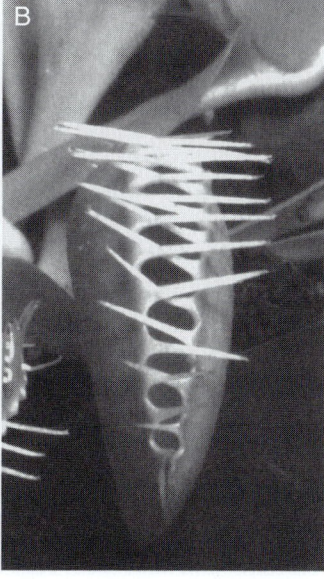

Fig. 12.1 The leaves of the Venus flytrap (*Dionaea muscipula*) close rapidly to trap insects. The inner surface of the open trap (A) contains sensitive trigger hairs which, when stimulated by touch, lead to rapid closure of the leaflets (B), trapping the prey inside.

movement, directional growth responses to directional stimuli are the major means by which plants adapt to a non-uniform environment. These growth responses are called **tropisms** and form the subject of the rest of this chapter.

12.3 | Tropisms

12.3.1 The nature of tropisms

All tropisms result from differential growth rates on opposite sides of the reacting organ. In nearly all cases this growth results from cell expansion – cell division has been implicated in only a few cases. Plants exhibit tropic responses to many environmental stimuli. Responses to light (phototropism) and gravity (gravitropism, formerly geotropism) have been most thoroughly studied, but plants also respond to temperature (thermotropism), touch (thigmotropism), chemicals (chemotropism), the presence of water (hydrotropism) or running water (rheotropism), oxygen (aerotropism) and even electrical fields (galvanotropism). In each case a directional stimulus, whether a light gradient, a gravity field or a temperature gradient, elicits a **directional response**. Plants also exhibit the phenomenon of **autotropism** in that the orientation of organs can be fixed in relation to other organs.

Different organs of the plant respond in different ways to these signals. Figure 12.2 shows developing maize seedlings (*Zea mays*), one of which was reorientated after germination. The coleoptile is **negatively gravitropic** as it grows vertically upwards whilst the root is

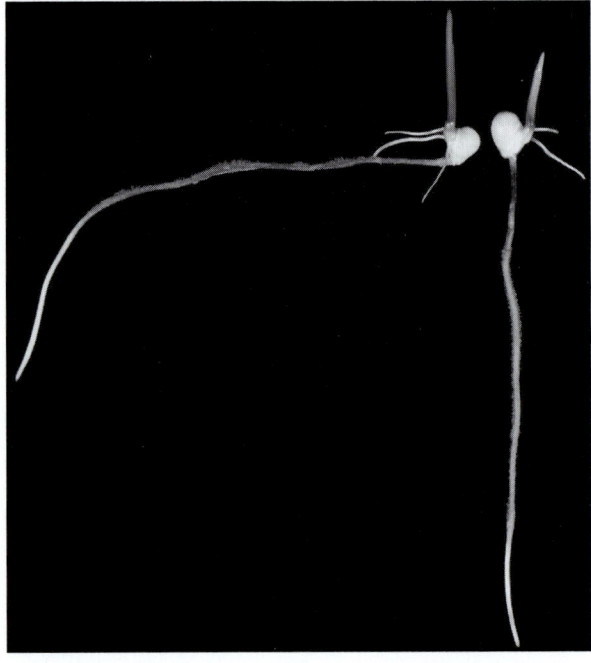

Fig. 12.2 Tropic responses of a maize seedling (*Zea mays*). The negative gravitropic responses of the coleoptile and positive gravitropic response of the root cause these organs to re-orientate themselves parallel to the direction of gravity when a maize seedling is placed horizontally.

positively gravitropic and grows vertically downwards. These same organs show opposite responses to light – the coleoptile will grow towards light (**positive phototropism**) and the root will grow away from light (**negative phototropism**). Our understanding of tropisms therefore has to account for the differing responses of organs to stimuli and must also allow for interactions between stimuli (in this case light and gravity).

This simplistic view of positive or negative responses is not a full representation of how a plant responds to the environment. Leaf orientation is a good example, as leaves must be held at an angle to the sun (and gravity) in order to gather light efficiently. The terms **diageotropic** and **plagiogravitropic** were introduced to describe respectively leaves held at 90° or some other angle to gravity (Fig. 12.3), and **orthotropic** to describe responses directly towards, or away from, the stimulus. It is now recognized that plant organs which respond to gravity do so by maintaining a fixed angle relative to the field known as the **gravitropic setpoint angle** (GSA). This single term allows us to describe the responses of different organs to gravity. The maize coleoptile has a GSA of 180°, the root has a GSA of 0° whilst the leaves have a GSA somewhere between the two. The ways in which a plant's organs respond to gravity are important determinants of its growth habit. Figure 12.4 shows some of the diversity seen in different plant species. Because tropisms result from differential growth rates, only growing organs can show a tropic

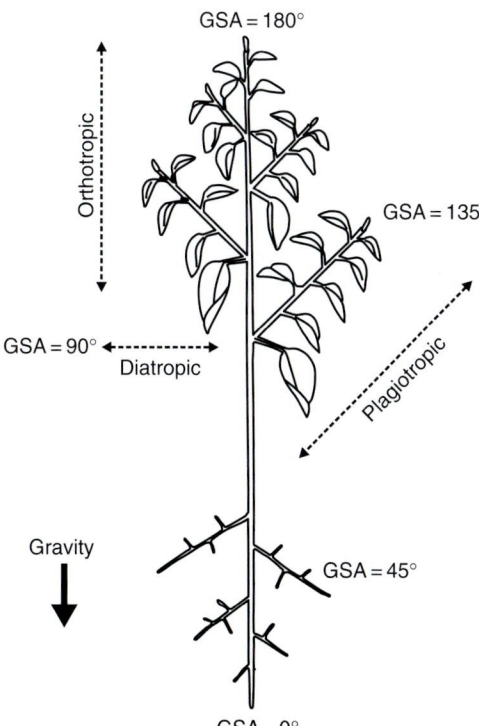

Fig. 12.3 The organs of a plant are held at different angles to gravity. Terms such as diatropic, plagiotropic and orthotropic have been used to describe the orientation of organs in relation to a directional stimulus – in this case gravity. More recently, the term gravitropic setpoint angle (GSA) has been introduced to describe the orientation relative to gravity more precisely.

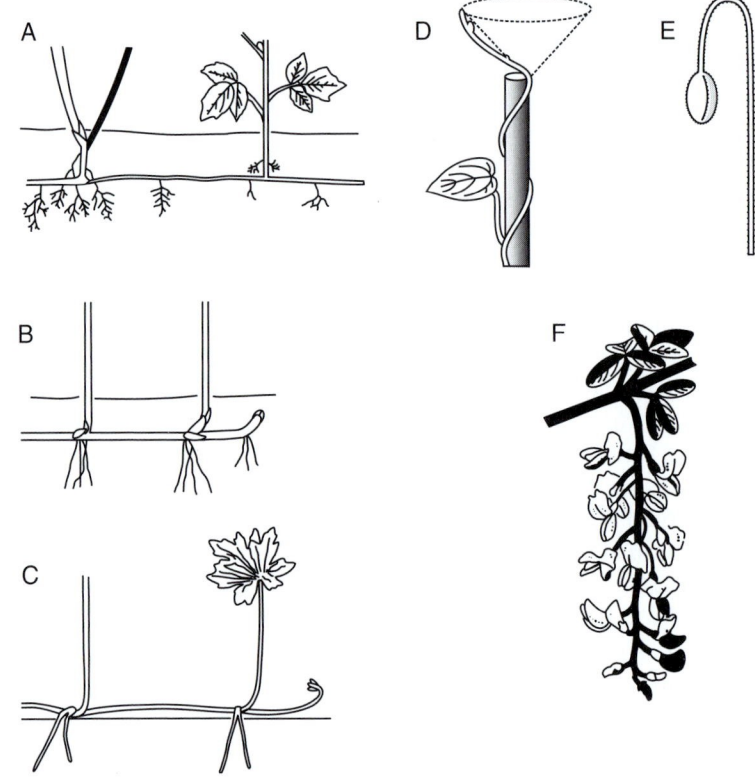

Fig. 12.4 Variations in growth habit controlled by gravitropic responses. (A) Diagravitropic root (GSA = 90°) of *Rubus idaeus*, raspberry; (B) diagravitropic rhizome of *Eleocharis*, spike rush; (C) diagravitropic runners of *Ranunculus repens*, creeping buttercup; (D) twining shoot of *Ipomoea purpurea*, morning glory, showing strong circumnutation and lateral gravitropism; (E) young peduncle of *Papaver*, poppy, showing positive gravitropism (GSA = 0°); (F) inflorescence of *Laburnum anagyroides* – the axis is non-gravitropic, and the inflorescence hangs down under its own weight. Redrawn from Larsen (1962).

growth reaction. During development, the sensitivity of an organ towards a particular stimulus increases to a maximum, then decreases and finally is lost when the organ reaches maturity and ceases to grow.

The interactions which occur between tropisms are of supreme importance in maintaining the most efficient orientation of plant organs. Leaf orientation, for instance, is controlled by a combination of tropic (or nastic) responses towards light and gravity, in conjunction with an autonomous tendency for unequal growth on the upper and lower sides. Light modifies the reactivity of the plant towards gravity. The mature nodes of *Tradescantia* grow upwards in darkness (GSA = 180°) but downwards under moderate irradiances (GSA = 0–90°). This modification of the gravitropic response by light is dependent upon photosynthesis as it is largely abolished in plants treated with photosynthetic inhibitors. In other species, e.g. tomato (*Lycopersicon esculentum*), both photosynthesis and phytochrome interact with GSA (Digby & Firn, 2002). Many plants produce runners which grow parallel to the soil surface (i.e. they are plagiogravitropic and exhibit a GSA of 90°), either on the soil surface or just below it. The stolons of *Circaea* and the runners of *Fragaria*, which normally grow close to the soil surface, turn upwards as if negatively gravitropic when the plants are kept in the dark. The rhizomes of many plants grow horizontally at a more or less fixed level below the soil surface and, if the level is disturbed, they will grow up or down at an angle until they reach their normal depth before resuming horizontal growth. The horizontal growth is generally regarded as a gravitropic response but it is not clear how the constant depth is maintained. Rhizomes of *Aegopodium podagraria* react to light by a positive gravitropism which causes the rhizomes to grow downwards if they come too near to the surface. What prevents growth to excessive depth is not clear but increasing CO_2 concentration at greater depths has been implicated; *Aegopodium* rhizomes respond to elevated CO_2 by turning upwards. For *Polygonatum multiflorum* rhizomes the control has been suggested to reside in a balance between negative gravitropism and photoepinasty – the rhizomes growing below the surface where the irradiance is just sufficient to induce a nastic reaction balancing the inherent negative gravitropism.

The sign (positive or negative) of the tropic response of an organ may change during its growth; this happens frequently in reproductive organs, where the flower bud, the open flower and the fruit may each have a characteristic orientation owing to a different gravitropic response of the pedicel (Fig. 12.5). The flower stalk of the peanut (*Arachis hypogaea*) (the gynophore) is at first negatively gravitropic, but after fertilization it becomes positively gravitropic and buries the fruit in the soil (see also Fig. 7.8). The phototropic reaction of the flower stalk of the ivy-leaved toadflax (*Cymbalaria muralis*) changes from positive to negative after fertilization; this results in fruits being pushed into crevices of the rocks or walls on which the plant grows.

Tropisms involve good examples of signal transduction pathways. An environmental signal is perceived by one or more receptors and generates a mobile signal within the plant which elicits a growth response. Sometimes the perceptive and receptive regions are separated by an appreciable distance. Undoubtedly auxin has an important role in these responses since the stimuli induce auxin concentration gradients in the growing regions, and, as we have seen in Chapter 7, this hormone is an important regulator of plant growth. However, many questions remain to be answered:

- What are the receptors for light and gravity (and other) signals? How can these signals interact?

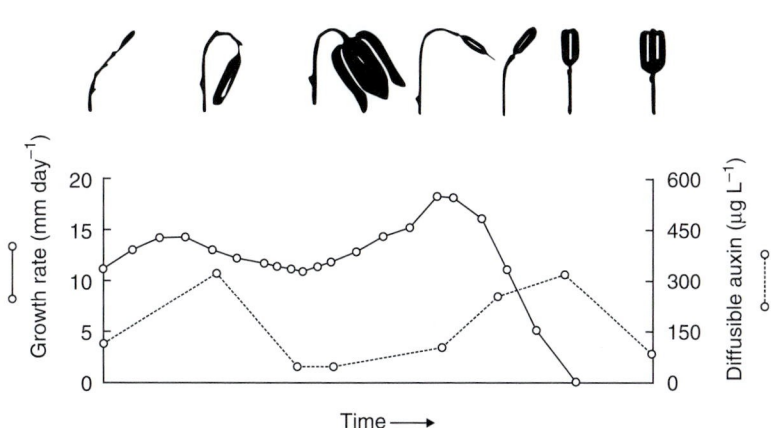

Fig. 12.5 Stages in the development of the flower of *Fritillaria meleagris*, showing the reversal of the gravitropic response of the flower stalk during the development of the flower and fruit. The continuous line shows the growth rate of the flower stalk in mm per day. The lower graph shows the amount of auxin diffusing from the flower and through the flower stalk. From Kaldewey (1957).

- How does the stimulation of these receptors lead to an asymmetrical distribution of auxin within the organ?
- What other factors, in addition to auxin, are involved in these responses?

We have answers to some of these questions but by no means all. In addition, some of the key experiments underlying well-established models of tropic responses have recently been questioned, so an element of caution is required. Inevitably, the study of tropic responses has been greatly aided by the isolation and characterization of plants with mutations in these responses, and these have helped to support or refute many theories. Likewise, modern technology, whether the availability of whole genome sequences or the use of space rockets for studying gravitropic responses, would have been unimaginable to early investigators of tropisms such as Charles Darwin, although his famous voyage on the *Beagle* was, in its time, probably just as daunting as a trip on the space shuttle is now.

12.3.2 The Cholodny–Went model

Early studies

Pioneer studies on plant responses to light and gravity led to the discovery of plant hormones. Charles Darwin, in 1880, reported upon his studies of the positive phototropic curvature of grass coleoptiles and of the hypocotyls of dicotyledonous seedlings and, finding that the bending failed to occur when the extreme tips of the organs were shaded by metal foil 'hats', concluded that some stimulus must be transferred from the tip to the curving region. Rothert (1894) extended these observations to a wider range of plant material. Then came the studies of Fitting (1907), Boysen Jensen (1910–1911) and Páal (1914–1919), which established that when the tip of a coleoptile is cut off and placed back in position the stimulus is not interrupted but passes across a moist cut surface or a gelatin layer. The stimulus failed to be transmitted when a thin sheet of mica, platinum foil or cocoa butter was placed between the tip and the responding part of the organ. Such observations indicated that the transmitted signal was most probably chemical (now known to be auxin).

If coleoptile tips were removed and then replaced asymmetrically on the stumps curvature occurred without unilateral illumination (Páal, 1918). The straight growth of *Avena* coleoptiles decreased severely or stopped when they were decapitated and could be restored by placing the tips back in position (Söding, 1925; Went, 1928; Fig. 12.6). The stimulus involved in the phototropic reaction was thus apparently also active in controlling normal growth. Parallel observations were made in the study of gravitropism, namely that the tip of a stimulated organ produced a growth-controlling stimulus which, under the influence of gravity, was transmitted along the lower side.

A Unilateral illumination results
in a positive phototropic curvature

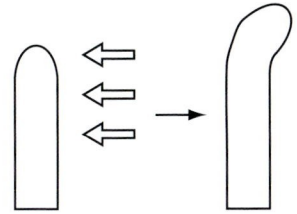

This response is abolished if
the tip of the coleoptile is shaded

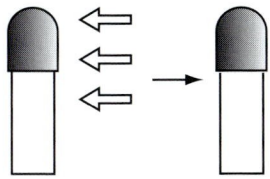

The signal can be transmitted
through a block of gelatin or agar

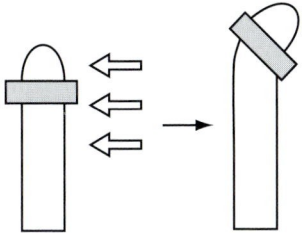

but not through a mica
or platinum sheet

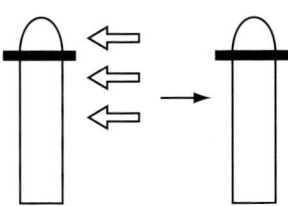

An asymmetrically placed agar
block containing auxin will cause
curvature in darkness

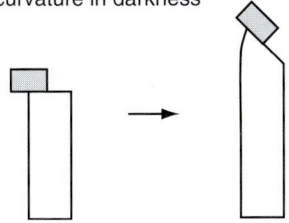

B When a coleoptile tip is unilaterally illuminated auxin is diverted to the
shaded side, but the total released remains unaltered.

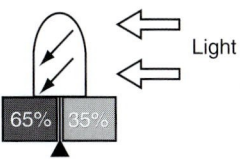

Gravity also results in an unequal distribution of auxin, in this case
to the lower side.

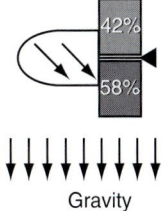

Unilateral illumination of the hypocotyl results in diversion of
auxin to the shaded side.

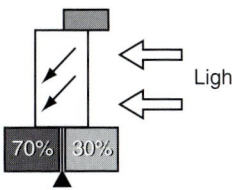

Fig. 12.6 A summary of the experiments leading to the
Cholodny–Went model of tropic responses.

Clear demonstration that a chemical capable of controlling growth was transmitted from the tip was achieved in 1928 by Went. He demonstrated that the chemical agent could be collected in agar blocks by diffusion from the cut surfaces of coleoptile tips, and that when these blocks were placed asymmetrically on decapitated coleoptiles, they caused curvature away from the side receiving the chemical messenger (Fig. 12.6). Further, by collecting the hormone released from the illuminated and shaded sides of a coleoptile tip exposed to directional light into separate agar blocks, Went was able to show that the total amount of hormone released from the tip was not altered significantly but that more hormone passed to the shaded side than the lit side.

These studies led to the development of the Cholodny–Went theory of tropisms. Although these researchers worked independently, their names have become irrevocably linked. In 1937, Went and Thimann defined the Cholodny–Went model as follows:

> Growth curvatures, whether induced by internal or external factors, are due to unequal distributions of auxin between the two sides of a curving organ. In the tropisms induced by light and gravity, the unequal auxin distribution is brought about by a transverse polarization of the cells which result in the lateral transport of auxin.

This definition omits to mention the role of the tip (root or shoot) in tropisms and, indeed, there are many examples of phototropic and gravitropic responses which occur in regions other than the tip. However, the omission most probably arose from the assumption that the importance of the role of the tip was obvious.

The model seeks to explain the tropic responses of both shoots and roots. Some aspects of this model remain controversial but it is worth examining its basic points as a starting point for further discussion. Recent studies, which have examined the molecular basis of auxin redistribution in roots undergoing gravitropism, have further supported the model.

Phototropic responses of the shoot

Auxin, synthesized in the shoot tip, moves basipetally via an active transport system. In darkness, or under uniform illumination, auxin is equally distributed on either side of the organ and stimulates growth equally on both sides (Fig. 12.7A). Upon directional illumination there is a redistribution of auxin towards the shaded side. Increased auxin concentrations stimulate growth on the shaded side whilst the reduced auxin concentrations on the illuminated side inhibit growth, forcing the organ to grow towards the light source. The positive effect of auxin on shoot growth causes the directional growth response.

Gravitropic responses of the root

Roots synthesize little auxin, rather it is delivered to the root cap via the vascular system. The root cap transports auxin laterally and it

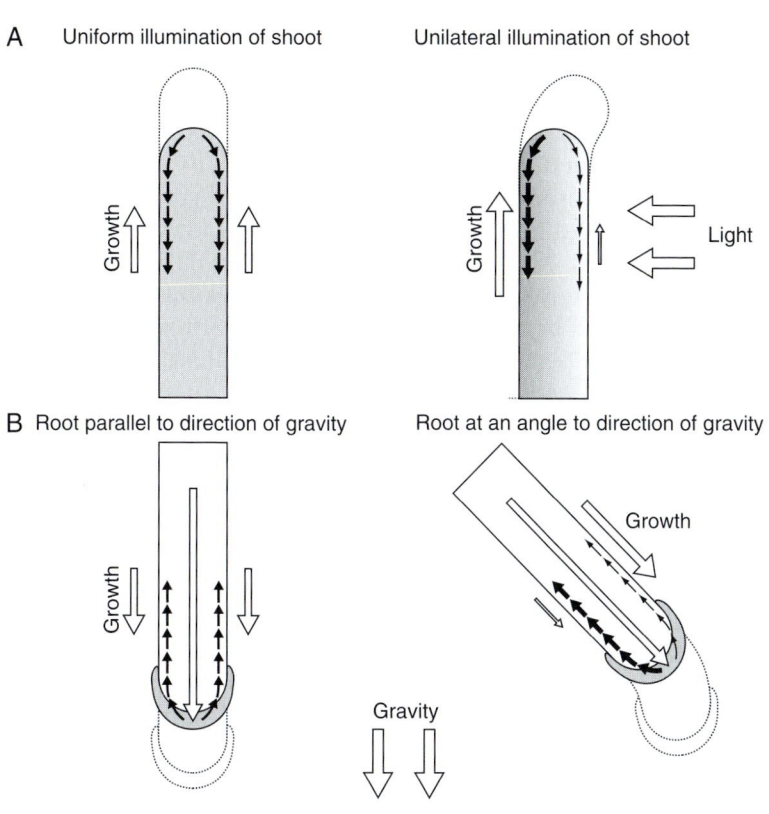

A Uniform illumination of shoot   Unilateral illumination of shoot

Growth

Growth

Light

B Root parallel to direction of gravity   Root at an angle to direction of gravity

Growth

Growth

Growth

Gravity

Fig. 12.7 Asymmetrical auxin gradients are thought to lead to tropic curvature. (A) In a shoot (or coleoptile) illuminated from above, auxin is transported from the tip and distributed uniformly. The growth rate of both sides of the shoot is the same and therefore it grows straight upwards. Illumination from one side leads to a redistribution of auxin within the shoot. Growth is stimulated on the shaded side and inhibited on the illuminated side. (B) In the root, auxin is redistributed in the root cap. When the root is placed at an angle to gravity, more auxin is transported to the lower side and less to the upper side. As auxin inhibits root growth, the root bends downwards.

then moves back up the root in the epidermal layers. Removal of the root cap abolishes gravitropic responses and stimulates root growth. Upon gravistimulation, an asymmetrical distribution of auxin develops. More auxin is delivered to the lower side, where it inhibits growth, whilst the reduced auxin content of the upper side stimulates growth. This differential growth causes the root to bend downwards (Fig. 12.7B). In this case the negative effect of auxin on root growth causes the directional growth response. The actual responses observed within roots are often more complex than a simple stimulation of growth on one side and inhibition on the other. Different parts of the root may respond at different times and the rate of elongation varies in a complex manner.

Predictions and problems
The simplicity of this model, and its agreement with many experimental observations, explains its durability. However, the strength of any model lies in its ability to generate testable hypotheses and, in the case of the Cholodny–Went model, not all of the predictions have been fulfilled. This has led to the model undergoing numerous modifications (although often retaining the same name). Some of these are discussed below but the reader is referred to Firn *et al.* (2000) for an excellent account of the development of the model and controversies and Yamamoto (2003) for a recent update.

DO AUXIN GRADIENTS EXIST WITHIN DIRECTIONALLY STIMULATED ORGANS?

The work of Went showed that a growth-regulating activity was asymmetrically distributed. However, auxin gradients have not always been found within gravistimulated organs and Hasegawa *et al.* (1989) proposed that a local accumulation of growth inhibitors on the illuminated side, rather than IAA, was responsible for phototropic curvatures in oat *(Avena sativa)*. Part of the difficulty in these experiments is measuring auxin concentrations on such a small scale. Went used bioassays (Section 7.2.2); modern analytical techniques greatly aid such studies, but highly localized and dynamic responses are still difficult to measure.

One useful approach has been to use transgenic plants which have been genetically modified so that they indicate where auxin is accumulating. McClure and Guilfoyle (1987) identified a series of soybean *(Glycine max (soja))* genes whose expression is rapidly up-regulated in response to auxin. If the promoter from one of these auxin-responsive genes is isolated and fused to a reporter gene such as β-glucuronidase (GUS), plants containing this construct will produce GUS activity in regions responding to auxin (Li *et al.* 1991). These cells can then be stained blue by incubating the tissue with X-glucuronide. The actively growing regions of transgenic tobacco plants *(Nicotiana tabacum)* stain blue, indicating the role of auxin in uniform plant growth. If the tobacco plants are laid on their side, GUS activity accumulates on the underside of the stems as expected and decreases on the upper side.

Recently this approach has been extended to produce a sensitive *in vivo* biosensor of auxin responsiveness (Ottenschläger *et al.* 2003). There is always the possibility that an auxin-responsive promoter is regulated by factors other than auxin. Therefore a simple synthetic promoter (called DR5) has been produced which contains auxin-responsive elements only, minimizing this problem. Also, the reporter gene GUS has been replaced by green fluorescent protein (GFP: Section 10.7.2) which can be visualized in living tissue by fluorescence microscopy without the need to kill and stain the plant tissue. Figure 12.8 shows that auxin responsiveness (and presumably, therefore, auxin itself) is uniformly distributed in the lateral root cap of vertically orientated *Arabidopsis* roots, but that an asymmetrical distribution occurs within a few hours of gravistimulation. This asymmetry can be generated by just 15 minutes of gravistimulation, but it takes a few hours for the GFP to be synthesized – a limitation of this technique. Another limitation is that the concentration of auxin must exceed a threshold before GFP expression can be detected, but other physiological responses may be triggered at lower concentrations. Yamamoto (2003) reports that the DR5 promoter is responsive to auxin concentrations in the range 0.3–9 μM (depending on whether the auxin is assumed to be uniformly distributed or just in those cells expressing the reporter gene) and that a

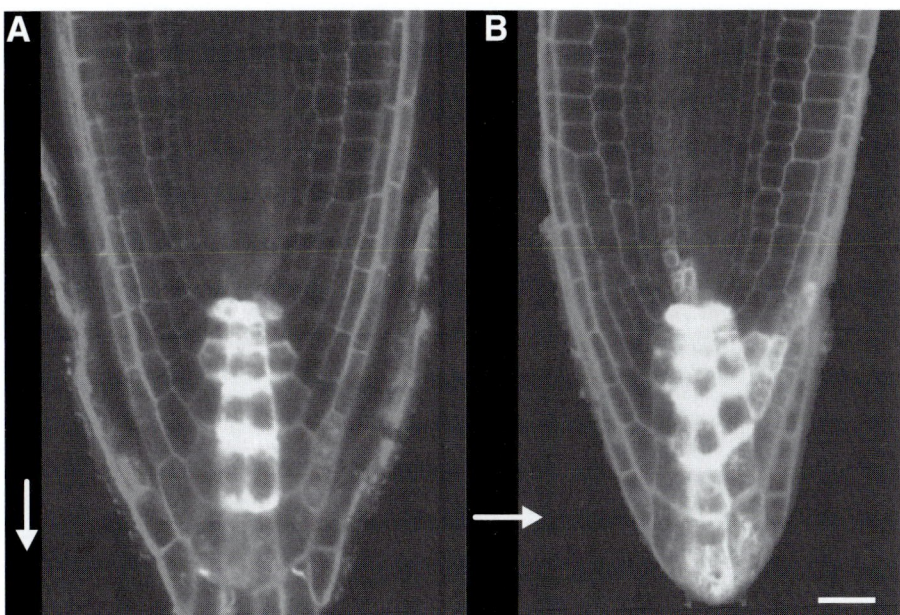

Fig. 12.8 Expression of the auxin-responsive gene DR5-GFPm in the lateral root cap of *Arabidopsis*. Transgenic *Arabidopsis* plants containing a synthetic auxin-responsive promoter, DR5, fused to green fluorescent protein (GFP), fluoresce green in response to auxin. The fluorescence has been imaged using confocal laser scanning microscopy and is shown as white. The cell walls were stained with propidium iodide and are shown as light grey. (A) GFP expression is restricted to the quiescent centre, columella initial and columella of roots grown in a vertical orientation. (B) After 3 h gravistimulation the entire lower half of the lateral root cap expresses GFP. Arrow = gravity vector; scale bar = 20 μm. Adapted from Ottenschläger *et al.* (2003). © 2003 National Academy of Sciences, USA.

sensitivity as much as 10 times greater may be required for studies of phototropic responses. Despite these limitations, the use of auxin-responsive reporter genes has proven invaluable in studying tropic responses.

DOES AUXIN MOVE ACROSS THE ORGAN?

As early as 1900 Copeland demonstrated that a bisected hypocotyl can still show a tropic response (see Firn *et al.* 2000), hence any transverse movement of auxin was likely to be local. Auxin transport inhibitors such as naphthylphthlamic acid (NPA) abolish auxin gradient formation and abolish gravitropic responses, strongly supporting the notion of auxin transport as an essential prerequisite for tropic growth responses. However, a number of research groups have shown that tropic responses are still seen in organs which have been treated with high concentrations of exogenous auxin, sufficient to swamp any internal gradient. However, these concerns have recently been addressed. Auxins, and auxin transport inhibitors, differ in their ability to penetrate cells. Ottenschläger *et al.* (2003) demonstrated that there is a good correlation between the treatments which abolish the formation of an asymmetrical auxin concentration gradient and the loss of gravitropic responsiveness.

Although our understanding of tropic curvatures is far from complete, great progress has been made using a mutational approach. Mutants in many species have been identified which lack one or more components of tropic responses, and the study of these mutants has led to the identification of many components of the signal transduction chain. For example, '*lazy*' and diageotropic mutants have been identified in many plant species including tomato, maize and several other cereals. Shoots of *lazy* mutants of maize and tomato grow initially upwards, but after a few days start to grow actively downwards. In contrast the shoot of the *diageotropica* mutant of tomato grows parallel to the soil surface, i.e. its GSA has changed from 180° to 90°. As well as searching for plants with mutations in tropic responses, great use has been made of plants with lesions in their responses to plant hormones – especially auxin. In many cases, plants which have been isolated in screens designed to look for altered tropic responses have independently been isolated in screens looking for altered hormone responses, emphasizing the importance of hormones in tropisms.

12.3.3 Mutants and tropisms

A complete description of all the tropic-response mutants identified to date would fill many pages, so only a few of the best characterized are described here. Many of these mutants have been best studied in *Arabidopsis*, although similar mutations have been described in numerous other species.

In a search for plants with mutations in root gravitropism, Bell and Maher (1990) isolated the *agr1* (altered gravitropism) mutant. This lacks root gravitropism but other gravitropic responses (those of the hypocotyl and inflorescence) are unaffected. The *AGR* gene has been isolated and found to encode a membrane transport protein. Mutations in this same gene have been found independently by groups examining polar auxin transport (*Atpin2*), root growth (*wav6–2*) and root responses to ethylene (*eir1*). This strengthens the suggestion that auxin transport is an important component of the root gravitropic response. Likewise, the *aux1* mutant, the roots of which can grow in inhibitory concentrations of the synthetic auxin 2,4-D, shows no gravitropism. The *AUX1* gene encodes a permease-like membrane transport protein, allowing auxin to *enter* cells, and is expressed in the root tip and epidermal cells of the elongation zone, but not in the root cap – the site of gravity perception.

If plants are treated with inhibitors of polar auxin transport, they develop a series of structural abnormalities including a 'pin-formed' naked inflorescence. A mutant of *Arabidopsis*, *pin1*, was identified which exhibited this phenotype in the absence of inhibitors and was found to be defective in polar auxin transport (Okada *et al.* 1991). The *PIN1* gene encodes an auxin efflux carrier protein required for the transport of auxin across the plasma membrane and *out of* the cell. It is now known that *Arabidopsis* contains a family of related auxin efflux carrier proteins. Recently, Friml *et al.* (2002) showed that mutations in one of these, AtPIN3, resulted in plants with reduced phototropic and gravitropic responses. The AtPIN3 protein is expressed in gravity-sensing cells, including the columella. In vertically growing roots the protein is symmetrically distributed within the columella cells. However, within a few minutes of gravistimulation the distribution becomes asymmetrical as the protein accumulates on the lateral plasma membrane of the columella cells. With continued stimulation this region of asymmetry expands to include the lateral root cap. This makes AtPIN3 an attractive candidate for the generation of auxin gradients within gravity-sensing cells.

It is now possible to combine the results from studies of the auxin efflux, influx and response to form a model, consistent with the Cholodny–Went theory, of the gravitropic responses of roots. Auxin is delivered to the columella, where it is distributed in the root cap by the PIN efflux carrier. Upon gravistimulation the efflux carrier is redistributed, generating an asymmetrical auxin gradient within the root. Auxin is taken up by the AUX1 influx carrier and delivered to the cells in the elongation zone. Inhibition of cell elongation by auxin causes the downward curvature of the root.

12.3.4 Gravity perception

Our discussion so far has focused on how the response to gravity signals occurs, rather than the means by which the signal is perceived. The majority of gravitropic studies have used seedling primary roots, seedling shoots or cereal coleoptiles, as these form amenable experimental systems.

In roots, gravity perception is confined to the root tip and, in many species, the root cap has been identified as the site of the gravity sensor. For example, removal of the root cap abolishes the root gravitropic responses of maize, pea (*Pisum sativum*), cress (*Lepidium sativum*) and lentil (*Lens culinaris*). A carefully focused laser beam can be used to destroy individual cells and such **laser ablation** studies have been used to demonstrate that *Arabidopsis* roots lose much of their gravitropic response when the central (but not lateral) cells of the columella root cap are destroyed. In shoots, perception is not confined to the tip. Decapitation of cereal coleoptiles or dicotyledonous shoots does not abolish their reactivity although their responses are often slowed.

Continuous stimulation is not required to elicit a gravitropic response. After a period of stimulation, curvature will occur even if the test object is subsequently placed on a klinostat – a slowly rotating table which results in the average gravitational force experienced in the plant in any one direction being zero. The minimum time during which the stimulus must be allowed to act to lead to subsequent curvature is known as the **presentation** or **perception** time. Under natural gravity and at room temperature this can vary from 10–30 s to 25 min. There is then a delay, the **latent** or **reaction** time, before a visible response is observed, although many different processes will be activated within the organ during this period.

The intensity of the gravitropic stimulation can be varied in a number of ways. If material is placed at different angles to the vertical, the force acting on it is then $g \sin \alpha$, where g is the force of

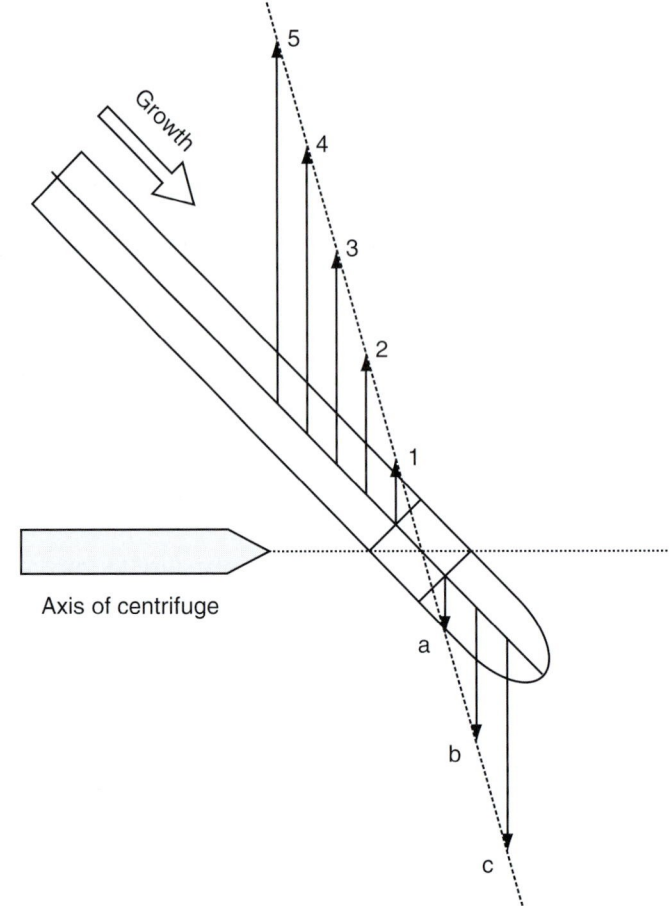

Fig. 12.9 Centrifugal stimulation of the root tip and base in opposite directions. The arrows a to c show the direction of the forces acting on the tip which projects beyond the centrifuge axis; the arrows 1 to 5 represent forces acting on the basal part; the lengths of the arrows are proportional to the force. If 1.5–2 mm of the tip is allowed to project beyond the axis, the root behaves as if the tip alone were stimulated, even though the forces acting on the base in the opposite direction are much stronger. From Larsen (1962).

gravity and α is the angle relative to the vertical (Fig. 12.9). Another means of controlling the gravity perceived by the plant is to rotate it. As described above, slow rotation on a klinostat has the effect of cancelling out the directional component of the gravitational force, providing that the rate of rotation is sufficiently slow to avoid centrifugal effects. Rotation at higher speed, in a centrifuge, allows the gravitational force experienced by the plant to be readily increased. A true reduction in the force of gravity is somewhat more challenging and requires flights on rockets on parabolic trajectories or into space. Such studies have shown that gravitropism exhibits **reciprocity**, i.e. the response observed is the product of the time of stimulation × the intensity of stimulation: a long period of weak stimulation causes the same response as a shorter period of more intense stimulation. For the oat coleoptile this relationship holds true for forces between 58.4 and $0.0014 \times g$, requiring exposure of 5 s or 68 h respectively. The threshold for the oat coleoptile is therefore approximately $300 \times g$ s at room temperature. Once this threshold has been exceeded the magnitude of the response generally increases with the intensity of the stimulus until a maximum is reached.

The role of statoliths

The primary perception of gravity must involve the movement of some entity in the sensitive cells – this is the only way in which the physical force can interact with the cell. In 1900, Haberlandt and Neměc independently put forward the statolith theory of gravitropism. This supposes that in the perceptive cells (the **statocytes**), mobile starch grains (the **statoliths**) move under gravity to lie in the lower parts of the cell. The cells have therefore acquired an 'up-and-downness', i.e. a polarization, which acts as a signal to trigger the response.

Statocytes are found within the gravity-sensing regions of the plant – the columella of the root cap and the starch sheath parenchyma cells that surround the vascular tissue of shoots and grass pulvini (Fig. 12.10) – and are absent from non-gravitropic organs. The starch which forms the statoliths differs from starch found within other tissues as it persists even under extreme starvation. There is some evidence that crystals of calcium oxalate may function as statoliths in some plants. Also the rhizoids of characean algae contain statoliths enriched in barium sulphate.

As well as their location in gravity-sensing regions, other evidence supports the role of statoliths in gravity perception. There is a good correlation between the density of the starch grains and the gravitropic response of the organ. Physiological treatments, such as

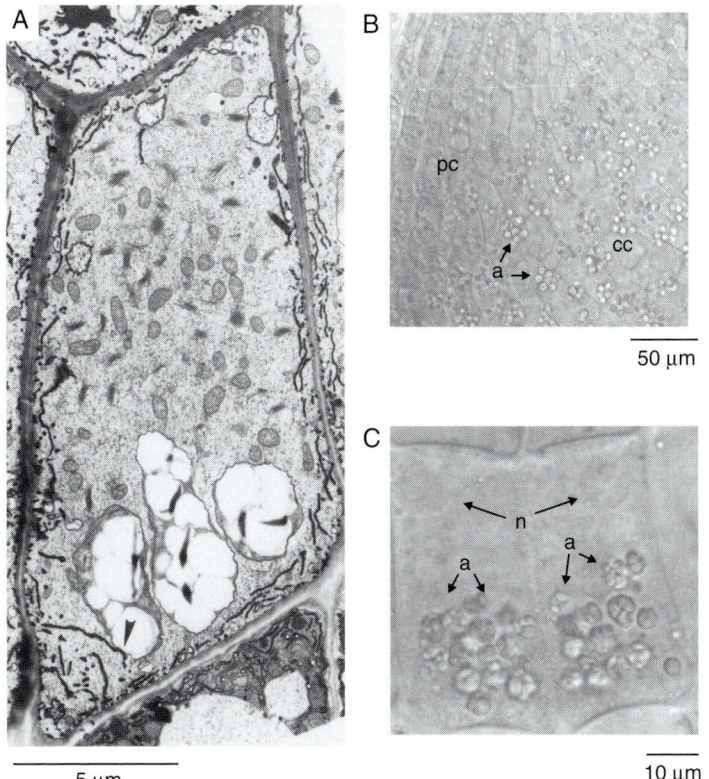

5 μm

50 μm

10 μm

Fig. 12.10 Statoliths and statocytes in root caps. (A) Electron micrograph of a central columella cell of *Arabidopsis*. The statoliths are white and have sedimented at the bottom of the cell. The endoplasmic reticulum has stained darkly and is located at the periphery of the cell. A point of contact between an amyloplast and the ER has been indicated with an arrowhead. (B) and (C) are brightfield images of the root cap of *Medicago truncatula* and *Linum usitatissimum*. (B) The central columella, cc, cells of the root cap contain many amyloplasts, a, whilst far fewer are found in cells of the peripheral root cap, pc. (C) At higher magnification the amyloplasts of the central columella cells are clearly visible and located at the base of the cell. The position of the nucleus, n, is indicated and is typically at the apex of these cells. (A) is from Sack (1997). (B) and (C) are from Collings *et al.* (2001). © Springer-Verlag.

exposure to low temperatures or treatment with gibberellins, which cause the statoliths to be lost, lead to a loss of gravitropism. Responsiveness is restored when the cells are allowed to recover their starch. The *lazy1* mutant of tomato lacks starch and is agravitropic. Treatment of this mutant with auxin restores both starch formation and gravitropism. The *shoot gravitropism 1 (sgr1)* mutant of *Arabidopsis* has agravitropic hypocotyls and these lack the endodermis containing sedimentable starch.

At one time the role of statoliths in gravity perception was challenged as it was reported that mutants of *Arabidopsis* which lacked starch, and hence lacked statoliths, were still able to show gravitropic responses. However, when these experiments were repeated under non-saturating gravitational conditions it was found that these plants were in fact less responsive. A careful examination of intermediate-starch mutants of *Arabidopsis* (Kiss *et al.* 1996) demonstrated that there was a good correlation between statolith density and graviresponsiveness, although the gravity-sensing mechanism appears to be somewhat 'over-built' in that the response of mutants with 50–60% starch is more similar to that of wild-type plants than to that of starchless mutants.

Of course, moving statoliths directly would provide excellent evidence for, or against, their role in gravity perception, but such experiments are somewhat challenging technically! Kuznetsov and Hasenstein (1996, 1997) exposed the root tips of *Arabidopsis* and barley (*Hordeum vulgare*) coleoptiles to extremely strong magnetic gradients which were sufficient to move the statoliths. In root tips this treatment led to the development of curvature in the direction of statolith displacement (as expected for a positively gravitropic organ) whilst the opposite was observed in the coleoptiles and hypocotyls (as expected for negatively gravitropic organs). In the characean alga *Chara*, which contains statoliths composed of $BaSO_4$, statoliths could be displaced using optical tweezers. This precise arrangement of laser beams uses the tiny, but perceptible, forces generated by the highly focused light to hold and move objects within cells (Leitz *et al.* 1995). Such experiments confirmed the role of statoliths in gravity perception in *Chara* rhizoids (Fig. 12.11).

Fig. 12.11 Optical tweezers can be used to move statoliths in *Chara* rhizoids. Before the experiment the statoliths are visible in the middle of the rhizoid. Switching on the laser causes the statoliths to be displaced. From Leitz *et al.* (1995). © Springer-Verlag, 1995.

10 µm

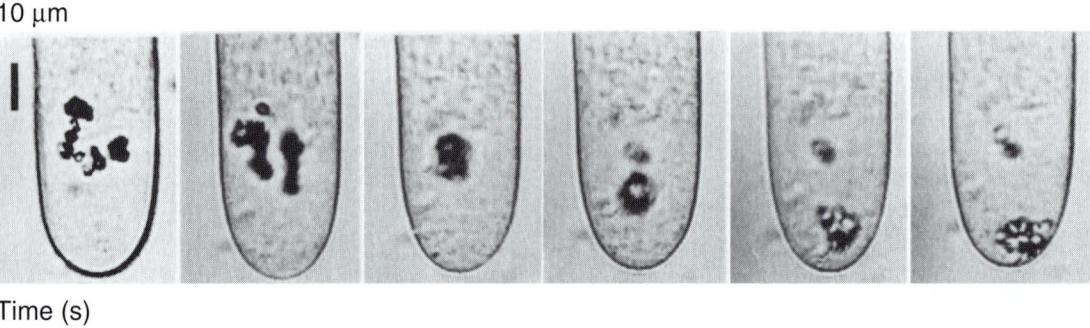

Time (s)

| 0 | 4 | 23 | 25 | 26 | 32 |

If the sedimentation of statoliths is the first step in gravitropism, how does this movement trigger a response? One theory suggests that the endoplasmic reticulum (ER) of the statocytes is an important component of the gravity perception pathway. The statocytes contain layers of ER arranged in a precise way near their distal cell walls. The nuclei of these cells tend to be located at the proximal end, presumably so as not to interfere with statolith sedimentation. The shape of these cells, and the arrangement of the ER, means that in a vertically orientated root the statoliths press equally against the ER on both sides of the root axis. When the root is reorientated so that it is horizontal the statoliths slide off the ER in the uppermost cells, and relieve the pressure upon it, but there is little effect on the lower cells (Fig. 12.12). This idea is supported by studies on a non-gravitropic mutant of pea (*Pisum sativum ageotropa*) in which the distal ER complex is missing in the root-cap statocytes. Other theories have proposed that the plasma membrane, rather than the ER, senses the movement of the statoliths. Of course, the statoliths do not need to contact the membranes directly to exert pressure upon it. The

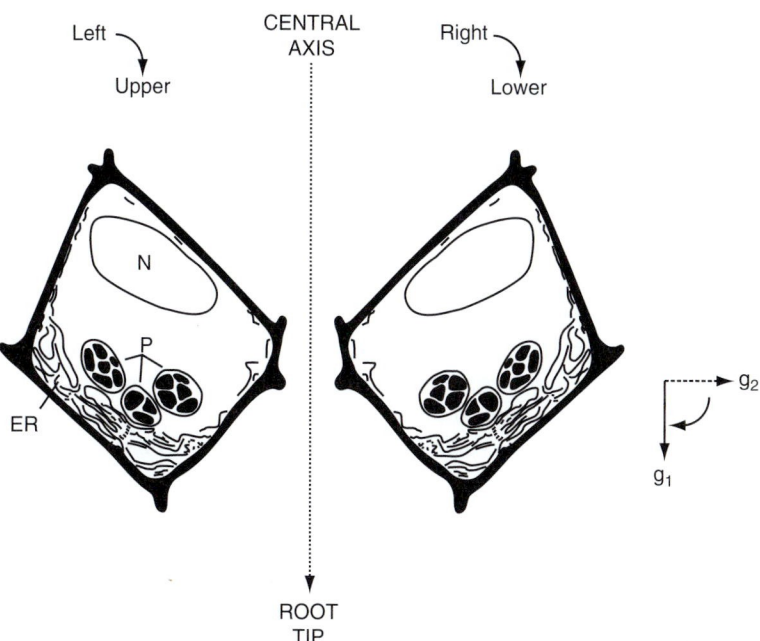

Fig. 12.12 Reorientation of a *Lepidium sativum* root. Diagrammatic representation, based on tracings from electron micrographs, of two cells on either side of the central axis of a *Lepidium sativum* root cap. When the root is orientated vertically (gravity acting in the direction g_1), the amyloplasts, P, press equally on the endoplasmic reticulum, ER, in the two cells. If the root is rotated 90° clockwise so that it is now horizontal (gravity acting in the direction g_2), the cell which was on the left of the root axis is now uppermost, whilst the cell to the right is lower. Sedimentation of the amyloplasts will greatly relieve pressure in the cell remaining uppermost, but in the lower cell the amyloplasts move less and still remain largely in contact with the ER. N = nucleus. From Sievers & Volkmann (1972).

interior of the cell is not a simple, fluid-filled sac but contains many cytoskeletal components such as microtubules and microfilaments. These may link the statoliths to membranes transmitting pressure changes in this manner. Collings *et al.* (2001) demonstrated that there are associations between actin filaments and statoliths in the columella cells of several species.

The molecules involved in the perception of statolith movement remain to be identified but membranes, whether plasma membranes or the ER, contain many ion channels, some of which are stretch-activated. Stretching a membrane alters the electrical gradient across it and many such channels sense, and respond to, such electrical signals. Inhibitors of stretch-activated channels, such as the cations gadolinium (Gd^{3+}) and lanthanum (La^{3+}), block gravitropism. A number of **second messengers** have been implicated in gravity signal transduction including calcium and inositol 1,4,5-triphosphate. Inhibitors of calcium-binding proteins such as calmodulin or Ca^{2+}-ATPases, which pump calcium across membranes, have been shown to block gravitropism. In addition, Plieth and Trewavas (2002) demonstrated that gravitropic stimulation generated transient changes in cytoplasmic Ca^{2+} concentration. However, the precise role of these compounds remains to be elucidated.

12.3.5 Phototropism

Phototropism has been extensively studied in the coleoptiles of cereals and in the epicotyls and hypocotyls of dicotyledonous species. In general, shoots are positively phototropic whilst roots are negatively phototropic or indifferent to light. However, the direction of the response can depend upon irradiance.

After a threshold irradiance has been exceeded, coleoptiles curve towards the source of illumination; this is the first positive response (Fig. 12.13). Over a limited range, the extent of this first curvature is related to the irradiance. However, as irradiance increases, a first negative and then a second positive curvature occurs. In some cases even a third positive curvature is seen although this is most likely to

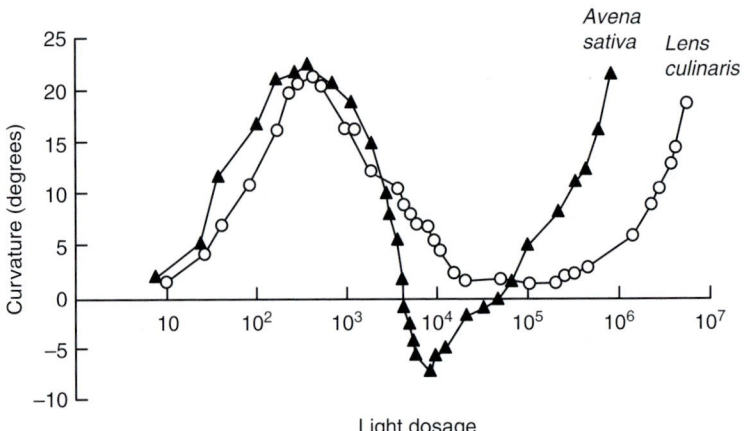

Fig. 12.13 Phototropic dose - response curves for etiolated coleoptiles of *Avena sativa* and etiolated epicotyls of *Lens culinaris*. Positive values are towards the light. Note that the light dosage is on a logarithmic scale. Adapted from Steyer (1967).

be a result of cell damage on the illuminated side leading to a loss of turgor. Not all organs show the same response. The first and second positive curvatures observed in etiolated epicotyls of lentil are separated by an indifferent zone. Light-grown radish seedlings are insensitive to directional illumination at low irradiances and begin to curve only in the second positive irradiance range.

The phototropic reaction has a number of features in common with gravitropism. There is a presentation and a reaction time, although these may be very short (a few seconds or less) especially at high irradiances. This probably reflects the fact that a photochemical reaction is almost instantaneous, whereas physical movement of the statoliths and structural rearrangement of membranes and filaments take a little time. Although a degree of reciprocity exists, this holds only over a limited range of irradiances; the duration of the exposure does affect the response observed. Phototropism is dependent upon the establishment of a light gradient across the reacting organ. Although a grass coleoptile is a delicate and translucent object, it is hollow, and light entering it is subject to significant internal reflection. The light gradients within an organ can be measured using an ultrafine fibre-optic and such studies have shown that unilateral illumination produces light gradients ranging from $1.5:1$ to $35:1$ depending upon the irradiance and the region illuminated. In more robust organs, absorption of light by the tissues forms a steep light gradient. A 1% difference in irradiance between the illuminated and shaded sides has been stated to be sufficient to elicit a response.

Light perception

As we have seen with phytochrome and cryptochrome (Chapter 10), the primary step in any light-mediated reaction must be the absorption of light by, and the activation of, one or more photoreceptors. Action spectra (Section 10.3.1) are a useful means by which to determine the spectral properties of the photoreceptor(s). This approach demonstrated that in phototropic responses blue light is active and the action spectrum has three peaks in the blue at 436 nm, 440 nm and 480 nm (Figure 12.14). These initial studies led to considerable controversy about the nature of the photoreceptor, as plants contain many different compounds with similar absorptive properties including flavins and carotenoids. Again, it has taken the isolation and characterization of mutants in these responses to identify some of the photoreceptors involved in tropic curvatures.

Phototropins

In 1995, Liscum and Briggs (1995) identified a number of mutants of *Arabidopsis* which had non-phototropic hypocotyls (*nph*). One of these, *nph1*, had a lesion in a gene, *NPH1*, which encodes a blue-light receptor involved in phototropic curvatures. Plants which lack this receptor do not exhibit phototropism when exposed to low fluence rates $(1 \ \mu mol \ m^{-2} \ s^{-1})$ of unidirectional blue light. The NPH1 protein, now

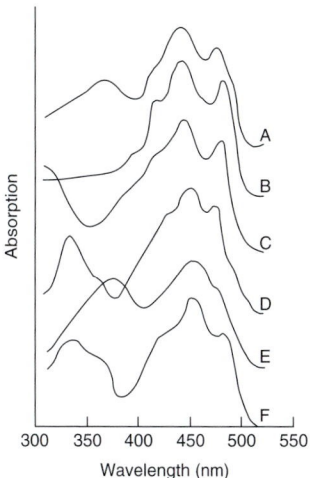

Fig. 12.14 The action spectrum of phototropism in the *Avena* coleoptile compared with flavins and carotenoids. (A) phototropic action spectrum; B, α-carotene in hexane; (C) 9–9' *cis*-β-carotene in hexane; (D) hexane extract of 50 *Avena* coleoptile tips; E, riboflavin in water; (F) 3-methyl-lumiflavin in benzene. From Thimann & Curry (1961).

known as a PHOTOTROPIN, PHOT1, is quite distinct from phytochrome or cryptochrome, but does share a number of motifs with signalling molecules identified in other organisms. It contains two conserved sequences, called LOV domains (light, oxygen, voltage), which have been identified in sensors in the fruit fly (*Drosophila melanogaster*) and also in vertebrates. In addition it is a serine/threonine kinase and phosphorylates itself when exposed to blue light. The protein binds a flavin chromophore which can undergo a light-stimulated conformational change (Fig. 12.15) with an associated alteration in its spectral properties. A number of related genes have been identified in the *Arabidopsis* genome and also in other plant species indicating that phototropins are encoded by a small multi-gene family, just as is the case with phytochrome and cryptochrome.

At first it was thought that PHOT1 was the only photoreceptor required for phototropic curvatures in *Arabidopsis*, but a number of lines of evidence suggest that others are involved, acting either independently or in conjunction with PHOT1. Although *nph1* (*phot1*) mutants lack phototropic curvatures in response to low fluence rates of blue light, exposure to higher fluence rates (100 μmol m^{-2} s^{-1}) for extended periods (12 h) leads to the same curvature as seen in wild-type plants. It has recently been demonstrated that a related gene, *NPL1* (*non-phototropic hypocotyl like*) encodes another phototropin required for this second phototropic response. This protein also contains a LOV domain and binds a flavin – hence it has been renamed PHOT2. In the double mutant, *phot1 phot2*, nearly all phototropic curvatures are eliminated (Sakai *et al.* 2001). These same photoreceptors are also required for the negative phototropic response of *Arabidopsis* roots.

These same two photoreceptors also control other aspects of plant responses to light (see Briggs & Christie 2002). When the chloroplast-containing mesophyll cells of leaves are illuminated with dim blue light, the chloroplasts accumulate at the uppermost part of the cell (adjacent to the periclinal wall), maximizing interception of light. However, if the same cells are exposed to strong blue light, the chloroplasts move to the anticlinal cell walls so that their self-shading is maximal (Fig. 12.16), probably a protective reaction since excessive illumination can cause photochemical damage (Chapter 2). Both of these responses are abolished in the *phot1 phot2* double

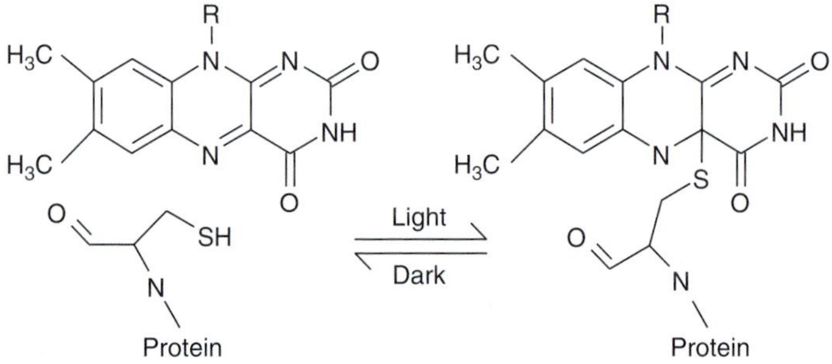

Fig. 12.15 Alterations in chromophore structure in phototropin in response to light. The flavin molecule which forms the chromophore in phototropin becomes covalently linked to a cysteine residue in the phototropin apoprotein upon illumination. A slow reversion occurs in darkness. Redrawn from Crosson & Moffat (2001). © 2001 National Academy of Sciences, USA.

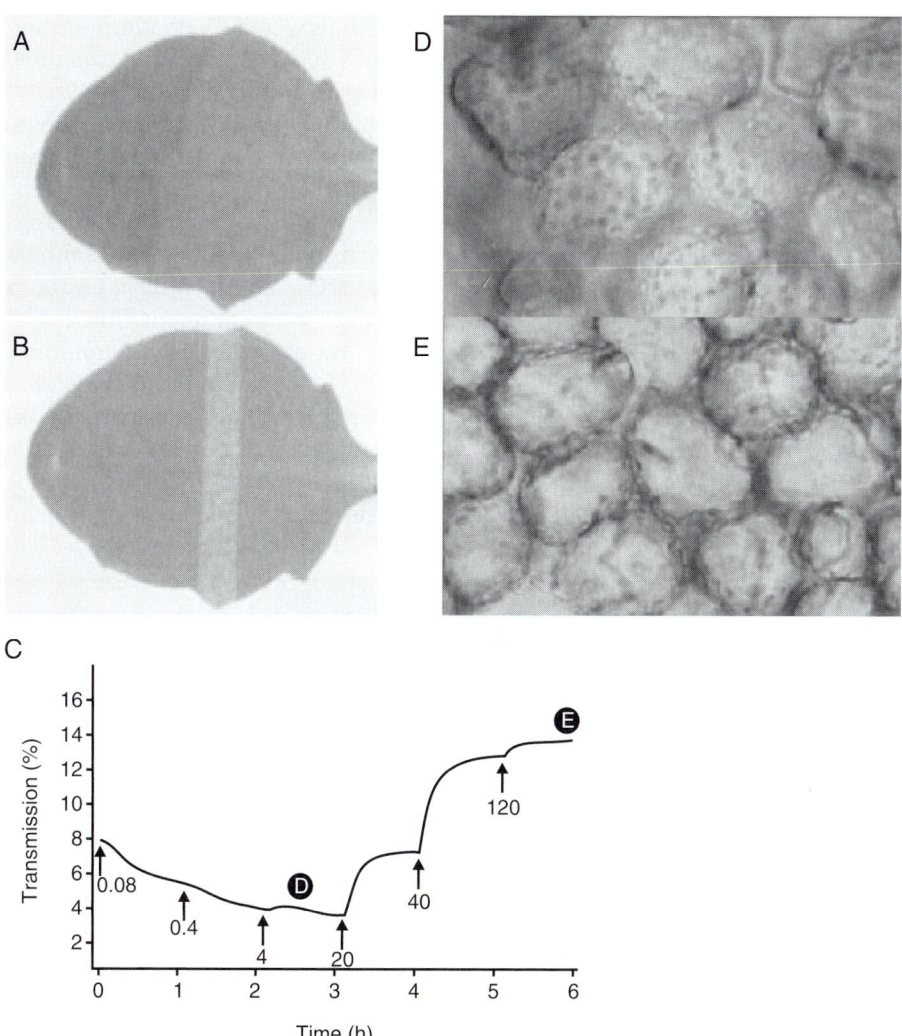

Fig. 12.16 Chloroplasts move in response to light. Leaves of *Arabidopsis* before (A) and after (B) exposure to strong white light. Illumination was confined to a small region of the leaf by covering it with an opaque plate in which a slit, 1 mm wide, had been cut. After 1.5 h illumination the transmission of the illuminated region has increased as chloroplasts have moved – this is apparent as the light grey stripe in (B). Chloroplast movement can be stimulated by blue light. (C) shows the changes in transmission as leaves of *Arabidopsis* are illuminated with increasing irradiances of blue light (μmol m^{-2} s^{-1}) at the times indicated by the arrows. At low irradiances (D) the chloroplasts cover the surface of the cell but after exposure to high light (E) they have moved to the periphery. (A) and (B) are from Kagawa *et al.* (2001). (C) – (E) are from Jarillo *et al.* (2001). © Nature Publishing Group.

mutant. Likewise these photoreceptors mediate stomatal opening induced by blue light (Kinoshita *et al.* 2001).

The role of other photoreceptors in phototropisms remains less clear. Ahmad *et al.* (1998) reported that *Arabidopsis* plants lacking either cryptochrome CRY1 or CRY2 still exhibited phototropic curvatures, but that in the *cry1cry2* double mutant the first (but not the second) phototropic curvature was abolished. However, Lascève *et al.* (1999) reported that *cry1cry2* plants did still exhibit a first positive phototropic response to blue light. The interactions between phototropin and cryptochrome at different irradiances has recently been examined by Whippo and Hangarter (2003). At low irradiances (1 μmol m^{-2} s^{-1}) cryptochrome enhances the phototropic response, but it has an inhibitory effect at high irradiances (100 μmol m^{-2} s^{-1}). Together, these photoreceptors balance the seedling response to light from different directions. It has been known for many years that exposing seedlings to uniform red light enhances their response to unidirectional blue light. Figure 12.17 shows the response of etiolated *Arabidopsis* seedlings to unidirectional blue light when exposed to different fluences of uniform red light. The etiolated seedlings were exposed to uniform, dim red light (1.6 μmol m^{-2} s^{-1}) for 1–1000 s and unidirectional dim blue light (0.5 μmol m^{-2} s^{-1}) for 4 h, after which the curvature was measured. Increasing the red fluence increased the curvature observed. This has all the hallmarks of a low-fluence phytochrome response (Section 10.3.2) as the enhancement caused by red light could

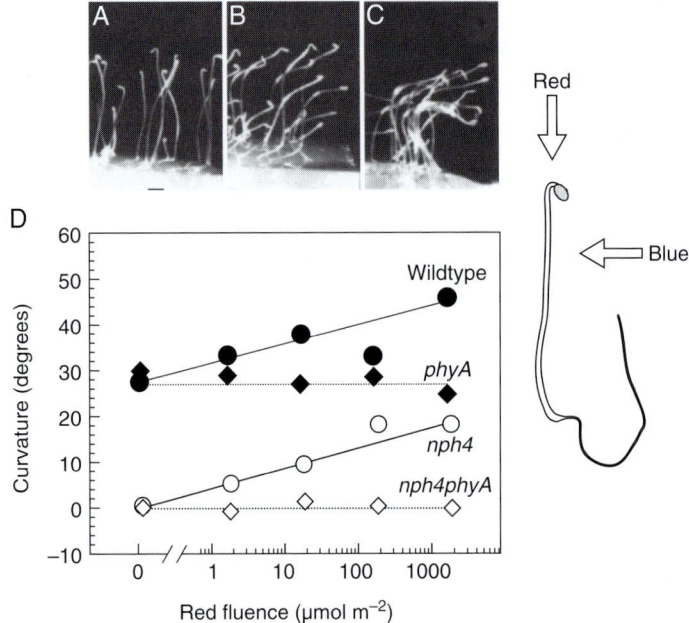

Fig. 12.17 Phytochrome A enhances blue-light-induced phototropism. Etiolated seedlings of *Arabidopsis thaliana* were exposed to weak (0.5 μmol m^{-2} s^{-1}) unidirectional blue light for 4 h and the curvature of the hypocotyl measured. The impact of red light from above was examined by exposing the plants to dim red light (1.6 μmol m^{-2} s^{-1}) for between 1 and 1000 s producing fluences between 1.6 and 1600 μmol m^{-2}. In wild-type plants exposure to unidirectional blue light alone produced a curvature of approximately 30°. Increasing fluences of red light increased the responsiveness of the seedlings (filled circles). *phyA* mutants were still responsive to blue light, but the enhanced response with additional red light was abolished (filled diamonds). The *nph4* mutant does not curve in response to blue light alone, but curvature does occur if plants are also exposed to red light. Again, this red-light-induced curvature is abolished in the *phyA* mutant. *phyB* mutants respond in the same way as wild-type plants (data not shown). From Stowe-Evans *et al.* (2001).

be abolished by subsequent exposure to far-red light. Unusually though, phytochrome A was implicated in this response, as it was no longer apparent in *phyA* mutants (Fig. 12.17). As seen in Chapter 10, phyA is normally associated with very-low-fluence or high-irradiance responses in which far-red light is the most important signal. These responses, unlike the low-fluence response, are not far-red reversible. The role of phytochromes in phototropic responses of etiolated, and de-etiolated, seedlings requires more detailed studies, but at least all the different pieces of the jigsaw appear to be in place.

As well as identifying the phototropins, the isolation of mutant plants has also led to the identification of other components of the signal transduction pathway. Other *nph* mutants (*nph2, nph3, nph4*) all operate downstream of *phot1* and a number of these have now been cloned. Some of these same mutations have also been independently identified in searches for mutants with altered root phototropism (*rpt* mutants) and other studies of plant physiology. Whilst the *phot1, nph2* and *nph3* mutants affected only phototropism, *nph4* is especially interesting as it also alters gravitropic responses. The *NPH4* gene has been cloned and found to encode a transcription factor (i.e. a gene which will regulate the expression of other genes) which had previously been identified as an auxin-responsive transcription factor, ARF7. This same gene is also mutated in plants which were identified on their altered responses following treatment with auxin, or inhibitors of auxin transport. As our understanding of the function of these genes and proteins increases, and especially their interactions with each other, we should hopefully soon be able to make the links between signal perception and the growth response seen in tropisms.

Complementary reading

Journal of Plant Growth Regulation, **21**(2) (2002). Issue largely devoted to tropisms, particularly gravitropism.

Boonsirichai, K., Guan, C., Chen, R. & Masson, P. H. Root gravitropism: an experimental tool to investigate basic cellular and molecular processes underlying mechanosensing and signal transmission in plants. *Annual Review of Plant Biology*, **53** (2002), 421–47.

Kagawa, T. & Wada, M. Blue light-induced chloroplast relocation. *Plant Cell Physiology*, **43** (2002), 367–71.

Sack, F. D. Plastids and gravitropic sensing. *Planta*, **203** (1997), S63–8.

References

Ahmad, M., Jarillo, J. A., Smirnova, O. & Cashmore, A. R. (1998). Cryptochrome blue-light photoreceptors of *Arabidopsis* implicated in phototropism. *Nature*, **392**, 720–3.

Bell, C. J. & Maher, E. P. (1990). Mutants of *Arabidopsis thaliana* with abnormal gravitropic responses. *Molecular & General Genetics*, **220**, 289–93.

Briggs, W. R. & Christie, J. M. (2002). Phototropins 1 and 2: versatile plant blue-light receptors. *Trends in Plant Science*, **7**, 204–10.

Collings, D. A., Zsuppan, G., Allen, N. S. & Blancaflor, E. B. (2001). Demonstration of prominent actin filaments in the root columella. *Planta*, **212**, 392–403.

Crosson, S. & Moffat, K. (2001). Structure of a flavin-binding plant photoreceptordomain: insights into light-mediated signal transduction. *Proceedings of the National Academy of Sciences (USA)*, **98**, 2995–3000.

Digby, J. & Firn, R. D. (2002). Light modulation of the gravitropic set-point angle (GSA). *Journal of Experimental Botany*, **53**, 377–81.

Firn, R. D., Wagstaff, C. & Digby, J. (2000). The use of mutants to probe models of gravitropism. *Journal of Experimental Botany*, **51**, 1323–40.

Friml, J., Wisniewska, J., Benkova, E., Mendgen, K. & Palme, K. (2002). Lateral relocation of auxin efflux regulator PIN3 mediates tropism in *Arabidopsis*. *Nature*, **415**, 806–9.

Hasegawa, K., Sakoda, M. & Bruinsma, J. (1989). Revision of the theory of phototropism in plants: a new interpretation of a classical experiment. *Planta*, **178**, 540–4.

Jarillo, J. A., Gabrys, H., Capel, J., Alonso, J. M., Ecker, J. R. & Cashmore, A. R. (2001). Phototropin-related NPL1 controls chloroplast relocation induced by blue light. *Nature*, **410**, 952–4.

Kagawa, T., Sakai, T., Suetsugu, N. *et al.* (2001). Arabidopsis NPL1: a phototropin homolog controlling the chloroplast high-light avoidance response. *Science*, **291**, 2138–41.

Kaldewey, H. (1957). Wuchsstoffbildung und Nutationsbewegungen von *Fritillaria meleagris* L. im Laufe der Vegetationsperiode. *Planta*, **49**, 300–44.

Kinoshita, T., Doi, M., Suetsugu, N., Kagawa, T., Wada, M. & Shimazaki, K. (2001). phot1 and phot2 mediate blue light regulation of stomatal opening. *Nature*, **414**, 656–60.

Kiss, J. Z., Wright, J. B. & Caspar, T. (1996). Gravitropism in roots of intermediate-starch mutants of *Arabidopsis*. *Physiologia Plantarum*, **97**, 237–44.

Kuznetsov, O. A. & Hasenstein, K. H. (1996). Intracellular magnetophoresis of amyloplasts and induction of root curvature. *Planta*, **198**, 87–94.

Kuznetsov, O. A. & Hasenstein, K. H. (1997). Magnetophoretic induction of curvature in coleoptiles and hypocotyls. *Journal of Experimental Botany*, **48**, 1951–7.

Larsen, P. (1962). In *Encyclopedia of Plant Physiology*, ed. W. Ruhland. Berlin: Springer. Vol. **17**, part 2, pp. 34–73.

Lascève, G., Leymarie, J., Olney, M. A. *et al.* (1999). *Arabidopsis* contains at least four independent blue-light-activated signal transduction pathways. *Plant Physiology*, **120**, 605–14.

Leitz, G., Schnepf, E. & Greulich, K. O. (1995). Micromanipulation of statoliths in gravity-sensing *Chara* rhizoids by optical tweezers. *Planta*, **197**, 278–88.

Li, Y., Hagen, G. & Guilfoyle, T. J. (1991). An auxin-responsive promoter is differentially induced by auxin gradients during tropisms. *The Plant Cell*, **3**, 1167–75.

Liscum, E. & Briggs, W. R. (1995). Mutations in the Nph1 locus of *Arabidopsis* disrupt the perception of phototropic stimuli. *The Plant Cell*, **7**, 473–85.

McClure, B. A. & Guilfoyle, T. (1987). Characterization of a class of small auxin-inducible soybean polyadenylated RNAs. *Plant Molecular Biology*, **9**, 611–23.

Okada, K., Ueda, J., Komaki, M. K., Bell, C. J. & Shimura, Y. (1991). Requirement of the auxin polar transport system in early stages of *Arabidopsis* floral bud formation. *The Plant Cell*, **3**, 677–84.

Ottenschläger, I., Wolff, P., Wolverton, C. *et al.* (2003). Gravity-regulated differential auxin transport from columella to lateral root cap cells. *Proceedings of the National Academy of Sciences (USA)*, **100**, 2987–91.

Plieth, C. & Trewavas, A. J. (2002). Reorientation of seedlings in the earth's gravitational field induces cytosolic calcium transients. *Plant Physiology*, **129**, 786–96.

Sack, F. D. (1997). Plastids and gravitropic sensing. *Planta*, **203**, S63–8.

Sakai, T., Kagawa, T., Kasahara, M. *et al.* (2001). *Arabidopsis* nph1 and npl1: blue light receptors that mediate both phototropism and chloroplast relocation. *Proceedings of the National Academy of Sciences (USA)*, **98**, 6969–74.

Sievers, A. & Volkmann, D. (1972). Verursacht differentieller Druck der Amyloplasten auf ein komplexes Endomembransystem die Geoperzeption in Wurzeln? *Planta*, **102**, 160–72.

Steyer, B. (1967). Die Dosis-Wirkungsrelationen bei geotropen und phototropen Reizung: Vergleich Von Mono- mit Dicotyledonen. *Planta*, **77**, 277–86.

Stowe-Evans, E. L., Luesse, D. R. & Liscum, E. (2001). The enhancement of phototropin-induced phototropic curvature in *Arabidopsis* occurs via a photoreversible phytochrome A-dependent modulation of auxin responsiveness. *Plant Physiology*, **126**, 826–34.

Thimann, K. V. & Curry, G. M. (1961). Phototropism. *In Light and Life*, ed. W. D. McElroy & B. Glass. Baltimore, MD: Johns Hopkins University Press, pp. 646–72.

van Doorn, W. G. & van Meeteren, U. (2003). Flower opening and closure: a review. *Journal of Experimental Botany*, **54**, 1801–12.

Went, F. W. & Thimann, K. V. (1937). *Phytohormones*. New York, NY: Macmillan.

Whippo, C. W. & Hangarter, R. P. (2003). Second positive phototropism results from coordinated co-action of the phototropins and cryptochromes. *Plant Physiology*, **132**, 1499–1507.

Yamamoto, K. T. (2003). Happy end in sight after 70 years of controversy. *Trends in Plant Science*, **8**, 359–60.

Chapter 13

Resistance to stress

13.1 | Introduction

Any factor that acts on an organism so as to impair its functions can be termed a **stress**. Plants growing in the field are habitually exposed to a number of environmental stresses, e.g. drought and frost. Being sessile organisms, plants cannot move away from a stressful situation. The ability to withstand environmental stresses therefore frequently becomes the limiting factor for plant growth, survival and geographical distribution. Plants in fact may possess remarkable powers of endurance. The vegetation of arctic regions can experience winter temperatures of $-70\,°C$, whilst in hot deserts over $50\,°C$ may be encountered, and even greater temperature extremes have been survived in the laboratory. On the other hand, some plants are killed by chilling at $10\,°C$: species vary tremendously in their resistance towards a particular stress. Studies of the reactions of plants under stress, and mechanisms of stress resistance, are of great practical importance, since agricultural yield is only too often drastically reduced by stressful external factors. The demands of an expanding human population have stimulated research into improving the stress resistance of crop species in order to extend the geographical range of a crop, or with a view to utilizing land areas previously regarded as too 'extreme' for cultivation, such as semi-deserts.

13.2 | Terminology and concepts

Stress is a very wide concept, and while the general idea is easily conveyed it is not so easy to decide where the limits should be drawn. Stress was briefly defined above as 'impairing function'. Any environmental condition below the absolute optimum impairs function to some extent. If the temperature optimum for photosynthesis of a temperate zone plant species is $25\,°C$, its rate of photosynthesis is lower at $20\,°C$, but one would hardly regard the plant as being cold-stressed at that temperature. The term stress is appropriate when there is *actual damage*, or *a specific stress response* is elicited. In some situations, what is considered a stress depends on individual judgement. Key terms and concepts applied in discussions of stress are explained below.

> **Stress resistance** (hardiness): the ability to endure an **externally** applied stress, e.g. the ability to survive a low external Ψ (water potential).
>
> **Stress avoidance**: the ability to prevent an externally applied stress from producing an equivalent **internal** stress in the plant, e.g. the ability to maintain a high cellular water content even when the external Ψ is low.
>
> **Stress tolerance**: the ability to survive an **internal** stress, e.g. the ability to survive a low cellular water content.

Resistance to a stress can thus be achieved by stress avoidance, by stress tolerance, or a combination of both to various degrees.

> **Hardening** (acclimation): the development of stress resistance, stimulated by subjection to mild and/or gradually increasing stress.

Some species are capable of hardening towards a particular stress, others are not. Hardening is a vital process, for plants potentially resistant towards some stress do not maintain the state of resistance continuously; for instance, perennials of climates with a cold winter are resistant towards subzero temperatures only during the cold season, undergoing annual cycles of hardening and de-hardening. Thus a *species* may be considered hardy towards a given stress, but an individual of that species may not be hardy at a particular time.

Throughout this chapter, reference is made to 'resistant' or 'tolerant' as contrasted with 'sensitive' plants and tissues, and to 'high' or 'severe' as opposed to 'low', 'moderate' or 'mild' levels of stress. These terms are relative, depending on the system in question and on the criteria used to assess resistance. An investigator comparing the effects of chilling at 5 °C on two plant species, and finding that species A survived without damage, while species B suffered some lesions to its leaves, might describe species A as chilling resistant and species B as sensitive, using leaf damage as the criterion. But a second worker comparing the effects of the same treatment on species B and species D, which was killed outright, could judge B to be resistant, on the criterion of survival. Literature should be consulted with such considerations in mind.

This chapter deals with stresses resulting from water deficit (including salinity), low temperature and high temperature. Additional stresses experienced by plants are excessive light, UV radiation, chemical pollutants (including atmospheric ozone and sulphur dioxide, and heavy metals in soil or water) and anaerobiosis. Lack of space precludes detailed consideration of these, but some reference is made in other chapters: excessive light and anaerobiosis are considered in Chapter 2 in connection with photosynthesis and

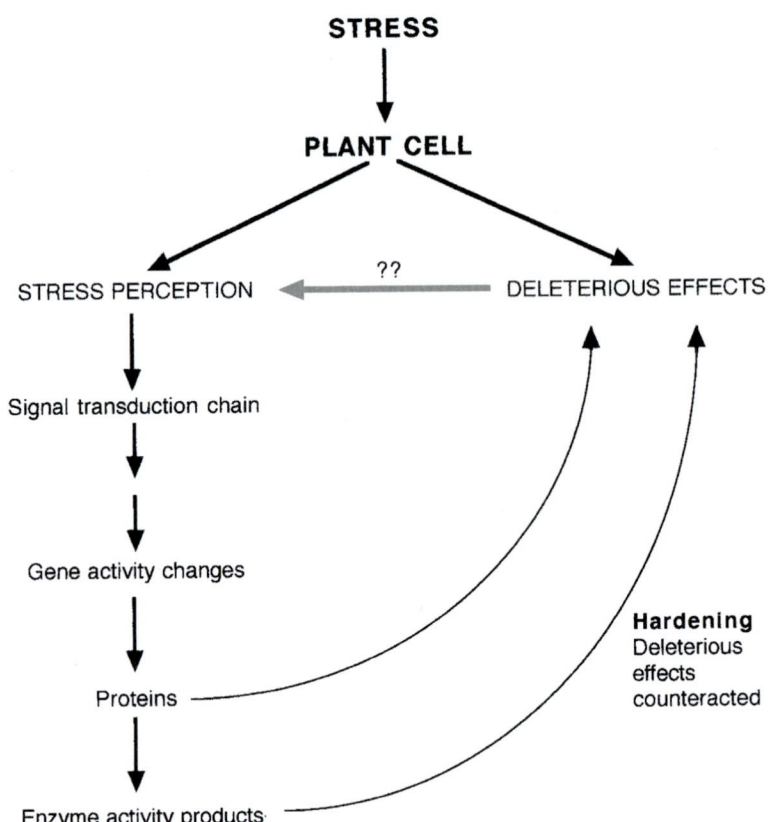

Fig. 13.1 Generalized scheme for the sequence of events initiated by stress on plant cells. Sensitive plants incapable of hardening suffer deleterious effects which can be fatal. In plants capable of hardening, the perception of stress, which may be via a damage effect, leads to a switch in gene activity and hence to changes in cell organization and physiology which counteract the deleterious effects (curved upward arrows).

respiration respectively, and heavy metals in Chapter 4, with mineral nutrition.

Hardening towards stress follows the same pattern in all cases, as illustrated in very general terms in Fig. 13.1. There is first **perception** of the stress; this may or may not be via the damage caused. This is followed by a **signal transduction chain**, also termed a signal cascade, leading to a changed pattern in gene transcription and in the synthesis of proteins that can counteract the deleterious effects of the stress, either directly or via their products if the proteins are enzymatic. Plant reactions to stress have proved to be extremely complex, with one stress able to evoke several reaction chains, with cross-reactions. On the other hand, the three factors, water deficit, low temperature and high temperature, might appear unrelated, but in fact their effects on plants show common features and they share (parts of) signal transduction chains.

13.3 | Water-deficit stress

Water-deficit stress (water stress, desiccation stress, drought) is defined as exposure to a low external Ψ. Except in the most humid environments, plants are likely to suffer some degree of water-deficit stress during their life cycle. For human populations, drought has been a scourge down the ages. Growth and photosynthesis, the two physiological activities most critical for crop yield, are among the first plant functions to be inhibited by water deficit. It is generally considered that the global losses in potential agricultural yield caused by water-deficit stress exceed losses from all other causes combined.

13.3.1 Water stress avoidance

Even under conditions of water shortage, some plants can reduce their rate of water loss to a very low level, so that a relatively high Ψ is maintained within their tissues and water stress is avoided. Xeromorphic characters which enable such water stress avoidance are described in Section 3.7. Water stress avoiders include numerous desert plants, e.g. cacti. The cells of many such plants cannot withstand dehydration to any marked degree. The supreme water stress avoiders are desert ephemerals, which spend most of their life span as dormant seeds. They germinate only after the occasional rainstorms and complete their growth cycle very rapidly while the water supply lasts. Less extreme examples are found in climates with a regular dry season (e.g. Mediterranean), where annuals yearly synchronize their growth period with the wet season, while perennials may lay up a water store during the rainy season in an underground organ for subsequent growth. Water stress avoidance is clearly of importance in arid habitats. In this chapter, however, emphasis is placed on the physiology of *tolerance*. The different types of resistance are not mutually exclusive; a plant growing in a xeric habitat may have a water-conserving avoidance mechanism as a first line of defence, but, when conditions become more extreme, plants may develop tolerance. For instance, many resurrection plants (see Section 13.3.2) are water stress avoiders, but are also the most desiccation-tolerant flowering plants.

13.3.2 Range of water stress tolerance; developmental changes

Most flowering plants pass through a stage of very low water content as mature seeds, which can dry out to equilibrium with atmospheric humidity, RH 20 to 50%. This produces seed water contents of 5 to 20%, corresponding to seed Ψ of -100 to -10 MPa. Experimentally, some seeds have survived drying to water contents below 1%. Not all seeds are highly resistant to desiccation, e.g. many species from the humid tropics have dehydration-sensitive seeds. Nevertheless the possession of 'dry' seeds is very widespread.

During germination, the desiccation resistance of seeds is gradually lost and, for the rest of their life, most flowering plants can never again endure desiccation to such low water contents. An exception is provided by the **resurrection plants**, which in the mature state tolerate extreme dehydration and revive rapidly when wetted. Over 100 species are known, belonging to several unrelated families, mainly from the hot dry regions of southern Africa (*Chamaegigas intrepidus*, *Craterostigma plantagineum*, *Myrothamnus flabellifolia*, *Talbotia elegans*, *Xerophyta* spp.) and Australia (*Borya nitida*, several grasses). During periods of drought, the leaves of these plants reach air dryness, with leaf Ψ values down to -160 MPa. The minimum leaf Ψ values survived in most mesophytic species fall between -3 and -22 MPa.

In perennials of temperate and arctic climates, water stress tolerance shows a regular seasonal fluctuation, being greatest in the winter months. During the winter, plants are subjected to water deficit if the soil freezes, and even if the soil contains liquid water at a high Ψ, the uptake capacity of the roots can be diminished (Section 3.6.1). Also, as explained in Section 13.4.3, freezing within the plant imposes a water-deficit stress.

13.3.3 Effects of water stress on plant cells and organs

Almost every process in plant cells is affected by water deficit (Fig. 13.2). This is not surprising, considering the ubiquitous involvement of water in cellular processes and structures, but makes it very difficult to distinguish between primary lesions directly resulting from water deficit and secondary effects consequent upon the primary damage, and to follow a chain of reactions from primary perception

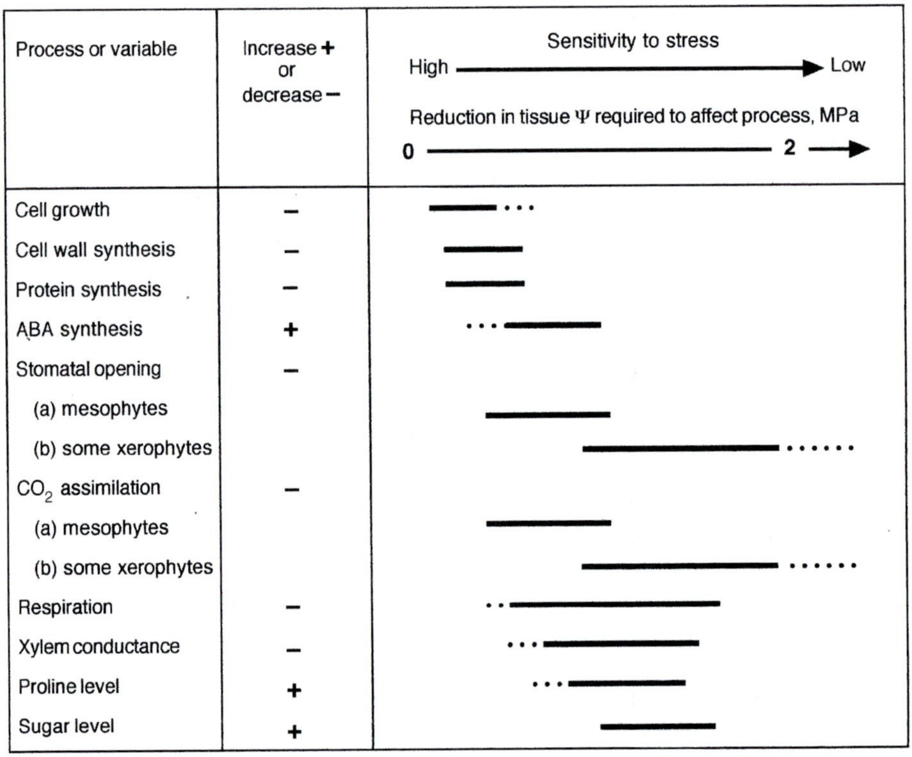

Fig. 13.2 Relative sensitivity of some plant characters to water-deficit stress. The horizontal lines represent the range of stress level, expressed as decrease in tissue Ψ, over which each process first becomes affected in a range of species. Dotted lines represent more tenuous data. The decrease in Ψ is from well-watered plants under mild evaporative demand. Adapted from Hsiao *et al.* (1976).

to observed effect, i.e. a specific damage process or a protective reaction. The effects of water-deficit stress are time-dependent, injury increasing with time.

Metabolic disturbances

Physiological effects of water stress depend on the severity of the water deficit (Fig. 13.2). **Growth** is extremely sensitive to water stress and growth rate may begin to decrease as soon as cellular water content falls below full saturation, before one would expect an impact on metabolic activities. This is probably a direct effect of the decrease in turgor, for cell expansion growth depends on the yielding of the cell wall under turgor pressure. **Photosynthesis** in mesophytes is affected at quite moderate levels of water deficit and rapidly declines to zero, more or less. Stomatal closure is an important factor here: in both mesophytes and xerophytes, the start of photosynthetic decline coincides with the closure of the stomata. There are also more direct effects of water deficit on the photosynthetic apparatus. When submerged aquatics, which do not rely on stomatal entry of CO_2, are dehydrated by immersion in a sugar solution, their rate of photosynthesis falls. Chloroplasts, isolated from water-stressed plants and assayed *in vitro* with an ample water supply, still show lowered photosynthetic activity. The reaction centre of photosystem II deteriorates at levels of water stress termed 'severe' and the amounts of some chloroplast proteins may fall. Decline or loss of photosynthetic activity, if prolonged, will lead to nutrient shortage and further limit growth. Suppression of CO_2 fixation leads to light-driven accumulation of ROS (reactive O_2 species) due to the surplus light energy absorbed (Box 2.1 and Section 2.4.2) and hence to oxidative damage, especially to membranes, further impairing the potential for photosynthesis. **Respiration** rate is less affected than photosynthesis, but uncoupling of ATP formation has been reported. The persistence of respiration after photosynthesis is inhibited hastens the build-up of nutrient shortages. Lowering of cellular water content leads to increased concentrations of solutes. This can have significant effects on metabolic reaction rates, which depend on substrate concentrations. Water stress results in increased tension in the xylem and promotes cavitation (Section 3.4.3, p. 80), aggravating the water shortage in the plant.

With slight or moderate water loss from tissues, the effects would be expected to be physiological and reversible, although over an extended period of time a combination of physiological effects as described above might lead to an overall metabolic disturbance sufficient to cause death. During more marked water deficit, however, structural damage becomes more critical.

Structural damage

With more severe dehydration, death results from structural disorganization at levels ranging from macromolecule structure to microscopically visible damage. When cells dry out, the protoplast is subjected to tension resulting from its contraction in volume and its adherence to the cell wall, physical stresses which can tear the plasmalemma. This type of damage is repeated during rehydration, when tensions build up between rehydrated outer layers and drier inner parts of the cell. Gradual desiccation and gradual rehydration can be endured better than sudden drying and rehydration. The extent of dehydration that a plant tissue can endure therefore depends on the particular conditions of treatment.

At the submicroscopic level, desiccation causes protein denaturation by withdrawal of water of hydration and by the effects of the increases in solute concentration. The tertiary structures of proteins are partly maintained by electric charges, and increased concentrations of charged molecules in the cells affect protein structure. This in turn affects enzyme activity and membrane integrity. The lipid regions of membranes, too, are affected by dehydration. The hydrophilic 'heads' of the membrane phospholipids are normally associated with water of hydration. For proper functioning, the lipids must be in a fluid state described as liquid crystalline; removal of hydration water can lead to a gelling of the lipids to a semisolid gel state with loss of functional integrity. Oxidative damage to membranes has already been mentioned.

13.3.4 Tolerance of water stress: hardening

To reach their maximal potential for tolerance, most species, including resurrection plants, need to be hardened by a few days of moderate water stress. Long-term growth with a low water supply can stimulate the development of xeromorphic features.

Perception of water stress

Plants can sense and react to very small changes in their water status, which should not have significant direct effects on cellular physiology. The turgidity and wall pressure of plant cells are, however, highly sensitive to changes in water content, diminishing rapidly in the initial stages of water loss (Fig. 13.3). The primary perception of incipient water stress may involve the effects of pressure on the plasma membrane, which contains proteins known as osmosensors. These are well known in yeast and bacteria, as two-component integral transmembrane proteins. At least one such osmosensor, ATHK1, has been characterized in *Arabidopsis*; the transcription of the *ATHK1* gene is up-regulated by water stress (Bray 2002). Pressure changes presumably affect the configuration of the sensor (receptor) part of the osmosensors, in the membrane, and this activates the component on the cytoplasm side, which acts as a kinase, to start a signal transduction chain via protein phosphorylation, leading to hardening. Figure 13.4 indicates some of the processes involved.

Gene activation and protein synthesis

Water-deficit stress results in a significant change in genetic activity. Numerous genes are switched on or up-regulated, while others are

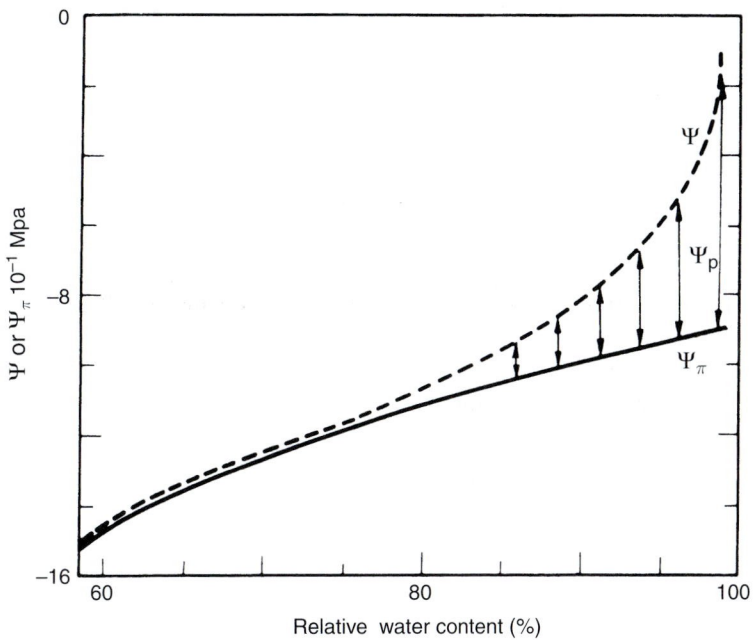

Fig. 13.3 Generalized relationships between water potential Ψ, osmotic potential Ψ_π and relative water content in leaves of herbaceous crop plants; 100% water content represents the maximum the tissue can hold, and maximal turgidity. The length of the arrows equals the turgor pressure Ψ_p, which can be seen to decrease rapidly in the early stages of water loss: over the first 5% loss in water content, the Ψ_p has halved. Adapted from Hsiao *et al.* (1976).

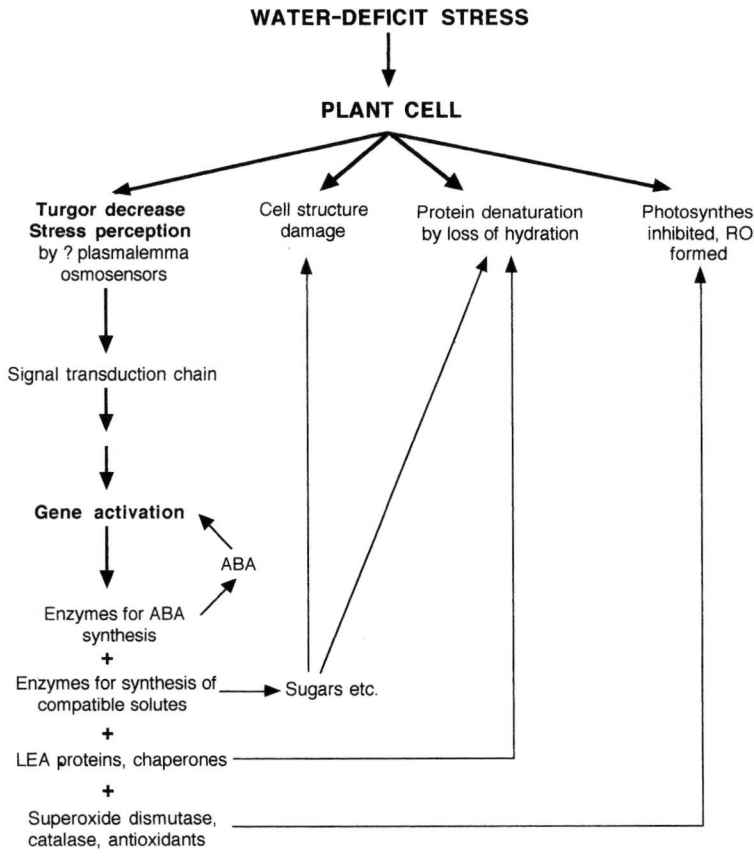

Fig. 13.4 Summary of the main events in development of water stress resistance. The perception of stress initiates signal transduction chains leading to the synthesis of proteins. These include enzymes for synthesis of ABA, which can then activate other genes. The new proteins, or their products, counteract (upward arrows) deleterious effects of water stress noted at the top of the diagram. ROS = reactive oxygen species.

switched off or down-regulated, with a corresponding change in the pattern of protein synthesis. One experiment with *Arabidopsis* identified over 44 up-regulated genes, for instance (Seki *et al.* 2001). Genes for the synthesis of the hormone ABA are up-regulated, leading to increased levels of this hormone. As discussed in Chapters 2 and 3, ABA transported into leaves induces stomatal closure in the early stages of water stress. But ABA also forms a part of the signalling chain for the controlling of other genes responsive to water stress (Fig. 13.4). Artificially applied ABA can substitute for water stress in the induction of many water-stress-responsive genes. But not *all* water-stress-responsive genes are induced by ABA. There obviously are several signal transduction chains involved in the response; in *Arabidopsis* at least four are inferred to exist, of which two involve ABA. Functions in combating the effects of water deficit can be assigned to quite a number, although far from all, of the coded proteins, as descibed below.

Compatible solutes and osmotic adjustment

A decrease in cellular Ψ_π (osmotic potential) typically occurs as a response to water deficit; this decrease is not just a concentration effect but results from synthesis of organic solutes, which accumulate in the cytoplasm (not the vacuole). The main solutes include sugars (sucrose, trehalose and umbelliferose); fructans (soluble polysaccharides of fructose units); polyols (sugar alcohols); cyclic polyols; amino acids, especially proline; quaternary ammonium compounds (QAC), often glycine betaine; and tertiary sulphonium compounds (Fig. 13.5). The accumulation of these solutes is controlled at the genetic level; e.g. genes encoding enzymes involved in the synthesis of proline are activated,

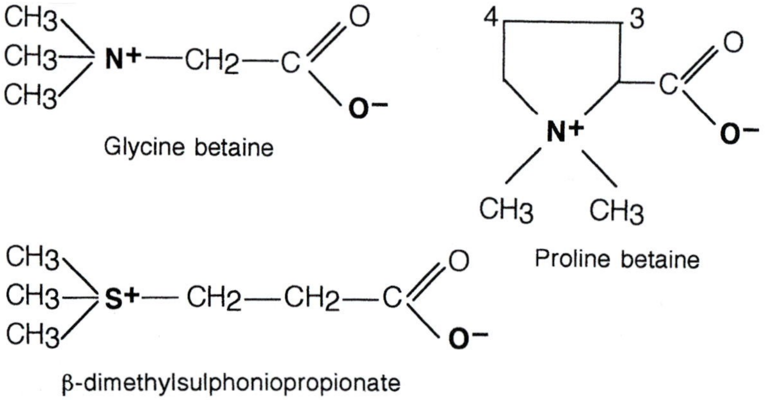

Fig. 13.5 The structure of some nitrogenous compatible solutes. These chemicals have charged atoms (shown in heavy type), but no *net* charge. Glycine betaine and proline betaine are examples of quaternary ammonium compounds (QAC), characterized by a fully methylated N atom with a permanent positive charge; other common plant QAC are alanine betaine, with an extra CH_2 group, also 3-hydroxyproline betaine and 4-hydroxyproline betaine, with an OH group on C atom 3 or 4. β-dimethylsulphoniopropionate (DMSP) is the most common plant tertiary sulphonium compound; this group of chemicals has a fully methylated S atom with a positive charge.

whilst those coding for enzymes of proline degradation are repressed. Transgenic individuals have been produced,which acquire the capability to synthesize a compatible solute not normally formed by the species, and which then gain increased water stress tolerance (Fig. 13.6).

The way in which such solutes function is still not completely clear. They are known as **compatible solutes**, since they are compatible with normal functioning of enzymes and other proteins: having no net electric charge, they do not disturb the ionic bonds maintaining tertiary and quaternary protein structure. The lowering of cellular Ψ_π, termed **osmotic adjustment**, enables the cell to hold water against a low external Ψ. Some species can maintain turgor in the face of an external Ψ of -4 MPa by this process. Compatible solutes, being in the cytoplasm, also maintain osmotic balance between the cytoplasm and the vacuolar sap, which becomes more concentrated as water is withdrawn from the cell. But sometimes the concentrations of these solutes are too low for a significant osmotic effect and other activities have therefore been proposed. Polyols can act as removers of oxidative radicals and some compatible solutes can stabilize molecular structures. Sugars can to some extent replace the water of hydration of proteins and prevent denaturation by water loss. In highly dehydrated plant tissues, sucrose participates in the formation of a glassy matrix, with tight H-bonding between molecules, in which proteins are protected from damage.

Protection against oxidative damage

There is an increase in transcription of genes coding for enzymes involved in the removal of ROS, e.g. SOD (superoxide dismutase) and catalase. The protective value of such enzymes is indicated by the fact that the water stress tolerance of plants can be improved by producing transgenic individuals which overexpress SOD. There are also reports of increases in the levels of antioxidants – β-carotene, ascorbate and α-tocopherol (vitamin E).

LEA genes and protein stabilization

There are also many genes activated coding for proteins for which the precise function is not elucidated. This includes *LEA* genes ('late embryogenesis abundant'), which are expressed during the normal desiccation occurring when seeds mature (Section 11.14.1), particularly the group of *LEA* genes coding for dehydrins (the D11 family of LEA proteins). Sequencing suggests that the majority of the LEA proteins should be highly hydrophilic. They have consequently been assigned protective functions in binding water and stabilizing membranes. Some are suggested to act as chaperones, proteins which bind to unfolded protein molecules and thereby prevent incorrect (denatured) folding and aggregation. Chaperone action is important in ensuring the correct folding of newly synthesized polypeptides in unstressed plants. In stressed cells, chaperones could react with proteins that have unfolded owing to loss of water of hydration, though direct proof of this is lacking.

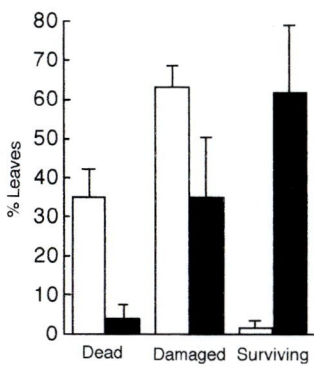

Fig. 13.6 Increased water stress resistance in transgenic tobacco (*Nicotiana tabacum*) expressing a yeast gene for synthesis of the sugar trehalose, not normally synthesized by tobacco. Both control and transgenic plants were deprived of irrigation for 15 days and the extent of stress injury was estimated in terms of the percentage of leaves that were dead, damaged, or surviving fairly undamaged. Open bars = control plants; black bars = transgenic plants; lines indicate SE. The concentrations of trehalose in the transformed plants were low, indicating that the effect in this instance was not osmotic. From Romero *et al.* (1997). © Springer-Verlag GmbH & Co. KG.

13.3.5 Salinity: a special case of water-deficit stress

Most of the water on the earth's surface is salt water containing NaCl at about 3% or $0.5 \, mol \, L^{-1}$ and with a Ψ of $c. -2.5 \, MPa$. Very few flowering plants are marine, but around sea coasts terrestrial saline habitats supporting an angiosperm flora occupy large stretches as salt marshes of temperate regions, and as mangrove swamps of the tropics. There exist also salt-rich inland areas – salt deserts and regions around salt lakes, e.g. the Dead Sea and the Great Salt Lake. Further areas have become saline as a result of irrigation. A common definition of a saline habitat is one where the concentration of NaCl equals or exceeds 0.5%. Plants that are able to grow normally at 0.5% (or more) NaCl are classed as **halophytes** as opposed to salt-sensitive **glycophytes**. Halophytes may be obligate or facultative. Nearly all agricultural crops are glycophytes and salinity limits the utilization of considerable expanses of land.

The ions of the salt do exert direct toxic effects, but the major component of salinity stress is water deficit. Halophytes accumulate high levels of Na^+ and Cl^-. This enables them to maintain a low internal Ψ and avoid water deficits; for instance *Salicornia europaea*, grown in a saline soil at $\Psi -1.6 \, MPa$, can maintain a shoot Ψ of -6.5 MPa. In the halophyte tissues, the ions are segregated away from the cytoplasm in vacuoles and cell walls; in the cytoplasm, high concentrations of ions would interfere with protein structure and enzyme activity owing to their electric charges. The ions in the cell wall and vacuolar compartments are osmotically balanced by corresponding cytoplasmic concentrations of compatible solutes, especially proline and QAC. In plants capable of hardening towards salt stress, such compounds are synthesized in response to exposure to salt. Other common halophytic characters are succulence, the very large vacuoles permitting a dilution of the salt, and the possession of salt glands. These glands are found on the aerial parts of many halophytes and eliminate excess Na^+ and Cl^- ions by secretion onto the epidermal surface.

Some species achieve a certain degree of avoidance of salinity by salt exclusion, i.e. by avoiding the uptake of excessive amounts. Varieties of barley (*Hordeum vulgare*) and the grass *Agrostis stolonifera* differ in their resistance to salinity; in saline media, the more resistant varieties take up less Na^+ and Cl^- than the more sensitive ones. In some plants salt is absorbed into the root system but very little is transported into the shoots. The extent of resistance that can be achieved by salt exclusion is, however, limited, permitting the plants to endure only moderate levels of salinity.

13.4 | Low-temperature stress

13.4.1 Reactions to low temperature in the field

Most plants do not possess any specific system of temperature regulation. In the centres of tree trunks, temperatures as much as $10 \, °C$

above the ambient have sometimes been recorded, but usually, even inside bulky organs, the temperature is close to air temperature except in direct sunlight. On cold nights, leaf and bud temperature often falls below the ambient. Some mechanisms for keeping the temperatures of plant organs above air temperature have been noted. For example, at high altitudes in Africa and South America, where night temperatures may fall below $-10\,°C$ throughout the year, some species produce leaf rosettes up to 0.5 m in diameter. The adult leaves on the outside of the rosette fold over nightly into a giant bud enclosing the apical meristem, and the temperature of the meristem remains above zero. But usually, when a plant tissue withstands a specified external temperature, this is also the actual temperature of the cells. Avoidance of low temperatures by sensitive tissues is achieved almost exclusively through appropriate timing of developmental stages. Actively growing tissues are the most susceptible to cold stress, and the growth periods of many plants are confined to warm seasons, leaving dormant, resistant, tissues to overwinter.

The geographical spread of many species is determined by their ability to survive low temperatures. Plant species native to climates with a cold winter are more resistant than species of warm climates, as might be expected. The northernmost limit of a perennial plant species in the northern hemisphere frequently coincides with a certain winter minimum isotherm. Whilst frost is not responsible for as much famine as drought, it still has a considerable impact on agriculture, limiting the areas for growing particular crops and causing large losses in produce when frosts occur atypically late in the spring or early in the autumn, when even frost-hardy species are in a sensitive state. The cold resistance of perennials of temperate and arctic zones undergoes large annual changes (Fig. 13.7), being very much greater in the winter months than in the summer. The hardening occurs in the autumn as a response to the falling temperature and shortening daylength and is associated with the onset of winter dormancy. The loss of hardiness in the spring follows from the rise in temperature in the spring. Artificially, hardening can be induced at any season by chilling (see below), and resistance can be destroyed in midwinter by warming. There is a clear correspondence between levels of resistance and the temperatures to which various plant organs are exposed in their normal environment. Underground structures are much less resistant than aerial parts and their resistance is more uniform over the year, reflecting the less extreme and less fluctuating temperatures of the soil. In severe winters greater resistance is developed than in mild ones.

13.4.2 Chilling stress, freezing stress and limits of resistance

Numerous species originating in warm climates cannot endure exposure to temperatures below $5\,°C$ or $10\,°C$, in some cases not even below $15\,°C$. This reaction is known as chilling injury or cold shock. The ability to survive temperatures from $10\,°C$ (or $15\,°C$) down to zero, is

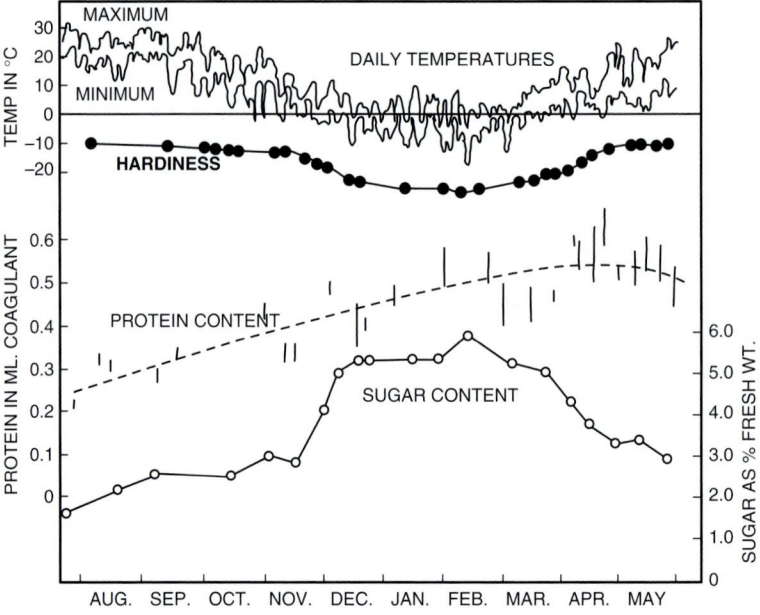

Fig. 13.7 Seasonal change in frost hardiness of the leaves of ivy (*Hedera helix*), a temperate-zone perennial. Hardiness is measured as the lowest temperature endured, with viability judged by the tetrazolium test for dehydrogenase enzyme activity. Actual temperatures during the period of measurement are given in the top traces (daily maximum and minimum). The content of soluble protein is also given, as ml precipitated by ethanol from 1 mL leaf extract, and the content of sugar as % of leaf fresh weight. Adapted from Parker (1962). © American Society of Plant Biologists; reprinted with permission.

chilling resistance. The ability to survive subzero temperatures is **freezing (frost) resistance**, which may be subdivided into freezing avoidance (no ice formed inside the plant) and freezing tolerance (survival of ice formation within the plant). Hardening towards freezing (frost hardening) is induced by exposure to temperatures of 0–10 °C for some days or weeks, and possession of chilling resistance is therefore a prerequisite for frost hardening.

The freezing resistance of a specimen is defined as the lowest temperature that it can survive. The effects of cold depend not only on the temperature reached, but upon rates of cooling and rewarming, and the time of exposure. Any value for low-temperature resistance therefore holds precisely only for the conditions under which it was determined. Examples of minimum temperatures which have been endured by flowering plants, or their organs, are given in Table 13.1. Since the lowest temperature recorded in regions of the earth where flowering plants grow is −60 to −70 °C, it can be seen that certain plant material can endure at least for short periods temperatures below those which plants will ever meet in nature, indeed close to the lowest temperature possible, absolute zero at −273 °C.

Table 13.1 Examples of low-temperature resistance of flowering plants. The 'time endured' refers to the particular experiments without implying that this is the longest that can be endured. For examples marked *, the temperature is the limit of resistance, in the other cases the temperature is the lowest tested. Where detached organs were used, long-term survival cannot be expected and viability was checked by tetrazolium tests for respiratory dehydrogenase activity, or by tests for normal membrane function (maintenance of semipermeability or capacity for plasmolysis). ND indicates no data available. From Street & Öpik (1984).

Plant material	Temperature (°C)	Time endured	Criterion of survival
Seeds, vacuum dehydrated	c. −273	2 h	Germination
	c. −273	10 h	Germination
	−190	6 weeks	Germination
Winter leaves, pine (*Pinus strobus*)	−196	5 min	Tetrazolium test, appearance for 4 weeks
Bark, mulberry (*Morus* sp.) twig	−196	160 days	Plasmolysis tests
Germinating seedlings, pea (*Pisum sativum*)	−183	1 min	Subsequent growth
Bark cells, false acacia (*Robinia pseudacacia*)	< −59	ND	Vital staining; cell permeability tests
*Winter buds, some temperate-zone deciduous trees	−21 to < −40	2 h	No visible damage up to 5 weeks later
*January leaves, ivy (*Hedera helix*)	−25	varied	Tetrazolium test, appearance
*Winter leaves, holm oak (*Quercus ilex*)	−13	ND	ND
*Summer leaves, holm oak	−6	ND	ND

13.4.3 Damage at low temperatures

Chilling damage

As stated, chilling-sensitive plants can suffer fatal cold shock at temperatures above zero but below a value depending on the species, most often about 10 °C. The main processes involved in chilling damage are indicated in Fig. 13.8. Membrane damage is a major factor in cold shock, which induces leakage from cells; this loss of semipermeability is often used as a measure of chilling injury. A physical effect of lowering of temperature is increased viscosity, or rigidity, of biological membranes. Below the critical temperature for cold shock, the membrane lipid bilayers of chilling-sensitive plants gel (Fig. 13.9) as the thermal motion of the lipid tails slows down. ROS accumulate, and (among other effects) also contribute to membrane gelling by peroxidation of membrane lipids. The accumulation of ROS is particularly high in photosynthetic tissues, for the rate of CO_2-fixing reactions is slowed drastically, but the photochemical reactions are not temperature-dependent and excess energy channels into ROS formation with consequent chloroplast damage. Low temperature inhibits translocation of photosynthate, causing end–product inhibition of photosynthesis and down-regulation of genes

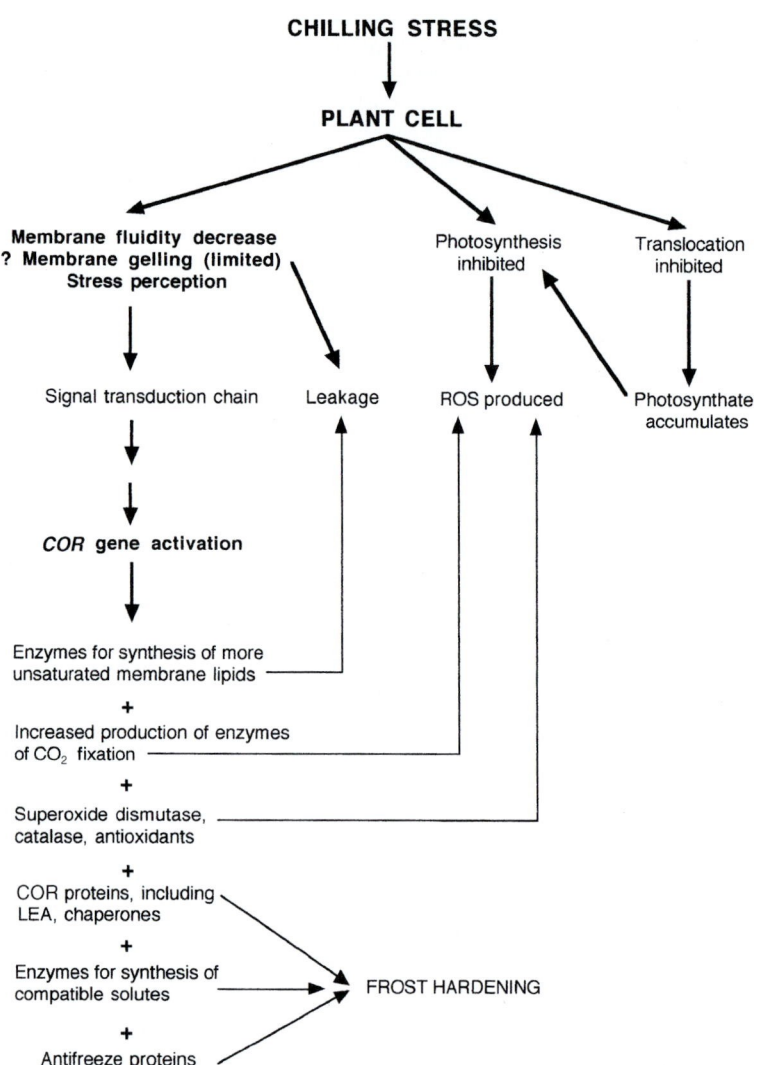

Fig. 13.8 Chilling stress and frost hardening of plant cells. The perception of stress leads to a signal transduction chain and gene activation. The new proteins include some that are active in counteracting (upward arrows) deleterious effects of chilling noted at the top of the diagram. Others, at the bottom, are listed as involved in frost hardening, but *COR* gene products and compatible solutes can also act in chilling tolerance. ROS = reactive oxygen species.

for components of the photosynthetic apparatus. This can lead to eventual loss of Rubisco protein; irreversible inactivation of Rubisco under chilling stress has also been reported, so that photosynthesis suffers severely, leading to nutrient shortages which further aggravate the chilling injury.

Freezing damage

Freezing damage depends on ice formation within plant tissues. Freezing can be **intracellular**, with the ice crystals forming inside the cells, or **extracellular**, with the ice confined to the apoplast, i.e. cell walls and extracellular (intercellular) spaces. In the field, extracellular freezing is much more common. Intracellular ice forms only when cooling is rapid, or when the tissue has supercooled. With slow cooling, ice formation always begins outside the cells and the lipid bilayer of the plasma membrane offers a high resistance to ice crystals

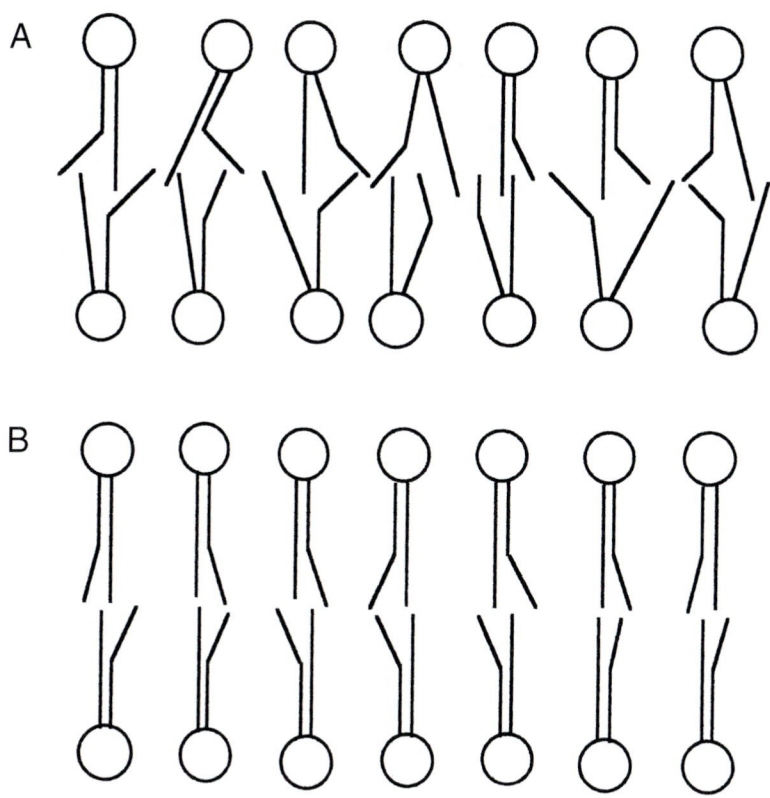

Fig. 13.9 (A) Diagrammatic cross-section of a bilayer of phospholipids in a fluid (liquid crystal) state; the hydrophobic tails of the lipids point towards the middle of the bilayer, but the arrangement is not very regular and it changes from moment to moment owing to thermal motion of the molecules. (B) Bilayer in a gel state, as at low temperature. The tails are now very regularly arranged and the structure is much more rigid, with little thermal motion.

growing into the protoplast. Although extracellular ice can distort and compress cells greatly, the cells remain intact and survive in hardy tissues. Intracellular freezing is fatal, the ice crystals disrupting the protoplasmic structure. In the laboratory, cells can survive 'vitrification' by very fast cooling to a glassy matrix with submicroscopic ice crystals, but vitrifying rates are never reached in nature.

The ice that forms in the apoplast is largely derived from the cytoplasmic and vacuolar water, which passes out to the growing ice crystals because the Ψ of ice (and of vapour pressure above ice) is lower than that of liquid water at the same temperature. Extracellular freezing therefore results in dehydration; (extracellular) freezing stress is basically a form of water-deficit stress, and desiccation damage accounts for a large part of freezing damage. Membrane damage, as discussed with reference to water-deficit stress, is fundamental also to freezing injury. Freeze-damaged cells typically leak on thawing. Freezing resistance often parallels desiccation resistance closely, i.e. the more resistant a tissue is to drying out (at higher temperatures), the more resistant it is also towards freezing. Nevertheless the removal of a given quantity of water by freezing does not have precisely the same effect as the removal of that amount at above-zero temperatures, and the fatal degree of dehydration has a different value when achieved by freezing and drying respectively.

One obvious reason for the difference is that at subzero temperatures metabolic reactions become very slow and metabolic disturbances would be expected to play a smaller role in freezing stress. Low temperature on the other hand weakens hydrophobic bonds and, in addition to the membrane damage that occurs on water withdrawal, membranes at low temperatures will lose stability because of the weakening of hydrophobic lipid–lipid and lipid–protein attractions.

13.4.4 Resistance to low-temperature stress

Chilling resistance, stress perception and frost hardening

In chilling-resistant plants, membranes do not undergo such damage at chilling temperatures as in sensitive plants. Chilling is indeed necessary for frost hardening (and for certain other developmental processes: vernalization, Section 11.3.5 and breaking of seed dormancy, Section 11.14.2). One postulate is that, in resistant plants, the lowering of temperature below a critical level causes a gelling only in limited areas, microdomains, of the membranes, resulting in less dramatic changes in membrane permeability, which now are sensed as the low-temperature signal. Within seconds of lowering the temperature, Ca^{2+} channel opening occurs in plasma membranes and this presumably sets off the signalling chain (Fig. 13.8). Profound changes take place in gene activity and protein content although all 'normal' protein synthesis is not repressed. Over 30 cold-induced genes (or gene families) are known and designated as the *COR* (COld Responsive or COld Regulated) genes/families. ABA can substitute for chilling in inducing acclimation, and many of the *COR* genes are responsive to ABA.

Cold treatment induces an increase in the degree of unsaturation of membrane phospholipids, and changes in membrane sterols and cerebrosides, making the membranes more fluid and functional at chilling temperatures. Chloroplast lipids are modified as well as those of the plasma membrane. Tobacco plants (*Nicotiana tabacum*) have been made more or less chilling sensitive by transferring to them genes of lipid metabolism from chilling-resistant or sensitive plants respectively. SOD activity increases, presumably helping to combat the increased production of superoxide radicals. Photosynthetic rate may undergo some recovery. In *Arabidopsis*, a sudden shift to temperatures below 10 °C results in a considerable inhibition of photosynthesis. But over a period of time the inhibition is relieved. The amount of the enzymes of C fixation increases, whether measured per unit weight, unit leaf area, or unit chlorophyll. When photosynthesis is measured at 5 °C, new leaves developed at 5 °C photosynthesize five times faster than leaves grown at 23 °C, and the photosynthetic rate of 5 °C-grown leaves at 5 °C is about the same as that of 23 °C-grown leaves at 23 °C (Fig. 13.10). The ability to translocate efficiently at low temperature can also be acquired during hardening.

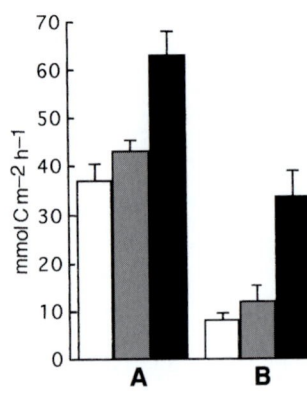

Fig. 13.10 Acclimation of *Arabidopsis* leaf photosynthesis rate to chilling temperatures. Photosynthesis, as rate of carbon assimilated per unit leaf area, was measured at 23 °C = (A), or 5 °C = (B). Open bars = plants grown at 23 °C throughout experiment; shaded = transferred to 5 °C for 10 days before measurement; black = transferred to 5 °C for 40 days before measurement, by which time a complete set of new leaves had grown. Each bar represents a mean of at least three measurements; the lines indicate the SD. At 5 °C the leaves develop a greatly enhanced potential for photosynthesis, associated with increases in enzymes of C fixation. Adapted from Strand *et al.* (1999). © American Society of Plant Biologists; reprinted with permission.

The physiological and structural changes described above and summarized in Fig. 13.8 help the plant to survive the chilling temperatures. At the same time, in potentially frost-hardy tissues, hardening towards subzero temperatures occurs, leading to freezing resistance by avoidance, tolerance, or both.

Freezing avoidance

Avoidance of freezing may be achieved by lowering of the freezing point by the synthesis of compatible solutes during hardening, mainly sugars (especially sucrose), polyols, and sometimes proline. The accumulation of these involves up-regulation of genes coding for enzymes of their synthesis, but the slowing down of their metabolic utilization, and slower translocation from source tissues, may also play a part. Freezing may additionally be inhibited by proteins with antifreeze activity coded by *COR* genes and secreted into the apoplast. These proteins bind to growing ice crystals, locally block the deposition of water molecules and break up the crystal's growth surface into numerous small areas, resulting in a higher free surface energy and requiring a lower temperature for more ice deposition, i.e. the proteins effectively depress the freezing point. The presence of antifreeze proteins also keeps down the crystal size.

The freezing point of cell contents is lowered by compatible solutes by a few degrees. Much lower temperatures can be reached without ice formation in tissues which become capable of **deep supercooling**. Ice crystals usually are initiated by nucleating substances or particles, on which water molecules align. In the absence of such 'nuclei' or 'seeds', water can supercool down to the **homogeneous nucleation point** at which ice forms spontaneously; for pure water this point is at $-38\,°C$, in solutions (and hence cells) it is some degrees lower. In many temperate-zone woody perennials deep supercooling occurs in xylem ray parenchyma and in the flower primordia inside the buds, while other tissues undergo and survive ice formation; the apple is an example of such a plant. The northern growth limit (in the northern hemisphere) of species which undergo deep supercooling is set by the $-40\,°C$ winter minimum isotherm. When supercooled cells do freeze, they freeze rapidly, intracellularly – and fatally. It is not known how deep supercooling is achieved. One would expect a cell, complex as it is, to contain some potential nucleating components. Possibly antinucleating compounds, which inhibit ice crystal formation, are produced during hardening.

Freezing tolerance

Freezing tolerance enables hydrated tissues to endure lower temperatures than supercooling, in some cases right down to $-196\,°C$ (Table 13.1). The freezing tolerance of a tissue is affected by the amount and state of water in its cells. The greatest cold resistance among flowering plants is encountered in dry seeds. Before subjection to the temperatures shown in Table 13.1, the seeds were dried in a vacuum, and the minute traces of water left were in a bound state

Box 13.1

Some bacteria (non-pathogenic) which inhabit plant surfaces, e.g. *Pseudomonas syringae*, synthesize proteins that can act as ice nucleating agents, and thereby cause frost damage to crops which otherwise would supercool by a few degrees – enough to survive spring/autumn frosts. This can be counteracted by spraying the plants with cultures of *P. syringae* engineered to lack the nucleating protein, and letting these outcompete the wild strain.

inaccessible to freezing. During the period of frost hardening of perennials, there is a tendency for a decrease in the total water content of the tissues, and an increase in the proportion of water bound to colloids, diminishing the amount of water available for freezing. Even so, many frost-hardy tissues still contain a very appreciable amount of freezable water.

Freezing tolerance depends on (1) ice formation being confined to the apoplast and (2) the ability of the protoplasts to survive the water deficit and compression by extracellular masses of ice. Extracellular freezing is promoted by increases in the permeability of the cells to water, making it easier for water to leave the cells, although this is not universal. The formation of compatible solutes is thought to stabilize protoplasmic colloids against dehydration damage, in addition to lowering the freezing point.

An appreciable number of the *COR* genes are homologous with the *LEA D11* or dehydrin group already mentioned in connection with water stress (Section 13.3.4) and their function in freezing tolerance is thought to be similar to that in desiccation tolerance in stabilizing the cells against the effects of water loss. Many of the *COR* genes can be induced by water stress, or by ABA. This emphasizes the close relationship between water stress and freezing stress.

13.5 | High-temperature stress

Heat injury is less of a problem in agriculture than drought or frost damage; high temperatures more frequently cause problems indirectly by induction of water stress. Photosynthesis, however, has a rather low temperature optimum in plants of temperate climates, only 20–30 °C, and therefore moderately high temperatures can already cause decreases in crop yield although no injury is apparent. In hot climates, high temperature does become a limiting factor for survival, but since the hottest regions are also the most arid, it is difficult to distinguish the effects of heat stress from water-deficit stress in such habitats.

13.5.1 Limits of survival of high temperature

The heat-killing temperature, the **thermal death point**, of a plant tissue is highly dependent on the time of exposure: with shorter times, higher temperatures can be endured. For example, cells of *Tradescantia discolor* have been killed in 7 min at 60 °C, in 4 h at 50 °C, and about 22 h at 40 °C. Dry seeds have survived heating above 100 °C for periods from a few hours to a few days. For hydrated organs of temperate-zone plants, the thermal death point generally lies between 45 and 55 °C for exposures of some hours – the length of period that daily maximal temperatures would last in natural habitats. Desert plants such as cacti have thermal death points of over 60 °C, whilst aquatic and shade plants may survive only up to about 40 °C.

These values might suggest that plants are safe from heat injury in temperate climates, where the air temperature is unlikely to exceed 40 °C. The temperatures of plant tissues may, however, be well above the ambient. In the sun, even thin leaves can warm up 6 to 10 °C above the air temperature in spite of transpirational cooling. Fleshy leaves have reached internal temperatures of 40–50 °C with the air temperature at 20–30 °C, and the temperature of the cambium on the sunny side of a tree has been recorded at 55 °C. Sunscald injury to the southern sides of trees, and the sunny sides of fruits, is a well-known phenomenon. Soil surface temperatures can rise far above air temperature on sunny days and seedlings are sometimes killed at soil level by overheating.

13.5.2 Basis of thermal injury

Death that occurs within a few seconds or minutes at very high temperatures can be attributed primarily to protein denaturation, leading to a catastrophic collapse of cellular organization. Coagulation of cell contents can be observed microscopically and many proteins are known to undergo denaturation at temperatures that correspond to the thermal death points of plants. The Q_{10} value of thermal killing of plants can be up to 2150 (!); the only chemical process known to have such a high Q_{10} is protein denaturation.

The slower heat injury observed at less extreme temperatures, when the time for development of damage is measured in hours or days, is a more complex phenomenon (Fig. 13.11). Protein denaturation, while slower than at the very high temperatures, occurs and leads to numerous disturbances in cell organization, including membrane organization, and metabolic abnormalities. Increased thermal motion of molecules results in increased membrane fluidity and excessive fluidity (like excessive rigidity at low temperatures) is deleterious. There is production of ROS. Photosynthesis is very sensitive to high temperature, with photosystem II being a primary site for damage, as in the case of water stress. The enzyme Rubisco activase, required to activate Rubisco, is also exceptionally heat-labile. Respiration rate by contrast increases with increasing temperature to near the lethal point and consequently there is danger of starvation. There is some evidence that particular essential metabolites may run short, leading to metabolic imbalances.

13.5.3 High-temperature resistance

Heat avoidance

Plants native to hot climates show some adaptations for heat avoidance, i.e. for keeping their temperatures below the air temperature, or at least for not heating up above it. Some of the xeromorphic features shown by plants of hot, dry climates serve to keep the temperature down – reflective surfaces and small leaves coolable by convection. Transpirational cooling lowers leaf temperatures, in some cases to 10 °C below air levels. Significant transpirational

Fig. 13.11 High-temperature stress and heat hardening of plant cells. The signal perception may include sensing of denatured proteins in the cytosol, as well as the effect on membranes. Upward arrows indicate the counteracting of deleterious effects of the high temperature by the products of the gene activation, i.e. the hardening process. The production of heat-stable enzymes is bracketed since it is not a major process in heat hardening of flowering plants. ROS = reactive oxygen species.

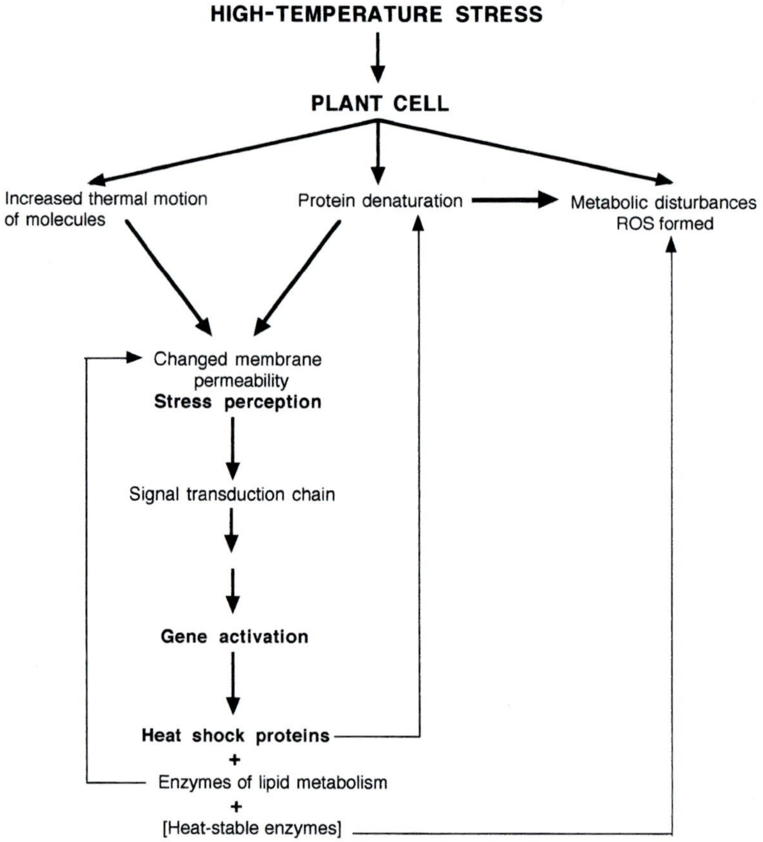

Box 13.2

SDS-PAGE is sodium dodecyl sulphate/polyacrylamide gel electrophoresis, a method for separating proteins in extracts. The extract is treated with SDS, an anionic (negatively charged) detergent, which binds to the proteins, forming negatively charged complexes. The mixture is then loaded on to strips of polyacrylamide gel and an electric field is applied. The negatively charged complexes migrate towards the positive pole, the smallest proteins, with the highest charge : mass ratio, moving fastest, and hence the proteins separate into bands according to molecular mass. The bands are detected by colour staining, or, if the proteins were radiolabelled, by autoradiography. Mixtures of proteins of known mass are run along parallel gel strips as calibration standards.

cooling, however, is possible only when there is an ample water supply, and many heat-resistant plants grow in dry habitats and are water conservers.

Heat hardening and heat tolerance (thermotolerance)

There is good evidence that tolerance to high temperature resides largely in the properties of cellular proteins and membrane lipids. When plant tissue is suddenly transferred to a temperature above a threshold, there is a drastic change in protein synthesis. The threshold depends on the species; for temperate zone plants it is in the 35–45 °C range, higher for plants of hot climates and lower for cool-climate species. The production of 'normal' proteins ceases, but within minutes transcription of new species of mRNA becomes detectable. The rapidity of the response suggests a perception mechanism that is more or less instantaneous, such as an effect on membrane fluidity and hence permeability (Fig. 13.11). The newly transcribed mRNA species code for numerous **heat-shock proteins**, HSP (Fig. 13.12). A gradual rise in temperature also results in HSP synthesis and hardening occurs. Organisms of other living kingdoms produce HSP under similar circumstances, and there is considerable homology between a number of HSP from unrelated organisms, e.g. flowering plants and *Drosophila*. But there are several families of HSP and these vary in importance in different organisms. The small HSP family (masses 17–30 kDa) is particularly abundant in flowering plants; these HSP form granules which are believed to provide surfaces on which partially heat-denatured proteins are held and protected from irreversible denaturing and aggregation. The HSP100 family (masses about 100 kDa) is also common in flowering plants; HSP100 are thought to act by solubilizing proteins which have aggregated. At least one member of this family, HSP101 from *Arabidopsis*, is of major importance. Heat-shock survival of plants engineered with lowered HSP levels is reduced in proportion to the amount of decrease in HSP101, and plants constructed to make HSP101 constitutively show increased thermotolerance without heat hardening (Gurley 2000). Other types of HSP

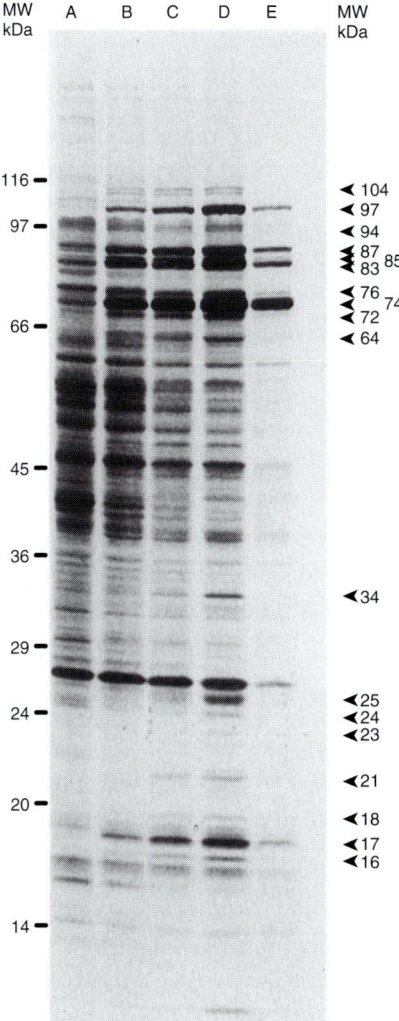

Fig. 13.12 Detection of heat-shock protein synthesis by SDS-PAGE separation (see Box 13.2). Seedlings of sorghum (*Sorghum bicolor*) were grown at 35 °C (optimal temperature for this cereal of warm climates), and samples were incubated for 2 h at temperatures of up to 47 °C in the presence of the amino acid methionine, radioactively labelled. The proteins synthesized during that period are detectable as radioactive bands on the gels. Temperatures of incubation were: (A) = 35 °C; (B) = 40 °C; (C) = 43 °C; (D) = 45 °C and (E) = 47 °C. With increasing temperature, synthesis of many proteins ceases, but new HSP appear. Molecular mass values of standards (run parallel to the extracts) are shown to the left of the gels; arrowed numbers on the right indicate molecular masses of recognizable HSP bands. From Howarth & Ougham (1993).

are believed to refold unfolded proteins. Thus many HSP appear to have chaperone activity (Section 13.3.4), as do some COR and LEA proteins. Another HSP, ubiquitin, tags proteins to which it binds for degradation, and it is thought that this is how heat-damaged proteins are marked for hydrolysis.

The production of HSP is a transient phenomenon and the synthesis of 'normal' proteins is resumed after a period; in soybean (*Glycine max* (soja)) treated at 40 °C this occurs after six hours, the original mRNA being preserved in an inactive form while the mRNA for HSP is being translated. On return to below the threshold temperature, the HSP synthesis ceases and thermotolerance is lost. Pearl millet (*Pennisetum glaucum*) given daily 50 °C heat shocks undergoes daily cycles of acclimation, with HSP synthesis and dehardening (Howarth 1991). Although the HSP are long-lived – e.g. one HSP from pea has a half-life of 52 days – new synthesis seems to be required each time the temperature is raised. The presence of previously synthesized HSP fails to maintain thermotolerance at lower temperatures.

Some enzymes from thermotolerant plants are more heat-stable than the corresponding enzymes from heat-sensitive species, and increases in the thermostability of some enzymes during heat hardening have been noted, but this is only occasional.

The properties of membrane lipids are also implicated in heat tolerance, the lipids from more tolerant or hardened tissues being more saturated. With a rise in temperature, membrane fluidity increases because of increased thermal motion of the molecules; in an excessively fluid membrane, membrane proteins are no longer held in the precise spatial arrangement needed for normal membrane function: in effect the membrane melts. A high level of lipid saturation raises the melting point and confers more rigidity on the membranes.

13.6 | Relationships between different types of stress resistance: cross-tolerance

The resistance to high temperature of a number of temperate-zone evergreens shows two annual maxima. One is in the summer as expected; the second coincides with the midwinter maximum of resistance to cold and water-deficit stress. Hardening against cold or water stress increases thermotolerance; hardening against freezing hardens against water-deficit stress and vice versa. Furthermore, during the midwinter maximum resistance towards freezing, water deficit and high temperature, plants also show high resistance towards other injurious factors such as toxic chemicals and lack of O_2. This is known as **cross-tolerance**. Individual resistance genes may be activated via more than one pathway, and one stress signal may activate several pathways, resulting in complex networks

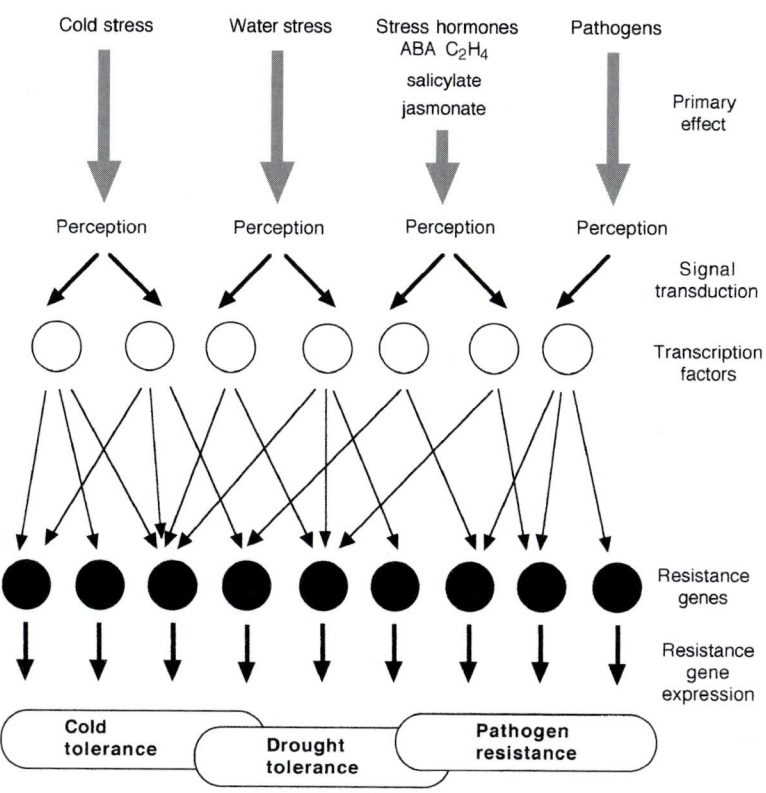

Fig. 13.13 Networks (crosstalk) between stress reactions: a highly diagrammatic representation of a small segment, including only a few stresses and signals. The signal transduction chains leading to transcription factor production involve activation of transcription factor genes. Resistance gene expression initiates reaction chains culminating in the development of resistance.

('crosstalk') of interactions, with which defence reactions against pathogens are also integrated (Fig. 13.13). Numerous individual reaction chains, transcription factors and resistance genes have been identified; the challenge is to understand how the network acts as a functional whole.

The likely physiological advantages of cross-tolerance can be recognized. Water-deficit stress and freezing stress both involve danger of desiccation of protoplasm, so it is not surprising that *COR* genes can be induced by water stress, and that many *COR* genes are related to, or even identical with, *LEA* genes active during seed dehydration and induced by water stress. Compatible solutes are efficacious in maintaining water status under water stress, in stabilizing structure in dehydrated tissue (including frozen) and in depressing the freezing point. Some HSP are induced by low temperature and water stress. Many HSP are also present at low levels, or as isoforms, in unstressed tissues. They appear, or increase in amount, at normal growth temperatures during certain developmental stages: embryogenesis, pollen development, seed germination, fruit maturation, and photoperiodic response. The heat shock response and heat hardening involve an intense activation of systems for maintenance of protein structure, which is very heat sensitive. But similar protein maintenance processes are required to meet other stresses, too; they are required to some extent even at lower temperatures – and utilize the same so-called HSP. Other features shared by varied stresses are the involvement of 'stress hormones', e.g. ABA, in signalling cascades, and an increase in ROS; the latter has actually been suggested to be the common starting point in the signal transduction chains for all types of stress. In *Arabidopsis* 113 genes inducible by ROS have been identified (Vranová *et al.* 2002). Indeed, ROS may be involved in the control of development in unstressed conditions.

Each type of stress nevertheless carries with it specific problems. While cold hardening improves heat resistance, the reverse is not true and the midsummer maximum of heat resistance coincides with the minimum of freezing resistance. Some mechanisms of protection against one type of stress are incompatible with resistance towards another; e.g. a high proportion of unsaturated lipids in membranes is favourable at low temperatures, but unfavourable at high temperatures. The concept of a general resistance of protoplasm towards all injurious agents, sometimes mooted, is an oversimplification.

13.7 | Development of stress-resistant crop plants

The production of stress-resistant varieties of crop plants by selection from varieties already present, and by cross-breeding, has been carried out over centuries. Now attention is focusing on the production of transgenic plants with resistance-conferring genes. It should be obvious from the preceding text that stress resistance is a very complex phenomenon under multigenic control. Nevertheless, there are numerous records of transfer of single 'tolerance genes' which have resulted in an improvement of the stress resistance of the recipient plant, as already occasionally noted; Table 13.2 summarizes some examples, but the problems for practical application are numerous. The increases in resistance are often insufficient to make a significant difference to survival under field conditions. However, in some cases it works. Tomato plants expressing a tonoplast Na^+/H^+ antiporter which pumps Na^+ ions into the vacuole (Table 13.2) grew well in 0.2 M NaCl, and the salt was moreover not accumulated in the fruits.

Some increases in gene activity in response to a stress may be a sign of injury rather than of acclimation, and activation of such genes is not desirable. The precise functions of the majority of stress-induced proteins are not known, and deductions made can be disputable. For example, there is a general assumption that the COR proteins are highly hydrophilic, and protect cellular structures against dehydration damage. But two particular COR polypeptides synthesized in transgenic *E. coli* exhibited only an average level of hydration and failed to protect artificial lipid vesicles against dehydration-induced gelling of the lipid. There is a pressing need for fuller information on the functions of the very numerous stress-induced proteins, for optimal planning of transformations.

Table 13.2 | Some examples of increases in plant stress resistance achieved by single gene transfers.

Gene transferred	Recipient	Resistance type increased	Reference
Gene for trehalose synthesis from yeast	Tobacco (*Nicotiana tabacum*)	Water stress	Holmström *et al.* (1996)
SabB for fructan synthesis from *Bacillus subtilis*	Tobacco	Water stress	Pilon-Smits *et al.* (1995)
HVA1 (*LEA* group3) from barley (*Hordeum vulgare*)	Rice (*Oryza sativa*)	Water stress	Xu *et al.* (1996)
Arabidopsis gene *AtNHX1* for Na^+ porter into vacuole	Tomato (*Lycopersicon esculentum*)	Salinity	Zhang & Blumwald (2001)
Gene for proline synthesis from *Vigna aconitifolia*	Tobacco	Salinity	Kishor *et al.* (1995)
Lipid-synthesis gene from chilling-resistant *Arabidopsis*	Tobacco	Chilling	Murata *et al.* (1992)
FeSOD gene from *Arabidopsis* over-expressed in plastids and mitochondria	Maize (*Zea mays*)	Oxidative stress; possibly chilling	Van Breusegem *et al.* (1999)
Gene for MnSOD from *Nicotiana* sp. over-expressed in plastids and mitochondria	Alfalfa (*Medicago sativa*)	Better field survival of winter	McKersie *et al.* (1999)

To get a full benefit from introduced 'resistance genes' it may be necessary to make other transfers. When genes for synthesis of glycine betaine were introduced into a species which normally does not accumulate the chemical, only a small rise in the betaine content and stress tolerance was produced. The workers suggested that this resulted from a lack of a precursor, choline; to obtain higher levels of the betaine, it might be necessary to up-regulate choline synthesis, too. The withdrawal of metabolic intermediates into synthesis of compatible solutes could have deleterious effects on other aspects of a plant's physiology. Betaines and proline are nitrogenous compounds and N tends to limit plant growth. Resistance to water stress through an engineered synthesis of N-containing compatible solutes might be bought at the price of a lower yield. Even the accumulation of carbohydrates, such as mannitol, might decrease the harvestable mass. Also, the change in chemical composition could affect the quality of produce. Such side effects of transfer of genes for resistance also need investigation before the transgenic plants are grown on the commercial scale.

Many 'resistance genes' are present in most flowering plants, but may be expressed in non-hardy species only in certain tissues, and not in response to stress – e.g. the *LEA* genes. Or differences between resistant and sensitive species/varieties may be quantitative rather than qualitative. Winter barley is cold resistant in comparison with spring barley; *COR* genes are expressed in both varieties in response to chilling, but to a larger extent in the winter variety. In such instances the genetic engineering must aim for control of gene expression, so that the required genes are expressed (1) in response to stress, (2) to an adequate degree and (3) in all relevant parts of the plant. Attention is being directed towards transgenic manipulation of the control mechanisms for switching on 'tolerance genes' (in contrast to introducing such genes), all the more since protective mechanisms and signal transduction chains are shared by several stresses, and multiple stresses are also common in the field (e.g. water deficit and low temperature in winter). One successful example is the introduction into *Arabidopsis* of a construct coding for a transcription factor DREB1A controlled by a stress-inducible promoter (Kasuga *et al.* 1999). DREB1A can activate numerous stress-tolerance genes. The transgenic plants exhibited markedly increased resistance towards water stress, salinity and low temperature, with increased activation of at least six genes involved in stress tolerance. It is probably in such manipulations that the key to successful engineering for hardiness lies.

Box 13.3

Constitutive expression of tolerance genes may be undesirable; it puts an extra drain on the plant's resources. With a single tolerance gene transfer this may not be too serious, but the *Arabidopsis* plants engineered to express the DREBIA gene, and hence numerous other stress-inducible genes constitutively, suffered severe growth inhibition. Hardened plants may be at a disadvantage in unstressed conditions; e.g. certain cold-hardy evergreen leaves translocated sugars out better at 10 °C than at 20 °C. Plants have been reported even to undergo daily cycles of hardening and dehardening towards frost and UV light.

Complementary reading

Bohnert, H. J., Nelson, D. E. & Jensen, R. G. Adaptations to environmental stresses. *The Plant Cell*, **7** (1995), 1099–111.

Campalans, A., Messeguer, R., Goday, A. & Pagès, M. Plant responses to drought, from ABA signal transduction events to the action of the induced proteins. *Plant Physiology and Biochemistry*, **37** (1999), 327–40.

Giorni, E., Crosatti, C., Baldi, P. *et al.* Cold-regulated gene expression during winter in frost tolerant and frost susceptible barley cultivars grown under field conditions. *Euphytica*, **106** (1999), 149–57.

Hernandez, J. A., Jiménez, A., Mullineaux, P. & Sevila, F. Tolerance of pea (*Pisum sativum* L.) to long-term salt stress is associated with induction of antioxidant defences. *Plant, Cell and Environment*, **23** (2000), 853–62.

Hughes, M. A., & Dunn, M. A. The molecular biology of plant acclimation to low temperature. *Journal of Experimental Botany*, **47** (1996), 291–305.

McNeil, S. D., Nuccio, L. & Hanson, A. D. Betaines and related osmoprotectants. Targets for metabolic engineering of stress resistance. *Plant Physiology*, **120** (1999), 945–9.

Munns, R. Comparative physiology of salt and water stress. *Plant, Cell and Environment*, **25** (2002), 153–61.

Murata, N. & Los, D. A. Membrane fluidity and temperature perception. *Plant Physiology*, **115** (1997), 875–9.

Pastori, G. M. & Foyer, C. H. Common components, networks and pathways of cross-tolerance to stress. The central role of 'redox' and abscisic acid mediated controls. *Plant Physiology*, **129** (2002), 460–8.

Queiroz, C. G. S., Alonso, A., Mares-Guia, M. & Magalhães, A. C. Chilling-induced changes in membrane fluidity and antioxidant enzyme activities in *Coffea arabica* L. roots. *Biologia Plantarum*, **41** (1998), 403–13.

Queitsch, C., Hong, S.-W., Vierling, E. & Lindquist, S. Heat shock protein 101 plays a crucial role in thermotolerance in *Arabidopsis*. *The Plant Cell*, **12** (2000), 479–92.

Ramanjulu, S. & Bartels, D. Drought- and desiccation-modulated gene expression in plants. *Plant, Cell and Environment*, **25** (2002), 141–51.

Tabaeizadeh, Z. Drought-induced responses in plant cells. *International Review of Cytology*, **182** (1998), 193–237.

Webb, M. S., Gilmour, S. J., Thomashow, M. F. & Steponkus, P. L. Effects of COR6.6 and COR15am polypeptides encoded by COR (cold-regulated) genes of *Arabidopsis thaliana* on dehydration-induced phase transitions of phospholipid membranes. *Plant Physiology*, **111** (1996), 301–12.

Wolkers, W. F., Tetteroo, F. A. A., Alberda, M. & Hoekstra, F. A. Changed properties of the cytoplasmic matrix associated with desiccation tolerance of dried carrot somatic embryos: an *in situ* Fourier transform infrared spectroscopic study. *Plant Physiology*, **120** (1999), 153–63.

Xin, Z. & Browse, J. Cold comfort farm: the acclimation of plants to freezing temperatures. *Plant, Cell and Environment*, **23** (2000), 893–902.

Yu, X.-M. & Griffith, M. Antifreeze proteins in winter rye leaves form oligomeric complexes. *Plant Physiology*, **119** (1999), 1361–9.

References

Bray, E. A. (2002). Abscisic acid regulation of gene expression during water-deficit stress in the era of the *Arabidopsis* genome. *Plant, Cell and Environment*, **25**, 153–61.

Gurley, W. B. (2000). HSP101: a key component of the acquisition of thermotolerance in plants. *The Plant Cell*, **12**, 457–60.

Holmström, K., Mäntyla, E., Welin, B., Mandal, A. & Palva, E.T. (1996). Drought tolerance in tobacco. *Nature*.

Howarth, C. J. (1991). Molecular responses of plants to increased incidence of heat shock. *Plant, Cell and Environment*, **14**, 831–41.

Howarth, C. J. & Ougham, H. J. (1993). Gene expression under temperature stress. *New Phytologist*, **125**, 1–26.

Hsiao, T.C., Acevedo, E., Fereres, E. & Henderson, D. W. (1976). Stress metabolism, water stress, growth and osmotic adjustment. *Philosophical Transactions of the Royal Society of London*, **B 273**, 479–500.

Kasuga, M., Liu, Q., Miura, S., Yamaguchi-Shinozaki, K. & Shinozaki, K. (1999). Improving plant drought, salt, and freezing tolerance by gene transfer of a single stress-inducible transcription factor. *Nature Biotechnology*, **17**, 287–91.

Kishor, P. B. K., Hong, Z., Miao, G.-H., Hu, C. A. A. & Verma, D. P. S. (1995). Overexpression of Δ1-pyrroline-5-carboxylate synthetase increases proline production and confers osmotolerance in transgenic plants. *Plant Physiology*, **108**, 1387–94.

McKersie, B. D., Bowley, S. B. & Jones, K. S. (1999). Winter survival of transgenic alfalfa overexpressing superoxide dismutase. *Plant Physiology*, **119**, 839–47.

Murata, N., Ishizaki-Nishizawa, O., Higashi, S., Hayashi, H., Tasaka, Y. & Nishida, I. (1992). Genetically engineered alteration in the chilling sensitivity of plants. *Nature*, **356**, 710–13.

Parker, J. (1962). Relationships among cold hardiness, water-soluble protein, anthocyanins and free sugars in *Hedera helix* L. *Plant Physiology*, **37**, 809–13.

Pilon-Smits, E. A. H., Ebskamp, M. J. M., Paul, M. J., Jeuken, M. J. W., Weisbeek, P. J. & Smeekens, S. C. M. (1995). Improved performance of transgenic fructan-accumulating tobacco under drought stress. *Plant Physiology*, **107**, 125–30.

Romero, C., Bellés, J. M., Vayá, J. L., Serrano, R. & Culiáñes-Maciá, F. A. (1997). Expression of the yeast *trehalose-6-phosphate synthase* gene in transgenic tobacco plants: pleiotropic phenotypes include drought tolerance. *Planta*, **201**, 293–7.

Seki, M., Narusaka, M., Abe, H. *et al.* (2001). Monitoring the expression pattern of 1300 *Arabidopsis* genes under drought and cold stresses by using a full-length cDNA microarray. *The Plant Cell*, **13**, 61–72.

Street, H. E. & Öpik, H. (1984). *The Physiology of Flowering Plants*, 3rd edn. London: Edward Arnold.

Strand, Å., Hurry, V., Henkes, S. *et al.* (1999). Acclimation of *Arabidopsis* leaves developing at low temperatures: increasing cytoplasmic volume accompanies increased activities of enzymes in the Calvin cycle and in the sucrose-biosynthetic pathway. *Plant Physiology*, **119**, 1387–97.

Van Breusegem, F., Slooten, L., Stassart, J.-M. *et al.* (1999). Overproduction of *Arabidopsis thaliana* FeSOD confers oxidative stress tolerance to transgenic maize. *Plant and Cell Physiology*, **40**, 515–23.

Vranová, E., Inzé, D. & Van Breusegem, F. (2002). Signal transduction during oxidative stress. *Journal of Experimental Botany*, **53**, 1227–36.

Xu, D., Duan, X., Wang, B., Ho, D. & Wu, R. (1996). Expression of a late embryogenesis abundant protein gene, *HVA1*, from barley confers tolerance to water deficit and salt stress in transgenic rice. *Plant Physiology*, **110**, 249–57.

Zhang, H.-X. & Blumwald, E. (2001). Transgenic salt-tolerant tomato plants accumulate salt in foliage but not in fruit. *Nature Biotechnology*, **19**, 765–8.

Appendix

A.I | NAMING GENES, PROTEINS AND MUTATIONS

A consistent nomenclature has been adopted for naming genes and proteins. A wild-type gene is written in italic as e.g. *CRY1* (for crypto-chrome) whilst mutated versions of it are written *cry1*. The first mutation discovered in the gene would be *cry1-1*, the second *cry1-2*, etc. In the case of cryptochrome (and many other proteins, including phytochrome) the receptor consists of a protein (the apoprotein) and a chromophore which together make the functional protein (holo-protein). The apoprotein is written as CRY1 whilst the holoprotein is written as cry1. The same mutants are often isolated by different research groups and are given different names. Once a class of genes has been relatively well studied they are sometimes renamed to avoid confusion.

A.2 | UNITS OF MEASUREMENT

The system of SI units, Système International d'Unités, was intro-duced in 1960. In this system, the basic units of mass, length and time are the kilogram (kg), metre (m), and second (s); a number of common units, e.g. litre and hour, are abandoned. Older units, however, still abound even in current scientific literature because of their conve-nience (and familiarity), and some are retained in this text. Note that there is no full stop after the abbreviations, and no 's' to denote the plural: thus we write 1 m and 10 m.

SI units	Previously employed units
Length Basic unit metre, m 1 μm, micrometre $= 10^{-6}$ m 1 nm, nanometre $= 10^{-9}$ m	Basic unit metre, m 1 μ, micron $= 10^{-6}$ m 1mμ, millimicron $= 10^{-9}$ m
Volume Basic unit cubic metre, m^3 1 dm$^3 = 10^{-3}$ m^3 1 cm$^3 = 10^{-6}$ m^3 1mm$^3 = 10^{-9}$ m^3	Basic unit litre L (or l) $= 10^{-3}$ m^3 1 L = 1dm^3 1 mL, millilitre $= 10^{-3}$ L = 1 cm^3 1 μL microlitre $= 10^{-6}$ L = 1 mm^3
Energy Basic unit joule, J	Basic unit erg $= 10^{-7}$ J 1 calorie, cal $= 4.18$ J

Temperature
 Basic unit kelvin, K (formerly degrees Kelvin, °K).
    1 K = 1 °C, but the kelvin scale starts at absolute
    zero, so that 0 K $= -273$ °C, and 0 °C $= 273$ K
	Degrees centigrade (Celsius), °C

Pressure Basic unit newton per square metre, N m$^{-2} =$ the pascal, Pa (1 N = 1 kg m s^{-2}) kPa, kilopascal $= 10^3$ Pa MPa, megapascal $= 10^6$ Pa	Bar or atmosphere, atm 1 bar $= 10^5$ Pa 1 atm $= 1.01325 \times 10^5$ Pa

Amount of substance
 Mole (symbol mol), amount of substance of a
    system which contains as many elementary
    units as there are carbon atoms in 0.012 kg of
    carbon-12. Can be applied to entities other
    than chemicals, e.g. used for 'amount' of light in
    terms of photons
	Gram-molecule (g-mole), the amount of a
	    compound whose mass in g is equal to its
	    molecular weight; numerically equals mole but
	    only applicable to chemical substances

Concentration Measured in mol m^{-3} (or mol dm^{-3}, mol cm^{-3} etc.)	Measured in molarity, M A 1 M (molar) solution contains 1 g-mole L$^{-1} = 10^3$ mol m^{-3}

Avogadro's number
The number of elementary units per mole (or g-mole) $= 6.02252 \times 10^{23}$ mol^{-1}

Mass of very small entities
Masses of atoms, molecules and molecular aggregates are minute and therefore generally not measured in
    SI mass units, but given relative to each other; starting with a value of 1 for the lightest atom, H. One can
    either say that the H atom has a relative molecular mass M_r of 1, or that it has a mass of 1 dalton, Da.

A.3 | PREFIXES FOR UNITS

Prefixes for large numbers		Prefixes for small numbers	
Prefix	Multiply by	Prefix	Multiply by
deka- da	10	deci- d	10^{-1}
hekto- h	10^2	centi- c	10^{-2}
kilo- k	10^3	milli- m	10^{-3}
mega- M	10^6	micro- μ	10^{-6}
giga- G	10^9	nano- n	10^{-9}
tera- T	10^{12}	pico- p	10^{-12}
peta- P	10^{15}	femto- f	10^{-15}
exa- E	10^{18}	atto- a	10^{-18}

Index

Page numbers in *italics* refer to figures and tables

ABA (abscisic acid) 21, 86, 105, *180*, 187, 198–199, 210, 311, *312*, 315, 352, 360, 362, 368
ABC model *see* floral development
abrin 308
Abrus precatorius 308
abscisic acid *see* ABA
abscisins 187
abscission 138, 188
absorption spectrum
  carotenoids 13, *13*, *337*
  chlorophylls 13, *13*
  flavins *337*
  phytochromes 249, 250, *250*, 263
ACC oxidase 310
ACC synthase 197, 310
acclimation *see* stress
accumulator species 106, 118
Acer platanoides see Norway maple
Acer saccharum see sugar maple
acetaldehyde 51, 52
acetyl CoA 43, 45, 315
acid growth theory 215–217
Acorus calamus 50, 52
action spectrum
  germination 248, *249*, 258, *258*
  photomorphogenesis *252*, 258
  photoperiodism 273
  photosynthesis 13
  phototropism 337, *337*
adenosine diphosphate *see* ADP
adenosine monophosphate *see* AMP
adenosine triphosphate *see* ATP
Adenostoma fasciculatum 69
ADP (adenosine diphosphate) 10, 15, 41, 54–55
ADP glucose pyrophosphorylase 304, 305
Aegopodium podagraria 323
Aequoria victoria 199, 265
aequorin 199

aeration of soil 88
aerobic respiration 39, 45, 50, 126
aerotropism 320
after-ripening 313
AG (AGAMOUS) gene 289–290
  agamous 288, 289
agaves 30, 150
age
  sc
  of plants 165, *230*, 277
ageing 164
agriculture 38, 181, 185, 206, 263, 299, 300, 308, 314, 355
  yield in 32, 34, 129, 344, 347, 362
Agrobacterium tumefaciens 186, 187, 213
Agrostis stolonifera 24, 354
Ailanthus altissima 134
Aizoaceae 29
alcohol *see* ethanol
alcohol dehydrogenase (ADH) 51, 52, 53, 105
alder 127
aldolase 41, 51
aleurone layer 302, 315
alfalfa 68, *369*
algae 11, 22, 122, 129
  gravitropism 333
  ion uptake by 112
  translocation in 155
Alisma plantago-aquatica 311
alkaloid 102
Allium cepa see onion
Allium sativum see garlic
allometric growth *see* growth
almond *304*
Alnus see alder
Aloe 95
Alternaria brassicola 202
alternative oxidase 47–48, 54
aluminium 106
Amanita muscaria 121

Amaranthaceae 29
amides 85, 137, 143
amino acids 23, 26, 39, 44, 123, 137, 138, 143, 352
  non-protein 103, 308
  in seeds 307, 308
  transport 85, 143, 155
ammonia 26, 128
ammonium 107, 113, 122–123, 126, 128
AMP (adenosine monophosphate) 10
amylase 52, 103, 315
amyloplast 304, *333*, *335*
anaerobic respiration 41, 49–53
anaerobiosis 49–53
Ananas comosus see pineapple
androecium 285
annual
  rhythms 175
  rings 71, 228
anoxia 50–51, 52
anther 285, 291, *292*, 298, 299
anthocyanin 247, 257, 259
antibiotic resistance 186
antibodies to hormones 193–194
antifreeze protein 361
antioxidants 16, 103, 353
antipodal cells 294
antiport 117, *117*
Antirrhinum majus 231, 282, 283, 285, 286, *286*
antisense repression 75, 143
AOS *see* reactive oxygen species (ROS)
AP (APETALA) gene 283–285
  apetala 288, 289
AP protein 290
aphids 137, 150
apical dominance 182, 206, 253
apical hook 188, 259, *259*
apical meristems *see* root, shoot
  meristems

apomixis 303
apoplast 72–76, *74*, 93, 108, 114, 215, 232, 358, 359, 361
apoplastic loading *see* phloem
apple 68, 93, 308, 309, 361
aquaglyceroporins 67
aquaporins 67, 75, 118, 298
aquatic plants 30, 50, 52, 105, 166, 349, 362
Arabidopsis thaliana (thale cress) 187, *188*, *231*, 234, *256*, *261*, 294, *302*, 331, 334, *339*, 340, *360*
   flower structure 285, 286
   flowering control 279–281, *280*, 282, 287, *290*
   genome size 196, 200, 209
   morphology *222*, *223*, *233*, *236*, 282
   mutagenesis 182, 196–197
   root structure 239, *239–241*, *240*, *242*, *333*
Araceae 48, 167
Arachis hypogaea see peanut
arbuscule 121, *121*
Arbutus unedo 90
area
   leaves 12, 34, *171*, *172*
   root systems 68, 241
argentum mutant, pea 232
Arisaema triphyllum 167, 207, *207*
Aristolochia brasiliensis 72
ARR4 protein 266
Arum maculatum spadix
   cell multiplication 213, *213*
   respiration 48, *48*, *213*
   thermogenesis 48
ascorbate (vitamin C) 16, 353
asparagine 137
aspartate 26, *26*, 126, 137
aspartic acid *see* aspartate
aspen 285
Asteraceae 29, 144, 303
Astragalus 106
ATHK gene 350
ATHK1 protein 350
atmospheric conditions and water uptake 78, 88–89, 91
ATP (adenosine triphosphate) 10, 22, 25, 39, 54–57, 115, 126, 138
   in cellular work 53, 153
   in membrane transport 115, 116
   requirement for CO_2 fixation 26, 29, 32

synthesis 45, 48–49, 349
   photosynthetic 15–16, 22
   respiratory 40–41, 45, 46–49
ATP : ADP ratio 54
ATPase 103
AtPIN3 protein 330–331
autotrophy 11, 122, 133
autotropism 320
AUX gene 330
   aux 330
AUX1 protein 331
auxanometer 162
auxin 124, 179–182, 186, 210, 211, 214–215, *216*, 218, 236–237, 242, 276, 308, 310, 323–324, 326–329, 334
   concentration 179
   conjugated 181
   effects 181
   mechanism of action 215–217
   receptors 195
   structure 181
   synthetic 181
   transport 181, 236, *236*, 326, *327*, 329, 330–331
auxin-binding protein 195
auxin-responsive
   genes *327*, 328–329
   transcription factors 341
Avena fatua see wild oat
Avena sativa see oat
avocado 309
Avogadro's number *374*
axillary bud *see* bud, axillary
Azospirillum 129
Azotobacter 128

Bacillus subtilis 369
Bacillus thuringensis 36
bacteria 11, 15, 185, 186, 190
   denitrifying 128
   growth rates 168
   nitrifying 128
   symbiotic 123
bacteroids *125*, 125–126
bamboo 167, *168*
Bambusa see bamboo
banana 207
barium sulphate in statoliths 333, 334
bark 133, 228, *357*
barley *26*, 36, 111, 113, *120*, 273, 295, *304*, 334, 370
barnase (RNase) 300
basipetal transport, auxin *see* auxin

bay laurel 90
bean *92*, *119*, 128, *135*, *167*, 302, 304, 305, 307
   see also Phaseolus
beech 72
beetroot 142
Bellis perennis see daisy
Bertholletia excelsa see Brazil nut
Beta vulgaris 142, 297
betaine 124
Betula sp. see birch
Betula pendula 314
Betula pubescens 91
bicarbonate 9, 26
bidirectional transport in phloem 141
Biloxi soybean 273, 275
bimolecular lipid layer, membrane 67, 103, 357, *359*
bioassay of hormones 181
bioinformatics 202
biological clock 174, 274
biomass 29, 37, 38, 54
biosphere 9, *10*, 11, 127
biosynthesis of macromolecules 11, 56
birch 86, *91*, 121, 122, 314
birch bolete 121
black cottonwood 254
blackberry 308
bleeding sap 77–78, 85
blue light *339*
   effect on stomata 21, 339
   in photomorphogenesis 255, 255–256
   in photoperiodism 280
   in tropisms 330–331, *340*
bog myrtle 127
boron 105, 112, 128
Borya nitida 348
Botrytis cinerea 168, *168*, 169
bound water 362
boundary layer 19, 88, 96
Bradyhizobium 123
Brassica sp. 298, 299, 308
Brassica napus oleifera see oilseed rape
Brassica napus rapifera see swede
Brassicaceae 103, 144
brassinolide *189*
brassinosteroids *189*, 191, 211, *256*, *257*, 266
Brazil nut 307
bristlecone pine 165
broad bean 168, *168*
brown-heart disease 128

Bromeliaceae 29
bromeliads 30
Bryonia 168
bud 361
   axillary 182, 221
   dormancy 183
bundle sheath *26*, 26–28, 230, 237

C *see* carbon
C_3 cycle 22–23, *23*, 26, 29, 237
C_4 cycle 26–27, *27*, 33, 35, 237, *237*
C_3 plants 22–24, *26*, 28, *28*, 29, 35, 37, 38
C_4 plants *26*, 26–28, *28*, 29–30, 35, 37, 38, 105, 129
Ca^{2+} *see* calcium ion
Cactaceae 29
cacti 95, 96, 347, 362
calcium 103, 336
   ion (Ca^{2+}) 103, 106, 112, 116, 118, 199, 263, 295–296
      channels 199, 295, 360
      cellular concentration 22, 117, 118, 199, 336
      oscillation 199, *200*
      pumps 117
      as second messenger 103, 336
   oxalate 112, 333
   pectate 103
   sensitive dye *see* Fura-2
   sensitive protein *see* Cameleon protein
callose 136, 151, 153, 292, 295, *296*
callus 206, 213, 218, 292
   cultures 184, 206
calmodulin 263, 336
Calvin cycle *see* C_3 cycle
CAM *see* Crassulacean acid metabolism
cambium
   activity 228
   cork 228, 242
   development 227
   fascicular 227
   interfascicular 227
   vascular 50, 227–228, 241
Cameleon protein 199, 201
Camellia sinensis see tea
campion 308
canola *see* oilseed rape
capillarity 77, 83, 85, 86
Capsella bursa-pastoris see shepherd's purse
carbohydrate 28, 133, 138, 281
   oxidation 39

photosynthetic product 23, 32
   respiratory substrate 39
   seed reserve 303
   see also starch; sugar; polysaccharide
carbon 9–11, 101
   cycle, in biosphere 9, *10*, 22
   global store 11, 32, 34
   global turnover 11
carbon dioxide (CO_2) 9–12
   concentration 12, 17, 19, 24, 25, 30, 32, 36, 234–235, 310, 323
   diffusion 18–19, 20, 28, 38, 96
   fixation *see* photosynthesis
   radioactive ($^{14}CO_2$) 24, 26, 134
carbonate 9, 11, 105
carbonic anhydrase 26, 105
β-carotene *13*, 353
carotenoid 14, 17, 18, 182, 309, 337
   absorption spectrum 13
carpel 285
Casparian strip 74, *74*, *75*, 110–111
castor bean 137, 139, 140, *293*, 304
catalase 16, *24*, 104, 353
cations
   adsorption 107, 107–108
   exchange 107
   in soil 107
cavitation 80–85, *84*, 349
cDNA *see* DNA
cell
   cultures 163, 189, 206
   cycle 168, 207–211, *208*, *210*
   division 163, 184, 189, 205–206, 207, 208, 209, 211, 213, *213*, 223–224, 227–228, *231*, 233–234, 239–241, 275, 301–302, 308, 311
      asymmetric 233–234, 241, 292, 301
   elongation 67, 181
   expansion 104, 163, 207, 213–218, 225, 232, 308, 311, 320
   number 213, *213*
   size 207
cell wall 64, 79, 83, 84, 107, 109–110, 155, *216*, 225, 308, 354
   composition 103, 104, 105
   elasticity 215
   enzymes of wall metabolism 218, 308, 309
   extensibility 62, *215*, 232
   growth 214, 214–218, *216*
   ingrowths 143, 301

plasticity 215, 217, 309
   protein in 215
   secondary 71
   secondary thickenings 20, 71
cellular differentiation 163–164, 218–219, 232, 241
   gradient-dependent 218–219
   in meristems 224
   polarity and cell division 218–219
   position-dependent 218, 224, 226
   totipotency 206
cellulase 214
cellulose 215, 218, 295
   microfibrils *21*, 214, 215, *216*
CEN (CENTRORADIALIS) gene 283
centrifugation 80, 332, *332*
Cercidium see paloverde
cereals 29, 34, 104, 183, 302, 304, 315, 336
Chamaegigas intrepidus 348
chamise *69*
channels *116*, *117*
   ion 22, 118, 199, 232
   stretch-activated 336
   water *see* aquaporins
chaperone protein 353, 366
chaperonin 139
Chara 333, 334, *334*
chemical potential of water (μ_w) 96
chemiosmosis 15
chemotropism 295, 320
Chenopodiaceae 29, 144
Chenopodium album 230, *230*, 246
Chenopodium rubrum 273, *274*
cherry 309
chilling injury 355, 357–358
chilling stress *see* stress, low temperature
chloride (Cl^-) 22, 105, 113, 115, 117, 199, 354
chlorine 105
chlorophyll 14, 16, 25, 31, 33, 104, 259, 309
   absorption spectrum 13, *13*, 22, 252
   synthesis 104, 224, 247, 257
chloroplast 18, 22, 24, 26, *26*, 41, 104, 211, 247, 257, 309, 349, 357, 360
   division 211–212, *212*
   genome 264
   movement 338, *339*
   structure 14, *15*, *24*, 211
Cholodny–Went model of tropisms *325*, 326–327, 331
chromatid 209

chromatography
gas–liquid (GLC) 193
gel 364, *365*
chromophore
cryptochrome 256
phototropin 338
phytochrome 249, 263, *264*
chromoplast 211, 309
chromoprotein 248
chromosome 208–209
circadian
clock 274, 280–281
rhythms 67, 77, 174–175, 273, 280–281, 319
Circaea lutetiana see enchanter's nightshade
circumnutation *see* nutation
citrate 43
Citrullus vulgaris 213
Cl⁻ *see* chloride
clay 87, 87–88
climacteric 309, 310
climbing plants 175
Clostridium 128
clover 170
CLV (CLAVATA) gene 225, 282
clv 224
CLV protein 225, 234, 239
co (constans) mutant 280
Cochiolobus heterostrophus 299
coenzyme A (CoA) 43, 45, 103
CO₂ *see* carbon dioxide
cobalt 105, 106
cocklebur *see Xanthium*
cocoa 128, 314
coconut 305, *306*
Cocos nucifera see coconut
cohesion–tension theory *see* water transport
colchicine 184, 214, 224, 292
cold stress *see* stress, low temperature
coleoptile 179–181, *180*, 195, 215, *215*, 320, 324–326, 331, 332, 334, *336*, 336–337
colloidal imbibition 62
colloids, soil 62
columella *240*, 241, 330, 331, 333
Commelina benghalensis 105
common toadflax 286, 288
companion cells *see* phloem
compatible solutes *352*, 354, 359–360, 361, 362, 367, 369
compensation point, photosynthesis 18

Compositae *see* Asteraceae
confocal microscopy 151
conifers 76, *76*, 155, 284
contact exchange 108
Convallaria majalis 308
conversion table, units of measurement 373
cooling effect of transpiration 95
COP1 protein 265
cop (constitutively photomorphogenic) 256–257, 265
Copernicia cerifera 93
copper 104
COR genes 360, 361, 362, 367, 370
COR protein 366, 368
cork 228
corotanine 190
corpus 223
cortex, root 72, *733.5253*
cotransport *see* symport
cotton 80, 138, 299
cotyledons *223*, 251, 302, 304, 315
countertransport *see* antiport
cowpea *125*
Crassulaceae 29
Crassulacean acid metabolism (CAM) 28–29, 33
Craterostigma plantagineum 348
creeping buttercup *322*
cress *166*, 331, *335*
cristae 47, *47*
crop 185, 354, 355
yields 29, 34, 37
cross-talk, signal transduction 367
crown gall disease 186, 213
CRY gene 255–256
cry 255, 259, 339
CRY protein *see* cryptochrome
cryptochrome 248, 255–256, *260*, 273, 339, 340
cry 256
CTR1 protein 198
ctr (constitutive triple response) 196
cucumber *166*, 170, *171*, *172*, 217, *217*, 255
Cucumis sativus see cucumber
Cucurbita maxima 138, *139*
Cucurbita pepo 166
Cucurbita sp. 169
Cucurbitaceae 134, 137, 138, 191
cuticle 12, 19, 21, 79, 90, *92*, 92–93, 96, 235
cuticular
resistance 19

transpiration 93
cutin 92
cyanide 47
cyanide resistant respiration 47
cyanobacteria 15, 22, 123, 212
CYC (CYCLOIDEA) gene 287
cyclic adenosine 5′ diphosphate ribose 199
cyclin 209–210, 231
cyclin-dependent kinase (CDK) 209–211
Cymbalaria muralis see ivy-leaved toadflax
Cyperaceae 29
cytochrome 104
chain 47, *47*, 54
oxidase 47, *47*, 49
P450 257
cytokinesis *see* mitosis
cytokinin 106, 148, *180*, 182, 184, 186, 210, 211, 257, 266, 308, 311

2, 4-D *see* 2,4-dichlorophenoxyacetic acid
daisy 17
dandelion 218
dark
acidification 28–29
CO₂ fixation 28
respiration *see* respiration, dark
Darwin, C. 324
date palm 167
day-neutral plants *see* flowering
death 52, 112, 127, 164, 349, 363
programmed 205
de-etiolation *see* photomorphogenesis
deep supercooling 361
defence
compounds 190, 307
reactions 190, 191, 298, 367
dehydrins 311, 353, 362
dehydrogenases 104
Delphinium ajacis 233
denaturation *see* macromolecules; proteins
denitrification 128
deoxyribonucleic acid *see* DNA
depression of freezing point 361
desaturase 305
desert plants 30, 64, 80, 314, 347, 362
desiccation
seeds 311
stress *see* stress, water deficit
det (de-etiolated) mutant 256–257, 265

development
  environmental control 165, 177, 179, 246, 277, 320–323
  general concepts 164
  phases 163–164, 165, 206, *206*, 213, 223, 232, 260, 270–271
  programmed 270
diatropic reactions 321–322
2,4-dichlorophenoxyacetic acid (2,4-D) *180*, 181, 330
dicotyledons 30, 136, 186, 247, 252, 254, 302, 324
  anatomy 223, 227
differentiation 169
  role in development 163–164
  see also cellular differentiation
diffusion
  CO_2 18–19, 20, 28, 38, 96
  resistance
    cuticular 19
    stomatal 19
    testa 50
  through membranes 116
  water vapour 88
dihydroxyacetone-3-phosphate 41, 45
β-dimethylsulphoniopropionate *352*
Dionaea muscipula see Venus flytrap
disulphide (SS) bridge 103
diurnal changes
  phloem sap composition 138
  plant water uptake 88
diurnal rhythms *see* circadian rhythms
DNA (deoxyribonucleic acid) 185–186, 208–209
  binding proteins 198, 200, 243, 289
  complementary (cDNA) 201
  content, nucleus 207
  errors in replication 279
  light-responsive elements (LRE) 265
  methylation and demethylation 276, 280, 303
  microarray *201*, 201–202
  replication 208–209
  transfer (T-DNA) 186
DNA polymerase 208–209
dog's mercury 17, 253
Donnan free space 109
dormancy
  bud 187
  seed *see* seed dormancy
dormin *see* ABA

dose–response curve
  IAA *181*
  phototropism *336*
double fertilization 301
DREB1A gene 370
DREB1A transcription factor 370
drought 32, 38, 68, 88, 187, 347, 348, 355
drought hardiness *see* stress, water deficit
dulcitol 137
dwarf plants 182–183, 262, 277
dwarfing genes 183, 257
dynamin 212

Echinochloa 51
Echinocystis 191
ecotypes
  metal-tolerant 130
  photoperiodic 272, 279
ectomycorrhiza *see* mycorrhiza
edge diffusion 21
egg cell 218, 293, 294
EIN protein 198
  ein (ethylene insensitive) 196
elasticity of cell wall 215
electrochemical potential gradient 15, 114–115, 116, 117
electromagnetic radiation 9
electron transport
  photosynthetic 14
  respiratory 45, 47, 47–48
electroporation 185
electroosmosis 154
elements 100
  beneficial 101, *102*, 106
  deficiencies 128–129
  essential 100–101, 101–106, *102*, 128
  functions in plants 101–106
  in soil 106–107
  toxic 130
Eleocharis see spike rush
ELISA *see* enzyme-linked immunoassay
embolism in xylem *see* cavitation
embryo 222, 310–311
  development *see* embryogenesis
  microspore 292
  polarity 294, 301
  sac, development 293–294, *295*, 301
    Polygonum type 293, 294, *294*
embryogenesis 221, 278, 301–302, *302*
  somatic 184

emf (embryonic flowering) mutant 279, *280*
enchanter's nightshade 17, 322
end-of-day far-red effect (EOD-FR) 253, 267
endodermis 72, *73*, 74, *74*, *75*, 110–111, 241, 334
endogenous rhythms
  of growth 174–175, *175*
  in photoperiodism 273
endomycorrhiza *see* mycorrhiza
endoplasmic reticulum (ER) *135*, 136, 139, 196, 198, 295, 305, 307, *333*, 335, *335*
endosperm 294, 301, 302, 303, 304, 315
  liquid 191
energy 9–11, 39, 53, 79
  and cellular work 56–57, 153, 154
  conservation 56
  conversion efficiency
    in photosynthesis 28
    in respiration 48–49
  flow through biosphere 9, *10*
  free *see* free energy
  per quantum *see* quantum
  release as heat 11, 16, 33, 56–57
  transduction 9, 13, 22, 57
enolase 51
entropy 57
enzyme-linked immunoassay (ELISA) 194
EOD-FR *see* end-of-day far-red effect
epicotyl *336*, 337
epidermis 301
  leaf 12, 20, 230, 232, 233
  root 69, 72, *73*, 92
Epilobium see willowherb
epinasty 188
epiphytes 30
ER *see* endoplasmic reticulum
EREBP (ethylene response element binding proteins) 198
Escherichia coli (E. coli) 185, 211, 368
EST *see* expressed sequence tag
ethanol 50, 52, 315
ethylene (ethene) 106, *180*, 187–188, 196–198, 202, 267, 276, 310, 330
  effects 187, *188*
  signal transduction chain *197*, 197–198
etiolation *see* photomorphogenesis
etioplast 211
eto (ethylene overproducing) mutant 196

ETR1 protein 198
   etr (*ethylene resistant*) mutant *188*, 196
Euphorbia corcollata 30
Euphorbia grandidens 30
Euphorbia maculata 30
Euphorbiaceae 29
eutrophication 129
evaporation of water 66, 68, 79, 87
evolution 297
   genetic systems 289
   land plants 12, 21, 60, 71, 92, 155, 303
   photosynthesis 25, 30, 37
exclusion limit, of plasmodesmata 136, 138–139, 144, 146
expansin 217, *217*, *217*, 225–226, *226*, 308
expansion growth *see* growth
expressed sequence tag (EST) *201*, 201
extensometer 217, *217*

Fabaceae 123, 144, 286, 311
FACE technology 37
FAD *see* flavin adenine dinucleotide
Fagus sylvatica see beech
Fallopia sachalinensis 106
false acacia 357
falsiflora mutant, tomato 284, *284*
far-red light 248–253
   in flowering 273
   in germination 258
   in phototropism 341
   see also phytochrome; photomorphogenesis
fats *see* lipids
fatty acids 41, 45, 92
   nomenclature 305
   in seeds *see* lipids, in seeds
feedback control 54, 119
fermentation 50
   alcoholic 51–53
   lactic 51
   malic 51
ferredoxin 103
fertilization 295, 296, 301, 308, 323
Fibonacci sequence 291
fibres 207, 209, 242
field capacity of soils 87
field pennycress 276
filiform apparatus 301
fire, and seed germination 313
fix genes (nitrogen fixation) 126
flavin 337, *337*

flavin adenine dinucleotide (FAD) 255
flavonoid 124
flax 304, *322*
FLC (*FLOWERING LOCUS C*) gene 279–280
floral
   development 281–291, *290*
     ABC model *287*, 288, 288–290
   inflorescence *286*
     development 282–285
     meristem 281–283
   meristem 270, 281–283, *284*, 291
   see also flower
florigen 274, 281
flower 270
   stalk, growth of 323, *323*
   structure 285–287, *286*
     actinomorphic 286, 288
     helical (spiral) 290–291, *291*
     whorled 285, 288–290
     zygomorphic 286, 288
     unisexual 296, 299
   see also floral
flowering
   juvenility and maturity 270, 277
   photoperiodism 272–275
     action spectrum 273
     circadian rhythms 273
     critical daylength 272–273
     daylength sensing 53, 274
     day-neutral plants 272, 274
     florigen 274
     hormonal control 276–277
     LDP, long-day plants 272–274, 279–281
     photoreceptors 273, 280
     SDP, short-day plants 272–274
     stimulus perception and transmission 274, *280*
   and plant size 277, 277–278
   time of year 272, 275
fluence 250
fluorescent dyes 76, 138, 139, 201
   calcium-sensitive 199
fly agaric 121
food chains 11
Fragaria ananassa see strawberry
Frankia 127
free energy (*G*) 11, 49, 57, 86, 96
   change (Δ*G*) 33, 45
   gradients 15, 19, 114
   of water 61, 61–62, 67
free space 109, *109*

freezing stress *see* stress, low temperature
French bean 20, *92*, *135*, 302, 307
FRI (*FRIGIDA*) gene 280
Fritillaria meleagris 323
frost 355, 362
   resistance *see* stress, low temperature
fructans 39, 41
fructose 39, 304, 309
fructose-1, 40, 42
fructose-6-phosphate 41
fruit 301
   development 308–310, *309*
   ripening 188, 309–310
   set 308
   size and ploidy 207
   structure 308
fucoxanthin *13*
fungi
   growth rate *168*, 168–169
   mycorrhizal 121–122
   pathogenic 104, 182, 299
Fura-2 199
fus mutant 256–257, 265
fusicoccin 215, *216*

g (gravitational force) 331–332, *335*
G_1 phase, cell cycle 208, 209–210
G_2 phase, cell cycle 209
GA_3 *see* gibberellic acid
GA-insensitive mutants *183*
galactose 144
galvanotropism 320
gamete
   female 291, 293, 301
   male 291, 292, 295, 301
gametophyte 298–299
   female 293, 294
garden balsam *168*
garlic 103
gas chromatography – mass spectrometry (GC–MS) 193, 193–194
gas–liquid chromatography (GLC) 193
gel electrophoresis 138
gene
   duplication 254, 289
   expression 35, 51, 67, 194, 198, 199–202, *201*, 238, 264, 265–266
   homeotic *287*, 287–290, *288*, 302
   homologous 289
   light regulation 264–266
   nomenclature 373

gene (cont.)
  promoter 147
  reporter 231, 265
  repression 143, 147
  transfer *see* genetic engineering
generative cell 292
genetic engineering 35, 129, 143, 261, 368–370
  genetically modified (GM) plants 185–187, 188, 206, 265, 301, 307
  recombinant DNA technology 185
  transgenic plants 35, 138, 185–187, 193–194, 226, 243, 254, 262, *262, 262,* 279, 285, 300, 328, 353, 368–370, *369*
genome
  of *Arabidopsis* 200, 209
  sequencing 200
  size 196, 200, 208, 209
genomics 200, 202
geotropism *see* gravitropism
germination *see* seed germination
GFP *see* green fluorescent protein
gi (gigantea) mutant 280
Gibberella fujikuroi 182
gibberellic acid (GA$_3$) *180,* 182, 210, 211, 276, 281, 308, 311–314, 315
gibberellins *180,* 182–184, *183, 192,* 334
  effects 183
  in flowering 276
  in germination 312, 313
  structure *180,* 182
girdling 133
glands 53, 354
GLC *see* gas–liquid chromatography
global photosynthesis 11, 32, 38
global warming 36, 38
globoid 307
glucose 40, 43, 48, 52, 137, 304, 309
glucose-1-phosphate 41
glucose-6-phosphate 41, 42
glucose-6-phosphate dehydrogenase 42
glucose phosphate isomerase 51
glucosinolates 308
β-glucuronidase (GUS) 231, 328
glutamate 26, 123, 137
glutamic acid *see* glutamate
glutamine 123, 137
glutathione 16, 103
glyceraldehyde phosphate dehydrogenase 41, 51

glycerol 45
glycine betaine 352, *352,* 368
Glycine max (soja) see soybean
glycolysis *40,* 40–42, 43, 45, 48, 51–52, 55
glycophyte 354
glyoxylate cycle 45, 315
glyoxysome 45, 315
GM plants *see* genetic engineering
Gossypium barbadense see cotton
grafting 139, 274
gramicidin 118
Gramineae see Poaceae
grand period of growth *see* growth
granum, of chloroplast *15*
grape 308, 309
grapevine 277
grass 20, *21,* 29, 104, 106, 119, 178, *180,* 254, 297, 303
gravitropic setpoint angle (GSA) *321,* 321–322, 330
gravitropism *320,* 320–336, *321, 322, 323, 327, 329,* 341
  perception 331–336, *332*
  presentation time 331
  reaction (latent) time 331
  statocyte *333,* 333–336
  statolith *333,* 333–336, *334*
    theory 333
  stimulus reciprocity 332
great fen ragwort 314
green fluorescent protein (GFP) 138, 146, 199, 265, 328
Green Revolution 183, 184–185
greenhouse effect 36, 38
grey speck disease 128
groundsel 165
growth
  allometric 173, *174*
  arithmetic 170
  by cell division 163, 223–224, 229–231, 239–241
  by cell expansion (elongation) 163, 225, 232
  conditions necessary 165–167
  correlation 86, 146
  definition 161, 164
  determinate 165, 278, 281, 283
  exponential 170–172, *171*
  grand period *172,* 172–173, 175
  heterogonic *see* allometric
  indeterminate 165, 277, 282, 283
  limiting factors 167
  linear 162, 168, *170*

localization 164–165
logarithmic *see* exponential
mathematical analysis 170–173
measurement of 162–163
movements
  nastic *see* nastic responses
  tropic *see* tropisms; gravitropism; phototropism
primary 205, 227
rate 29, 37, 54, 69, 88, 96, 167–173, 181
  absolute 168, 169
  net assimilation rate (NAR) 173
  relative (RGR) 54, *168,* 168–170, *169,* 170, 173
  regulating substances *see* hormones
rhythms 119, 174–175, 174–176, 175
secondary 136, 206, 227–228, 241
temperature limits 166, *166, 167*
guard cells *see* stomata
GUS *see* *β*-glucuronidase
guttation 77
gynoecium 285, 293
gynophore *194,* 323

H *see* hydrogen
H$^+$ *see* hydrogen ions
H$^+$ ATPase 215–217
  see also proton pump
halophytes 80, 354
haploid plants 184, 292
hardiness *see* stress
Hatch–Slack cycle *see* C$_4$ cycle
heartwood 71
heat
  loss from plants 34, 56–57
  production 48, 56–57
  radiation into space *10,* 11
  shock 364, 366
    proteins (HSP) *365,* 367
  stress *see* stress, high temperature
heavy metals 130
Hedera helix see ivy
Helianthus annuus see sunflower
Helleborus niger 291, *291*
hemicellulose 215, 295
henbane 272
herbaceous plants 68, 75, *76,* 78, 144, 169, *169*
herbicide resistance 185, 186
herbivores 103, 189, 190, 191, 308, 318
heteroglycan 215
heterozygous plants 299, 303, 312

hexokinase 41

high-irradiance response (HIR) *see* phytochrome

high-performance liquid chromatography (HPLC) 192, *192*

high-photosynthesis plants 28

high-temperature stress *see* stress, high temperature

HIR (high-irradiance response) *see* phytochrome

histones 208

holly 83

holm oak *357*

homeotic mutants 287–290

homogeneous nucleation point, ice 361

homozygous plants 184, 297, 299, 312

honesty 276

hop 169

Hordeum sativum (vulgare) see barley

hormones 106, 178, *180*, 206, 207
  analytical methods 192–194, *194*
  bioassays 192
  classes of 179
  commercial uses 181
  concentrations in plants 178–179, 191, 310, 311–314
  conjugation 181
  definition 178
  interactions between 181, 184, 187, 190, 210, 311–314, *313*
  receptors 178, 187, 197–198
  synthetic 178
  transport 86, 139, 178
  see also under individual hormones

horticulture 206, 270

HSP *see* heat shock proteins

Humulus lupulus see hop

hy mutants *see PHY* genes

HY5 protein 265

hybrid 185
  F_1 299–301
  vigour (heterosis) 299, 303

hydration
  in germination 315
  membranes 350
  shells, of molecules 110, 116, 350, 353

hydraulic conductivity 67, 75

Hydrilla verticillata 30

Hydrodictyon 112

hydrogen 101, 126
  bonds (H bonds) 79, 143, 353
  ions (H^+) 15, 107, 117, 154

hydrolases 294, 315

hydrolysis, seed reserves 39, 315–316

hydrolytic enzymes *see* hydrolases

hydrophilic proteins 353

hydrophobic bonds 360

hydrotropism 69, 320

hydroxyl ions (OH^-) 107–127

Hyoscyamus niger 272, 273, 275

hyperaccumulation of elements 130

hypocotyl 181, 188, 217, *217*, 324, 334

hypophysis 301

hypoxia 50, 52, 88

IAA (indole-3-acetic acid) 179, *180*, 181, *181*, 194, 242
  see also auxin

ice formation 81, 356, 358, 361
  extracellular 358–359
  homogeneous nucleation point 361
  intracellular 358–359
  nucleation 361, 362

ice plant 30

imbibition 315

Impatiens balsamina 168

Impatiens hawkeri 168

indole-3-acetic acid *see* IAA

inflorescence development *see* floral development

inositol hexaphosphate 307

insects 36, 106, 137

integuments 294, 301

internode 221
  elongation 249, 253

invertase 304, 309

ion
  accumulation 108, 114–115, *119*
  channels 22, *116*, *117*, 118, 199, 232
  concentrations, plants 112–114, 114–115, 118, *119*, *120*
  exchange 107, 109
  transport
    apoplastic 110–112
    long-distance 85, 110–112
    across membranes 112–118, *115*, *117*
  uptake 107–110, *109*, 118–121
    by cells *108*, 108–110
    by roots 107–108, *120*, *120*

Ipomoea caerulea 233

Ipomoea purpurea 293, *322*

iron 104, 126

iron–sulphur proteins 103, 104

ironwood 314

irradiance 13, 16–18, 24, 234–235, *247*, 250, 336–337, 340
  units 16

irrigation 354

isozymes
  definition 41
  glycolysis 41, 51
  PPP and C_3 cycle 42

ivy 223, *356*, *357*

ivy-leaved toadflax 323

jasmonate 202

jasmonic acid *189*, 190–191

jojoba 305

Juglans nigra see walnut

juvenility 223, 232, *233*, 270, 277

K *see* potassium

K^+ *see* potassium ion

kinases *see* protein kinases

kinetin *180*, 184

klinostat 331, 332

Klopstockia cerifera 93

Knightia excelsa 73

KNOX1 gene 238

Kranz anatomy 26, *26*, 237

Krebs cycle 43–46, *44*, 49, 315

Laburnum anagyroides 322

lactate 51

lactate dehydrogenase 51, 52

lactic fermentation 51

Lactuca sativa see lettuce

LAI *see* leaf area index

Lamiaceae 286

Laminaria 155

late embryogenesis abundant
  genes *see LEA* genes
  proteins *see* LEA proteins

latent
  heat of evaporation 79
  time 331

lateral
  buds 188
  roots 70, *242*, 242–243

Laurus nobilis see bay laurel

LDP (long-day plant) *see* flowering

leaching of minerals 112, 128

LEA genes 311, 353, 362, 367, 370

LEA proteins 353, 366

lead 130

leaf 78
  abscission 138, 188
  area index (LAI) 34

leaf (cont.)
cell division 229–231, *231*
compound 237–238, *238*
development 228–238, *229*, *231*, *232*, *233*
stomatal patterns 233–235, *235*
developmental gradients 228, 229, 231
growth 146, 147, *171*, *172*
initiation 221
internal area 12, 18, 79
meristems 229
mesophyll 230, 232, 237
mosaic 18
movements 67, 319
orientation 321–322
as photosynthetic organ 12, 13, 18
phyllotaxis 222, *223*
primordia *222*, 225–226, *226*, *229*, 230
senescence 188
shade 18, 230, *230*
structure 12
dicot 230, 233, 236, 237
Kranz anatomy 26, *26*, 237, *237*
monocot 228, 236, 237
xeromorphic 95–96
sun 18, 230, *230*
vasculature 12, 227, *236*, 236–237
leakage
from damaged membranes 357
in phloem 145, 153
Leccinum scabrum 121
leghaemoglobin 126
legume 123, 307, 308
Leguminosae *see* Fabaceae
Lens culinaris see lentil
lenticels 228
lentil 331, *336*, *337*
Lepidium sativum *166*, 331, *335*
lettuce 248, 249, *249*, *252*
leucoplast 211
LFR (low-fluence response) *see* phytochrome
LFY (*LEAFY*) gene 283–285
LHP *see* light-harvesting pigment
lianas 76, *76*
light 33, *247*
compensation point *17*, 18
as energy source 9–11, 16, 33
fluence 250
influence on growth 167
as information medium 246
irradiance 13, 250

saturation, photosynthesis 16, *17*
light-harvesting pigment (complex), LHP(C) 14, 18
light-sensitive seeds *see* seed, photoblastic
lignin 79, 96
Ligustrum lucidum 80
Liliaceae 292, 314
Lilium longiflorum 209, *212*
lime 72, 136
limiting factors
agricultural yield 32, 129, 347, 362
growth 167
photosynthesis 17, 19, 30–32, 34, 349
Linaria vulgaris see common toadflax
Linnaeus, C. 288
Linum usitatissimum see flax
lipase 45
lipids 10, *15*, 43
conversion to carbohydrate 315
membrane 315, 357, *359*, 360, 366
as respiratory substrate 39, 45
in seeds 39, 45, 305, 315, *352*
fatty acids, saturated 305, *306*
fatty acids, unsaturated 305, *306*
oils 39, 41, 305, *306*
lipoxygenases 315
Litchi sinensis 276
loading of phloem *see* phloem
lodging 183
Lolium perenne 54
long-day plant (LDP) *see* flowering
low-fluence response (LFR) *see* phytochrome
low-photosynthesis plants 28
low-temperature stress *see* stress, low temperature
Lunularia annua see honesty
lupin 106
Lupinus albus see lupin
Lycopersicon esculentum see tomato

M phase, cell cycle 210
Macrocystis 155
macromolecules 102
denaturation 349, 350
phloem transport 138–139
synthesis 9
energetics 56
macronutrients 101, *102*
Macroptilium atropurpureum 120
MADS-box genes 289
magnesium 104

maize (sweetcorn) 26, 29, 35, 52, 53, 67, 68, 69, *73*, 83, *119*, 166, *166*, 167, 237, *237*, 239, 248, 278, *278*, 281, 294, 295, 299, 302, 304, *304*, 307, 320, *320*, 330, 331, *369*
tassel 278, *278*, 299
major intrinsic protein (MIP) 67
malate 22, 26, *26*, *27*, 28, 44, 51, 52, 199
male sterility 299–300, *300*
malic acid *see* malate
malic enzyme 44, 51
maltase 315
Malus domestica see apple
manganese 104, 128
Mangifera indica see mango
mango 303
mangrove 314, 315
swamp 354
mannitol 137, 155
maple 230
marrow 169
Maryland Mammoth tobacco 271
mass flow, phloem 148–154
mass spectrometry 193, 193–194
mathematical analysis of growth *see* growth
matric potential (Ψ_m) 62, 86
mature stage 232, 270, 275
meadow buttercup 297
Medicago sativa see alfalfa
Medicago truncatula 333
megaspore 294
megasporocyte 293
meiosis 292, 294, 299
membrane 357–360, *359*, 360, 363, 364, 366
leakage 315, 357, *359*
permeability 52, 62, 66–67, 75, 103, 106, 116, 357, 360
transport
auxin 236
ions *115*, 116
sugars 143
water 66–67, 75
see also plasma membrane; tonoplast; transport proteins
Mendel, G. 182
Mercurialis perennis see dog's mercury
meristemoids 233
meristems 50, 163, 164, 205–206, 213, 219, *222*
determinate 278, 281, 283
fate of cells in 278

indeterminate 277, 282, 283
secondary 206, 227, 228
see also cambium; floral, leaf, root, shoot meristems
Mesembryanthemum crystallinum see ice plant
mesocotyl *180*
mesophyll 12, 26–28, 142–144, *143, 144,* 230
palisade 12
resistance 19
spongy 12, *223*
mesophytes 349
messenger RNA (mRNA) *see* RNA
metabolomics 202
metal
contamination 130
tolerance 130
3-methyl-lumiflavin *337*
microfibrils *see* cellulose
microfilaments 336
micronutrients 101, *102,* 130
as enzyme activators 104–105
microorganisms 11, 88, 104–105, 107, 128, 168
micropyle 294, 295, 301
microspore 292
embryo 292
mother cell 292
microtubules 209, 214, 292, 336
middle lamella 103, 309
millet 299
Mimosa pudica see sensitive plant
mineral deficiencies 31
mineral ions 106
Minuartia verna 130
mistletoe 314
mitochondria 24, *24,* 43, 45, 49, 50, 136, 299, 301, 315
cristae 47, *47*
division 212, 315
electron transport 45, 47, *47*–48
structure *47*
mitosis 168, 209, 276
asymmetric 292
see also cell division
mitotic spindle 209
mobilization of food reserves 313, 315–316
molybdenum 105, 122, 126, 128
monocotyledons 30, 136, 186, 247, 254, 302
anatomy 223, 224, 227
Monterey pine 276

morning glory *293, 322*
morphogenesis 165, 205
organ initiation 221, 225–226
see also cellular differentiation; photomorphogenesis
Morus see mulberry
mRNA *see* RNA
Mucor stolonifer 168
mulberry 296, *357*
Münch hypothesis *see* phloem
mung bean *21, 47, 92*
mustard *166,* 251
mutagenesis 196–197, 263, 289
mycorrhiza 121–122
ectomycorrhiza 121
endomycorrhiza 121
vesicular–arbuscular (VA) 121
Myrothamnus flabellifolia 348

N *see* nitrogen
Na⁺ *see* sodium ions
NaCl *see* salt
NAA *see* naphthaleneacetic acid
NAD⁺ (nicotinamide adenine dinucleotide) 41, 43
NADH 40, 46, 47, 48, 49, 51, 123
NADP⁺ (nicotinamide adenine dinucleotide phosphate) 14–16, *15,* 42, 55
NADPH 16, 22, 32, 40, 43, 44, 45, 46, 47, 49, 55
naphthaleneacetic acid (NAA) *180,* 181
naphthylphthlamic acid (NPA) 329
nastic responses 318–320
growth movements 319
turgor movements 318–319
NEEDLY gene 285
networks
of interacting genes 225, 239
of signal transduction chains *367*
New Guinea balsam 168
nickel 105, 130
Nicotiana tabacum see tobacco
nif genes (nitrogen fixation) 126
Nitella 112
nitrate 107, 113, 128, 129, 242–243
assimilation 16, 105, 122–123, 137
fertilizer 129
reductase 105, 122–123
uptake 116, 119
nitrite 122
reductase 122–123
Nitrobacter 128

nitrogen 31, 102, 122, 242
assimilation 122–123
cycle *127,* 127–128
fixation 105, 123–127, 129
nitrogenase 126, 127
Nitrosomonas 128
NMR *see* nuclear magnetic resonance
nod genes (nodulation) 126
nodes 221, 322
nodule, root 123–126, *124, 125*
Norway maple *166*
NPA *see* naphthylphthlamic acid
NPH gene 337, 341
nph 341
see also phot mutants
NPH protein *see* PHOT proteins
NPL1 gene 338
nucellus 293
nuclear magnetic resonance (NMR) 139, 150
nuclear volume 207, *207*
nucleus 208, 265, 335
polar 294
sperm 292, 295
vegetative 295
Nuphar luteum 51
nutation 175, *322*
nutrition
human 100, 303, 305, 307
plant 11, 128–129

O₂ *see* oxygen
oak *72,* 122, 169, 296, 314
oat 20, 36, 128, 215, *216,* 332, *336, 337*
oil *see* lipid
oil body (oleosome) 45, 305
oilseed rape 300, *300,* 304, *304,* 305, *306*
Olea europaea see olive
olive *306*
Olneya see ironwood
onion 103
optical tweezers 334, *334*
optimum temperature
growth 166
photosynthesis 28, 31, 362
vernalization 275
Orchidaceae 29, 286
orchids 30, 121, 294, 303
organ
formation 225–226
initiation 221
orientation *321*

organelle division 163, 211–212, *212*, 315

organic acids, terminology *10*

Oryza sativa see rice

osmoregulation *see* osmotic adjustment

osmosensors 350

osmosis 22, 62, 79, 81, 145, 199, 232

osmotic
  adjustment 353
  potential (Ψ_π) 62, *63*, *65*, 66, 86, 148, *351*, 352
  pressure 62, 77

ovary 285, 301, 308

ovule 285

ovum 218

oxaloacetate 26, 43, 44, 315

oxidation
  carbohydrates 39
  lipids 39, 45
  β-oxidation 45, 315
  proteins 39, 45

oxidative stress 52, 349, 353

2-oxoglutarate 43, 44

oxygen (O_2) 39, 44, 46–47, 102
  concentration 25, 50–51, 52, 310
  evolution 15, 104, 105
  uptake 23, *48*

P *see* phosphorus

P-protein *135*, 136, 139, 151, 152, *152*, 153, 154

palm 134, 136, 150, 153, 227

paloverde 314

Papaver see poppy

Papaveraceae 298

PAR *see* photosynthetically active radiation

parthenocarpy 183, 308

parthenogenesis 303

partial molal volume of water (V_w) 96

particle bombardment 186

partitioning of metabolites 36, 146–148

patch-clamping 113

pathogenesis-related proteins 190

pathogens 93, 178, 186, 188, 190, 202, 213, 218, 367

pea *125*, *166*, 182, 232, 237, 255, 275, 286, 302, *304*, 331, 335, *357*, 363

peanut *194*, *306*, 323

pear 308

pearl millet 362

pectin 92, 105, 215, 295, 309

pectinase 214

Pennisetum glaucum 362

pentose phosphate pathway (PPP) 41, 42, 42–43, 45, 49

pentoses 43

PEP *see* phosphoenolpyruvate

PEP carboxylase 26, 26–28, 35, 44

peptide hormones *189*, 189–190

perennials 71, 85, 136, 138, 169, 175, 355, 361

perforation plate 71, *72*

pericycle 72, *73*, 111, 241

periderm 69, 228, 242

Perilla ocymoides 275

permanent wilting point (PWP) *87*, 87–88

permeability *see* membrane

peroxidase 104

peroxisome 24, *24*

petal *see* floral development; flower

petiole 319

Petunia 284, 298

PFD *see* photon flux density

P_{FR}
  absorption spectrum *250*
  interconvertibility with P_R 248–249, *249*, *264*
  see also phytochrome

PGA *see* 3-phosphoglycerate

pH
  cell wall 217
  phloem sap 138
  xylem sap 85

Phaseolus coccineus 167

Phaseolus vulgaris 20, 92, *135*, 302, 307

phi (Φ) *see* phytochrome

Phleum pratense 106

phloem 50, 82, 85, 134–136, *135*, 136, 142, 228, 241
  companion cells *135*, 136, 138, 142–144, 145, 154, 155
  loading 142–145, *145*, 149
    apoplastic 142–143, *143*, 144–145
    symplastic 143–145, *144*
  parenchyma 142, 143
  sap composition 137–139, 150, 153, 154, 190, 238
  SE–CC complex *135*, 136, 142, 142–143
  sieve tube 134–136, 149–155, 236
    cells 134–136, *135*, 139
    plates 134, *135*, 151–152, *152*, 153–155
    pores 134–136, 151–153, *152*, 153–155

translocation 274, 303, 357
  direction 140–142, *141*
  Münch mass flow hypothesis 148–154, *149*
  rate 139–140, 155
  unloading 145–146, 149, 153

Phoenix dactylifera see date palm

phosphatase 198, 211

phosphate, inorganic (Pi) 15, 23, 103, 107, 111, 113, 114, 119, 122

phosphoenolpyruvate (PEP) 26, 28, 41, 44

phosphoenolpyruvate carboxylase *see* PEP carboxylase

phosphofructokinase 41, 55

6-phosphogluconate dehydrogenase 42

3-phosphoglyceraldehyde 41

3-phosphoglycerate (PGA) 22, 24, 26, 41

phosphoglycerokinase 41

phosphoglycolate 24, 25, 26, 35

phospholipase 103

phospholipids, membrane 103, 360

phosphorus 103
  radioactive (^{32}P) 111, 208

phosphorylation
  oxidative 47
  photosynthetic 15
  of proteins 198, 209, 263, 350
  substrate level 41, 43, 45

phot mutants (*nph*) 337–339, 341

PHOT proteins 338

photoautotrophy 11

photoblastic seeds *see* seed, photoblastic

photochemical reactions 14, 16, 31, 33

photoepinasty 323

photoinhibition of photosynthesis 16

photometric units 16

photomorphogenesis
  action spectrum 248–249, *252*
  de-etiolation 247, *248*, 253, *256*, 256–257, 259–260, 265
  etiolation 247, *248*, *256*, 256–257, 260, 265
  hormones in 266–267
  photoreceptors 248, 255, *260*
    interactions *259*, 259–260
  see also phytochrome

shade
  avoidance 253, 260–262, *261*, *262*, 267

tolerance 253, 260–263, *262*
signal transduction pathways 255, 257, 259, 263–266, *266*
photon 13, 14, 16
  energy content 33
  flux density (PFD) 16–18, 25, 28
photooxidation 16
photoperiodism *see* flowering
photophosphorylation *see* phosphorylation, photosynthetic
photoreceptors
  photomorphogenesis 248
  photoperiodism 273, 280
  phototropism 337–341
  stomatal movement 339
photorespiration 24–26, *25*, 28, 35–36
  nitrogen cycle 26
photosynthesis 9, 57, 322, 363
  action spectrum 13, 347
  CO$_2$ fixation 16
    C$_3$ cycle 22–23, *23*, 26, 29, *237*
    C$_4$ cycle 26–27, *27*, 33, 35, 37, *237*, *237*
    CAM 28–29, 33
  energy conversion efficiency *14*, 32–34, 33–34
  evolution 25, 30, 37
  global 11, 32, 38
  light
    absorption 12–13, 16, 34
    utilization 13–16, 33
  limiting factors 17, 19, 30–32, 34, 349
  pigments 12–13
  productivity 33–34
    improvement 34–36
  rate 16, *17*, *18*, 21, 23, 28, 29, 30–32, 35, 37, 38, 138, 360, *360*
photosynthetically active radiation (PAR) 12, 16, 33, 34, 247
photosystem I 14, 15
photosystem II 14, 15, 349, 363
phototropin 248, 337–341, *338*
hototropism 320, 323, 324–326, *327*, *336*, 336–341, *340*
  action spectrum 337, *337*
  irradiance effects 336–337
  perception 337–341
  photoreceptors 337–341
    see also phototropin
  presentation time 337
  reaction time 337
  reciprocity 337
phragmoplast 209

PHY (phytochrome) genes 254–265, 259, 261, 265
  hy mutants 254–255
phyllotaxis 222, *222*, *222*, 223, *226*, 232
phytin 307
phytochrome 248–252, *258*, *260*, 322
  absorption spectrum 250, *250*, 258, 263
  in flowering 273
  in germination 249, 258–267, 314
  HIR, high-irradiance response 251–252, *252*, 253, 259, 260, 265
  LFR, low-fluence response 249–250, 253, 258, *258*, 260, 273, 340
  localization 265
  multigene family 254–255
  phi (Φ) 251, *251*
  photoreversibility 249
  photostationary state 251
  P$_R$ – P$_{FR}$ conversion 248–249, *249*, 250, 252, *264*
  in phototropism 319
  structure 248, 263, *264*
  VLFR, very-low-fluence response 250–251, 258, *258*, 265
phytoextraction 130
phytomer 221, 277, *277*, 283
Pi *see* phosphate, inorganic
PI (PISTILLATA) gene *288*, *290*
PI protein 290
PIF3 protein 263, 265, *266*
PIN1 gene 330
  pin 189, 330
PIN protein 330–331
pine 122, *357*
  see also Pinus
pineapple 30
Pinus longaeva 165
Pinus radiata 284
Pinus strobus 357
Pisum sativum 125, *166*, 281, 302, 331, 335, *357*
Pisum sp. *see* pea
pith cells 165
pits, cell wall 71, 83, 84
plagiotropic reactions 321–322
plant
  breeding 54, 184–185, 305, 307
  communities 38
  distribution 29, 344, 355
Plantago major 54
plasmalemma *see* plasma membrane

plasma membrane (plasmalemma) 22, 74–75, 108–110, 113, 114, *115*, 136, 142, 199, 215, 232, 335, 336, 358
  intrinsic proteins (PIP) 67
  permeability 66–67
  receptors 195, 196
  stress effects 349, 350
plasmid 185–186
plasmodesmata 26, 74, *74*, *135*, 136, 138–139, 142–144, 146
  exclusion limit 136, 138–139, 144, 146
plasmolysis 63, 64, *65*, 66
plasticity of cell wall 215, 217, 309
plastid 41, *135*, 136, 211, 305
  division 211
  see also chloroplast
plastochron 221, 237
Poaceae 20, 29, 302, 304
polar nuclei 294
polar transport 330–331
polarity 218–219, 294, 333
pollen 297–299
  formation 106, 291–293, *292*, 299–300
  germination 295, 297
  grains 184, *280*, 285, 292, 293, *293*, *300*
  sac 285, 292
  tube 295, 295–296, *296*, 298
    growth rate *168*, 168–169, 295
  wall 292, 295
pollination 48, 292, 294, 295–301
  cross 296, 299
  self 299
Polygonatum multiflorum 323
polymer trap 144, *144*
polyols (sugar alcohols) 137, 352, 353, 361
polyploidy 126, 206–207, *207*, 209, 224
polysaccharides 10, 39, 215
poplar 144
poppy 298, 308, *322*
Populus deltoides 144
Populus tremula *see* aspen
Populus trichocarpa 254
porters 116, 117, *117*
post-transcriptional gene silencing 190
potassium 103
  ion (K$^+$), 22, 104, *109*, 109, 113, 114, *115*, *120*, 116–121, 138, 154, 198

potato 143, 213, 298
PPP *see* respiratory pentose phosphate pathway
P-protein *135*, 136, 139, 151, 152, *152*, 153, 154
P_R
  absorption spectrum *250*
  interconvertibility with P_{FR} 248–249, *249*, *264*
  see also phytochrome
prairie hollyhock *293*
pressure bomb 66, 80, 82
pressure potential (Ψ_p) 62–64, *63*, *65*, 66
pressure probe 66, 81–82, 150
primordium 221–223, *223*, 225
privet 80
production *28*, *32*, 54
  primary 11
  secondary 11
prokaryotes 47, 122, 123, 211–212
proline 352, 354, 361
proline betaine 147,165
proplastid 211
protein 39, 45, 138
  body 136, 307
  denaturation 363, 365
  dephosphorylation 209
  kinases 198, 199, 263
  nomenclature 373
  phosphorylation 198, 209, 263, 350
  seed reserve 304, 307
  synthesis 56, 352, 364
proteinase inhibitors 189
proteomics 202
proton 15, 47, 114, 215
  gradient 117
  pump 107, 116–117, 142, 154
  see also hydrogen ions
protoplasmic streaming 154
protoplast 74, 185, 214
  fusion 185
Prunus dulcis see almond
Pseudomonas syringae 190, 362
psi (Ψ) *see* water potential
pterin 256
pulvinus 319, 333
pumpkin *166*, 304, *304*
pumps 116, 116–117, *117*
  calcium (Ca^{2+}) 117
  ion 22
  proton (H^+) 117
PWP *see* permanent wilting point
Pyrus communis see pear

pyruvate 41, 43, 51
pyruvate decarboxylase 51
pyruvate dehydrogenase 41, 53
pyruvate kinase 41, 55
pyruvic acid *see* pyruvate

Q_{10} (temperature coefficient) 31, 166, 363
QAC *see* quaternary ammonium compound
quantum 11, 14, 15, 16, 57
  energy and wavelength 33
  requirement, photosynthesis 33
  yield, photosynthesis *14*, 33
quaternary ammonium compound (QAC) 352, *352*, 354
Quercus ilex see holm oak
Quercus robur see oak
quiescent centre 239, *240*

radicle 311
radioactive compounds, as tracers 61, 85, 111, 113, 134, 136, 139, 141, 208, 223, 239
radioimmunoassay of hormones (RIA) 193–194
radish 337
raffinose 137, 144, 145
Ranunculus acris 297
Ranunculus repens 322, *322*
rape *see* oilseed rape
Raphanus sativus see radish
raspberry 308, *322*
rays 71, *73*, 228
reaction time 331
reactive oxygen species (ROS) 16, 17, 25, 349, 353, 357, 368
reciprocity, in stimulus perception 250, 332
recombinant DNA technology *see* genetic engineering
red goosefoot 273
red light 248–252
  in flowering 273, 280
  in germination 248–252, 258
  in phototropism 340, *340*
  see also phytochrome
relative growth rate (RGR) *see* growth rate, relative
relative humidity (RH) 89
reproductive phase 270–271, 281
respiration 9, 10, 11, 18, 29, 38, 115, 166, 315
  and activity 53–57, 115, 153

aerobic 39, 45, 50, 126
anaerobic 49–53, 88, 315
ATP yield 40–41, 48–49, 52, 54
climacteric 309, 310
cyanide-resistant 47
dark 23–24, 39
growth 53
in the light 39
  see also photorespiration
maintenance 53, 153
rate 23, 47, *48*, 53–55, 153, *167*, *213*, 349, 363
thermogenic 48, *48*, *213*
respiratory pathways 45–46, *46*
  glycolysis *40*, 40–42, *43*, 45, 48, 51–52, 55
  Krebs cycle 43–46, *44*, 49, 315
  pentose phosphate pathway (PPP) 41, 42, 42–43, 45, 49
  terminal oxidation 40, *44*, 46–48, *47*
respiratory quotient (RQ) 50, 123
respiratory substrate 39, 52
resurrection plants 347, 362
retinoblastoma protein 210
reverse transcriptase 201
RGR *see* growth rate, relative
RH *see* relative humidity
rheotropism 320
Rhizobium 123–127, *124*, *126*
rhizoid 322, 334, *334*
rhizome 50, 52, 322–323
Rhus ovata see sumach
rhythms *see* annual; circadian; endogenous
riboflavin *337*
ribonuclease *see* RNase
ribonucleic acid *see* RNA
ribulose-1,5-bisphosphate 22, 24
ribulose-1,5-bisphosphate carboxylase – oxygenase *see* Rubisco
rice 35, 36, 52–53, 143, 182, 183, *183*, 285, 299, *304*, *369*
ricin 308, 311, 314
Ricinus communis see castor bean
ringing experiments 133
ripeness to flower 270
RNA (ribonucleic acid) 201
  messenger (mRNA) 75, 138, 190, 200, 315, 364, 366
  micro (miRNA) 190, 232
  in phloem sap 138, 146, 190, 238
  short interfering (siRNA) 190
  as signalling molecule 139
RNase 298, 300

Robinia pseudacacia see false acacia
root 78
    apex 69, *70*, 146
    apical meristem (RAM) 206, 213,
        222, 239–241, 301
    cap 239, *240*, 241, 326, 328, *329*,
        331, 333, *333*, 335
    cortex 110, 113, 121, 125
    development 239–243
        response to nutrients 242
        stereotyped 239
    gravitropism 320, 326–327, *327*,
        330–336, 331
    growth 69, 88, 146, *168*, *174*, *181*, 241
    hairs 68–70, *70*, 107, 124, *240*, 241
    lateral 70, 242, *242*–243
    surface area 68, 241
    structure 72–75, *73*, *75*, 239–241,
        240, 242
    systems 68, *69*, 75, 95
    see also ion uptake; water uptake
root pressure 77–78, 83, 85, 88
ROS *see* reactive oxygen species
Rosa see rose
rose 270
RQ *see* respiratory quotient
rubidium 106, 116, *120*
Rubisco (ribulose-1,5-bisphosphate
        carboxylase-oxygenase) 22,
        24–28, 29, 35, 147, 358
    cellular concentration 36
    CO_2 affinity 25
    localization, C_4 plants 26, 237, *237*
    modification 35
    O_2 affinity 25, 35
Rubisco activase 363
Rubus idaeus see raspberry
rye 68, 70, *168*, 275, 328

S-factors in pollination 297–298
S locus, self-incompatibility 297–298
    alleles 297–299
S phase, cell cycle 208, 209, 210
Saccharum officinarum see sugar cane
Salicornia europaea 354
salicylic acid *189*, 190–191
salinity stress *see* stress
Salix babylonica 144
salt 64, 354, 368
    glands 115, 354
    marsh 64, 354
    water 354
sapwood 71, 83
scald disease 128

sclerophylls 76
Scrophulariacae 286
scutellum 302
sdd (*stomatal density*) mutant 234
SDP (short-day plant) *see* flowering
SDS–PAGE separation 364, *365*
SE–CC complex *135*, 136, 142, *142*–143
sea water *see* salt water
Sebertia acuminata 130
Secale cereale see rye
second messengers 103, 336
secondary compounds 92, 102, 103,
    247
secondary thickening 136, 227–228,
    241
secretion 81, 111, 115
Sedum 30, 95
seed
    bank, in soil 314
    coat (testa) 50, 301, 311, 313, 315
    defence compounds 307
    desiccation tolerance 310, 311, 314,
        315, 347
    development 310–312
    dormancy 311–314, *312*
        breaking 312–314, *313*
    germination 50, 52, *180*, 187, 249,
        250, 253, 315
        imbibition 50, 315
        inhibitors 311
    nutrient reserves 39, 302, *304*
        commercial applications 305
        in human and animal nutrition
        303
        mobilization 313, 315–316
        synthesis 303–308, 311
        see also lipids; protein; starch
    photoblastic 272–273, 314
    recalcitrant 314
    survival
        of cold *357*, 362
        of heat 362
seismonasty 318, *319*
selectable marker 186–187
selenium 106, 118
self-incompatibility 296–299
    gametophytic 297, *297*, 298–299
    sporophytic 297, *297*, 298, 299
semipermeability 62
Senecio jacobaea 68
Senecio paludosus see great fen ragwort
Senecio vulgaris see groundsel
senescence 106, 112, 170, 172, 188
sensitive plant 318

SEP (SEPALLATA) genes 289–290, *290*
SEP protein 290
sepal *see* floral development; flower
sesame 304, *304*
Sesamum indicum see sesame
SH (sulphydryl) group 103
shade
    avoidance 253, 260–262, *261*, *262*,
        267
    leaves 18, 230, *230*
    plants 17, *17*, 230, 362
    tolerance 253, 260–263, *262*
shepherd's purse 311
shoot 75
    apex *194*
    apical meristem (SAM) 205,
        221–226, *222*, *223*, *226*, 230,
        270, 277–279, *278*, 281, *284*,
        302
    bud 182
    growth 148, *168*, *181*
    tropic reactions 326, *327*, 336
Shorea 314
short-day plant (SDP) *see* flowering
SI units 19, 373
sickle-leaf disease 128
sieve tubes *see* phloem
signal transduction pathways
    cross-talk 367, *367*
    ethylene action *197*, 197–198
    photomorphogenesis 255, 257,
        259, 263–266, *266*
    stress reactions *351*, 358, 360
signalosome 265, *266*
Sildacea malviflora see prairie hollyhock
silicon 104, 112, 119
Simmondsia chinensis see jojoba
Sinapis alba see mustard
sink, for metabolites 140, 146–148,
    303, 309
skotomorphogenesis 256–257, 266
sleep movements 319
snapdragon 282
    see also Antirrhinum
SOD *see* superoxide dismutase
sodium 105
    ions (Na^+) 113, 115, 117, 118, 354
soil
    aeration 88
    calamine 130
    clay 87, 87–88, 107, 107–108
    colloids 62, 86, 107
    conditions and water uptake 64
    mineral ions in 86, 106–107

soil (cont.)
  sandy 68, 86
  serpentine 130
  solution 107
  temperature 88
  water potential 62, 64, 77, 78, 80, 84, 87, 87–88
  water-yielding capacity 78
Solanaceae 144, 189, 292, 298
Solanum tuberosum see potato
sorbitol 137
sorghum 255, 267, 299, *365*
Sorghum vulgare (bicolor) see sorghum
source to sink transport 140–142, *141, 147*, 149, *152, 155*
southern corn leaf blight 299
soybean 150, 271, 286, 304, *304*, 307, 328, 366
sperm 292, 295, 301
spike rush *322*
spinach 276
Spinacia oleracea see spinach
spindle fibres 209
sporophyte 292, 298
sporopollenin 293
SS (disulphide) bridge 103
stachyose 137, 144
stamen 285, *292*
staminal filaments 285, 291, *292*
  growth 168, *168*
starch 22, 39, 41, 52, 281, 333–334
  grains 39, 304, *305*
  seed reserve 304–305, 315
  sheath 333
  synthesis 304–305
starvation 52, 333, 363
statocyte 333–336
statolith 333–336
  theory 333
stem
  growth *174*, 253
    primary 227
    secondary 227–228
  structure 75
    dicot 227
    monocot 227
  water storage 95
stigma 285, 295, 297–298
STM (*shoot meristemless*) gene 282, 302
  stm 223, *224*
STM protein 225
stomata 12, 20–22, 88, 90–92, 96
  aperture 20, *21*, 28, 38, *90*, 90–92, *91*, 187, *200*

density (frequency) 19, 20, 37, 90, *233*, 234, 235, *235*
development 233–235
  environmental control 233, 234–235
guard cells 20, 22, 198–199, 233–234
index 234, 235, *235*
movements 19, 21, 28, 29, 37, 90, 118, 198–199, 349, 352
structure 20, *21, 92*
stomatal resistance 19
stratification 313
strawberry 77, 207, 308, 322
strawberry tree *90*
stress
  concepts and definitions 344, 344–345, *346*, 355–356
    avoidance 345
    hardening (acclimation) 345, *346*
    perception 346
    resistance (hardiness) 345
    signal transduction 346
    tolerance 345
  cross-tolerance 346, 366–368, *367*
  engineering for resistance 368–370
  general resistance 187, 368
  high temperature 362–366, *364*
    avoidance 363–364
    damage 363
    hardening *364*, 364–366
    resistance 363–366
    thermal death point 362–363
    thermotolerance 364–366
  low temperature 349, *367*
    avoidance 354–355, 361
    damage 357–360
    hardening 355, *358*, *360*
    resistance 355–356, *356*, *357*, 360–362
    tolerance 361–362
  salinity 354
  water deficit 21, 30, 32, 88, 92, 346–354, *348*, 359–360
    avoidance 347
    damage 348–350
    hardening 350–353
    resistance 36, *351*, *353*
    tolerance 347–348, 350–353
stretch-activated channels 336
strontium 106
style 285, 295, 297
suberin 74, 228
succinate 43, 45, 47, 315

succinic acid *see* succinate
succulence 12, 28, 95, 354
sucrose 39, 41, 45, 210, 218, 303, 304, 309, 352, 353, 361
  as signal molecule 147, 147–148
  transport in phloem 137, 142–145
  transporter (SUT) 142, 143, 146
sucrose synthase 304, 309
sugar 9, 10, 39, 186, 352, 353
  alcohols *see* polyols
  transport 86
  see also sucrose
sugar beet 297
sugar cane 29, 35
sugar maple 86
sulfokin 189, *189*
sulphate 113, *120*, 137
sulphur 103, 128
sulphydryl (SH) groups 103
sumach 313
sun 9
  leaves 18, 230
  plants *17*
sunflower 17, 83, 106, *293*, 299, 305, *306*
sunlight 9, 12, 16, 18, 57, 267
  spectral composition 13, *247*, *252*, 252–253, *253*
sunscald injury 363
supercooling, deep 361
superoxide
  dismutase (SOD) 16, 105, 353, 360
  radical 16, 25, 48, 360
suspensor 301
SUT *see* sucrose transporter
swede 128
sweetcorn *see* maize
symbiosome 125
symbiotic nitrogen fixation *see* nitrogen
symplast 74, 74–76, 110
symplastic loading *see* phloem
symport 117, *117*
synergids 301
synthetic auxin 181
Système International d'Unités (SI) 373
systemin 189, *189*

T-DNA *see* DNA, transfer
Talbotia elegans 348
tapetum 292, 300
Taraxacum officinale see dandelion
target tissue 179, 181
tassel *see* maize
Taxus baccata see yew

tea 106, 128
tea-yellows 128
temperature
  for growth *167*
  and photorespiration 24
  and photosynthesis 17, 25, 28, 31, 357, 363
  regulation in plants 48, 354–355
  and respiration *167*
  see also optimum temperature
temperature coefficient *see* Q_{10}
tendril 277
tension
  in cell walls 64
  in xylem 78, 79–85, 111
tension wood 228
terminal oxidation *see* respiration
Terminator Technology 311
tertiary sulphonium compounds 352, *352*
testa (seed coat) 50, 285, 301, 311, 313
tetracycline 226
tetrapyrrole 249, 255, 263
tetrazolium test *357*
TFL (*TERMINAL FLOWER*) gene 283–285
thale cress *see Arabidopsis thaliana*
thermal death point 362–363
thermochemical reactions 31
thermocouple psychrometer 65
thermotolerance *see* stress, high temperature
thermotropism 320
thigmonasty 318, *319*
thigmotropism 320
Thiobacillus denitrificans 128
Thlaspi arvense 276
Thlaspi caerulescens 130
threshold stimulation 332, 336, 336–337
thylakoid 14, 15, *15*, 17, 26, 31
thymidine 208, 223
Tilia see lime
tissue culture 184–185, 186
tmm (*too many mouths*) mutant 234
tobacco 75, 139, 143, 225, 226, *226*, 230, 255, 261, 262, *262*, *262*, 267, 271, 277, *277*, 284, 295, 298, 300, *300*, 328, *353*, 360, *369*
  Maryland Mammoth 271
tomato 80, 143, 189, 225, 237, 238, *238*, 254, 255, 256, 276, 298, 308, 309, 311, 322, 329, 368, *369*
  falsiflora mutant *284*
  FlavrSavr *284*

fruit development 308–310, *309*
α-tocopherol 353
tonoplast 66–67, 74, 75, 110, 113, 114, *115*, 368
  intrinsic protein (TIP) 67
totipotency 206
toxic elements 130
toxins, in plants 103, 308
tracheids *see* xylem
Tradescantia sp. 322
Tradescantia discolor 362
transcription 18, 121, 265, 364
  factors 200, 210, 243, 266, 281, 285, 289, 370
transfer cell 71, 115, 143
transfer DNA *see* DNA
transformation 185, 186
transgenic plants *see* genetic engineering
translation 51
translocation *see* phloem translocation
transmitting tissue 295
transpiration *see* water transpiration
transport *see* membrane transport; phloem transport; xylem transport
transport proteins *116*, 116–118, *125*, 129, 236, 330, 368
trees 68, 75, *76*, 77, 78, 79, 80, 83, 121, 138, 144, 165, 285, 314
trehalose 352
trichoblast 241
trichome 226, 235
Trifolium repens 170
triose sugars 22
triple response, to ethylene 188, *188*, 196, 198
Triticum sativum (*vulgare*) *see* wheat
tRNA *see* RNA
tropical
  grasses 29
  rainforest 30, 77, 246
  species 144, 175, 272, 314, 347
tropisms 320–324
  see also gravitropism; phototropism
tunica 223
turgor 62, 63, 199, 232, 350, 353
  movements 20, 67, 104, 319
  pressure (Ψ_p) 62, 148–149, 150–151, 350, *351*

ubiquitin 211, 366
ultraviolet radiation 93

umbelliferose 352
uniport 118
units of measurement, SI 373
unstirred layer 19, 88, 96
UV-A/blue light photoreceptors 248, 252, 255
  see also cryptochrome
UV-B photoreceptors 256

VA mycorrhiza *see* mycorrhiza
vacuole 28, 39, 74, 110, 113, 114, 115, 214, 307, 354, 368
Valonia 113
van Niel equation 32
Vanilla fragrans 30
vapour pressure, water 64, 78, 88, 96
vascular
  bundles 26, 75, *135*, 142
  cambium 50, 227–228, 241
  differentiation 218, 301
  tissue 133, 218, 227
    see also phloem; xylem
vegetative cell, of pollen 292
Venus flytrap 318, 319, *319*
verbascose 137
vernalin 276
vernalization 275–276, 279–280, 313
very-low-fluence response (VLFR) *see* phytochrome
vesicular–arbuscular (VA) mycorrhiza *see* mycorrhiza
vessels *see* xylem
viability, seed 314
Vicia faba see broad bean
Vigna aconitifolia 369
Vigna radiata see mung bean
Vigna unguiculata see cowpea
Viola calaminaria 130
viruses 138, 190
Viscum album see mistletoe
vivipary 187, 311, 315
VLFR (very-low-fluence response) *see* phytochrome
VRN (*vernalization*) genes 280

wall *see* cell wall
wall pressure *see* turgor pressure
water 32
  bound 362
  chemical potential, μ_w 96
  conservation 29, 30, 95–96, 347
  content, plant tissues 60, 62, 68, 311, 347, *351*, 362
  culture 100

water (cont.)
  deficit stress *see* stress
  free space (WFS) 109
  heavy 60
  loss from plants 12, 21, 28, 79, 115, 350
  metastable state 81
  properties 60, 79, 88
  relations
    of cells *63*, *65*, 66–67
    of whole plants 68
  supply and growth 167, 349
  transpiration 68, 75, 78, 79, 94–95
    cuticular 93
    pull 78, 79, 81, 85
    rate 76, *76*, 88–92, *89*, *90*, *91*
    stomatal 90–92
    stream 68, 84
  transport 61
    across membranes 66–67
    cohesion-tension theory 78–85
    motive forces 78
    root pressure 77–78, 83, 85, 88
    route in plant 70, 72, 74, 84
  uptake 88, *89*, 93, 214
    by roots 69–70, 86–88
  vapour pressure 64, 78, 88, 96
  see also water potential
water lily, yellow 51
water melon 213
water plantain 311

water potential (Ψ) 61–66, *63*, *65*, 68, 77, 346–348
  atmospheric 68, 78, 88, 96
  definition 61, 96
  forces determining 61–62
  gradients 61, 66, 67, 68, 76, 79, 80, 88
  measurement 64–66
  of soil *87*
  values, plants 62, *63*–64, 78, 80, 137, *351*
wax *94*, 95–96, 305
wax palms 93
weeds 29, 165, 169, 314
wheat 34, 36, 52, *94*, 166, *166*, *168*, 183, *183*, *304*
wild oat 314
willow 144, 218–219, 296
  see also Salix
willowherb 314
wilting *63*, 64, *65*, 81, 87, 87–88, 92
wind 80, 89, 295
wood *76*, 81, 228
  see also xylem
woody plants 85, 183, 361
  growth of 71, 169, *169*
work 9, 53, 56–57
wounding responses 186, 188, 189, 190, 213
WUS protein 225, 239
  wus mutant 224

Xanthium pennsylvanicum (*strumarium*) 272–275, 311
xanthophyll *13*
xeromorphic characters 95–96, 347, 350, 363
Xerophyta sp. 348
xerophytes 95–96, 349
X-ray microanalysis 95–96, 349
xylem 79–86, 236
  metaxylem 207, *207*
  protoxylem 146, 241
  sap composition 77–78, 82, 85
  structure 71–72, *72*, *76*
  tracheids 71, *72*, 76, 84–85
  transport 70–74, 76, 94, 108
    of solutes 85–86, 110, 111–112
  vessels 71, *72*, *73*, *73*, 76, 81, 82, 83–85
xyloglucan 215, *216*, 218

yeast *353*
yew 296
yield
  agricultural *see* agricultural yield
  photosynthetic *see* photosynthesis
  quantum *see* quantum yield

Zea mays see maize
zeatin *180*
Zebrina pendula 20
zinc 105, 128
zygote 301